P9-APY-552

ADVANCES IN CHEMICAL PHYSICS

VOLUME LII

Advances in
CHEMICAL PHYSICS

EDITED BY

I. PRIGOGINE

University of Brussels
Brussels, Belgium
and
University of Texas
Austin, Texas

AND

STUART A. RICE

Department of Chemistry
and
The James Franck Institute
The University of Chicago
Chicago, Illinois

VOLUME LII

AN INTERSCIENCE® PUBLICATION

JOHN WILEY & SONS

NEW YORK · CHICHESTER · BRISBANE · TORONTO · SINGAPORE

An Interscience® Publication

Copyright © 1983 by John Wiley & Sons, Inc.

All rights reserved. Published simultaneously in Canada.

Reproduction or translation of any part of this work
beyond that permitted by Section 107 or 108 of the
1976 United States Copyright Act without the permission
of the copyright owner is unlawful. Requests for
permission or further information should be addressed to
the Permissions Department, John Wiley & Sons, Inc.

Library of Congress Catalog Number: 58-9935
ISBN 0-471-86845-0

Printed in the United States of America

10 9 8 7 6 5 4 3 2 1

CONTRIBUTORS TO VOLUME LII

STANISLAV BISKUPIC, Faculty of Chemistry, Slovak Technical University, Bratislava, Czechoslovakia

JOHN M. BROWN, Department of Chemistry, The University of Southampton, Southampton, England

MIROSLAV HARING, Computing Centre, Slovak Academy of Sciences, Bratislava, Czechoslovakia

DAVID R. HERRICK, Chemistry Department, University of Oregon, Eugene, Oregon

FLEMMING JØRGENSEN, Chemical Laboratory V, The H. C. Ørsted Institute, Copenhagen, Denmark

VLADIMÍR KVASNIČKA, Faculty of Chemistry, Slovak Technical University, Bratislava, Czechoslovakia

VILIAM LAURINC, Faculty of Chemistry, Slovak Technical University, Bratislava, Czechoslovakia

C. Y. NG, Department of Chemistry, Iowa State University, Ames, Iowa

ROBERT J. RUBIN, National Bureau of Standards, Washington, D.C.

GEORGE H. WEISS, National Institutes of Health, Bethesda, Maryland

INTRODUCTION

Few of us can any longer keep up with the flood of scientific literature, even in specialized subfields. Any attempt to do more, and be broadly educated with respect to a large domain of science, has the appearance of tilting at windmills. Yet the synthesis of ideas drawn from different subjects into new, powerful, general concepts is as valuable as ever, and the desire to remain educated persists in all scientists. This series, *Advances in Chemical Physics*, is devoted to helping the reader obtain general information about a wide variety of topics in chemical physics, which field we interpret very broadly. Our intent is to have experts present comprehensive analyses of subjects of interest and to encourage the expression of individual points of view. We hope that this approach to the presentation of an overview of a subject will both stimulate new research and serve as a personalized learning text for beginners in a field.

ILYA PRIGOGINE
STUART A. RICE

CONTENTS

ADVANCES IN CHEMICAL PHYSICS

VOLUME LII

NEW SYMMETRY PROPERTIES
OF ATOMS AND MOLECULES

DAVID R. HERRICK

Chemistry Department
University of Oregon
Eugene, Oregon 97403

CONTENTS

I. INTRODUCTION

It is safe to say that the modern language (if not most of our understanding) of the electronic structure of atoms and molecules is derived from the shell structure of representative one-electron systems, and the independent-particle model with a single configuration of orbitals. More subtle electron correlation effects from the Coulomb repulsion between electrons are neglected in a first approximation, but may be estimated on the computer using perturbation theory or variational approaches including configuration interaction. The subject of this article is a relatively small, but growing class of systems for which both theory and experiment suggest a fundamental breakdown of the single configuration picture. The breakdown is attributed to electron correlation in the form of a particularly strong interaction between excited configurations that are otherwise degenerate, or nearly degenerate. Strong coupling of these excitations produces new types of spectra, or even familiar spectra in a seemingly unfamiliar setting. The intrinsic multiconfigurational structure of the levels affects the behavior of these systems in radiative and collisional processes or in external fields.

Specific problems of interest are the long-range dipole coupling of electrons to rotating polar molecules or hydrogenlike atoms in collisions or photodetachment (Section III), doubly excited states of two-electron atoms (Section IV), and valence states of molecules predicted by the Pariser–Parr–Pople model (Section V). The common idea is that these systems offer concrete examples of complex spectra that could originate in approximately separable, collective degrees of freedom for electrons, linked to coupled representations of Lie groups for the underlying orbital basis. In this article we shall focus on basic questions that need answers if one is to grasp the sense and direction of the work: (1) What are the reasons for introducing these groups? (2) What do the spectra look like? (3) What physical pictures of electron correlation can we infer from the new classifications of spectra?

The traditional role of group theory in physics and chemistry is largely interpretive. Groups for the manifest geometrical, rotational, spin, and permutation–inversion symmetries are exploited for state labels and selection rules for matrix elements of tensor operators. The familiar Lie groups for exact symmetries are SU(2) for spin, and R_3, or SO(3), for proper spatial rotations. The accidental degeneracy of the hydrogen atom is another exact symmetry, and Fock[1] and later Bargmann[2] described this with a higher rotation group R_4, or SO(4), for four-dimensional rotations in momentum space. The idea of higher groups for approximate symmetry was introduced by Wigner[3] with SU(4) for the spectra of light nuclei. Racah[4] introduced approximate symmetry for atoms when he described five components of a d orbital with R_5 for rotations in five dimensions, and used this to derive "seniority" labels for two ^2D terms that arise in a d^3 configuration of electrons. This generalizes to a group R_{2l+1} for configurations with higher integer orbital angular momentum. The coupled states in a seniority classification diagonalize a simple "pairing" interaction between electrons, instead of the actual Coulomb repulsion, and hence the quantization is very approximate. Pairing is generally thought to be more realistic in nuclei, and Flowers[5] extended the seniority concept there using the symplectic group Sp_{2j+1} with half-integer values of j for the $j-j$-coupling model. A common feature of both atomic and nuclear seniority is that the groups are subgroups of a larger group of unitary transformations defined on the orbitals. One therefore has two types of seniority chains, $U_n \supset R_n \supset R_3$ or $U_n \supset Sp_n \supset R_3$, according as the number of orbitals n is *odd* or *even*. Properties of these higher groups have been documented in great detail by several different approaches with an eye toward the physical applications.[6-11] Gilmore[12] presents some of the deeper properties of Lie groups and the associated Lie algebras.

The modern way to describe seniority for approximate symmetries is with so-called quasispin groups, whose Lie-algebra generators change the number of particles in a system. Quasispin groups are complementary to seniority groups in the sense that the pairing interaction can be represented with quadratic Lie-algebra invariants of either group, and generators of the two groups commute. The associated group for R_n seniority is R_3^Q for three-dimensional quasispin,[8,13,14] and for Sp_n seniority there is R_5^Q for five-dimensional quasispin.[11,15-17]

The basic problems in the orbital structure theory of atoms, nuclei, and molecules are first constructing the configuration interaction matrix of the energy, and then diagonalizing it. The former is complicated in large systems because of the large number of configurations possible, and so one is faced with the logistics of merely labeling configurations for expedient setup of the energy matrix. Subgroups of U_n like the seniority–rotation chain do not generally provide the complete set of labels that is needed, and so alternative approaches derived from a canonical subgroup chain $U_n \supset U_{n-1} \supset \cdots \supset U_1$ and symmetric group representations have been considered.[18-39]

Transformation of the configuration-interaction matrix to bases adapted to the exact symmetries block-diagonalizes the energy, and invariants of the groups are exact constants of the motion. The idea of approximate symmetries is that other groups could approximately block-diagonalize the energy even further within each exact symmetry block, and invariants of the groups would be approximate constants of the motion. Implicit in the labeling of energy levels with approximate symmetries, then, are pictures of the underlying correlation of particles whose motions are coupled, as well as modeling of the spectra. The basic pictures which account for many features in nuclear spectra have been different groups linked to collective rotations and vibrations of particles.[40-44] Elementary particles[45] offer another example of spectra that are linked to group representations.

Elementary properties of some of the groups of interest in the present article are summarized in Section II. Correlation symmetries for electron–molecule and electron–atom dipole coupling involve mixing of different values of l and conservation of angular momentum about internal symmetry axes. The correlation symmetry group in both cases is SO(4), and the spectra look like representations of $SO(4) \supset SO(3) \supset SO(2)$ or $SO(4) = SU(2) \times SU(2)$, depending on the coupling strength. One reason for describing the molecular dipole problem first is that it is the simplest example of a correlation symmetry, resulting from internal angular momentum coupling. The atomic SO(4) is derived from the hydrogen-atom degeneracy. In two-electron atoms the Coulomb repulsion operator breaks the degeneracy, and coupled representations of SO(4) are exploited to find supermultiplet spectra[46,47] which display essential features of a rotor–vibrator picture.[48,49] Groups for correlation in molecular polyenes are derived from the unitary group U_n for orbitals on different atoms, taking into account the basic particle–hole exchange symmetry of the model. This leads to seniority subgroups very similar to the ones for the atomic and nuclear shell models, although the physical interpretations are different. In polyenes, for example, we find two types of pairing symmetry, one for short-range correlation and one for long-range correlation effects. The recent[50] pseudorotation symmetry of linear polyenes is found as a special case in this more general approach to molecular correlation.

The investigation of approximate correlation symmetry is necessarily semiempirical to some extent, because at some stage comparisons have to be made with wave functions and energies from configuration interaction calculations. The methods and models described here should therefore be viewed as prototypes for future extensions, and several suggestions are offered. As these ideas develop, it is hoped the methods will continue to apply in regions of the spectra where other methods fail.

II. PROPERTIES OF RELEVANT LIE ALGEBRAS

A. Angular Momentum

Since the main interest is with commutation relations, invariant operators and quantum numbers for irreducible representation spectra, no careful distinction will be made between the groups SU(2) and SO(3) for angular momentum.[51] There are three Lie-algebra generators j_x, j_y, and j_z which satisfy the commutation relations

$$[j_x, j_y] = ij_z, \qquad [j_y, j_z] = ij_x, \qquad [j_z, j_x] = ij_y \qquad (2.1)$$

or in terms of the generators $j_\pm = j_x \pm ij_y$ and $j_0 = j_z$,

$$[j_0, j_\pm] = \pm j_\pm, \qquad [j_+, j_-] = 2j_0 \qquad (2.2)$$

The quadratic operator which commutes with the Lie-algebra generators is the square of the angular-momentum vector, $\mathbf{j}^2 = j_x^2 + j_y^2 + j_z^2$ or $\mathbf{j}^2 = j_+ j_- + j_0(j_0 - 1)$. States in each unitary irreducible representation are labeled $|jm\rangle$ with two quantum numbers for two diagonal operators $\mathbf{j}^2 = j(j+1)$ and $j_z = m$ of the subgroup chain SO(3)⊃SO(2). In general the values of j are positive integers or half-integers, and $m = j, j-1, j-2, \ldots, -j$ for a total of $2j+1$ states in each representation $[j]$. The usual convention for matrix elements of the generators is[52]

$$\langle jm \pm 1 | j_\pm | jm \rangle = [(j \mp m)(j \pm m + 1)]^{1/2}$$
$$\langle jm | j_z | jm \rangle = m \qquad (2.3)$$

When the representations of two commuting angular momentum groups are coupled so that the total angular momentum vector is $\mathbf{J} = \mathbf{j}_1 + \mathbf{j}_2$, the irreducible representations with $\mathbf{J}^2 = J(J+1)$ diagonal are described by the Clebsch–Gordan series

$$[j_1] \times [j_2] = [j_1 + j_2] + [j_1 + j_2 - 1] + \cdots + [|j_1 - j_2|] \qquad (2.4)$$

Individual states with $J_z = M$ diagonal are constructed on the product basis $|j_1 m_1, j_2 m_2\rangle \equiv |j_1 m_1\rangle |j_2 m_2\rangle$ with the orthogonal transformation

$$|(j_1 j_2)JM\rangle = \sum_{m_1, m_2} |j_1 m_1, j_2 m_2\rangle \langle j_1 m_1, j_2 m_2 | JM \rangle \qquad (2.5)$$

where the Clebsch–Gordan coupling coefficient is usually represented in terms of the more symmetrical 3-j symbol

$$\langle a\alpha, b\beta | c\gamma \rangle = (-1)^{a-b+\gamma}(2c+1)^{1/2}\begin{pmatrix} a & b & c \\ \alpha & \beta & -\gamma \end{pmatrix} \quad (2.6)$$

Coefficients on the right-hand side of (2.5) vanish unless $m_1 + m_2 = M$, and unless the triangle condition $|j_1 - j_2| \leq J \leq j_1 + j_2$ is satisfied. The generalization of (2.5) to the coupling of $3, 4, 5, \ldots$ independent angular-momentum vectors leads to the higher-order coupling coefficients involving 6-j, 9-j, 12-j,... symbols.

An important aspect of the coupling scheme in (2.5) is that it is designed specifically to account for the usual types of angular momentum that arise in physical problems, such as spin or spatial rotations. Operators satisfying the commutation relations (2.1) may also arise in entirely different physical circumstances. In these cases it may be appropriate to consider different types of coupling schemes. One example is the approximate pseudorotation symmetry of linear polyenes, which is described in Section V. In the case of neutral polyenes the usual angular-momentum coupling is followed in order to construct correlated wave functions for valence electrons. In the case of just two electrons, however, it is found to be more appropriate to couple components of the two independent pseudorotation vectors $\mathbf{j}_1$ and $\mathbf{j}_2$ to give a resultant vector $\mathbf{K}$ with

$$K_x = j_{1x} - j_{2x}, \quad K_y = j_{1y} - j_{2y}, \quad K_z = j_{1z} + j_{2z} \quad (2.7)$$

The precise physical reasons for this are the subject of Section V.F. 6. It is straightforward to construct states which diagonalize $\mathbf{K}^2 = K(K+1)$ and K_z, using properties of the usual states in (2.5) under the reflection $(x_2, y_2, z_2) \to (-x_2, -y_2, z_2)$. The result is

$$|(j_1 j_2)KM\rangle = \sum_{m_1, m_2} |j_1 m_1, j_2 m_2\rangle\langle j_1 m_1, j_2 m_2 | KM\rangle(-1)^{j_2 - m_2} \quad (2.8)$$

Individual terms on the right-hand side of (2.8) differ from those in (2.5) only by a simple phase factor $(-)^{j_2 - m_2}$. This is the only nonstandard coupling scheme that we shall have occasion to refer to.

B. SO(4) Representations

The general analysis of the Lie-algebra structure of the group SO(4) is facilitated by the fact that there is a representation in which the group factors into a product of two commuting angular-momentum groups, SO(4) =

SU(2)$\times$SU(2).[53] If the two independent angular-momentum vectors are $\mathbf{a}$ and $\mathbf{c}$, the basic states for each irreducible representation are labeled $|a\alpha, c\gamma\rangle$ with four quantum numbers for the diagonal operators $\mathbf{a}^2 = a(a+1)$, $a_z = \alpha$, $\mathbf{c}^2 = c(c+1)$, and $c_z = \gamma$. Generally speaking then, one can use the language of SO(4) whenever there is coupling of two angular momenta.

The more conventional way of representing states is related to a subgroup chain SO(4)$\supset$SO(3)$\supset$SO(2). Lie-algebra generators for this picture are $\mathbf{l} = \mathbf{a} + \mathbf{c}$, which generates the SO(3) algebra, and $\mathbf{b} = \mathbf{a} - \mathbf{c}$. The commutation relations are

$$[l_x, l_y] = il_z, \qquad [l_y, l_z] = il_x, \qquad [l_z, l_x] = il_y$$

$$[l_x, b_y] = ib_z, \qquad [l_y, b_z] = ib_x, \qquad [l_z, b_x] = ib_y \qquad (2.9)$$

$$[b_x, b_y] = il_z, \qquad [b_y, b_z] = il_x, \qquad [b_z, b_x] = il_y$$

Each irreducible representation of SO(4) is labeled with two indices $[p, q]$ that are related to the SU(2) quantum numbers

$$p = a + c, \qquad q = a - c \qquad (2.10)$$

Individual states in each representation are labeled $|pqlm\rangle$, with four quantum numbers for eigenvalues of the four commuting operators

$$\begin{aligned} \mathbf{l}^2 + \mathbf{b}^2 &= p(p+2) + q^2 \\ \mathbf{l} \cdot \mathbf{b} &= q(p+1) \\ \mathbf{l}^2 &= l(l+1) \\ l_z &= m \end{aligned} \qquad (2.11)$$

The triangle rule for the coupling of $\mathbf{a}$ and $\mathbf{c}$ to give $\mathbf{l}$ shows that each SO(4) representation breaks up into a series of SO(3) representations $[l]$,

$$[p, q] = [p] + [p-1] + \cdots + [|q|] \qquad (2.12)$$

and the total number of states is described by the dimension formula $\dim[p, q] = (p+1)^2 - q^2$, or simply $(2a+1)(2c+1)$.

Since the commutation relations of $\mathbf{b}$ establish this operator as a tensor of rank 1 with respect to the SO(3) group, selection rules for matrix elements are $\Delta l = 0, \pm 1$. It is sufficient to specify the nonzero matrix elements of the

z-component in terms of

$$\langle pql+1m|b_z|pqlm\rangle$$

$$=\left[\frac{(p+2+l)(p-l)\left[(l+1)^2-q^2\right]\left[(l+1)^2-m^2\right]}{(2l+1)(2l+3)(l+1)^2}\right]^{1/2}$$

$$\langle pqlm|b_z|pqlm\rangle=\frac{mq(p+1)}{l(l+1)} \qquad (2.13)$$

The expressions are symmetrical in exchange of the indices m and q because these are related to quantum numbers for the projection of the angular-momentum vector $\mathbf{l}$ along two different axes. In the case of ordinary angular momentum for three-dimensional rotations, m labels the quantization in the laboratory z axis, while q labels the quantization with respect to the internal vector $\mathbf{a}-\mathbf{c}$.

Coupling of two independent representations of SO(4) is straightforward because it is essentially the problem of coupling four SU(2) angular momenta. Irreducible components of the product representation are labeled $[P,Q]$ with $P=a+c$ and $Q=a-c$. The Clebsch–Gordan series is

$$[p_1,q_1]\times[p_2,q_2]=\sum_{a,c}[a+c,a-c] \qquad (2.14)$$

with possible values of the indices a and c restricted by the SU(2) triangle rules $|a_1-a_2|\leqslant a\leqslant a_1+a_2$ and $|c_1-c_2|\leqslant c\leqslant c_1+c_2$. There is a certain degree of arbitrariness in specifying a set of coupled states for the irreducible basis because the coupling of SU(2) vectors can be carried out in several different ways. This is analogous roughly to the interpretation of atomic states with total angular momentum quantum numbers J and M_J with either $L-S$ or $j-j$ coupling.

Biedenharn[53] has described coupling for the canonical representation of SO(4) generated by $\mathbf{L}=\mathbf{l}_1+\mathbf{l}_2$ and $\mathbf{A}=\mathbf{b}_1+\mathbf{b}_2$. Working in a coupled SO(3) basis, his coupled SO(4) states are

$$|PQLM\rangle_A=\sum_{l_1,l_2}|p_1q_1p_2q_2(l_1l_2)LM\rangle\langle p_1q_1l_1,p_2q_2l_2|PQL\rangle \qquad (2.15)$$

with the coupling coefficient defined in terms of a 9-j symbol

$$\langle p_1q_1l_1,p_2q_2l_2|PQL\rangle=\left[(P+Q+1)(P-Q+1)(2l_1+1)(2l_2+1)\right]^{1/2}$$

$$\times\begin{Bmatrix} a_1 & a_2 & \tfrac{1}{2}(P+Q) \\ c_1 & c_2 & \tfrac{1}{2}(P-Q) \\ l_1 & l_2 & L \end{Bmatrix} \qquad (2.16)$$

The diagonal quadratic SO(4) invariants in this basis are $\mathbf{L}^2 + \mathbf{A}^2 = P(P+2)$ $+ Q^2$ and $\mathbf{L} \cdot \mathbf{A} = Q(P+1)$.

Another representation of SO(4) that is useful has the Lie algebra generators $\mathbf{L} = \mathbf{l}_1 + \mathbf{l}_2$ and $\mathbf{B} = \mathbf{b}_1 - \mathbf{b}_2$. Coupled states for this picture are

$$|PQLM\rangle_B = \sum_{l_1, l_2} |p_1 q_1 p_2 q_2 (l_1 l_2) LM\rangle \langle p_1 q_1 l_1, p_2 - q_2 l_2 | PQL \rangle (-)^{l_2}$$

(2.17)

This is similar to (2.15) except for the phase factor and $q_2 \to -q_2$ which are related to the transformation $\mathbf{b}_2 \to -\mathbf{b}_2$ in the Lie-algebra generator. The diagonal SO(4) invariants for the basis (2.17) are $\mathbf{L}^2 + \mathbf{B}^2 = P(P+2) + Q^2$ and $\mathbf{L} \cdot \mathbf{B} = Q(P+1)$, although this P and Q are different from the labels P and Q in (2.15).

Other SO(4) coupling schemes are possible, such as $\mathbf{l}_1 + \mathbf{l}_2$ and $\mathbf{l}_1 + \mathbf{b}_2$, but explicit details are omitted. The coupling scheme of greatest physical significance would have to be identified in the particular problem of interest.

C. Hydrogen-Atom SO(4)

The SO(4) structure of the hydrogen atom is very well established.[1,2,54,55] The notation and interpretations introduced here will be used repeatedly in subsequent sections. Working in atomic units ($e = m_e = \hbar = 1$) the Hamiltonian for electronic states is

$$H = \tfrac{1}{2} p^2 - \frac{1}{r}$$

(2.18)

Two constants of the motion are the angular-momentum vector $\mathbf{l} = \mathbf{r} \times \mathbf{p}$ and the Runge–Lenz vector[56]

$$\mathbf{a} = \tfrac{1}{2}(\mathbf{l} \times \mathbf{p} - \mathbf{p} \times \mathbf{l}) + \hat{\mathbf{r}}$$
$$= \mathbf{p}(\mathbf{r} \cdot \mathbf{p}) - \mathbf{r}\left(p^2 - \frac{1}{r}\right)$$

(2.19)

For bound states the vectors $\mathbf{l}$ and $\mathbf{b} = (-2H)^{-1/2} \mathbf{a}$ satisfy the commutation relations in (2.9), and hence the discrete portion of the spectrum may be cast into irreducible representations of SO(4).

The classical interpretation is that $\mathbf{l}$ is perpendicular to the plane of the elliptical Kepler orbit, while $\mathbf{a}$ lies in the plane along the semimajor axis.[57] This is depicted in Fig. 1 for an orbit with nonzero angular momentum. The classical vectors are related to the energy by the formula $a^2 = 1 + 2Hl^2$, and $\varepsilon = |\mathbf{a}|$ is the eccentricity of the orbit. The greatest distance of the electron

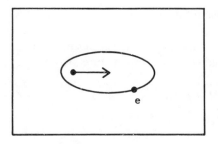

Fig. 1. Classical Kepler orbit for bound electron (e). The arrow indicates the direction of the Lenz vector, originating at the nucleus.

from the nucleus is $r_{max} = (1 + |a|)(-2H)^{-1}$, and the distance of closest approach is $r_{min} = (1 - |a|)(-2H)^{-1}$.

The quantum-mechanical interpretation of the Lenz vector is due to Pauli,[56] who showed that $a^2 = 1 + 2H(l^2 + 1)$. In terms of $\mathbf{l}$ and $\mathbf{b}$ the energy is

$$-2H = (\mathbf{l}^2 + \mathbf{b}^2 + 1)^{-1} \qquad (2.20)$$

which is related to the SO(4) invariant in (2.11). The vectors $\mathbf{l}$ and $\mathbf{b}$ are orthogonal ($\mathbf{l} \cdot \mathbf{b} = 0$), and so the particular SO(4) irreducible representations $[p, q]$ of interest have $q = 0$. This means that (2.20) becomes $-2H = (p + 1)^{-2}$, and hence $p = n - 1$ where n is the usual principal quantum number.

The basis $|p0lm\rangle$ for the representation SO(4)$\supset$SO(3)$\supset$SO(2) corresponds to the separated hydrogen-atom wave function in spherical polar coordinates, $R_{nl}(r)Y_{lm}(\theta, \phi)$. A radial basis consistent with the phase convention for matrix elements of the SO(4) generator in (2.13) is

$$R_{nl}(r) = N_{nl} \left(-\frac{2r}{n} \right)^l \exp\left(-\frac{r}{n} \right) F\left(l + 1 - n, 2l + 2, \frac{2r}{n} \right)$$

$$N_{nl} = \frac{1}{(2l+1)!} \left[\frac{(n+l)!}{(n-l-1)!2n} \right]^{1/2} \left(\frac{2}{n} \right)^{3/2} \qquad (2.21)$$

where $F(a, b, z)$ is the confluent hypergeometric function.[58] An important point in subsequent applications of the hydrogenic basis to problems with mixing of different-l states is the phase of the radial function at large r, where the probability density is greatest. The form of the functions in (2.21) at large r is

$$R_{nl}(r) \sim C_{nl} \left(-\frac{2r}{n} \right)^{n-1} \exp\left(-\frac{r}{n} \right)$$

$$C_{nl} = \left[\frac{(n-l-1)!}{(n+l)!2n} \right]^{1/2} \left(\frac{2}{n} \right)^{3/2} \qquad (2.22)$$

and so all of the functions with the same principal quantum number have the same asymptotic phase.

The alternate representation of SO(4) as SU(2)×SU(2) is related to the basis $|a\alpha, c\gamma\rangle$ with two equal angular momenta, $a = c = \frac{1}{2}(n-1)$. Whereas the spherical basis had the operator $\mathbf{l}^2$ diagonal, this basis has the operator b_z diagonal and hence individual states contain admixtures of different l components. The transformation between the two bases is[59]

$$|a\alpha, a\gamma\rangle = \sum_l |p0lm\rangle\langle a\alpha, a\gamma|lm\rangle \qquad (2.23)$$

where the summation over l includes values in the range $|m| \leqslant l \leqslant n-1$. Physically, the basis (2.23) is related to the separation of hydrogen-atom wave functions in parabolic coordinates, $F(\xi)G(\eta)\exp(im\phi)$, with $\xi = r + z$ and $\eta = r - z$. Bethe and Salpeter[60] give further details of parabolic wave functions for splitting of the hydrogen degeneracy in a Stark electric field, $V = Fz$. The first-order splitting is obtained by diagonalizing V in the degenerate spherical basis, and was described by Pauli[56] using the operator replacement $\mathbf{r} \leftrightarrow 1.5n^2\mathbf{a}$. There is a similar relationship[61,62] with the unit vector of the electron in a fixed-n basis, $\hat{\mathbf{r}} \leftrightarrow \mathbf{a}$. In terms of SO(4) generators these are

$$\mathbf{r} \leftrightarrow 1.5n\mathbf{b}, \quad \hat{\mathbf{r}} \leftrightarrow n^{-1}\mathbf{b} \qquad (\text{constant } n) \qquad (2.24)$$

and the first-order Stark splitting is described by the operator $V_1 = 1.5nFb_z$.

The spectrum of the SO(4) generator b_z thus provides a direct representation of the linear Stark effect in hydrogen. It is convenient to describe the states with the SO(2) label m and the electric-field quantum number k, which are related to α and γ in the SU(2)×SU(2) basis by

$$m = \gamma + \alpha, \quad k = \gamma - \alpha \qquad (2.25)$$

The quantization for principal quantum number $n = 5$, including a total of 25 states, is depicted in Fig. 2. This is linked to the actual first-order splitting of the level by the formula

$$V_1 = -1.5nFk \qquad (2.26)$$

The complete hierarchy of discrete hydrogenic Stark basis functions is

$$n = 1, 2, 3, \ldots$$
$$m = n-1, n-2, \ldots, 1-n \qquad (2.27)$$
$$k = n - |m| - 1, n - |m| - 3, \ldots, 1 + |m| - n$$

DAVID R. HERRICK

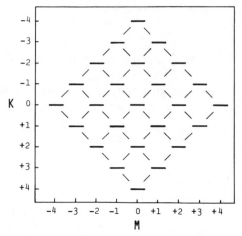

Fig. 2. First-order Stark energy levels for the $n = 5$ shell of the hydrogen atom, illustrating labeling of states with SO(4) quantum numbers $l_z = m$ and $b_z = k$. Diagonal lines connect states with the same value of $\alpha = \frac{1}{2}(m - k)$ or $\gamma = \frac{1}{2}(m + k)$ for SO(4) = SU(2)×SU(2).

The hierarchy of spherical states, on the other hand, is

$$n = 1, 2, 3, \ldots$$
$$l = 0, 1, \ldots, n - 1 \tag{2.28}$$
$$m = l, l - 1, \ldots, -l$$

It is necessary to understand the basic structure of these two different quantization schemes in order to appreciate the structure of more complicated spectra in Sections III and IV. An important point is that the Stark quantum number k carries information about the θ dependence of the wave function relative to the field axis $z = r \cos \theta$. This is related to the fact that the matrix of the operator $\cos \theta$ is diagonal within each shell, by virtue of the operator replacement in (2.24). The diagonal term for Stark states $|nkm\rangle$ is

$$\langle nkm | \cos \theta | nkm \rangle = -\frac{k}{n} \tag{2.29}$$

In other words, states with $k > 0$ have angular correlation (i.e., l mixing) so that the region of highest probability for the electron is in the range $90° < \theta \leqslant 180°$, while states with $k < 0$ have correlation which favors the region $0° \leqslant \theta < 90°$. As a concrete example of this consider the Stark functions with $n = 2$ and $m = 0$. There are two states, with $k = +1$ and -1, which may be expressed in the usual orbital notation as

$$\psi_{k=+1} = \frac{2s - 2p_z}{\sqrt{2}}$$

$$\psi_{k=-1} = \frac{2s + 2p_z}{\sqrt{2}}$$

These are the ordinary sp-hybridized one-center orbitals for chemical bonding in linear molecules. There is a similar linear hybridization of orbitals in the Stark basis for higher n, although this generally includes higher values of l.

There is an interesting property of the hydrogen-atom Stark wave functions which does not seem to have been explained in the literature, and which illustrates nicely the idea of *approximate symmetry*. This is related to the lifetime of states against radiative decay in the Stark basis (2.23). Exact symmetry properties of the SO(4) states lead to the sum rule[63]

$$\sum{}' B(nkm) = \frac{T(n)}{n} \qquad (2.30)$$

where $B(nkm)$ is the transition probability from the Stark state $|nkm\rangle$, and $T(n)$ is the total transition probability from all n^2 states in the manifold to states in lower shells. The prime in (2.30) indicates the summation extends over all states with the same value of $\alpha = \frac{1}{2}(m - k)$, or $\gamma = \frac{1}{2}(m + k)$, and the value of the sum is independent of this value of α or γ. These are the states connected by diagonal lines in Fig. 2, for example. There are no obvious group-theoretical grounds for expecting any additional symmetry properties of the transition probabilities. Direct numerical calculations[64,65] reveal, however, that $B(nkm)$ is very nearly independent of the quantum number k when $|m| > 0$. Individual transition probabilities deviate from this empirical rule by only a few percent in most cases. This is an example of how a semi-empirical approach can uncover approximate symmetries that are otherwise hidden. The precise reasons for the accidental degeneracies are not easy to see, because the explicit form of $B(nkm)$ involves summations of matrix elements over many different shells. One useful result of the symmetry is a simplification of the sum rule in Eq. (2.30) that leads to a simple empirical formula for Stark radiative lifetimes.[63]

D. Unitary Groups for Orbital Bases

A very general approach for analyzing many-electron spectra derived from a finite n-dimensional orbital basis is based on representations of the unitary groups $U(2n)$ and $U(n)$ in a second-quantization picture. This is briefly outlined, but most details of specific subgroup structure are left for applications in Section V. Judd[7,8,66] has described the second-quantization approach including tensors for seniority subgroup chains of the unitary groups.

The basic approach for spin-$\frac{1}{2}$ particles starts with a set of creation operators $X_{i\sigma}^\dagger$ and annihilation operators $X_{i\sigma}$ for a set of orthonormal one-electron orbitals $i = 1, 2, \ldots, n$ with spin component $\sigma = \pm\frac{1}{2}$. These satisfy the usual fermion anticommutation relations

$$X_{i\sigma}^\dagger X_{j\mu}^\dagger + X_{j\mu}^\dagger X_{i\sigma}^\dagger = 0$$

$$X_{i\sigma} X_{j\mu} + X_{j\mu} X_{i\sigma} = 0 \tag{2.31}$$

$$X_{i\sigma}^\dagger X_{j\mu} + X_{j\mu} X_{i\sigma}^\dagger = \delta_{ij}\delta_{\sigma\mu}$$

related to the fundamental antisymmetry of the N-particle Slater determinant basis

$$X_{i_1\sigma_1}^\dagger X_{i_2\sigma_2}^\dagger \ldots X_{i_N\sigma_N}^\dagger |0\rangle \tag{2.32}$$

defined relative to the vacuum state $|0\rangle$.

The $4n^2$ operators $E_{i\sigma j\mu} \equiv X_{i\sigma}^\dagger X_{j\mu}$ are elementary Lie-algebra generators of $U(2n)$ with commutation relations

$$\left[E_{i\sigma j\mu}, E_{k\eta m\alpha} \right] = \delta_{jk}\delta_{\mu\eta} E_{i\sigma m\alpha} - \delta_{mi}\delta_{\alpha\sigma} E_{k\eta j\mu} \tag{2.33}$$

The n^2 Lie-algebra generators of the group $U(n)$ are

$$E_{ij} = \sum_\sigma X_{i\sigma}^\dagger X_{j\sigma} \tag{2.34}$$

with

$$\left[E_{ij}, E_{km} \right] = \delta_{jk} E_{im} - \delta_{mi} E_{kj} \tag{2.35}$$

Applications of these groups to physical problems involve the identification of subgroup chains which provide state labels for the exact and approximate symmetries of the system. The generality of this approach is seen in the fact that the hydrogen-atom symmetry group for each principal quantum number n can be obtained as a subgroup $U(n^2) \supset SO(4)$, along with many other possibilities.

One difficulty with the general unitary approach is that it cannot be used to describe Lie algebra structure like that in Eq. (2.7), or the operator **B** preceding (2.17), since these generators break the antisymmetry of the wave function. This is a subtle point, but one of importance because these are precisely the Lie algebras of physical interest in two-electron atoms and molecules.

III. SO(4) SYMMETRY OF ELECTRON–DIPOLE CORRELATION

Correlation effects originating in a long-range charge–dipole interaction offer one example of orbital configuration interaction that is central to the usual coupled-channel theories of electron–molecule[67] and atom–molecule[68] collisions. The problems described in this section illustrate two fundamentally different ways in which the group SO(4) provides a useful description of the dipole coupling channels. The first example is the coupling of an electron to a rotating symmetric-top molecule which has a permanent dipole moment. Engelking[69] proposed the basic set of coupled channels for this system as a model for anomalous threshold scaling laws in photodetachment experiments, and the SO(4) interpretation is due to Herrick and Engelking.[70] The second example is the long-range coupling of an electron to an excited hydrogen atom, which is the basic model for the interaction of an ion with the near-degeneracy manifold of a Rydberg atom. This has been described by Seaton,[71] Gailitis and Damburg,[72] Burke,[73] and others.[74,75] The SO(4) quantization of atomic dipole channels followed here is that of Herrick[62,76] and Herrick and Poliak.[77] The difference between the two problems is that in the molecular case the SO(4) structure originates in angular-momentum coupling only, whereas the atomic SO(4) is the symmetry of the hydrogen atom. There is obviously no permanent dipole moment in the hydrogen atom, and the correlation is linked to l mixing in each shell by the Coulomb repulsion between two electrons, and hence the formation of an induced dipole.

A. Electron–Molecule Channels

The problem is one of identifying a set of correlated channels which take into account long-range coupling of an electron to a rotating polar molecule by the interaction $V = -\mu \hat{\mathbf{R}} \cdot \hat{\mathbf{r}} / r^2$. The notation (atomic units) is

$\mathbf{r}$ = electron position

$\hat{\mathbf{r}} = \mathbf{r}/r$

$\mathbf{l}$ = electron angular momentum

$\hat{\mathbf{R}}$ = unit vector of molecule symmetry axis

$\mathbf{J}$ = molecule angular momentum

μ = molecule dipole moment $(1 \text{ au} = 2.54 \times 10^{-18} \text{ esu cm})$

$\mathbf{L} = \mathbf{J} + \mathbf{l}$

and the Schrödinger equation of the system is

$$(T_e + H_{rot} + V - E)\psi = 0 \qquad (3.1)$$

T_e represents the kinetic energy of the electron, and H_{rot} is the energy of the molecule treated as a symmetric top, or a linear rotator with a component of internal angular momentum along the symmetry axis. There are a total of six degrees of freedom. Three of these are for rotations of the molecule, and the quantization is

$$(H_{rot} - \varepsilon_{JK})|JKN\rangle = 0 \qquad (3.2)$$

with labels from the diagonal operators $\mathbf{J}^2 = J(J+1)$, $\mathbf{J} \cdot \hat{\mathbf{R}} = K$ for rotations about the molecule symmetry axis, and $J_z = N$ for rotations in a laboratory frame. Neglecting spin, the electron has two angular degrees of freedom and one radial degree of freedom in spherical polar coordinates. The angular states are $|lm\rangle$, with $\mathbf{l}^2 = l(l+1)$ and $l_z = m$.

The idea of a coupled-channel approach for solving (3.1) is an expansion in terms of a complete set of states $|i\rangle$ for the angular degrees of freedom,

$$\psi = \sum_i \frac{|i\rangle F_i(r)}{r} \qquad (3.3)$$

Working in a basis in which H_{rot} is diagonal, the basic form of the coupled radial equations obtained by this method with a long-range electron–dipole interaction is

$$\left(\frac{d^2}{dr^2} + k_i^2\right) F_i = \frac{1}{r^2} \sum_j F_j \langle j|\Lambda|i\rangle \qquad (3.4)$$

Here $k_i^2 = 2(E - \varepsilon_{JK})$, and

$$\Lambda = \mathbf{l}^2 - 2\mu \hat{\mathbf{R}} \cdot \hat{\mathbf{r}} \qquad (3.5)$$

represents a type of "generalized angular momentum" resulting from centrifugal and dipole contributions to the $1/r^2$ coupling interaction.

The matrix of the operator Λ in (3.4) is blocked according to the exact symmetries of the problem. The uncorrelated single-configuration channels are the product states $|JKN, lm\rangle \equiv |JKN\rangle|lm\rangle$, and ordinary angular-momentum coupling of J and l leads to states $|(JK, l)LM\rangle$ with $\mathbf{L}^2 = L(L+1)$ and $L_z = M$ diagonal. K is also a good quantum number because the operator $\mathbf{J} \cdot \hat{\mathbf{R}}$ commutes with both Λ and H_{rot}. Values of J and l, on the other hand, are mixed by the dipole coupling according to selection rules $\Delta J = 0, \pm 1$ and $\Delta l = \pm 1$.

The SO(4) symmetry is related to the diagonalization of Λ within blocks of constant J, so that only the angular momentum of the electron is mixed. This is a model for leading-order threshold photodetachment of the electron from a rotating molecular anion.[69] The $\Delta J = 0$ part of the operator Λ is

$$\Lambda_{JK} = \mathbf{l}^2 - 2\bar{\mu}\,(\mathbf{J} \cdot \hat{\mathbf{r}}) \tag{3.6}$$

where the effective dipole parameter $\bar{\mu} \equiv \mu K / J(J+1)$ depends on both J and K. There is no l mixing when $K = 0$, because $\bar{\mu}$ vanishes in this case. Strong coupling between channels with the same molecule quantum number J can occur only when there is a nonzero component K along the symmetry axis. The matrix elements for constant J are

$$\langle (JK, l)LM | \Lambda_{JK} | (JK, l)LM \rangle = l(l+1)$$

$$\langle (JK, l+1)LM | \Lambda_{JK} | (JK, l)LM \rangle \tag{3.7}$$

$$= -\bar{\mu} \left[\frac{(L+J+l+2)(J-L+l+1)(L-J+l+1)(L+J-l+1)}{(2l+1)(2l+3)} \right]^{1/2}$$

One consequence of the tridiagonal structure of this matrix is that the eigenvalues are the same for $+\bar{\mu}$ and $-\bar{\mu}$, and hence the channel eigenvalues of Λ_{JK} and Λ_{J-K} are the same also.

The SO(4) connection is evident in the similar forms of the matrix elements of Λ_{JK} and the SO(4) Lie algebra generator b_z in Eqs. (3.7) and (2.13). The result is that eigenvalues of Λ_{JK} may be computed in the SO(4) basis $|pqlm\rangle$ with

$$\begin{aligned} \Lambda_{JK} &= \mathbf{l}^2 - \bar{\mu}b_z \\ p &= J + L \\ q &= J - L \\ m &= 0 \end{aligned} \tag{3.8}$$

Equation (3.8) illustrates the intrinsic SO(4) symmetry of the electron–molecule dipole coupling channels. This is a result of pure angular-momentum coupling, although with respect to an internal symmetry axis of the problem. The operators l_z and b_z (this z should not be confused with the laboratory axis), for example, are SO(4) representations of $(\mathbf{l} \cdot \hat{\mathbf{r}}) = 0$ and $2(\mathbf{J} \cdot \hat{\mathbf{r}})$, respectively. The basis for an SU(2)$\times$SU(2) representation of the SO(4)

18   DAVID R. HERRICK

group, obtained by diagonalizing b_z, thus corresponds to a rotating-frame picture in which $\mathbf{J}\cdot\mathbf{r}$ is diagonal. Rotating frames are used frequently in scattering problems,[67,68] and it is interesting that they can be described with SO(4).

The physical interpretation of the dipole symmetry is that the total angular momentum $\mathbf{L}$ is conserved about a collective electron–molecule axis $\mathbf{J}+\bar{\mu}\hat{\mathbf{r}}$. When $\bar{\mu}$ is small the approximate symmetry is SO(4)⊃SO(3)⊃SO(2) for a spherical representation in which $\mathbf{l}^2$ is diagonal. When $\bar{\mu}$ is large the approximate symmetry is SO(4) = SU(2)×SU(2) for a representation in which b_z is diagonal. The intermediate coupling regime has $\mathbf{L}\cdot(\mathbf{J}+\bar{\mu}\hat{\mathbf{r}})$ diagonal, and this is related to the operator Λ_{JK} by

$$\Lambda_{JK} = J(J+1) + L(L+1) - 2\mathbf{L}\cdot(\mathbf{J}+\bar{\mu}\hat{\mathbf{r}}) \qquad (3.9)$$

One physical consequence of the dipole channels and conservation of angular momentum about the internal SO(4) symmetry axis is the occurrence of anomalous threshold behavior in the radial part of the problem. The radial equations for each J decouple in the dipole channel basis,

$$\left(-\frac{d^2}{dr^2} + \frac{\lambda(\lambda+1)}{r^2} - k_{JK}^2\right)F_\lambda = 0 \qquad (3.10)$$

where $\Lambda_{JK} \equiv \lambda(\lambda+1)$ is the effective centrifugal angular momentum. When $\bar{\mu}=0$ this is just the usual term $l(l+1)$. When $\bar{\mu}\neq 0$, however, λ differs from

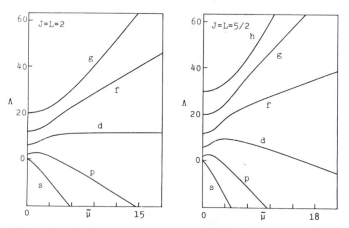

Fig. 3. The curves show eigenvalues of the generalized centrifugal angular momentum $\Lambda = \lambda(\lambda+1)$ versus the effective dipole parameter $\bar{\mu}$ for coupling of an electron to the rotating molecule; $s, p,\ldots$ label the electron's angular momentum ($\lambda = l$) when $\bar{\mu}=0$.

l and may even become negative or complex if the coupling is sufficiently strong. Some examples of the eigenvalues of Λ_{JK} are shown in Fig. 3 as a continuous function of $\bar{\mu}$. The point is, negative values of λ give rise to fractional exponents in the general threshold formula[78] $\sigma \propto k^{2\lambda+1}$.

B. Electron–Hydrogen Channels

The same coupled-channel approach and radial equations in (3.4) continue to apply when the three rotational degrees of freedom of the symmetric top are replaced by the three coordinates of the bound electron in a hydrogenlike atom. The difference is the form of the channel states and the quantization of the effective centrifugal operator Λ. The position and angular momentum of the outer particle are $\mathbf{r}$ and $\mathbf{l}$, and the position and angular momentum of the bound particle are $\mathbf{r}_1$ and $\mathbf{l}_1$. The basic uncorrelated single-configuration channel states are $|Nl_1m_1\rangle|lm\rangle$, where N is the principal quantum number of the orbital for the bound electron. $L = \mathbf{l}_1 + \mathbf{l}_2$ is the total angular momentum, and single-configuration channels with $\mathbf{L}^2 = L(L+1)$ and $L_z = M$ conserved are $|(Nl_1, l)LM\rangle$.

Interchannel coupling that mixes different values of N, l_1, and l originates in the Coulomb repulsion operator $1/|\mathbf{r}-\mathbf{r}_1|$, and the attraction of the outer electron to the nucleus of the atom. At large r the leading-order charge-dipole interaction is

$$V = \frac{\mathbf{r}_1 \cdot \hat{\mathbf{r}}}{r^2} \tag{3.11}$$

This is combined with the $1/r^2$ centrifugal potential of the outer electron to obtain the operator Λ for use in Eq. (3.4):

$$\Lambda = \mathbf{l}^2 + 2\lambda(\mathbf{r}_1 \cdot \hat{\mathbf{r}}) \tag{3.12}$$

The parameter λ has been introduced here to indicate the form of the operator in the more general case of an atom with charge Z, and an outer ion with charge Z_2 and mass M_2. The form of the coupling parameter is $\lambda = -M_2Z_2/Z$. This reduces to $\lambda = 1$ for electron–hydrogen coupling, for which $Z = 1$, $Z_2 = -1$, and $M_2 = 1$ in atomic units.

The SO(4) symmetry part of Λ is associated with the $\Delta N = 0$ part of the dipole interaction. This is similar to the investigation of Λ for constant J in the electron–molecule channels. Here the constant-N part of Λ is simplified by the replacement of the bound-electron coordinate with the SO(4) generator $\mathbf{b}_1$ as in Eq. (2.24). The result is an effective dipole channel operator

$$A = \mathbf{l}^2 + 3N\lambda(\mathbf{b}_1 \cdot \hat{\mathbf{r}}) \tag{3.13}$$

which is the analogue of the operator Λ_{JK} in the molecular problem. The atomic dipole channels have a more complicated structure, however, because now there is mixing of angular-momentum quantum numbers for both the outer electron and the bound electron. Diagonalization of A leads to a quantization of channels involving labels which account for only four degrees of freedom: A, L, M, and N. These are the exact symmetry blocks for the coupled radial equations in (3.4) when Λ is replaced by A. Since there are a total of five degrees of freedom, there must be another quantum label for the system. Quantization of the fifth degree of freedom is linked to diagonalization of the operator[76]

$$W = \mathbf{L} \cdot (3N\lambda\hat{\mathbf{r}} - 2\mathbf{b}_1) \qquad (3.14)$$

representing a component of the total angular momentum along a collective internal axis of the system. W commutes with A and $\mathbf{L}$, and so a complete set of labels for the atomic dipole channels is N, A, W, L, and M.

Except for a few special cases the spectra of the channel invariants are determined on the computer. The interpretation of these results is complicated by the large number of channels that can occur, growing as N^2 for higher values of L and N. Furthermore, there are many crossings of eigenvalues as the value of the parameter λ is varied; this makes it difficult to carry over interpretations of dipole structure from one atom to another, for example. It is also difficult to understand the electron correlation in the computed wave functions for channels, since these contain complicated admixtures of many different values of l and l_1.

The interpretation of channel spectra and correlation simplifies considerably if one adopts an SO(4) approach. The problem is reduced to one of classifying structure according to one of the representations SO(4)⊃SO(3)⊃ SO(2) or SO(4) = SU(2)×SU(2). The result is that when $\lambda \simeq 0$ the dipole angular momentum is $A \simeq l(l+1)$ and the spectrum has the approximate symmetry SO(4)⊃SO(3)⊃SO(2) for a representation with Lie-algebra generators $\mathbf{L} = \mathbf{l}_1 + \mathbf{l}$ and $\mathbf{C} = \mathbf{b}_1 + \mathbf{l}$. When λ is large, on the other hand, the approximate symmetry is SU(2)×SU(2) from a projection of SO(4) generators of the bound electron along the frame of the outer particle: $A \simeq 3N\lambda(\mathbf{b}_1 \cdot \hat{\mathbf{r}})$ and $W \simeq 3N\lambda(\mathbf{l}_1 \cdot \hat{\mathbf{r}})$. An understanding of channels for the intermediate values of λ, including ones for the e–H and e–He$^+$ systems, is thus gained by understanding how to correlate states for the two different SO(4) limits.

The labeling scheme that has been adopted uses quantum numbers from SU(2)×SU(2) in the high-λ limit. These are K and Q defined as

$$\mathbf{l} \cdot \hat{\mathbf{r}} = -Q, \qquad \mathbf{b}_1 \cdot \hat{\mathbf{r}} = -K \qquad (3.15)$$

The correspondence between these channel labels and the SO(4) labels k and

m in (2.25) for the one-electron Stark spectra is

$$K \leftrightarrow k, \qquad Q \leftrightarrow m \qquad (3.16)$$

A hierarchy of labels for all five degrees of freedom is then

$$
\begin{aligned}
&N = 1, 2, 3, \ldots \\
&Q = N-1, N-2, \ldots, 1-N \\
&K = N-|Q|-1, N-|Q|-3, \ldots, 1+|Q|-N \qquad (3.17) \\
&L = |Q|, |Q|+1, \ldots \\
&M = L, L-1, \ldots, -L
\end{aligned}
$$

The physical interpretation of channels at high λ is linked to the representation of Stark states for the hydrogen atom. The actual spectra of A for the electron–hydrogen channels are in fact quite similar to the first-order Stark splittings. This is shown in Fig. 4 for the case $N = 5$, including only states $L = |Q|$ for which there is no overall rotational excitation. The values $Q = 0, \pm 1, \pm 2, \pm 3, \pm 4$ thus correspond to S-, P-, D-, F-, and G-wave dipole channels in the usual spectroscopic notation. Comparison of Figs. 4 and 2

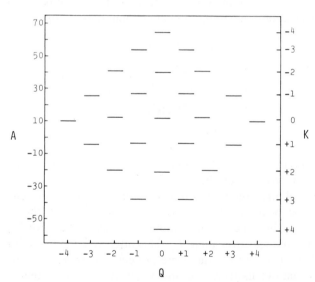

Fig. 4. Eigenvalues of the generalized centrifugal angular momentum A for electron–hydrogen dipole coupling channels with principal quantum number $N = 5$ and total angular momentum $L = |Q|$. Here K and Q label states in an SU(2)×SU(2) picture of Stark levels in the strong-coupling limit.

shows the very high degree of $SU(2) \times SU(2)$ symmetry of the hydrogen dipole channels. The first-order Stark pattern is broken only slightly, and may be estimated with perturbation formulas from the high-λ limit as

$$A \simeq -\bar{u}K + \tfrac{1}{2}\left[N^2 - 1 - K^2 - 3Q^2 + 2L(L+1)\right]$$
$$\qquad - \tfrac{1}{4}K\bar{\mu}^{-1}\left[N^2 - 1 - K^2 - 15Q^2 + 8L(L+1)\right] + \cdots \qquad (3.18)$$
$$W \simeq -Q\left\{\bar{\mu} + 2K + \bar{\mu}^{-1}\left[N^2 - 1 - K^2 - 3Q^2 + 2L(L+1)\right] + \cdots\right\}$$

in terms of the parameter $\bar{\mu} \equiv 3N\lambda$.

In addition to the dependence on the Stark-like quantum numbers K and Q, the hydrogen dipole channels depend on the value of the total orbital angular momentum L. The perturbation formula for each channel (with fixed values of K and Q) suggests a form

$$A \simeq a_0 + a_1 L(L+1) \qquad (3.19)$$

This is essentially the formula for the energy of a rigid rotor, though the two parameters a_0 and a_1 do depend on K and Q. The rotor structure is evidently linked to yet *another* representation of SO(4) in the channel basis. In this picture N and Q are preserved, but K is replaced by a new label $d \equiv \tfrac{1}{2}(N - 1 + K + |Q|)$ that is derived from representations of SO(4)$\supset$SO(3)$\supset$SO(2). The second hierarchy of channel quantum numbers is

$$N = 1, 2, 3, \ldots$$
$$d = 0, 1, \ldots, N - 1$$
$$Q = d, d - 1, \ldots, -d \qquad (3.20)$$
$$L = |Q|, |Q| + 1, \ldots$$
$$M = L, L - 1, \ldots, -L$$

Note that the role of d here is similar to that of l for spherical representations of the hydrogen atom in (2.28).

The physical interpretation of the quantum number d is not entirely clear, but the analogy with the hydrogen-atom functions suggests that the value of $N - d - 1$ could be linked to the number of nodes in the wave function with respect to the interelectron angle θ_{12}. This interpretation is supported by the discovery[77] that d controls the correspondence between channel quantum numbers in the two limits $\lambda = \infty$ and $\lambda = 0$. It is expected that the number of nodes is conserved as λ is varied, but no one has yet investigated the actual θ_{12} dependence of the channel wave functions explicitly. The formula relating l for the weak-coupling channels $A \simeq l(l+1)$ at $\lambda = 0$ to the labels at high values of λ is

$$l = \max(L, d) - K \qquad (3.21)$$

The relationship of all this to the L dependence of the channel spectra is that at high L, the spectrum of A for *each* value of λ becomes

$$A \simeq l(l+1) \qquad (3.22)$$

where l is given in (3.21). The simple form of this result shows that a rotor interpretation of the spectra, like that in (3.19) for example, becomes exact as the value of L grows larger. Moreover, there is a degeneracy of channels which have the same value of l. For additional details of the degeneracy, pictures of the rotorlike channel spectra including breaking of the degeneracy at low L, and pictures of the λ dependence of spectra, see Reference 77. The λ dependence is important because it is linked to the degree of angular correlation of the electrons by the formulas

$$\frac{dA}{d\lambda} = 3N^2 \langle NKQLM, \lambda | \cos\theta_{12} | NKQLM, \lambda \rangle$$
$$\frac{dW}{d\lambda} = 3N \langle NKQLM, \lambda | \mathbf{l}_1 \cdot \hat{\mathbf{r}} | NKQLM, \lambda \rangle \qquad (3.23)$$

At high values of λ this becomes simply $\cos\theta_{12} \simeq -K/N$, $\mathbf{l}_1 \cdot \hat{\mathbf{r}} \simeq -Q$. The most favorable correlation, when the electrons are on opposite sides of the nucleus, occurs when $K = N - 1$ and $Q = 0$.

Nikitin and Ostrovsky[79] have derived some additional details of the L dependence of dipole channels, based on a perturbation expansion of the generalized angular momentum in the high-L regime. Greene[80] has discussed implications of the degeneracy of A for $\pm Q$ in terms of possible asymmetry of photofragment distributions. The relationship of the spectra of the channel operator A to the actual energy spectra of bound states below threshold will be discussed in Section IV.

C. Comparison of Molecular and Atomic Channels

Although they are very different physical problems, the fact that the electron–molecule and electron–atom dipole channels both have SO(4) symmetry suggests that the channel structure could become identical in some limits. This is found to occur when the atomic channels are restricted to $L = 0$, which corresponds to S-wave states for the two-electron system. In this case the angular momenta of the two electrons satisfy $\mathbf{l}_1 + \mathbf{l} = 0$, and the operator A in Eq. (3.13) becomes

$$A = \mathbf{l}_1^2 + 3N\lambda(\mathbf{b}_1 \cdot \hat{\mathbf{r}}) \qquad (3.24)$$

Comparison with Eq. (3.8) shows that the SO(4) structure of the atomic channel A and the molecular channel operator Λ_{JK} is the same after we make the following correspondence of quantum numbers and parameters for the two problems:

$$\textit{Molecule } (J = L) \quad \textit{Atom } (L = 0)$$

$$l \quad \leftrightarrow \quad l_1 = l$$

$$J \quad \leftrightarrow \quad \tfrac{1}{2}(N-1)$$

$$|\bar{\mu}| \quad \leftrightarrow \quad 3N|\lambda|$$

The parameter $\bar{\mu} = 3N\lambda$ was also used in the perturbation expansion of A in Eq. (3.18).

As an example of this correspondence between the two problems, the molecular channels for $L = J = 2$ and $L = J = \tfrac{5}{2}$ shown in Fig. 3 are also representations of the atomic channels $L = 0$ for $N = 5$ and $N = 6$, respectively. There is a similar physical interpretation of states at high or low $\bar{\mu}$ as representations of $SO(4) = SU(2) \times SU(2)$ or $SO(4) \supset SO(3) \supset SO(2)$. The physical difference is that typical molecular dipole moments correspond to values of the coupling parameter with $\bar{\mu} \cong 0$, while the values for low-Z atoms like H ($\lambda = 1$) or He$^+$ ($\lambda = \tfrac{1}{2}$) would lie in the other limit $\bar{\mu} \gg 1$. The fact that $\bar{\mu}$ and λ may have different signs in different problems is not important in the channel spectrum, which is the same for $+\bar{\mu}$ and $-\bar{\mu}$. The molecular channels are thus more like $SO(4) \supset SO(3) \supset SO(2)$ spectra, while the low-Z atomic channels are more like $SO(4) = SU(2) \times SU(2)$.

One consequence of the different values of $\bar{\mu}$ for the two problems is the different type of photodetachment threshold behavior that can occur. In the electron–molecule problem the small value of $\bar{\mu}$ means that most of the channels have cross section scaling $\sigma \propto k^x$ with $x > 1$, while only the lowest channel eigenvalue for the state with $J = L = 0$ and $l = 0$ could have $x < 0$. This is the basis of Engelking's interpretation of observed thresholds.[69]

In the electron–hydrogen system the values of $\bar{\mu}$ are so large that many low-L channels can have eigenvalues $A < 0$. When $A < -\tfrac{1}{4}$ the effective radial potential $V \sim A/2r^2$ is attractive enough to support an infinite number of bound states,[81] and the predicted threshold scaling is more abrupt, and oscillatory.[72,78] An additional complication is the possibility of resonance structure in the cross section, due to the short-range part of the potential that is neglected in the dipole approximation. Hyman et al.[75] have illustrated the dipole channel classification of computed photodetachment cross sections for the final state of hydrogen in the $N = 2$ shell.

D. Semiclassical SO(4) Symmetry

Though diagonalization of the operator A in Eq. (3.13) gives the exact quantum-mechanical angular correlation in the dipole coupling model, additional insight to the SO(4) symmetry of the system is gained if the outer particle is treated classically. In the case of hydrogen the assumption is a heavy ion passing by in a straight-line trajectory. The l-mixing of the bound electron is described with time-dependent quantum mechanics, and in a first approximation[82, 83] the dipole model predicts transitions $\Delta l = \pm 1$. Here we shall outline a derivation of the exact solution of the dipole coupling model when only transitions between degenerate levels are allowed. The result is that the overall evolution of the bound electron within the manifold is described by an SO(4) group rotation.

The notation is now $\mathbf{l}$ and $\mathbf{b}$ for generators of the hydrogen SO(4) Lie algebra. The trajectory of the outer particle is $\mathbf{r} = (-vt, 0, a)$ with velocity v, impact parameter a, and time t. We denote by N the principal quantum number of the manifold for the bound electron, and dipole coupling within this is described by

$$V = 1.5N(\mathbf{b} \cdot \mathbf{F}) \tag{3.25}$$

with components of the vector $\mathbf{F}$ defined as

$$F_x = -\frac{|Z|vt}{(a^2 + v^2t^2)^{3/2}}, \qquad F_y = 0, \qquad F_z = -\frac{|Z|a}{(a^2 + v^2t^2)^{3/2}} \tag{3.26}$$

Z is the charge of the outer particle.

The l mixing obtained from V amounts to a Stark mixing in a rotating field. The equation of motion is

$$i\frac{\partial U}{\partial t} = 1.5N(\mathbf{F} \cdot \mathbf{b})U \tag{3.27}$$

with $U = 1$ initially. Equation (3.27) assumes the field-free levels are degenerate. The exact solution is found by taking into account the SO(4) symmetry of the problem, which is evidently that U is a rotation operator for the Lie group. If X is a Lie-algebra generator, the corresponding Lie group generator is $\exp(iX)$ in the quantum-mechanical notation. The effect of a group rotation on another generator Y is

$$\exp(iX)\,Y\exp(-iX) = Y + i[X, Y] + \frac{i^2}{2!}[X, [X, Y]] + \cdots \tag{3.28}$$

In order to use this for hydrogen we define an angle α by the formula $\tan\alpha = F_z/F_x$. Over the course of the outer particle's trajectory $\alpha = 0$ initially, and $\alpha = \pi$ radians in the final state. After making the transformation from t to α, Eq. (3.27) is

$$i\frac{\partial U}{\partial \alpha} = w(b_x\cos\alpha + b_z\sin\alpha)U$$
$$= w\exp(i\alpha l_y)\,b_x\exp(-i\alpha l_y)\,U \qquad (3.29)$$

with the new parameter $w \equiv 3N|Z|/2a\upsilon$. Since w is independent of time, the exact solution is

$$U = \exp[i\alpha l_y]\exp[-i\alpha(l_y + wb_x)] \qquad (3.30)$$

or

$$U = \exp(i\alpha l_y)\exp(i\gamma b_z)\exp\left(-\frac{i\alpha l_y}{\cos\gamma}\right)\exp(-i\gamma b_z) \qquad (3.31)$$

Here the angle γ is related to w by the formula $w = \tan\gamma$.

The final result is obtained by taking the limit $\alpha \to \pi$, and

$$U = \exp(-i\gamma b_z)\exp(-i\beta l_y)\exp(-i\gamma b_z) \qquad (3.32)$$

where the two angles may be expressed simply as

$$\cos\gamma = \frac{1}{(1+w^2)^{1/2}}$$
$$\beta = \pi\left[(1+w^2)^{1/2} - 1\right] \qquad (3.33)$$

The point is that the overall time evolution of the hydrogen atom in the field of the passing ion boils down to a product of SO(4) group rotations in Eq. (3.32). Biedenharn has described some properties of SO(4) rotation-matrix elements.[53] One feature of (3.32) is that it includes infinite-order time dependence. The usual first-order result is obtained as the linear term in an expansion with respect to the parameter w,

$$U \simeq 1 - 2iwb_z + \cdots \qquad (3.34)$$

and this describes dipole transitions because of the selection rule $\Delta l = \pm 1$ for the Lie-algebra generator b_z.

IV. DOUBLY EXCITED TWO-ELECTRON ATOMS

A. Introduction

After scaling $H \to H/Z^2$ and $r \to r/Z$, the Hamiltonian for two-electron atoms (atomic units) is

$$H = -\tfrac{1}{2}\left(\nabla_1^2 + \nabla_2^2\right) - \frac{1}{r_1} - \frac{1}{r_2} + \frac{\lambda}{r_{12}} \tag{4.1}$$

with $\lambda \equiv 1/Z$. Operators which commute with H are the total angular momentum $\mathbf{L} = \mathbf{l}_1 + \mathbf{l}_2$ and spin $\mathbf{S} = \mathbf{s}_1 + \mathbf{s}_2$. There are altogether eight degrees of freedom, including spin. Four of these are removed by block-diagonalizing the energy according to quantum numbers L, M, S, M_S for the mutually commuting operators $\mathbf{L}^2 = L(L+1)$, $L_z = M$, $\mathbf{S}^2 = S(S+1)$, and $S_z = M_S$. Another exact symmetry is the total parity Π for inversion of electrons through the origin.

The four degrees of freedom remaining can be described in a variety of ways, none of which exactly separates the energy. They are essentially related to the two radial coordinates r_1 and r_2, and interparticle correlation contained in the two operators $\mathbf{l}_1 \cdot \mathbf{l}_2$ and $\hat{\mathbf{r}}_1 \cdot \hat{\mathbf{r}}_2 \equiv \cos\theta_{12}$. Correlation in the energy originates entirely in the Coulomb repulsion operator, $1/r_{12}$.

Including antisymmetrization, the single-configuration representation of states is $|(Nl, nl')LMSM_S\rangle$, with $N - l - 1$ and $n - l' - 1$ labeling the number of nodes in each radial function. Subsequent notation will assume $n \geqslant N$. The quantization is exact in the limit $\lambda = 0$, where the energy is that of two independent electrons in hydrogen-atom orbitals,

$$E = -\frac{1}{2}\left(\frac{1}{N^2} + \frac{1}{n^2}\right) \tag{4.2}$$

Doubly excited configurations have both electrons in excited shells, $n \geqslant N \geqslant 2$, and lie above the first ionization limit. The hydrogen atom symmetry is broken when $\lambda > 0$, and there is mixing of different values of the approximate symmetry labels N, n, l, and l'. The mixing of l and l' when $\Delta N = 0$ and $\Delta n = 0$ is particularly strong, because these configurations are nearly degenerate in the single-configuration picture. There is also autoionization due to coupling of the doubly excited configurations with the continuum.

Theoretical interest in the configuration mixing structure of doubly excited states stems from the puzzling[84,85] set of channels of ^1P resonances in the photoabsorption spectrum of helium[86] in the range 60–65 eV just below threshold for $He^+(2s$ or $2p) + e$. Two series of levels, $(2s, np)$ and $(2p, ns)$, are expected on the basis of dipole selection rules from the ground $(1s, 1s)$

state, but only one series of levels stood out in the spectrum. A qualitative reason was[87] that electron correlation in the excited states is described better with configuration-mixed channels $(2s, np) \pm (2p, ns)$. The interpretation of radial functions suggested that $r_1 - r_2$ was smaller in the $+$ channel, and this would lie higher in energy. Radiative excitation would go preferentially to the $+$ channel because the radial motion of the electrons is in phase, and could overlap strongly with the ground state. Actual wave functions for doubly excited states are far more complicated than this, since both angular and radial degrees of freedom are involved. The basic theoretical problem has been one of drawing a picture of the correlation that is easily understood.

The picture we are suggesting is derived from an interpretation of the configuration-mixing correlation with an approximate SO(4) symmetry group. Alper and Sinanoğlu[88] and Wulfman[89] were the first to investigate SO(4) for electron correlation. In the case of two electrons, they concluded that the Lie-algebra generator $A = b_1 + b_2$ and the coupled basis in Eq. (2.15) described an approximate symmetry. Later, this interpretation was shown to be incorrect.[90-94] After more careful investigation of the operator $1/|r_1 - r_2|$ in the light of SO(4) representation of $r_1 - r_2$ in constant-energy shells, it was found[61,95] there *is* an approximate symmetry, related to the generator $B = b_1 - b_2$ and the coupled basis in Eq. (2.17). The result that B generates the more physical approximate symmetry has been interpreted as a type of "maximal symmetry breaking" of exchange and parity,[47] and in terms of the fundamental symmetry of a generalized pairing interaction for two electrons in the same shell.[96,97] Most of the current interpretations of correlation are derived from analyses of configuration-interaction wave functions with the coupled SO(4) basis. It was through studies like this that the SO(4) channel structure of doubly excited states was discovered, for example.[98,99] Recent work has focused on the so-called "intrashell" part of the spectrum, with $n = N$, and this forms the basis of a supermultiplet picture of the spectra,[46,47] including the interpretation of states as rotor–vibrator levels.[48,49] Additional areas of interest are the generalization to systems of variable dimensionality[100] and atoms with more than two electrons.[101]

B. Approximate Channel Decoupling

The initial description of channel decoupling for two-electron atoms was by Herrick and Sinanoğlu.[98] Working with configuration-interaction wave functions, they found that interactions with $\Delta N = 0$ are blocked approximately according to the selection rules

$$\Delta|Q| = 0, \qquad \Delta(P - n + 1) = 0 \tag{4.3}$$

Here P and Q are the SO(4) labels for the configuration-mixed basis (2.17), with invariants $\mathbf{B}^2 + \mathbf{L}^2 = P(P+2) + Q^2$ and $|\mathbf{L} \cdot \mathbf{B}| = |Q|(P+1)$. The use of the label $|Q|$ instead of Q is related to the fact that $\mathbf{L} \cdot \mathbf{B}$ mixes states of different parity and exchange; simple $\pm$ admixtures of $+|Q|$ and $-|Q|$ states have definite parity. For convenience, the label $T \equiv |Q|$ is used, and the quantum number K is defined as $K \equiv P - n + 1$. The meaning of K and Q is essentially the same as the dipole-channel labels K and Q in Section III.B, and so the basic SO(4) structure of the channels is described with SU(2)$\times$ SU(2) and the hierarchy in Eq. (3.17).

The selection rules in (4.3) can be rationalized in terms of matrix elements of the operator $\cos \theta_{12}$, which gives the degree of correlation of the two electrons. When $\cos \theta_{12} = +1$, the two electrons are on the same side of the nucleus, and the electron–electron repulsion is high. When $\cos \theta_{12} = -1$, the two electrons are on opposite sides of the nucleus, and their repulsion energy is low. Evaluation of matrix elements of $\cos \theta_{12}$ when there is no exchange is simplified by the operator replacement $\cos \theta_{12} \rightarrow (\mathbf{b}_1 \cdot \mathbf{b}_2)/Nn$ for the part $\Delta N = 0$, $\Delta n = 0$. In the SO(4)-channel basis this rearranges to[62]

$$\cos \theta_{12} \rightarrow -\frac{K}{N} + \frac{N^2 - 1 - K^2 - Q^2 + 2\mathbf{l}_1 \cdot \mathbf{l}_2}{2Nn} \tag{4.4}$$

The second term contains a part $\mathbf{l}_1 \cdot \mathbf{l}_2$ that mixes different SO(4) states, but this is small when $n \gg N$ for constant L. The result is $\cos \theta_{12} \simeq -K/N$ in the SO(4) basis. This is depicted in Fig. 5, where pairs of classical Kepler orbits are shown for the two extreme cases corresponding to $K/N = \pm 1$. The interpretation of angular correlation is similar to that described for dipole channels in Eq. (3.23) in the strong-coupling limit, where it was also found

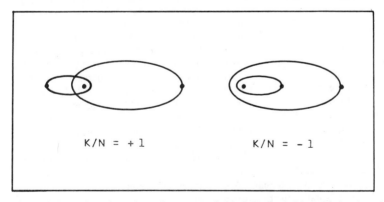

$$K/N = +1 \qquad\qquad K/N = -1$$

Fig. 5.   Relative orientation of nondegenerate classical Kepler orbits with $\mathbf{l}_1 + \mathbf{l}_2 = 0$, illustrating two extremes with favorable ($K/N = +1$) and unfavorable ($K/N = -1$) angular correlation.

that $\cos \theta_{12} \to - K/N$. In the full configuration-interaction matrix (constructed with hydrogen-atom orbitals which diagonalize the one-electron energies), the Coulomb repulsion mixing of the type $\Delta N \neq 0$, $|\Delta K| \gg 0$ is expected to be small, because of a corresponding small overlap of the part of the wave functions depending on θ_{12}.

The actual decoupling found in the configuration-interaction wave function is much better than could be expected from (4.4) alone. An example of the (K, T) symmetry classification is shown in Table I for low-lying states of the helium Rydberg series for $^1P^o$ levels in the shell $N = 3$. A similar table constructed on the basis of configuration assignments (l, l'), on the other hand, would show very strong configuration mixing for each level. The empirical rule for the strongest channel in photoexcitation from the ground state is[98] $(K, T) = (N - 2, 1)$. The physical picture in these quantum numbers is electrons on opposite sides of the nucleus, and one unit of angular momentum excited about the symmetry axis.

The explicit SO(4) coupling coefficients which diagonalize $\mathbf{B}^2$ and $\mathbf{L} \cdot \mathbf{B}$ for $L = 1$, $N = 3$ are shown in Table II, where they are displayed as functions of n. Another example is the configuration mixing for the shell $N = 2$ (all L) in Table III. It is interesting to note that the series with $K = N - 2$ has a mixing coefficient for $(2s, np)$ that goes to zero as n approaches infinity. This would destroy the simple picture of $(2s, np) + (2p, ns)$. It is not known if this

TABLE I

Channel Classification of $^1P^o$ Levels of Helium

below Threshold for $\mathrm{He}^+(\mathrm{N} = 3) + e$ [a]

n^*		Fraction of total normalization in (K, T) [b]				
Theory[b]	Expt.[c]	$(2, 0)$	$(1, 1)$	$(0, 0)$	$(-1, 1)$	$(-2, 0)$
2.109	2.125	0.0003	0.9951	0.0009	0.0035	0.0002
2.806		0.9980	0.0016	0.0004	0.0000	0.0000
3.011		0.0000	0.0222	0.0028	0.9273	0.0476
3.245	3.240	0.0026	0.9686	0.0065	0.0205	0.0017
3.347		0.0001	0.0081	0.9857	0.0046	0.0014
3.779		0.9942	0.0054	0.0004	0.0000	0.0000
4.190		0.0009	0.1391	0.0147	0.8266	0.0187
4.260	4.25	0.0056	0.8390	0.0075	0.1423	0.0055
4.411		0.0001	0.0154	0.9666	0.0124	0.0055
4.741		0.0012	0.0005	0.0072	0.0171	0.9740
4.765		0.9892	0.0092	0.0003	0.0000	0.0013

[a] The effective quantum number of the outer electron is n^*.

[b] Configuration-interaction prediction.

[c] Experiment, Reference 86.

TABLE II

SO(4) Configuration-Mixing Coefficients for Configurations $(3l, nl')$ with Total Angular Momentum $L=1$

(K,T)	$(3s, np)$	$(3p, ns)$	$(3p, nd)$	$(3d, np)$	$(3d, nf)$
$(+2,0)$	$-\left[\dfrac{(n-1)(n+2)(n+3)}{3n(n+1)^2}\right]^{1/2}$	$\left[\dfrac{(n+2)(n+3)}{6n(n+1)}\right]^{1/2}$	$-\left[\dfrac{(n-1)(n-2)(n+3)}{3n(n+1)^2}\right]^{1/2}$	$\left[\dfrac{(n-1)(n+2)(n+3)}{15n(n+1)^2}\right]^{1/2}$	$-\left[\dfrac{(n-1)(n-2)(n-3)}{10n(n+1)^2}\right]^{1/2}$
$(+1,1)$	$\left[\dfrac{8(n+2)}{3n(n+1)^2}\right]^{1/2}$	$\left[\dfrac{(n+2)(n-1)}{3n(n+1)}\right]^{1/2}$	$\dfrac{n+5}{n+1}\left[\dfrac{n-2}{6n}\right]^{1/2}$	$\left[\dfrac{(n+2)(3n-1)^2}{30n(n+1)^2}\right]^{1/2}$	$\left[\dfrac{(n-2)(n+3)(n-3)}{5n(n+1)^2}\right]^{1/2}$
$(0,0)$	$-\dfrac{n^2-5}{(n^2-1)\sqrt{3}}$	$\left[\dfrac{2}{3(n^2-1)}\right]^{1/2}$	$\left[\dfrac{16(n^2-4)}{3(n^2-1)^2}\right]^{1/2}$	$-\left[\dfrac{4(n^2+1)^2}{15(n^2-1)^2}\right]^{1/2}$	$\left[\dfrac{2(n^2-4)(n^2-9)}{5(n^2-1)^2}\right]^{1/2}$
$(-1,1)$	$\left[\dfrac{8(n-2)}{3n(n-1)^2}\right]^{1/2}$	$\left[\dfrac{(n-2)(n+1)}{3n(n-1)}\right]^{1/2}$	$\dfrac{n-5}{n-1}\left[\dfrac{n+2}{6n}\right]^{1/2}$	$-\left[\dfrac{(n-2)(3n+1)^2}{30n(n-1)^2}\right]^{1/2}$	$-\left[\dfrac{(n+2)(n-3)(n+3)}{5n(n-1)^2}\right]^{1/2}$
$(-2,0)$	$-\left[\dfrac{(n+1)(n-2)(n-3)}{3n(n-1)^2}\right]^{1/2}$	$-\left[\dfrac{(n-2)(n-3)}{6n(n-1)}\right]^{1/2}$	$\left[\dfrac{(n+1)(n+2)(n-3)}{3n(n-1)^2}\right]^{1/2}$	$\left[\dfrac{(n+1)(n-2)(n-3)}{15n(n-1)^2}\right]^{1/2}$	$-\left[\dfrac{(n+1)(n+2)(n+3)}{10n(n-1)^2}\right]^{1/2}$

TABLE III
SO(4) Configuration-Mixing Coefficients for configurations
$(2l, nl')$ with Total Angular Momentum L

(K,T)	$(2s, nL)$	$(2p, nL-1)$	$(2p, nL+1)$
$(+1,0)$	$\left[\dfrac{(n-L)(n+L+1)}{2n^2}\right]^{1/2}$	$-\left[\dfrac{(n+L)(n+L+1)L}{(2L+1)2n^2}\right]^{1/2}$	$\left[\dfrac{(n-L)(n-L-1)(L+1)}{(2L+1)2n^2}\right]^{1/2}$
$(0,0)$	$\left[\dfrac{L(L+1)}{n^2}\right]^{1/2}$	$\left[\dfrac{(n+L)(n-L)(L+1)}{(2L+1)n^2}\right]$	$\left[\dfrac{(n+L+1)(n-L-1)L}{(2L+1)n^2}\right]^{1/2}$
$(-1,0)$	$\left[\dfrac{(n+L)(n-L-1)}{2n^2}\right]^{1/2}$	$\left[\dfrac{(n-L)(n-L-1)L}{(2L+1)2n^2}\right]^{1/2}$	$-\left[\dfrac{(n+L)(n+L+1)(L+1)}{(2L+1)2n^2}\right]^{1/2}$

really happens in the hydrogenic basis, because most of the calculations have
been performed with low-n orbitals, usually $n \leqslant 7$. One interesting point is
that the SO(4) mixing coefficients taken in the limit $n \to \infty$ reduce to the same
coefficients that diagonalize the dipole channel operator A (Section III.B) for
strong coupling.[62]

There is another approximate symmetry implicit in the SO(4) channel decoupling, which is related to a separation of channels into two classes that
we shall label $\nu = \pm 1$. The definition is

$$\nu = \Pi(-)^{T+S}$$
$$= \Pi P_{12}(-)^T \qquad (4.5)$$

where P_{12} is the symmetry of the wave function under exchange of only the
spatial coordinates of the two electrons. It is possible to do this because the
exact wave function for two electrons separates exactly into spin and spatial
parts. The label ν only describes an approximate symmetry, because T is only
an approximate symmetry label itself.

The physical difference between the two classes $\nu = +1$ and $\nu = -1$ is related to the corresponding symmetry of the SO(4) wave functions under exchange when $n = N$ (Section IV.D). The result is that the possible Rydberg
levels in each class are

$$\begin{aligned} \nu &= +1: \quad n = N, N+1, N+2,\ldots \\ \nu &= -1: \quad n = N+1, N+2, N+3,\ldots \end{aligned} \qquad (4.6)$$

The fact that only "plus" channels include intrashell states (with $n = N$)
means their electrons can occupy the same region of space, and hence have
higher repulsion energy. "Minus" channels, on the other hand, are killed off

when $n = N$ because they are forbidden by the Pauli exclusion principle when spin is taken into account. Their electrons are therefore not allowed to occupy the same region of space and, everything else being equal, would be expected to have lower repulsion energy than their "plus" counterparts. Basically, the labels $\nu = +1$ and -1 are a generalized SO(4) version of the "plus" and "minus" in the original classification of $(2s, np) \pm (2p, ns)$.

The different correlation of electrons in the two classes of doubly excited states also manifests itself in the autoionization process He** $\to$ He$^+ + e$, and $\nu = +1$ states tend to decay more rapidly than their $\nu = -1$ counterparts.[98] An approximate selection rule for ΔN follows from the fact that the SO(4) channel indices always have[62]

$$(-)^T = (-)^{N-1+K} \tag{4.7}$$

Since spin and parity are conserved in autoionization, the approximate symmetry of ν suggests that ΔT is *even*, and from (4.9) it is obvious that $\Delta(N + K)$ will be *odd*. On physical grounds, autoionization involving a strongly correlated initial state should tend to preserve the values of K (angular correlation) and T (angular momentum about the symmetry axis) as much as possible. In transitions $N \to N-1$, for example, the corresponding changes in the channel symmetry would be $\Delta T = 0$, $\Delta K = \pm 1$. This result was found to be in good agreement with close-coupling predictions of the coupling of $N = 3$ and $N = 2$ levels in H$^-$, and it was found that the strongest transitions were $\Delta T = 0$, $\Delta K = -1$.[62] The physical interpretation of this is that angular momentum about the symmetry axis and the angle between the electrons are approximately conserved during autoionization. This is one example of how the physical pictures we associate with the approximate symmetry labels for two-electron correlation can lend new insight to problems that are usually treated only by direct numerical approaches.

Very little is known at this point about the dependence of the energies of intershell levels ($n \neq N$) on the quantum numbers K, T, and L, though it is expected to be similar to the dependence of the dipole channel invariant A on these labels. As a first example of the types of spectra that could be found, we have arranged in Fig. 6 the set of helium levels with $N = 3$, $n = 4$, $L = T$, and $\nu = +1$ and -1. Recall that $L = T$ is the lowest possible value of L for a channel. There are two striking observations to be made about these pictures. First is the result that the spectrum for the states $\nu = +1$ is virtually identical in shape to that of the states $\nu = -1$, at least until you get out your ruler. The levels with $\nu = +1$ all lie 0.2 eV higher in energy than their -1 counterparts. This is evidently a result of the fact that "plus" states allow the electrons to come together more, and hence have higher repulsion energy. "Minus" wave functions are better correlated in the region $r_1 \simeq r_2$, θ_{12}

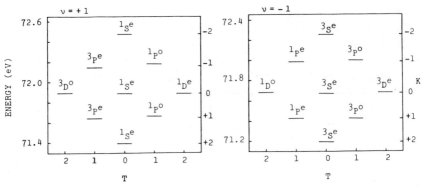

Fig. 6.   Configuration-interaction estimate of doubly excited helium spectra for $N = 3$, $n = 4$, $L = T$, and $\nu = \pm 1$.

$\simeq 0°$ where the Coulomb repulsion is singular, and have lower energies. The fact that each of the two spectra have an approximate left–right mirror symmetry is related to the fact that $(\mathbf{L} \cdot \mathbf{B})^2$ is a very good constant of the motion, with approximate twofold level degeneracy when $T \neq 0$. The actual splitting of the two levels would be the two-electron analogue of the general class of molecular doubling phenomena, such as Λ doubling, Ω doubling, and l doubling in linear molecules. The precise physical origin of doubling is of course different in each case. The analogy has led us to call the splitting of two atomic levels that are otherwise degenerate in an SO(4) picture "T doubling" (cf. Section IV.F).

The second striking feature of Fig. 6, if you compare it with Figs. 2 and 4, is that the $n = 4$ spectra look just like Stark levels for SO(4) = SU(2) × SU(2) in hydrogen. This is unexpected at this low n, because the hydrogen atom SO(4) symmetry with the Runge–Lenz vector is broken in helium, and the numerical configuration-interaction energies include some infinite-order effects when interpreted with a λ perturbation expansion of the energy about $\lambda = 0$. The implication of degeneracies like those in Fig. 6, which include different values of L, is an approximate separation of the wave function and energy. It is results like this that lead us to ask if, perhaps, atomic physics has overlooked something about two-electron atoms. There is no obvious set of coordinates that could have predicted these spectra without the benefit of a computer, although a reinvestigation of one-electron parabolic coordinates for two-electron atoms would seem to be in order. Some additional insight to the possible physical origin of the implied separability is given in Section IV.I, where a set of generalized parabolic coordinates for two electrons is proposed.

The levels in Fig. 6 only represent the lowest rotational state in each channel, $L = T$. When levels with $L > T$ are included the pictures become more complex, since one is trying to interpret what is basically a four-dimensional picture, with K, T, and L the variables for the energy. Both the intrashell studies (cf. Section IV.D) and dipole channels suggest that a way to interpret L-excitation spectra is to break them up into separate classes of levels according to the label $d = \frac{1}{2}(N - 1 + K + T)$ which was introduced in Eq. (3.20) for the subgroup chain SO(4)⊃SO(3)⊃SO(2). There the rule for angular momentum was $L \geq T$, with no upper bound because the angular momentum of the outer electron was unrestricted. In the SO(4) coupling picture for Rydberg states, on the other hand, the outer electron's angular momentum is restricted by $0 \leq l \leq n - 1$. In the coupled SO(4) model, this becomes a restriction on the total angular momentum, with $T \leq L \leq P$ for each value of $P = n - 1 + K$. This is nothing more than a triangle selection rule for the 9-j coupling coefficient in (2.17).

Each set of levels with the same label d is an example of one type of "supermultiplet" in our work. The supermultiplet with $d = 2$ in the shell $N = 3$, $n = 4$ of helium is shown in Fig. 7. The lowest level in each (K, T) symmetry channel has $L = T$. Note that L excitation of overall rotation of the two electrons increases the energy. The T doubling separates levels into two classes $\eta = \pm 1$, with

$$\eta = \Pi(-)^L \tag{4.8}$$

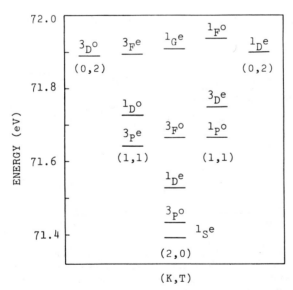

Fig. 7. Configuration-interaction estimate of L dependence of doubly excited helium levels in the supermultiplet $d = 2$ for $N = 3$, $n = 4$, and $\nu = +1$. There are not enough data available to show $^1F^e$, $^1G^o$, $^3H^o$, $^3G^e$, and $^3F^o$ levels for the highest value of L in each channel.

and levels with $\eta = +1$ have slighty higher energies than their -1 counterparts.

The SO(4) structure of L excitation for other channels is largely unexplored, because of a lack of data for higher-L levels. Nikitin and Ostrovsky[102] considered states when $n \gg N$ and $L \gg n$, and concluded that a different SO(4) coupling scheme might be better there. The states they proposed have constants of the motion l_2^2 and $\mathbf{l}_2 \cdot (\mathbf{l}_1 + \mathbf{b}_1)$, with 2 the outer electron. We have described a similar coupling for the dipole-channel operator W in the weak-coupling ($\lambda = 0$) limit.[76] The weak coupling is not useful for the strong correlation we are considering for low-L channels.

Analyses of the K, T symmetry for decoupled channels makes it very easy to sort out the complicated λ dependence of the doubly excited states. Examples are shown in Fig. 8 for $^1P^o$ and $^3P^o$ levels in the shell $N = 3$. Each level has an effective Rydberg quantum number n^* that is related to the binding energy $\varepsilon \equiv E + \frac{1}{2}N^{-2}$ of the outer electron by the formula

$$\varepsilon = -\frac{1}{2}\left(\frac{1-\lambda}{n^*}\right)^2 \tag{4.9}$$

$$n^* = (1-\lambda)(-2\varepsilon)^{-1/2} \tag{4.10}$$

This can be expanded in a power series in λ,

$$n^* = n + n_1^*\lambda + n_2^*\lambda^2 + \cdots \tag{4.11}$$

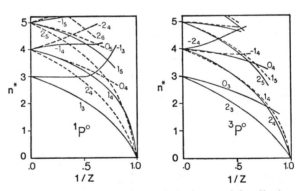

Fig. 8.   The curves show configuration-interaction estimates of the effective quantum number of the outer electron in doubly excited states when the inner electron is in the $N = 3$ shell, using continuous values of $\lambda = 1/Z$. The channel classification is that of Reference 98, with individual curves labeled as K_n. Values of T not shown in the figures are $T = 0$ for $K = 0, \pm 2$ and $T = 1$ for $K = \pm 1$.

just like the usual expansion of the energy itself,

$$\varepsilon = -\tfrac{1}{2}n^{-2} + \varepsilon_1\lambda + \varepsilon_2\lambda^2 + \cdots \tag{4.12}$$

At $\lambda = 0$, then, the curves in Fig. 8 approach a limiting value n that labels the independent-particle states. The other limit $\lambda = 1$ has $n^* = 0$ for levels bound below threshold. Not all the levels are bound in H^-, because the long-range potential in each dipole channel is $V \sim \tfrac{1}{2}Ar^{-2}$, and this supports an infinite number of states only when $A < -\tfrac{1}{4}$. The key point about Fig. 8 is the way it shows the breaking of the SO(4) hydrogen-atom degeneracy as λ increases. Levels with $K < 0$ mostly have $n^* > n$, corresponding to higher repulsion for the unfavorable angular correlation. Levels with $K > 0$ mostly have $n^* < n$, corresponding to lower repulsion for the more favorable angular correlation.

The Stark-like energies for helium in Fig. 6 suggest most of the splitting is due to K. An approximate formula can be derived if we consider only the first-order contribution in (4.11). The result is

$$n_1^* = \varepsilon_1 n^3 - n \tag{4.13}$$

where ε_1 is the eigenvalue of the Coulomb repulsion operator $1/r_{12}$ in the degenerate basis at $\lambda = 0$. A simple dipole approximation neglecting exchange when $n \gg N$ suggests the approximate leading-order scaling formula

$$n^* - n \propto -\lambda K \tag{4.14}$$

This is essentially the formula for a first-order Stark splitting of the degeneracy, only now the coupling parameter λ plays the role of the field intensity in Eq. (2.26). The effective quantum number has a linear dependence on λ in the weak-coupling limit (i.e. high Z), because of the degeneracy of independent hydrogenic states at $\lambda = 0$. Other examples of Stark-like splittings are shown in Reference 98, and in the recent work of Ho.[103, 104]

The actual splitting of levels is more complicated than the Stark formula (4.14), and includes exchange effects. It is reasonable to expect that features of the dipole-channel operator A will also be reflected in the actual energy spectra. This would be the case exactly if the energy of the outer electron could be described with an effective one-dimensional radial potential $V(A, r)$, for example. This type of approach has been used as a semiempirical method of predicting levels in H^-.[73, 105] Qualitative features of the spectra at each value of λ could be approximated with the model potential

$$V = \tfrac{1}{2}Ar^{-2} - (1-\lambda)r^{-1} \qquad (r \geq r_0)$$
$$= \text{constant or infinity} \qquad (r < r_0) \tag{4.15}$$

where it is assumed that the cutoff r_0 depends only on two-electron quantum numbers for the short-range part of the interaction. These are N, S, and Π. The point of considering (4.15) here is that the solutions of the one-dimensional channel radial equation would have built into them the SO(4) structure of the channel invariant A. For each fixed value of r_0 this could lead to accidental degeneracies of the type we have found, at least to the extent that degeneracies also occur in the spectra of A. These can be highly degenerate, as we showed in Fig. 4. Seaton[106] and Bely[107] have described properties of radial functions for long-range potentials like (4.15) in connection with quantum-defect theory.

The quantum defect we consider for doubly excited states is $\delta \equiv n^* - n$, and this generally has a linear dependence on λ at small values of λ, for example $\delta \propto -\lambda K$ with the approximate formula (4.14). The dipole model (4.15) with constant r_0 might not reproduce this linear behavior at small λ, because the spectrum of A has a *second-order* dependence near $\lambda \simeq 0$. An example is the solvable version[81] when $r_0 = 0$, which predicts

$$\delta = -\tfrac{1}{2} + \left(\tfrac{1}{4} + A\right)^{1/2} \tag{4.16}$$

Another way of writing this is $A = \delta(\delta + 1)$. Equation (4.16) is a valid approximation of the spectrum only when $A > -\tfrac{1}{4}$, a criterion that is satisfied by every dipole channel over some range of λ, since $A \geqslant 0$ at $\lambda = 0$. The condition is always satisfied at very high angular momentum, and Nikitin and Ostrovsky[102] derived a leading-order formula

$$\delta = -\frac{9}{8}\left(\frac{\lambda N}{L}\right)^2 K \tag{4.17}$$

This has the same linear K dependence we find with the approximate low-L formula (4.14), but quadratic λ dependence.

In the case of H$^-$ the Coulomb attraction term in (4.15) vanishes because $\lambda = 1$, and the general form of the energy at high n can be found for the physically interesting case when $A < -\tfrac{1}{4}$. The result of Gailitis and Damburg[72] is a level-ratio formula that is independent of the short-range boundary r_0,

$$\frac{\varepsilon_n}{\varepsilon_{n-1}} = \exp\left(-\frac{2\pi}{\mathrm{Im}(\delta)}\right) \tag{4.18}$$

This illustrates how the SO(4) structure of the dipole channels carries over to the energy spectrum, since δ is the generalized angular-momentum quantum number in the dipole picture.

C. General Properties of Intrashell Spectra

The lowest levels in the spectrum of channels with $\nu = +1$ are called "intrashell" because they would correspond to configurations with $n = N$ in the independent-particle model. We find them interesting because:

1. Their electron correlation is closer to that of atomic and molecular valence shells, with $r_1 \simeq r_2$.
2. The lowest levels in each shell have been of most experimental interest.[108–110]
3. As $N \to \infty$ their correlation should be similar to that which controls the leading-order threshold behavior for two leaving electrons[111] in processes like the double photoionization of helium or double photodetachment of the hydrogen negative ion:

$$\text{He}(1s^2) + h\nu(79.0\ \text{eV}) \to \text{He}^{2+} + 2e$$

$$\text{H}^-(1s^2) + h\nu(14.4\ \text{eV}) \to \text{H}^+ + 2e$$

The actual identification of levels as "intrashell" or "intershell" is problematical, because one cannot identify them solely on the basis of energies. An analogous problem in molecular spectroscopy is the identification of levels as "valence" or "Rydberg." The most logical approach in theoretical calculations is to vary the charge parameter $\lambda = 1/Z$ to obtain results like those in Fig. 8. Levels with $n = N$ are easily identified in the weak-coupling limit. If this method is not followed, it is sometimes difficult to identify intrashell levels by following the usual route of looking for an expected, "dominant" valence configuration in the wave function.

A simple example is the $(2s, 2p)^1P^o$ state, which is the only $^1P^o$ configuration in the shell $N = 2$. The basic approach of Herrick and Sinanoğlu[98] uses a correlated "charge wave function" that is derived from a general many-electron approach to electron correlation in open-shell systems.[112] More details are described in Reference 47. For the state at hand,

$$\psi \doteq c_0(2s, 2p) + c_1(2s, Fp) + c_2(2p, Fs) + c_3(2p, Fd) \qquad (4.19)$$

where c_i is a linear configuration-mixing coefficient, obtained by diagonalizing the energy, and F denotes a radial function for correlation in which one of the electrons occupies an outer orbital that is always orthogonal to the valence shell. Coefficients estimated by expanding the radial functions in a hydrogenic basis ($n = 2$–7) are given in Table IV for helium ($\lambda = \frac{1}{2}$) and the hydrogen negative ion ($\lambda = 1$). The valence structure of the state is evident in the independent-particle limit, where all coefficients except c_0 vanish. The valence structure is still evident in helium, though the charge correlation now makes up twenty-three percent of the total normalization. As λ increases further, the correlation gets stronger, and at $\lambda = 1$ we see that virtually none of the original $(2s, 2p)$ structure is left. The corresponding Herrick-Sinanoğlu classification with (K, T) labels, on the other hand, remains very strong, and this is shown on the right-hand side of the table.

It is interesting to note the way the mixing coefficients of the configurations $(2s, Fp)$, $(2p, Fs)$, and $(2p, Fd)$ in Table IV closely resemble the ones obtained by coupling two SO(4) orbital representations with $N = 2$ and $n = 3$. From Table III the mixing for $K = 0$, $T = 1$ is

$$\chi_F = 0.471(2s, Fp) + 0.771(2p, Fs) + 0.430(2p, Fd) \qquad (4.20)$$

and this is similar to the mixing in Table IV when $\lambda = 1$. The reason we are surprised by this close agreement is that the expansion of each radial function $F_l(r)$ in terms of hydrogen-atom radial functions $R_{nl}(r)$ involves mostly the values $n = 5-7$, and *not* $n = 3$. What this seems to imply is that the approximate channel decoupling for intrashell correlation may involve just *two* SO(4) representations: one for the valence intrashell orbitals themselves, and another one for the orbitals F_{lm}. These are called "internal" and "semiinternal" orbitals in the many-electron approach of Sinanoğlu.[112] It will be interesting to see in the future if this idea could extend to other doubly excited states, since it would reduce a very large configuration-interaction problem to just a 2×2 interaction matrix between the internal SO(4) state and the semiinternal SO(4) state. On the basis of the number of nodes expected in the functions $F_l(r)$, it seems reasonable to guess that the effective quantum number for the semiinternal SO(4) representation would be $n = N + 1$, at least for intrashell levels.

The structure of intrashell double excitation energies has been clarified greatly by the recent studies with group-theoretical supermultiplets,[46,47] the rotor-vibrator model,[48,49] and pairing symmetry.[96] These will be described

TABLE IV
Configuration-Mixing Structure of the $N = 2$, $^1P^o$
Intrashell Doubly Excited State

λ^a	$(2s, 2p)$	$(2s, Fp)$	$(2p, Fs)$	$(2p, Fd)$	SO(4)^b
0	1	0	0	0	100%
$\frac{1}{2}$	0.878	-0.182	-0.310	-0.316	98%
1	-0.077	0.486	0.684	0.539	92%

$^a = 1/Z$.
^{b}Percentage of total wave-function normalization with $K = 0$, $T = 1$.

in Sections IV.D, E, and F. The motivation behind all this work is that the analysis of intrashell states from the point of view of decoupled channels indicated that the spectra behaved as if there really were two constants of the motion, $\mathbf{B}^2$ and $\mathbf{L} \cdot \mathbf{B}$, in addition to $\mathbf{L}^2$. For example, states having the same values of the quantum numbers P, T, and L, but different parity, were nearly degenerate, indicating an approximate invariance with respect to rotations about the symmetry axis $\mathbf{B}$. Moreover, the splitting of this twofold degeneracy was always such that the level with $\Pi = (-)^{L+1}$ was lower in energy. Other near-degeneracies were found,[98] but it was not known then how to explain them. All this is puzzling, of course, in the sense that there seemed to be a very high degree of symmetry even at low Z, where the hydrogen-atom symmetry is severely broken.

Empirical fits of intrashell energies at low Z indicate the repulsion energy is correlated strongly to $\mathbf{B}^2$. A formula for qualitative orderings and a good fit to low-lying levels is[61]

$$E - E_0 = \left(\alpha + \beta \mathbf{B}^2 \right)^{-1} \qquad (4.21)$$

E_0 is the zero-order energy at $\lambda = 0$, and the constants α and β for each shell have an approximate linear dependence on λ. Formulas like Equation (4.21) assume the exact eigenstates of the problem are SO(4) states $|N, PQLM\rangle$, so that an operator like $\mathbf{B}^2$ can be replaced by its eigenvalue. The fact that the energy is empirically a strong function of $\mathbf{B}^2$ has a classical interpretation with the relative orientation of Kepler orbits, indicated in Fig. 9. The actual form of an SO(4) operator $\mathbf{B}$ that could explain the high symmetry at low Z is not known, but it is expected to look something like coupled Runge–Lenz vectors.

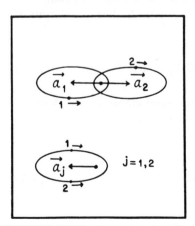

Fig. 9.   Degenerate classical Kepler orbits with $l_1 + l_2 = 0$, illustrating two extreme cases of favorable ($|a_1 - a_2| = $ max) and unfavorable ($a_1 - a_2 = 0$) angular correlation. The higher-energy configuration has the electrons paired in the same orbit. (The quantum-mechanical version of pairing symmetry is shown in Fig. 16.)

A more careful fit[99] of the helium intrashell energy levels $N = 2, 3,$ and 4 ($L \leqslant 2$) showed the spectra could be correlated better as a function of

$$X = \mathbf{B}^2 + \delta \mathbf{L}^2 \tag{4.22}$$

with $\delta \simeq 0.1$. In terms of invariant operators for the symmetry-breaking chain SO(4)⊃SO(3) this is $\mathbf{L}^2 + \mathbf{B}^2 - (1-\delta)\mathbf{L}^2$, which shows the strong rotational dependence of X a little more explicitly. The fact that X is a good measure of electron correlation may be seen in the following comparison of its eigenvalues with a formula[47] for average values of $\overline{\cos \theta_{12}} \equiv (\mathbf{b}_1 \cdot \mathbf{b}_2)/N^2$ in the SO(4) basis:

$$X = (N + K)^2 - 1 - (1 - \delta)L(L+1) + T^2 \tag{4.23a}$$

$$\overline{\cos \theta_{12}} = \frac{4N^2 - 3(N+K)^2 + 2L(L+1) - 3T^2 - 1}{8N^2} \tag{4.23b}$$

Classically $\overline{\cos \theta_{12}}$ is related to the average angle between two Lenz vectors. The two formulas in Eq. (4.23) have a similar dependence on K, T, and L that shows the angular correlation of electrons is more favorable (lower repulsion) at higher values of X. The range of values in each shell is $0 \leqslant X \leqslant X_m$, with $X_m = 4N(N-1)$. The correspondence is simply $\overline{\cos \theta_{12}} = (4N^2 - 4 - 3X)/8N^2$ when the symmetry breaking parameter is $\delta = \frac{1}{3}$.

The actual formula used to fit the helium energies was found by analyzing the X dependence of $(E - E_0)^{-1}$. This is related to the resolvent $(z - H)^{-1}$ when $z = E_0$ in the basis which diagonalizes the Hamiltonian. Although either the high- or the low-energy part of the spectrum for each shell could be fitted with a linear formula, it is necessary to introduce a quadratic dependence on X in order to fit both parts. The repulsion energy is then fitted approximately with the formula[99]

$$E - E_0 = [\alpha + \beta X + \gamma X(X_m - X)]^{-1} \tag{4.24}$$

where α, β, and γ are functions of N and, for other atoms, Z. The N dependence for the shells $N \leqslant 4$ in helium is approximately

$$\alpha = \tfrac{9}{20}N(N+1)$$
$$\beta = 0.25 - 0.00225(N-1)^{-1} \tag{4.25}$$
$$\gamma = 0.01/N$$

More accurate expressions could be found with the recent extensions to in-trashell levels $N \leqslant 5$ in He and H$^-$, but the important thing right now is the fact that the simple formula in Eq. (4.24) contains most of the information about correlation in doubly excited atoms, and we wish to extract this in a way that is easily understood.

The fit is shown in Fig. 10. Only levels of the symmetry $\Pi(-)^L = +1$ are displayed there; the -1 states have energies only slightly lower than these. The curve shows the formula (4.24) for continuous values of X. The highest energy level in each shell has $X = 0$, and corresponds to the totally symmetric SO(4) representation $[0,0]$. The lowest level in each shell is the ^1S state of the representation $[2N-2,0]$. These have $K = 1 - N$ and $K = N - 1$, respectively, and relative energies are consistent with the average values of $\cos\theta_{12}$ in (4.23) and with the picture of classical orbits in Fig. 9. The ground state of helium has $N = 1$, and the exact energy is $E - E_0 = 1.096$ in the scale of Fig. 10. In order to compute the corresponding ground-state repulsion energy with the empirical formula, it is necessary to use $K = 1 - N$ in the formula for X. The result is $E - E_0 = 1.111$, which is only 0.4 eV above the exact value. The fact that the ground state is the lowest member in a series of levels $N = 1, 2, 3, \ldots$ with $K = 1 - N$ is noteworthy because these are the states that are hardest to describe theoretically. Accurate energies of the doubly excited states in this series would require the same complicated machinery (Hylleraas coordinates, etc.) as the ground state. The lower-energy doubly excited states in each shell, on the other hand, have a very favorable angular correlation, and can be described very well with only the relatively simple "charge wave function" configuration mixing of the type shown in

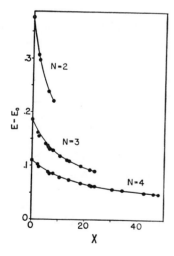

Fig. 10. Doubly excited energy spectrum ($\bullet$) of helium intrashell states of parity $(-)^L$, relative to $E_0 = -4/N^2$ au. The curves are empirical fits from Eq. (4.24) using continuous values of X.

Eq. (4.19). The idea is, there are both short- and long-range angular-correlation effects.

There is an interesting symmetry of the empirical formula (4.24) that has not been discussed before, and that illustrates nicely what is going on in the spectrum. Possibly this is an example of the type of symmetry that could be found for the hamiltonian (4.1) itself. The symmetry is related to the fact that the two extremes of the spectrum for each shell, $X \simeq 0$ and $X \simeq X_m$, are controlled by two simple poles that correspond to zeros of the denominator in the energy formula. These are called X_+ and X_-, and are described generally by the formula

$$X_{\pm} = \frac{1}{2}\left(X_m + \frac{\beta}{\gamma}\right) \pm \frac{1}{2}\left[\left(X_m + \frac{\beta}{\gamma}\right)^2 + \frac{4\alpha}{\gamma}\right]^{1/2} \qquad (4.26a)$$

The empirical form of the parameters α, β, and γ for low-N intrashell states suggests the singularities lie on the real X axis, and would have the following scaling at very high N:

$$X_+ \simeq X_m + 13.75N$$
$$X_- \simeq -11.25N \qquad (4.26b)$$

We see from this that, with increasing N, the singularities move away from the region of physical interest and there is a general ordering $X_- < 0 \leqslant X \leqslant X_m < X_+$. This is evident in Fig. 10 by the way the curves for higher N are becoming less steep.

The symmetry we are talking about is that the empirical energy formula is represented by a product of two simple poles, and this may be written a sum of the two poles:

$$E - E_0 = \gamma\left[(X - X_-)(X_+ - X)\right]^{-1}$$
$$= C(X - X_-)^{-1} + C(X_+ - X)^{-1} \qquad (4.27)$$

where $C = \gamma(X_+ - X_-)^{-1}$. This form captures the spirit of what all the recent work on doubly excited states with supermultiplets, the rotor–vibrator model, and pairing seems to suggest about a possible separation of the energy. The first pole lies closer to the physical part of the X axis and controls most of the energy, including short-range intrashell repulsion for a close encounter of the two electrons at $r_1 \simeq r_2$ and $\theta_{12} \simeq 0°$. The second pole lies farther away from the physical part of the X axis, and contributes mostly to low-energy repulsion that is related to favorable intrashell angular correlation when $r_1 \simeq r_2$ and $\theta_{12} \simeq 180°$. Most of the correlation, however, is described by the first term in Eq. (4.27), and if the second term is neglected we still recover the basic formula (4.21). Future work with fitting double excitation energies may wish to use a modified version of (4.28) in which the second coefficient C is different from the first, so that there are four basic parameters X_-, C, X_+, and C'. Branch singularities are another possibility.

From the point of group theory it is interesting to inquire about the significance of the pole at $X_- < 0$. To get there, we would have to make a transformation of Lie algebra generators from $\mathbf{L}$ and $\mathbf{B}$ for SO(4), to $\mathbf{L}$ and $i\mathbf{B}$ for the noncompact group SO(3,1), for example. The factor $i = \sqrt{-1}$ changes a sign in the commutation relations (2.9), but does not otherwise alter our physical interpretation of $\mathbf{B}^2$. The possibility that SO(3,1) might be linked to two-electron correlation was noted by Sinanoğlu and Herrick.[99]

The real significance of the group SO(3,1) for doubly excited atoms may well lie in the fact that coupled representations SO(3,1)$_1 \times$ SO(3,1)$_2$ for continuum electrons with the same energy could hold the key to the problem of threshold behavior in processes where two electrons escape. The coupling coefficients are linked to those for SO(4) in (2.17) by analytic continuation of the principal quantum number $n \to \pm in$. Just as coupled SO(4) representations describe correlation in the negative-energy doubly excited states, the coupled SO(3,1) states could account for correlation of two electrons as they recede from the nucleus. The logic of this becomes clear after one understands the physical picture behind intrashell correlation, supermultiplets, the rotor–vibrator model, and pairing.

D. Intrashell Supermultiplets

The empirical energy formula (4.24) and the associated curves in Fig. 10 suggested a smooth dependence of the correlation energy on the group-theoretical variable $X = \mathbf{B}^2 + \delta\mathbf{L}^2$. The idea of "supermultiplets" is to resolve the spectra further by casting out subportions of states according to quantum numbers for SO(4) = SU(2) × SU(2) or SO(4)⊃SO(3)⊃SO(2) that are linked to the internal degrees of freedom of the electrons. We have already described the basis for supermultiplets in Eqs. (3.17), with K and Q, and (3.20), with d and Q. Full details of the supermultiplets, including extensive pictures of their spectra for shells $N = 2$, 3, 4, and 5, are described in recent work.[46,47]

The first type of supermultiplet resolves the spectra according to angular correlation implicit in the (K, T) symmetry channels first, and then builds upon this with rotational excitations. Including the discrete symmetry labels

ν and η defined in Eqs. (4.5) and (4.8), the hierarchy of intrashell levels is

$$\nu = +1, \qquad \eta = \pm 1$$
$$N = 1, 2, 3, 4, \ldots$$
$$T = \begin{cases} N-1, N-2, \ldots, 0 & (\eta = +1) \\ N-1, N-2, \ldots, 1 & (\eta = -1) \end{cases}$$
$$K = N-1-T, N-3-T, \ldots, T+1-N$$
$$L = T, T+1, \ldots, N-1+K$$
$$\Pi = \eta(-)^L, \qquad (-)^S = \eta(-)^{L-T}$$

(4.28)

This accounts for a total of $\frac{1}{3}N(2N^2+1)$ energy levels in each shell, illustrating the exposive N^3 growth of this problem at high N. When the electrons in each channel are in their lowest rotational state, $L = T$. Herrick and Kellman[46] defined a new label $I = L - T$ which could describe the relative degree of rotational excitation in each channel. The "I supermultiplets" are then the collection of levels with the same value of I, arranged according to K and T. Figure 11 shows the I-supermultiplet structure of the level $N = 3$ in helium. The largest supermultiplet has $L = T$, and this may be compared to similar supermultiplets for $N = 3$, $n = 4$ in Fig. 6. There is only one of these diamond-shaped spectra for $n = 3$, because of the symmetry restriction that $\nu = +1$ for intrashell SO(4) states. Another example is the supermultiplet $I = 0$ for the level $N = 4$ in helium, shown in Fig. 12. It is clear from these spectra that there is quite a bit of SU(2)×SU(2) symmetry for this part of the correlation. The angular correlation grows less favorable as K decreases,

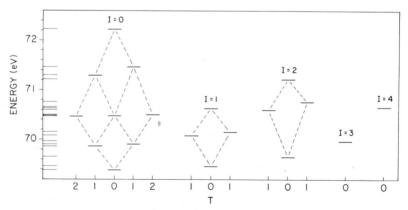

Fig. 11. I-supermultiplet classification of computed doubly excited states of helium in the $N = 3$ shell.

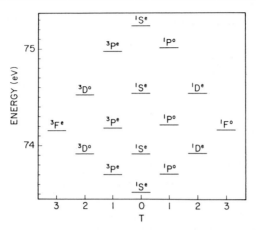

Fig. 12. Supermultiplet classification of computed doubly excited states of helium in the $N = 4$ shell, $I = 0$ only.

and hence the higher energy. Semiempirical interpretations like this do not tell us what is causing the degeneracies, however—only how to recognize them in a systematic fashion.

The second type of intrashell supermultiplet is derived from the SO(4)⊃ SO(3)⊃SO(2) channel picture with the label d. The idea is that each of the $^1S^e$ levels represents an intrinsic state, above which excitations are built with T and L. The complete hierarchy of intrashell levels in this scheme is

$$\nu = +1, \qquad \eta = \pm 1$$
$$N = 1, 2, 3, 4, \ldots$$
$$d = N - 1, N - 2, \ldots, 0$$
$$T = \begin{cases} d, d-1, \ldots, 0 & (\eta = +1) \\ d, d-1, \ldots, 1 & (\eta = -1) \end{cases} \qquad (4.29)$$
$$L = T, T+1, \ldots, 2d - T$$
$$\Pi = \eta(-)^L, \qquad (-)^S = \eta(-)^{L-T}$$

In the SO(4) picture with coupled hydrogen-atom orbitals, the label d is the larger value, $d = \max(a, c)$, of the SU(2) angular-momentum quantum numbers for $[\frac{1}{2}(\mathbf{L} + \mathbf{B})]^2 = a(a + 1)$ and $[\frac{1}{2}(\mathbf{L} - \mathbf{B})]^2 = c(c + 1)$. The quantum number is thus related to the SO(4) labels P and $T \equiv |Q|$ by the formula $d = \frac{1}{2}(P + T)$. The hierarchy is also described with representations of the group O(4), each of which splits up into two parts

$$[P, T] \rightarrow [P, T]^+ + [P, T]^-$$

when $T \neq 0$. Here the notation is $[P,T]^\eta$. Generators of the group O(4) include both proper SO(4) rotations and a reflection operation for $\eta \rightarrow -\eta$. When $T = 0$, this is $[P,0] \rightarrow [P,0]^+$. The full hierarchy of supermultiplets, $\{d\}$, and O(4) multiplets, $[P,T]$, is contained in the following reduction of the product of one-electron representations:

$$[N-1,0]_1 \times [N-1,0]_2 = \{N-1\} + \{N-2\} + \cdots + \{0\}$$
$$\{d\} = [2d,0] + [2d-1,1] + \cdots + [d,d] \qquad (4.30)$$

The result of all this for the shell $N = 3$ in helium is shown in Fig. 13. The "intrinsic" state for each d supermultiplet is the lowest level, and in each case this has $^1S^e$ symmetry. Excitations within each supermultiplet correspond to two types of rotational excitation: (1) rotations about the SO(4) symmetry axis ($\Delta T > 0$), and (2) overall rotation ($\Delta L > 0$). Neither of these is expected to introduce additional nodes into the wave function with respect to θ_{12}, so the basic topology of these nodes is preserved within each d-supermultiplet. This is why a $^1S^e$ state is called "intrinsic", because its wave function determines the number of nodes for intrashell angular correlation. Figure 13 shows that the energy of the intrinsic states increases with decreasing d, so it is reasonable to conclude that the number of nodes for θ_{12} is $N-1-d$, at least in the region of configuration space that determines the level structure. This is analogous to the fact that the number of nodes in a radial function $R_{Nl}(r)$ is $N-1-l$. The same physical interpretation of d has been reached in the dipole-channel problem, though it is difficult to prove these assertions without actually looking at plots of the electron density distributions.

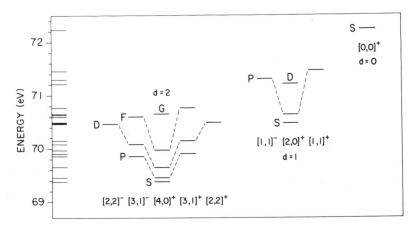

Fig. 13. d-supermultiplet classification of computed doubly excited states of helium in the $N = 3$ shell.

Rehmus, Kellman, and Berry[113] have illustrated some of the features of intrashell angular distributions computed with an SO(4) basis of hydrogen-atom orbitals.

Overall, the value of the supermultiplets lies in their power to efficiently resolve the entire intrashell spectrum into smaller clusters that are easier to understand. This is hard to appreciate until you actually sit down and try to sort out the data some other way.

E. Rotor–Vibrator Model

Perhaps the most appealing idea that has come out of the studies of intrashell levels is that they could originate in collective rotations and vibrations of electrons with respect to a quasirigid linear geometry for the intrinsic state.[48,49] This is represented by $X—Y—X$, where Y is the nucleus and X is an electron. A classical molecular structure like this has a total of four vibrational degrees of freedom, depicted in Fig. 14. Of these, it was proposed that the two bending modes could be similar to the motions of correlated electrons when they occupy the same shell. The picture is appealing because most of us know how to sketch out rotational–vibrational level structure on the back of an envelope. At the very least, then, it is a convenient mnemonic device. The remarkable result is that it works much better than one would have expected on the grounds of the usual independent-particle assumptions.

The essential idea is that we can associate the level structure in each shell with a zero-order energy formula

$$E = B_e\left[J(J+1)-l^2\right] + \omega_2(v_2+1) \tag{4.31}$$

Here J is the molecular rotation quantum number, and v_2 and l are the molecular symbols for quantization of the ν_2 bending mode: v_2 is the angular bending label, while l describes vibrational angular momentum about the

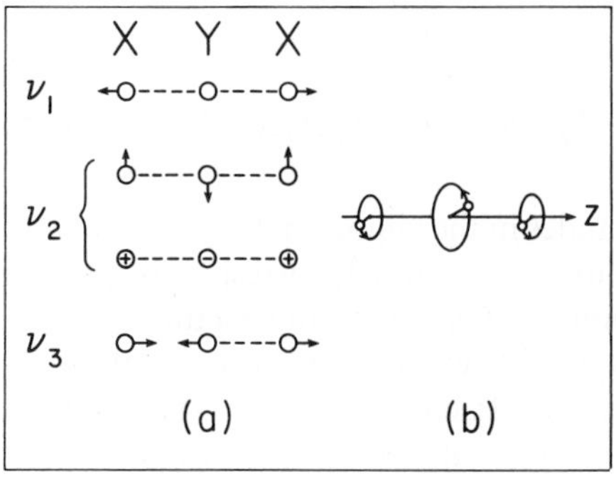

Fig. 14. (a) Normal modes of a linear symmetric XYX molecule. (b) Coupling of degenerate bending modes, which give a component of vibrational angular momentum along the molecular symmetry axis (z).

symmetry axis of the molecule as a two-dimensional vibrator (cf. Fig. 14). The relationship of these quantum numbers to the corresponding SO(4) symmetry labels provides us with an appropriate "picture" of just what the electrons are doing as they execute their correlated motion in each state. There are two molecular pictures, related to the two types of supermultiplets.

The I supermultiplets are described approximately by the energy formula (4.31) if the correspondence between atomic and molecular quantum numbers is

$$N - K - 1 \leftrightarrow v_2$$

$$T \leftrightarrow l \tag{4.32}$$

$$L - T = I \leftrightarrow R = J - l$$

and

$$E = B_e R(R+1) + \omega_2(v_2 + 1) + B_e l(2R + 1) \tag{4.33}$$

The quantum number R is that of a rigid dumbbell rotor, while v_2 is for the bending vibration. When $J = 0$, for example, the second term includes the exact $(v_2 + 1)$-fold degeneracy of a two-dimensional vibrator, and could account for the degeneracies found in the largest I supermultiplets, at least to the extent that B_e/ω_2 is small so that the contribution of the last term can be neglected.

The d supermultiplets are described approximately by the energy formula (4.31) if the correspondence between quantum numbers is

$$N - d - 1 \leftrightarrow n_2$$

$$T \leftrightarrow l \tag{4.34}$$

$$L - T \leftrightarrow R$$

and

$$E = 2\omega_2\left(n_2 + \tfrac{1}{2}\right) + l\left[\omega_2 - B_e(2R+1)\right] + B_e R(R+1) \tag{4.35}$$

Here $n_2 = \tfrac{1}{2}(v_2 - l)$ is the number of nodes in the vibrational part of the wave function with respect to the XYX bending angle, which corresponds to θ_{12} in the doubly excited atoms. The partitioning of energy in this picture is different because the first term in (4.35) is that of two one-dimensional vibrators, each with the same excitation number n_2. Each value of $n_2 = 0, 1, \ldots$ thus characterizes one of the intrinsic states for angular correlation. On top of this intrinsic spectrum there is built the energy for two types of rotational excitation. The second term in (4.35) is mostly excitation of vibrational angular momentum, while the third term represents overall rotational excitation only. Further interpretation of these two points of view is given in Reference 49.

The value of these pictures lies in the inferences we can draw. Consider, for example, excitation of $^1P^o$ doubly excited states by radiative excitation from the ground state, $(1s, 1s)\,^1S^e$. The SO(4) interpretation was that the strongest allowed transition goes to the intrashell channel $K = N - 2$, $T = 1$. The molecular picture of the excited state, then, is derived from the correspondence $v_2 = N - K - 1 = 1$, and $l = T = 1$. In this doubly excited state there is one degree of vibrational excitation, corresponding to one unit of vibrational angular momentum. This is the type of motion illustrated in Fig. 14b. In the SO(4) picture this is a component of angular momentum about the symmetry axis **B**, which represents roughly the average vector between the positions of the two electrons. It is interesting to note that if we assume the photodetachment process $H^- + h\nu \rightarrow H^+ + 2e$ is described by the same propensity rule in the limit $N \rightarrow \infty$, then the molecular model predicts the electrons coming off in opposite directions, but with a component of angular momentum about the axis defined by $\theta_{12} = 180°$. We are not aware that this possibility has been recognized for threshold experiments. Because parity is conserved in the excitation, the final state would be an admixture of two electronic states with different helicity $Q = \pm 1$. Observation of anything different from a 50% mixture of these two types would be evidence for parity violation. Now, this is of course pure speculation built from a crude model of the doubly excited atom, but it illustrates the whole point—namely, that the rotor–vibrator model allows us for the first time to think about electron correlation in ways that do not require a computer.

Inclusion of higher-order molecular terms in the energy for anharmonicity, coupling, rotational distortion (the atom *shrinks* with increasing L; it doesn't stretch), and other effects all lead to reasonable interpretations of the correlation when fits are made to the intrashell spectra. The possibility was also suggested[49] that autoionization breakup could be initiated by Coriolis coupling, which excites the asymmetric stretch mode ν_3 in Fig. 14. The asymmetric stretch would be coupled to the S_N–2 pathway proposed by Rehmus and Berry[114] for autoionization.

The accuracy of the rotor–vibrator model remains to be fully tested, and in particular there is no general theory which could predict *a priori* what the "molecular" constants are. Nevertheless, computed spectra seem to grow more like those of a rotor–vibrator at low Z, just where the correlation is strongest and hence the linear $X — Y — X$ "geometry" more reasonable. Figure 15 shows the rotor-like excitation[48] in the largest O(4) multiplet $[2N - 2, 0]$ of the shell $N = 3$, displayed as a series of spectra for different values of

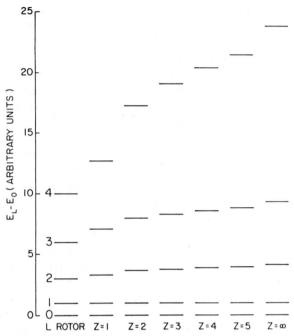

Fig. 15. Comparison of relative energies of rigid-rotor states and computed levels in the O(4) multiplet $[4,0]^+$ for $N = 3$, $Z = 1$–5, ∞. Energy is in arbitrary units with $E(^3P^o) - E(^1S^e)$ scaled to one unit. The energies for helium ($Z = 2$) may be seen in Fig. 13 as well.

the nuclear charge Z. The possible values of the angular momentum for this are $L = 0, 1, 2, \ldots, 2N - 2$ including a cutoff for the actual intrashell restrictions on angular momentum. The obvious conclusion is that these computed doubly excited spectra are looking more like rigid-rotor spectra at low Z. This is an example of what we mean when we say that the approximate symmetry of two-electron atoms seems to be stronger at low Z. Similar behavior is found with increasing N.[48,49]

Although the actual dependence of wave functions on θ_{12} has not yet been fully explored for higher values of N, a recent investigation[115] of Hylleraas-type wave functions for the $N = 2$ shell of doubly-excited helium has suggested the rotor–vibrator picture is essentially correct. Gaussianlike distributions of electron density were found peaked at $r_1 = r_2$ and $\theta_{12} = 180°$, lending support to the idea of an intrinsic structure for electron correlation.

F. Collective Modes and Pairing on a Spherical Shell

1. The Model

The model of two electrons on the surface of a sphere is interesting because it has angular correlation symmetry properties similar to the intrashell states of doubly excited atoms. The fixed radius of the sphere represents electrons confined to a single atomic shell, and this scales roughly as $r \simeq N^2$. The Hamiltonian is

$$H = \tfrac{1}{2}r^{-2}\left(l_1^2 + l_2^2\right) + \lambda r^{-1}V(\theta_{12}) \tag{4.36}$$

where $V(\theta_{12})$ is the angular potential. The linear $X-Y-X$ geometry is associated with a minimum in the potential when $\theta_{12} = 180°$. There are a total of four angular degrees of freedom, and the single-configuration classification is $|(l, l')LM\rangle$ when spin is neglected. Physical values of the coupling parameter $(\lambda > 0)$ lead to configuration mixing in the wave function. The angular dependence of the Coulomb repulsion when $r_1 = r_2 = r$ is represented by

$$V(\theta_{12}) = \left(2 - 2\cos\theta_{12}\right)^{-1/2} - \tfrac{1}{2} \tag{4.37a}$$

$$\simeq \tfrac{1}{8}\left(1 + \cos\theta_{12}\right) \tag{4.37b}$$

$$\simeq \tfrac{1}{16}\left(\pi - \theta_{12}\right)^2 \tag{4.37c}$$

where only the leading-order terms from the expansion about $\theta_{12} = 180°$ are indicated. Another way of writing the energy is

$$H = \tfrac{1}{2}r^{-2}\Lambda, \qquad \Lambda = l_1^2 + l_2^2 + 2\lambda r V(\theta_{12}) \tag{4.38}$$

where Λ represents a generalized angular momentum for correlation. Configuration mixing in the wave function grows stronger with increasing λ and r. The reason it increases with r is that the effective shell mass of the electrons in (4.36) is increasing. The result of this is that when $r \to \infty$ the system will "freeze" into the linear geometry for minimum potential energy.

2. Rotor–Vibrator Constants

The rotor–vibrator model is associated with collective excitations of the strongly coupled electrons with respect to the linear geometry. The picture is similar to the valence-force model for rotations and vibrations of linear YX_2 molecules[116] if the harmonic potential (4.37c) is assumed. The corresponding energy formula was given in Eq. (4.31). The rotation–vibration constants determined from the hamiltonian in Eq. (4.36) are

$$B_e = \tfrac{1}{4}r^{-2}, \qquad \omega_2 = \tfrac{1}{2}\lambda^{1/2}r^{-3/2} \tag{4.39}$$

The N dependence of B_e and ω_2 has been estimated[47] by fitting computed energies for the lowest rotational and vibrational excitations in the shells $N = 2-5$ for He and H^-. The results are shown in Table V, where we also display the corresponding effective shell radius determined from Eq. (4.39). The

TABLE V
Rotor–Vibrator Constants[a]

Atom	B_e	ω_2	$r = \frac{1}{2}(B_e)^{-1/2}$	$r = (\lambda/4\omega_2^2)^{1/3}$
He	$0.042\,N^{-4.38}$	$0.173\,N^{-3.38}$	$2.44\,N^{2.19}$	$1.61\,N^{2.25}$
H^-	$0.058\,N^{-4.29}$	$0.234\,N^{-3.25}$	$2.08\,N^{2.15}$	$1.02\,N^{2.17}$

[a] From lowest $^1S^e \rightarrow {}^3P^o$ rotational excitation and lowest $^1S^e \rightarrow {}^1P^o$ vibrational excitation in doubly excited shells, $N = 2$–5.

empirical scaling laws are roughly consistent with the scaling $B_e \propto N^{-4}$, $\omega_2 \propto N^{-3}$, and $\omega_2/B_e \propto N$ expected for a hydrogenlike shell radius $r \propto N^2$. The shell radius implied by the numerical data is more like $r \simeq 2N^{2.2}$. The empirical ratio for the relative partitioning of vibrational and rotational energy is $\omega_2/B_e \simeq 4N$ in both He and H^-.

Although results like this should not be taken very seriously until a more rigorous theoretical foundation for their applicabilty is developed, they do suggest that the rotor–vibrator picture of electrons on a sphere is consistent with the computed spectra. The larger shell radius (or $X - Y$ "bond" length) in the empirical results could be a result of radial correlation that is causing the independent-particle shells to expand by the additional factor $N^{0.2}$. If the empirical scaling is accurate at high N, the two-electron system would then begin to have a rotation–vibration energy separation like that found in molecules when $N \simeq 500$, where $\omega_2/B_e \simeq 2000$. Results like this suggest that a quantum-mechanical or semiclassical expansion of the intrashell correlation could be developed at very high N.

3. Long-Range Correlation Symmetry

The approximate SO(4) symmetry of the model is related to an assumption that the quantum-mechanical angular-momentum restriction $0 \le l \le N - 1$ for each electron carries over to the problem of two electrons on a sphere. A second assumption is that $\cos\theta_{12}$ in Eq. (4.37b) can be replaced with a product of two SO(4) generators, $(\mathbf{b}_1 \cdot \mathbf{b}_2)N^{-2}$. This is possible in low-lying states when $l \ll N$ because the matrix element of $\mathbf{b}$ in Eq. (2.13) reduces to that for $\hat{\mathbf{r}}$. The result for Eq. (4.38) is

$$\Lambda = 2N^2 - 2 + \tfrac{1}{4}\lambda r - \left(\mathbf{b}_1^2 + \mathbf{b}_2^2 - 2\mu\mathbf{b}_1 \cdot \mathbf{b}_2\right) \tag{4.40}$$

with the parameter $\mu \equiv (\lambda r/8N^2)$.

There are two exact SO(4) symmetries for Λ, when $\mu = \pm 1$. The case $\mu = -1$ is nonphysical in the present problem, but would correspond to an exact symmetry involving the coupled SO(4) generator $\mathbf{b}_1 + \mathbf{b}_2$. The more physical choice for electrons is $\mu = +1$, and SO(4) coupling with $\mathbf{B} = \mathbf{b}_1 - \mathbf{b}_2$ diagonalizes Λ:

$$\begin{aligned}
\Lambda(\mu=1) &= 4N^2 - 2 - \mathbf{B}^2 \\
&= 4N^2 - 2 - P(P+2) + \left[L(L+1) - T^2 \right] \\
&= 4N(v_2+1) + \left[J(J+1) - l^2 \right] - (v_2+1)^2 + 1 \quad (4.41)
\end{aligned}$$

This shows the explicit form of the eigenvalues in terms of the SO(4) symmetry labels $[P, T]$ on the one hand, and the related set of rotor–vibrator quantum numbers $v_2 = 2N - 2 - P$ and $l = T$. The energy formula is almost like the zero-order rotor–vibrator formula in Eq. (4.31), illustrating how similar results are obtained by the two different approaches. For low vibrational levels and high N, the two formulas (4.41) and (4.31) become identical.

The SO(4) symmetry of the electron correlation is only exact when $\mu = +1$, and is broken at other values of μ. The symmetry is for a shell radius $r = 8ZN^2$, where Z is the corresponding nuclear charge for the coupling parameter $\lambda = 1/Z$. This shell radius is unreasonably large for systems in which Z is large. At low Z, however, it becomes reasonable for He and H$^-$ when compared with the values in Table V. The point is, the model shows how the SO(4) symmetry of intrashell electron correlation could actually grow *stronger* at low Z, just the opposite of what we expect naively from the idea of broken Runge–Lenz SO(4) symmetry for the hydrogen atom. The long-range Coulomb potential in the full hamiltonian in Eq. (4.1) is attractive only for values of the nuclear charge $Z > \frac{1}{4}$. The SO(4) symmetry radius for $Z = \frac{1}{4}$ is $r = 2N^2$. Further details of the SO(4) symmetry model are given in Reference 47. One result is that the rotor–vibrator constants predicted by Eq. (4.41) at high N have $\omega_2 / B_e = 4N$. This is the same value found empirically for the computed spectra of He and H$^-$.

4. Short-Range Correlation Symmetry

There is another type of SO(4) symmetry for the intrashell model if we take into account the short-range part of the electron–electron interaction.[96,97] The idea is that if the short-range interaction is diagonalized on a restricted configuration basis $|(l, l')LM\rangle$ with $0 \leqslant l, l' \leqslant N-1$, then the highest-energy state that comes out is a *pairing state* in the Racah sense with seniority for the group U(N^2). The interesting result is that the SO(4) subgroup of this turns out to be the same SO(4) group we just described for the long-range intrashell angular correlation. In other words, the same SO(4) group is describing symmetries for the long-range and short-range parts of the problem, with the result that the total energy could have a very high degree of symmetry. What is unusual about this correlation pairing is that it occurs for the highest state. Physics is used to thinking about pairing symmetries in terms of the lowest-energy state of the system, at least when the interactions are attractive. What makes the electron correlation problem different is that the interaction is a repulsion between the electrons.

Our original reason for adopting a pairing approach derived from the old chemical notion for valence-shell electron repulsion that, if one chooses orbitals that maximize the repulsion between two electrons in the same orbital, the repulsion between different orbitals will be lowered. In the case of one electron on a valencelike spherical shell, Pauling[117] showed that hybridized linear combinations of atomic orbitals could be constructed by maximizing the bond strength (value of the orbital when $\theta = 0$) subject to constant normalization on the sphere. In the case of a hydrogenlike shell restriction $0 \leqslant l \leqslant N-1$ this leads to functions

$$\phi = \sum_{l=0}^{N-1} (2l+1)P_l(\cos\theta) \tag{4.42}$$

where $P_l(\cos\theta)$ is a Legendre polynomial.[58] The functions for values of $N = 2, 3, \ldots$ are just $sp^3, sp^3d^5, \ldots$ hybridized orbitals in the direction $z = \cos\theta$.

Similar reasoning for two electrons on a spherical hydrogenlike shell with $0 \leqslant l, l' \leqslant N-1$ leads to the "pairing state" that maximizes the pair strength (value of the function when $\theta_{12} = 0$) subject to constant normalization on the sphere. For $L = 0$ this is

$$\psi = \sum_{l=0}^{N-1} (2l+1)P_l(\cos\theta_{12}) \tag{4.43}$$

which is the two-electron version of Eq. (4.42). In the case $N = 2$, for example, the pairing state is[97]

$$\psi(K = -1) = \tfrac{1}{2}(2s,2s) - \tfrac{1}{2}\sqrt{3}\,(2p,2p) \tag{4.44}$$

The index $K = -1$ is included in Eq. (4.44) because the pairing state for $N = 2$ turns out to be identical with the SO(4) S state $K = -1$ in Table III. The reason for this will be explained shortly. The interesting point for $N = 2$ is that the other S state can be obtained from the requirement of orthogonality. The result is

$$\psi(K = +1) = \tfrac{1}{2}\sqrt{3}\,(2s,2s) + \tfrac{1}{2}(2p,2p) \tag{4.45}$$

That the wave function $\psi(K = +1)$ indeed has the more favorable angular correlation becomes more apparent when it is recalled that the standard convention for angular-momentum coupling[51,52] gives $(2p, 2p) \propto -\cos \theta_{12}$ for $L = 0$. The result is a density maximum when $\theta_{12} = 180°$. $\psi(K = +1)$ vanishes when $\theta = 0°$ because all the density has been put into the pairing state, which has the maximum pair strength.

When the procedure is generalized to states for each angular momentum $L = 0, 1, 2, \ldots, 2N - 2$ in a hydrogenlike shell model, the resulting states with maximum pair strength (MPS) at $\theta_{12} = 0°$ are[96]

$$\psi_L = \sum{}' |(l, l')LM\rangle [(2l+1)(2l'+1)]^{1/2} \begin{pmatrix} l & l' & L \\ 0 & 0 & 0 \end{pmatrix} \rho_L^{-1/2}$$

$$\rho_L = \sum{}' (2l+1)(2l'+1) \begin{pmatrix} l & l' & L \\ 0 & 0 & 0 \end{pmatrix}^2$$

(4.46)

The prime on the summation is a reminder that the summation for a hydrogenlike shell is restricted to the values $0 \leqslant l, l' \leqslant N - 1$. There is only one MPS state for each value of L, and all other wave functions orthogonal to the MPS states vanish when $\theta_{12} = 0°$. Because of the Pauli principle the MPS states all have *singlet* spin coupling ($S = 0$), and further investigation of the exchange and parity symmetry of the functions shows that $\eta = \Pi(-)^L = +1$. All states with $\eta = -1$ then, necessarily vanish when $\theta_{12} = 0°$. The physical significance of this difference becomes clear when it is recalled that T doubling for two-electron atoms is a splitting of two otherwise degenerate levels with different symmetries $\eta = +1$ and -1. In the SO(4) picture the degeneracy was for rotations about the symmetry axis **B**, while in the rotor–vibrator picture this was a twofold degeneracy for vibrational angular momentum about the symmetry axis of the linear $X — Y — X$ structure. We now see that the short-range part of the intrashell angular potential can split the degeneracy, with the result that the level with $\eta = +1$ would lie higher in energy in each T-doubled pair of states. This is the same ordering of levels found in the configuration-interaction studies. The fact that the T doubling is found to be small for most low-lying states in each shell is an indication that their wave functions do not contain large admixtures of the MPS state. This is consistent with the fact that the low-lying intrashell levels have most of their electron density near $\theta_{12} = 180°$. Short-range repulsion is not the only factor contributing to T doubling, however, as this can also occur on account of the electrons' motions as they rotate and vibrate near $\theta_{12} = 180°$.[49]

Another physical interpretation of the MPS states is that they are eigenfunctions of a δ-function pair interaction within the shell. The eigenvalues for the repulsion energy are just the normalization constants ρ_L defined in Eq. (4.46), and the ordering is generally $\rho_0 > \rho_1 > \rho_2 > \cdots > \rho_{2N-2}$, with ρ_0

$= N^2$, $\rho_1 = N(N-1)$, and

$$\sum_{L=0}^{2N-2} (2L+1)\rho_L = (\rho_0)^2 = N^4 \tag{4.47}$$

The repulsion energy of all other (non-MPS) states is zero.

The highest-energy MPS state, which has $^1S^e$ symmetry, is the totally symmetric state for a coupled two-electron representation of the group $U(N^2)$ for unitary transformations of one-electron states $|Nlm\rangle$ that have the hydrogenlike quantization.[96] For electrons on a sphere the pairing state [cf. Eq. (4.43)] is

$$\psi_0 = \sum_{l=0}^{N-1} \sum_{m=-l}^{l} Y_{lm}^*(\theta_1,\phi_1) Y_{lm}(\theta_2,\phi_2) \tag{4.48}$$

We have left out a normalization factor, and Y_{lm}^* denotes the complex conjugate of the spherical-harmonic orbital. There is a very subtle point about Eq. (4.48) that is easy to miss, and ultimately this is related to the subtlety underlying the choice of $\mathbf{B} = \mathbf{b}_1 - \mathbf{b}_2$ instead of $\mathbf{A} = \mathbf{b}_1 + \mathbf{b}_2$ for SO(4) coupling. The point is that if $G_1 \equiv \exp(ig_1)$ and $G_2 \equiv \exp(ig_2)$ are group elements of $U(N^2)$ for electron 1 and electron 2, then the product that leaves the function in Equation (4.48) invariant is

$$G_{12} = G_1 \Theta^\dagger G_2 \Theta \tag{4.49}$$

where Θ is the operator for time reversal. At the Lie-algebra level the proper coupling of generators for the pairing symmetry is

$$g_{12} = g_1 - \Theta^\dagger g_2 \Theta \tag{4.50}$$

With this coupling the Lie-algebra generators are antisymmetric under simultaneous application of conjugation (C), particle exchange (P), and time reversal (T). The Lie-group generators are symmetric under the CPT operation.

Now, the reason for considering all this is that the MPS states have an approximate symmetry that is described by an SO(4) subgroup of $U(N^2)$, and this is identified by using the criteria in Eqs. (4.49) and (4.50). The result, as we have already said, is a set of coupled SO(4) generators $\mathbf{L} = \mathbf{l}_1 + \mathbf{l}_2$ and $\mathbf{B} = \mathbf{b}_1 - \mathbf{b}_2$. The different signs are related to the fact that angular momentum is axial vector, whereas the operator $\mathbf{b}$ for a hydrogenlike shell is a polar vector like the Runge–Lenz vector. The only state with exact SO(4) symmetry is the one with $L = 0$. MPS states with $L > 0$ only have approximate SO(4) symmetry for electron correlation, but this is found to be very strong. The example in Table VI shows MPS and SO(4) configuration mixing coefficients for the shell $N = 3$. Also shown there are mixings computed by diagonalizing the full Coulomb repulsion operator $1/r_{12}$ in a configuration basis with hydrogen-atom radial functions. Overall, the results illustrate (1) the approximate SO(4) symmetry of the MPS states, and (2) the approximate MPS and SO(4) symmetries of the Coulomb repulsion.

The MPS classification of intrashell energies in the helium shell $N = 3$ may be seen in Fig. 13 by looking only at the highest energy level in each multiplet described by the formulas

$$[2d,0]^+, \qquad L = 2d, \qquad d = 0,1,\ldots,N-1$$
$$[2d-1,1]^+, \qquad L = 2d-1, \qquad d = 1,2,\ldots,N-1$$

(4.51)

The former is seen at the top of the primary rotorlike series at the center of

TABLE VI
Configuration-Mixing Structure of the
Highest Level in the $N = 3$ Shell[a]

	$^1S^e$		
	$3s3s$	$3p3p$	$3d3d$
MPS	.333	−.577	.745
HYD	.331	−.586	.739
SO(4)	.333	−.577	.745

	$^1P^o$	
	$3s3p$	$3p3d$
MPS	.577	−.817
HYD	.634	−.774
SO(4)	.667	−.745

	$^1D^e$		
	$3s3d$	$3p3p$	$3d3d$
MPS	.657	.509	−.556
HYD	.571	.563	−.597
SO(4)	.408	.500	−.763

[a] Predicted by the method of maximum pair strength (MPS), SO(4) theory, and diagonalization of $1/r_{12}$ in a hydrogenic (HYD) basis.

each d supermultiplet, while the latter is seen at the top of the second rotor-like series just to the right of the main one. The pairing state stands alone, at the highest energy in the shell, as the totally symmetric supermultiplet $d = 0$. The ordering of the energies E_L for the levels with approximate MPS symmetry is $E_0 > E_1 > E_2 > \cdots > E_{2N-2}$, just like the ordering of δ-interaction repulsion energies.

5. Illustration of Intrinsic Wave Functions

We now consider what the wave functions of the intrinsic SO(4) states with $L = 0$ look like in the model with two electrons on a sphere. The S-wave SO(4) states are basically Jacobi polynomials, which for the purposes of computation are expressed in terms of the Gauss hypergeometric function as[96]

$$\chi_d = \left(d + \tfrac{1}{2}\right)^{1/2} \binom{N+d}{2d+1} x^d F(d+1-N, d+1+N; 2d+2; x) \quad (4.52)$$

with $d = 0, 1, 2, \ldots, N-1$. The variable is $x \equiv \tfrac{1}{2}(1 - \cos\theta_{12})$, and $\binom{c}{a}$ is the binomial coefficient. The reader may wish to compare the function in Eq. (4.52) with that of the hydrogen-atom radial function in Eq. (2.21). The respective roles of d and l are similar, and in the two-electron wave functions the value of $N - d - 1$ gives the number of nodes in the θ_{12} distribution. This is also the number of nodes (n_2) in the corresponding vibrator wave function, illustrating the same topology for the SO(4) and rotor–vibrator models. Figure 16 shows the function χ_d for the values $N = 2$–4, which is enough to see the general trend. These are plotted from the perspective of an observer at the center Y of a horizontal axis $X_1 - Y - X_2$, with electron 1 to the left and electron 2 to the right. The curves then represent relative values of the wave function for different bending angles. A verticle axis drawn through each plot would represent a configuration $\theta_{12} = 0°$ in which one electron is sitting right on top of the other ("bond" angle zero). This way of plotting wave functions for θ_{12} makes the pictures look like the orbitals we're used to seeing. Each "orbital" is symmetric with respect to rotations about the symmetry axis ($T = 0$), just like a σ molecular orbital. The resulting three-dimensional figure would represent the intrinsic structure of the SO(4) state, having its own effective moment of inertia for rotations. Specifically for $N = 3$: the state $d = 2$ has electrons localized on opposite sides of the nucleus, $\theta_{12} \simeq 180°$, just like the ground state of a linear vibrator. We don't count the vertical axis as a node in this type of representation, so the state $d = 2$ is nodeless. The state $d = 1$ is like the vibrator wave function with $v_2 = 2$, and has one node. The peak probability is still the linear configuration though, and this is where we have placed the electrons. The pairing state has $d = 0$.

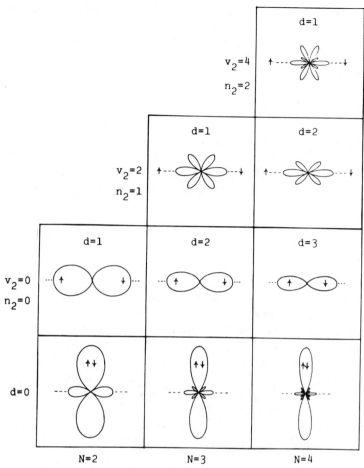

Fig. 16. Angular distributions of intrashell SO(4) states for two electrons on a sphere when the one-electron angular momentum is restricted by $0 \leqslant l \leqslant N-1$ and the total angular momentum is $L=0$. Quantum numbers are v_2 and n_2 for a vibrator, and d for intrinsic states in the SO(4) supermultiplet picture. The pairing wave function is $d=0$ (cf. Fig. 9).

One suggested application of the SO(4) angular functions is in expansions of the full wave function, including radial coordinates.[96] This is similar to the coupled-channel approaches for one electron, which lead to coupled radial equations like those in Eq. (3.4). The corresponding equation for coupled two-electron radial functions for $L=0$ is

$$\left(\frac{\partial^2}{\partial r_1^2} + \frac{\partial^2}{\partial r_2^2} + 2E - V_{dd} \right) F_d(r_1, r_2) = \sum_{j \neq d} V_{dj} F_j(r_1, r_2) \qquad (4.53)$$

where V_{dj} is a matrix element of the effective potential between two angular channels.

Some details of the diagonal and off-diagonal coupling terms are described in Reference 96. When $r_1 = r_2$, the diagonal potentials V_{dd} in the SO(4) basis have the form

$$V_{dd} = 2A_d r^{-2} - 4(1 - \lambda\sigma_d)r^{-1} \qquad (4.54)$$

The constant A_d for each channel represents the effective centrifugal angular momentum for the simultaneous approach of both electrons toward the nucleus, while σ_d is the effective screening constant. The Coulomb potential is either attractive ($\lambda\sigma_d < 1$) or repulsive ($\lambda\sigma_d > 1$). Values for $N = 3$ are

$$\begin{aligned} A_0 &= 4, & \sigma_0 &= 1.313 \\ A_1 &= 3, & \sigma_1 &= 0.395 \\ A_2 &= 1, & \sigma_2 &= 0.278 \end{aligned} \qquad (4.55)$$

The resulting potentials for radial channels reflect the angular correlation in Fig. 16. Electrons in high-d channels can move in closer to the nucleus, with less screening, than can their counterparts in low-d channels. It is clear from this why the energy of the pairing state ($d = 0$) is the highest in each shell. The screening parameter for the channel $d = 2$, on the other hand, is much closer to the classical value of 0.25 for two electrons with $\theta_{12} = 180°$. The critical nuclear charge at which the Coulomb potential first becomes repulsive in the diagonal matrix elements is $Z^* = \sigma_d$, so that only the intrinsic S-wave states with higher values of d are expected to remain stable at low Z. This agrees with the Z dependence of supermultiplets shown in Reference 47.

The actual radial channels for Eq. (4.53) would have to be found by numerically diagonalizing the full potential at each value of r_1 and r_2. Regions where SO(4) symmetry is high are characterized by nodes in the off-diagonal matrix elements evaluated in the SO(4) basis. The lines of nodes for matrix elements V_{01} and V_{12} for $N = 3$ are shown in Fig. 17. The nodes are a result of the cancellation of a positive Coulomb repulsion by an attractive "centrifugal" part. The matrix element V_{02} is purely repulsive. The curves are for H$^-$, and from Table V the empirical rotor–vibrator shell radius is $r \simeq 11$–22 au. In Fig. 17 this lies near the region of high SO(4) symmetry, offering another indication of higher correlation symmetry of low-Z two-electron atoms.

The assumption made for SO(4) was that the angular momentum of each electron is restricted to the values $0 \leqslant l \leqslant N - 1$. This is not very realistic for

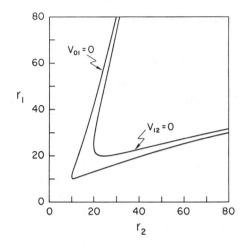

Fig. 17. Lines of nodes for off-diagonal radial coupling in Equation (4.53) for $N = 3$ states of H^-

the higher-energy states in each shell of the actual atoms, and some mixing with higher values of l is expected. The result would be a splitting of the pairing state, to allow the electrons to separate for lower repulsion. Similar correlation would occur in the radial part of the wave function, to allow in–out separation, like an antisymmetric stretch mode v_3 in Fig. 14. In practical applications with the SO(4) angular basis then, it might prove more efficient to use an *effective* principal quantum number $N^* > N$. Most of the short-range effects would then be dumped into wave functions with low values of d, while the higher values of d would represent the physical levels in each shell N. Functions with $N = 4$ in Fig. 16, for example, could represent the $N = 3$ shell if $d = 0$ were neglected. Some results of computed wave functions of electrons on a sphere suggest a reasonable approximation may be $N^* \simeq N + 1$.[118]

6. Implications for Experiments

Our analogy of pair strength with chemical bond strength suggests that doubly excited He** formed in collisions of slow, heavy ions M^{z+} with He involves a complex $[He^{2+} —:— M^{z+}]^{**}$ with electrons between the two centers. He** formed by adiabatic breakup would most likely have electrons in states of different L with maximum pair strength (toward M^{z+}). Evidence for this could be intrashell vibrational "cascade" transitions $\Delta K = +1$, with possible formation of high-L states. From the highest level in Fig. 12, for example, one sequence is $^1S^e \rightarrow {}^1P^o \rightarrow {}^1D^e \rightarrow {}^1F^o$ with photon emission at each step. Beam-foil spectra could also show this. Another possibility is vibrational excitation $\Delta K = -1$ by secondary photons after primary excitation of $^1P^o$ levels from ground states of He or H^-. In Fig. 12 this would be

$^1P^o \rightarrow {}^1D^e$ (or $^1S^e$) as a first step up. High vibrational states of H^- are expected to predissociate.[47]

G. Dimensionality Scaling

The introduction of a hypothetical variable dimension D to the problem of electron correlation has revealed some interesting symmetries that are exact for nonphysical values of D, but that shed light on the approximate correlation symmetries of real systems. The idea is to consider a generalized hamiltonian in which the position and momentum are described by the D-dimensional vectors $\mathbf{r} = (r_1, r_2, \ldots, r_D)$ and $\mathbf{p} = (p_1, p_2, \ldots, p_D)$, but the basic r^{-1} form of the nuclear attraction and electron repulsion terms is retained. The Schrödinger equation is then recast in a form so that D appears explicitly, and this is treated as an adjustable parameter for probing electron correlation in the wave function and energy. There are three types of symmetry that have been found by this approach, related to interdimensional degeneracies for many-electron systems, configuration interaction at low D, and the form of the energy at high D.

Herrick and Stillinger[119] investigated variable dimensionality for two-electron atoms, and found that with suitably scaled units the wave function and energy of the ground $(1s,1s)\,^1S$ state in five dimensions ($D = 5$) are identical to those of the doubly excited $(2p,2p)\,^3P$ state in ordinary three dimensions ($D = 3$). This is an exact degeneracy for the exact nonrelativistic energy. There is no established physical interpretation of the degeneracy,§ but similar types of degeneracy have been found in several other systems as well.[120]

One consequence of dimensionality scaling is that electron correlation in the limit $D \rightarrow 1$ becomes just like that of particles interacting by short-range δ-function potentials. The reason for this is that the form of the Coulomb potentials in the scaled Hamiltonian is $(D-1)r^{-1}$, and in the limit $D \rightarrow 1$ this vanishes everywhere except at $r = 0$. Accurate variational and perturbation estimates of the δ-function model, the $(1s,1s)\,^1S$ state, and the $(2p,2p)\,^3P$ state in three dimensions have established the basic scaling of the energy at low D for all physical values of the coupling parameter $\lambda = 1/Z$.[119]

The configuration-mixing symmetry was found in our investigation of variable D for electron correlation in the $N = 2$ shell with coupled representations of the group $SO(D + 1)$.[100] When $D = 3$ this is the $SO(4)$ group for physical doubly excited states. The Lie algebra generators include $\frac{1}{2}D(D-1)$ operators $l_{jk} = x_j p_k - x_k p_j$ for generalized angular momentum, and the D

---

§One possibility is mass decay from five to three dimensions, $He(D = 5) \rightarrow He^{**}(D = 3)$, though a coupling mechanism for this to occur is not known.

components of a generalized Runge–Lenz vector $\mathbf{b} = \mathbf{a}(-2H)^{-1/2}$. Coupled states for two electrons with total angular momentum $L = 0$ constructed by diagonalizing the operator $\mathbf{B}^2 = (\mathbf{b}_1 - \mathbf{b}_2)$ in the shell $N = 2$ are

$$\psi_- = \left[\frac{1}{D+1}\right]^{1/2}(2s,2s) - \left[\frac{D}{D+1}\right]^{1/2}(2p,2p)$$

$$\psi_+ = \left[\frac{D}{D+1}\right]^{1/2}(2s,2s) + \left[\frac{1}{D+1}\right]^{1/2}(2p,2p)$$

$$(4.56)$$

with $(\mathbf{B}^2 - 2D - 2)\psi_+ = 0$, and $\mathbf{B}^2\psi_- = 0$. These are the generalized versions of the wave functions in Eqs. (4.44) and (4.45), and in a model with electrons on a D-dimensional sphere ψ_- would correspond to the pairing state.

The preceding $SO(D+1)$ representation of electron correlation predicts very strong configuration mixing at low D, but at high D the independent-particle model is recovered as $\psi_+ \to (2s,2s)$ and $\psi_- \to (2p,2p)$. Naive intuition says this makes good physical sense, because there is more room for the electrons to escape from each other at higher D (think of a feuding couple locked in a large house, rather than a studio apartment, for example). The actual configuration-interaction mixing coefficients obtained by diagonalizing the operator $1/r_{12}$ in the hydrogen orbital basis are shown in Table VII, and suggest the interpretation is correct. Note that the $SO(D+1)$ symmetry of these mixings is *exact* at $D=1$. It is broken at higher values of D, though

TABLE VII
Dimensionality Dependence of Configuration Mixing in the Lower
Energy State in the Shell $N = 2$

D	$1/r_{12}{}^a$		$SO(D+1)^b$	
	$(2s,2s)$	$(2p,2p)$	$(2s,2s)$	$(2p,2p)$
1	0.707	0.707	0.707	0.707
2	0.825	0.565	0.816	0.577
3	0.880	0.476	0.866	0.500
4	0.909	0.416	0.894	0.447
5	0.928	0.373	0.913	0.408
∞	1.0	0.0	1.0	0.0

[a] Obtained from diagonalization of the 2×2 matrix of the Coulomb repulsion operator $1/r_{12}$.
[b] Obtained from the function ψ_+ in Eq. (4.56) with $SO(D+1)$ symmetry.

only slightly. This illustrates how we can associate approximate symmetries in the real problem with exact symmetries in a related nonphysical problem.

The reason for the exact symmetry at $D = 1$ is related to the short range of the interactions there. Basically, the pairing symmetry for the δ-function model is exact. The symmetry is actually more general than just for intrashell mixing, and it was found that the matrix of $1/r_{12}$ in the limit $D \to 1$ decouples *exactly* into noninteracting blocks of channels $(2s, ns) + (2p, np)$ and $(2s, ns) - (2p, np)$. A similar decoupling occurs for the channels $(2s, np) + (2p, ns)$ and $(2s, np) - (2p, ns)$ with one unit of total angular momentum. The latter channels are the one-dimensional analogues of the plus–minus states that were originally proposed[87] as an approximate symmetry for ^1P channels in helium. They are exact symmetry states for the hydrogenic mixings in one dimension because the radial functions of the $2s$ and $2p$ orbitals are *identical* there, and hence the wave function separates exactly into a radial part times an angular part. The "minus" states vanish at the origin, and have zero repulsion energy because of the δ-function limit. The "plus" states do not vanish at the origin and hence carry all of the repulsion energy.

There is a neat interpretation of all this using the form of the Runge–Lenz operator at $D = 1$.[100] The appropriate form of the operator for one electron on the x axis is

$$\mathfrak{a} = \mathrm{sgn}(x) \qquad (4.57)$$

where we are using the form without energy normalization. In other words, $\mathfrak{a} = +1$ when $x > 0$, and $\mathfrak{a} = -1$ when $x < 0$. One-electron states which diagonalize $\mathfrak{a}$ are the Stark-like mixings $ns \pm np$. The exact channel decoupling for two-electron configurations then involves states ψ_K with $K = \pm 1$ and

$$\mathfrak{a}_1 \mathfrak{a}_2 \psi_K = -\mathrm{sgn}(K) \psi_K \qquad (4.58)$$

This K is the one-dimensional analogue of the quantum number K introduced by Herrick and Sinanoğlu (cf. Section IV.B) to describe channel decoupling for doubly excited states when $D = 3$. It is interesting to see how the approximate symmetry there could be regarded as a broken symmetry of the hypothetical one-dimensional problem.

One possibility for future extension of this work is that the two-electron problem with $D = 2$ could be used to investigate the connection between approximate-symmetry supermultiplets and rotor–vibrator structure. Since $D = 2$ is closer to the exact symmetry at $D = 1$, the approximate symmetry might reveal itself in the spectrum in a more striking—and simpler—fashion. The reason is that the symmetry group for correlation when $D = 2$ is

SO(3), and everyone is familiar with that type of coupling. Calling the two coordinates x and y, the generators of the Lie algebra for one electron are the two-dimensional angular momentum $l_z \equiv xp_y - yp_x$, and the two Runge–Lenz vector components b_x, b_y. Diagonalization of the correlation operator $\mathbf{B}^2 = (\mathbf{b}_1 - \mathbf{b}_2)^2$ for two electrons corresponds to finding the irreducible representations of the coupled SO(3) group with Lie algebra generators $\mathbf{J}$ defined as

$$J_x = b_{1x} - b_{2x}, \qquad J_y = b_{1y} - b_{2y}, \qquad J_z = l_{1z} + l_{2z} \qquad (4.59)$$

This is just the type of coupling we talked about in Eq. (2.7), and the appropriate states are

$$|(j_1 j_2)JM\rangle = \sum_{m_1, m_2} |j_1 m_1, j_2 m_2\rangle (2J+1)^{1/2} \begin{pmatrix} j_1 & j_2 & J \\ m_1 & m_2 & -M \end{pmatrix} (-)^{j_1 + m_1}$$

$$(4.60)$$

The quantum number j is related to the principal quantum number n of the two-dimensional hydrogen atom by the formula $j = n - 1$. The quantum number M labels the total physical angular momentum of two electrons in the plane, while J labels the approximate correlation symmetry. The quantum number for channel decoupling when $n > N$ is $K \equiv J - n + 1$, and a quantum number v for the intrashell bending vibration is $v \equiv 2N - 2 - J$. It is reasonable to expect that, if the rotor–vibrator picture continues to work for $D = 2$ as well as $D = 3$, the energy spectra for the intrashell doubly excited states could have an approximate form

$$E = BM^2 + \omega\left(v + \tfrac{1}{2}\right) \qquad (4.61)$$

The physical picture for this is a flapping, propeller mode for collective excitations of a linear $X - Y - X$ structure. This type of correlation could occur for two electrons on a surface, or trapped in a two-dimensional conduction layer near a charge site.

The third type of symmetry found with dimensionality scaling involves the form of the energy at very high D. This has been investigated very recently by Mlodinow and Papanicolaou,[121] whose results were interpreted by Witten.[122] The interesting point here is that at very high D it is possible to scale units in a different way so that the effective mass of the electrons becomes infinite, leaving only an effective potential energy in the Hamiltonian. This "freezes out" a rigid geometry for the electron pair, with $\theta_{12} =$

$95.30°$ for the ground state. This is similar to the way our model of two electrons on a sphere freezes at $\theta_{12} = 180°$ for the lowest intrashell states—excluding the ground state. The fact that the $(1s, 1s)$ ground state is the lowest member in a series of pairing states $N = 1, 2, 3, \ldots$ suggests to us that the bending angle $\theta_{12} \simeq 95°$ may well characterize the shape of our pairing structure if one includes all of the configuration interaction, and not just intrashell mixings as our simple model assumed.

H. Extension to First-Row Atoms

It is not known for sure how much of the two-electron correlation symmetry survives in a many-electron environment, since the only study so far has been for first-row atoms.[101] The difficulty is that the configuration mixing is no longer described by irreducible representations of the SO(4) group. What was found was an approximate particle–hole symmetry of intrashell and intershell correlation energies with respect to occupancy of the $2s-2p$ shell. The degeneracy occurs between states having the same degree of ionization. A possible interpretation of this may be found with the approximate pairing symmetry for the Coulomb repulsion between two electrons. An interesting extension of the idea of diagonalizing the operator $\mathbf{B}^2 = (\mathbf{b}_1 - \mathbf{b}_2)^2$ for two electrons was the discovery that intrashell configuration mixing in a hydrogenic basis could be described by diagonalizing a pairwise sum of the operators for N_e electrons,

$$\Lambda(N_e) = \sum_{i<j} \left(\mathbf{b}_i - \mathbf{b}_j\right)^2 \qquad (4.62)$$

The physical picture this implies is a system of Runge–Lenz vectors with "harmonic" coupling.

I. Collective Electron Coordinates

We have tried to convince the reader up to this point that the double excitation spectra of two-electron atoms are telling us about an approximate separability, and that this is probably related to representations of Lie groups. Obviously much more investigation of levels from the point of view of decoupled symmetry channels is needed before our understanding of the symmetry is complete. Enough of the spectrum is already clear, though, that we can begin to ask the difficult question: What features of the hamiltonian in Eq. (4.1) are causing all this to happen?

The answer to this question lies, of course, in the identification of a suitable set of coordinates that could describe the types of collective electronic spectra that we have found. Fano[84,85] has described electron correlation for two-electron atoms in terms of hyperspherical radial coordinates $r = (r_1^2 +$

$r_2^2)^{1/2}$ and $\tan\alpha = r_1/r_2$. Although angular correlation is implicit in the decoupled channels computed by that approach, this is cast aside in the final picture, which describes correlation from the point of view of radial wave functions only. Our results, on the other hand, suggest that the approximate separability of the problem should be described with collective coordinates that include angular correlation at the outset.

One clue for constructing such a coordinate system is the factorization of the empirical formula for intrashell repulsion in Section IV.C into a product of singularities. This could well be a manifestation of a similar type of factorization of the Coulomb repulsion operator, $1/r_{12}$. The results are most conveniently expressed in terms of a modified set of hyperspherical coordinates involving one radial coordinate R, and two angles β and χ, defined by

$$R = r_1 + r_2 \qquad (R \geqslant 0)$$
$$\sin\chi = (r_1 - r_2)R^{-1} \qquad (-\tfrac{1}{2}\pi \leqslant \chi \leqslant \tfrac{1}{2}\pi) \qquad (4.63)$$
$$\beta = \tfrac{1}{2}(\pi - \theta_{12}) \qquad (0 \leqslant \beta \leqslant \tfrac{1}{2}\pi)$$

Single ionization channels when r_1 or $r_2 \to \infty$ correspond to the angle $\chi \to \pm\tfrac{1}{2}$, while $\chi = 0$ when $r_1 = r_2$. The value $\beta = 0$ corresponds in usual angles to $\theta_{12} = 180°$, and $\theta_{12} = 0°$ when $\beta = \tfrac{1}{2}\pi$. The proposed collective coordinates for electron correlation are

$$\eta = R(1 + \cos\chi\sin\beta),$$
$$\zeta = R(1 - \cos\chi\sin\beta) \qquad (4.64)$$

The appealing feature of η and ζ is that they look like the usual parabolic coordinates for the Stark effect in hydrogen, $\eta' = r(1 + \cos\theta)$ and $\zeta' = r(1 - \cos\theta)$. The generalized two-electron "parabolic" coordinates in (4.64) could well be linked to the Stark-like $SU(2) \times SU(2)$ energy spectra we have found. The total potential energy for the hamiltonian (4.1) is

$$V = -\frac{8}{(\eta + \zeta)(\cos\chi)^2} + \frac{\lambda}{(\eta\zeta)^{1/2}} \qquad (4.65)$$

The repulsion singularity occurs when $\zeta = 0$. Another singularity when $\eta = 0$ occurs only for nonphysical values of the angles, although this could be important in approaches which treat the autoionizing resonances with complex coordinates. A fuller description of two-electron atoms with coordinates like those in (4.64), including appropriate limits for two-electron escape, should be very interesting. For levels below threshold, another possibility is that wave functions built from generalized parabolic coordinates could give a more natural description of gaussianlike distributions for θ_{12} that are part of a true rotor–vibrator picture. The same types of angular distributions for the two electrons should be expected in momentum space if the gaussian picture is correct.

V. VALENCE-SHELL CORRELATION IN POLYENES

A. Introduction

In this section we shall describe very recent work[50, 123] and present some new results on a general approach that identifies subgroup chains of the unitary group U_{2n} (cf. Section II.D) for approximate symmetries of correlation in polyenes. The symmetries are approximate in two senses. First of all, we are considering spectra predicted by a Pariser–Parr–Pople (PPP) model hamiltonian for valence-shell π electrons. The spectra are nontrivial, and are difficult to compute for very large systems, but nevertheless only represent rough approximations of spectra one could obtain from a full-scale *ab initio* approach. We offer no new insight that justifies the widely used PPP model, only a discussion of some of its interesting symmetry properties for electron correlation. The second point is that these symmetries are exact only in certain limits of the model, with values of the underlying Coulomb repulsion parameters that are not truly accurate by chemical standards. They are close enough, though, that we gain an enormous amount of qualitative insight about the real PPP spectra in a very general sense, rather than being concerned with numerical details about one or two particular states. Although we shall speak directly about the problem of π-electron correlation, it is clear these ideas could be extended at the group-theoretical level to other shells or other molecules, including features such as the actual nonseparability of σ and π energies.

All chemists are familiar with the Hückel model of molecular orbitals, which assumes a constant resonance attraction of the electron between neighbor atoms. The PPP model extends the Hückel picture by including the effects of (1) nonuniform resonance and one-center core attraction energies, and (2) Coulomb repulsion between electrons approximated with complete neglect of differential overlap.[124] The model gives a good qualitative picture of valence-shell spectra of planar unsaturated hydrocarbons, and has been investigated recently by extended configuration-interaction methods.[125–130] Linear polyenes, $C=C-C=C\cdots$, are particularly interesting as prototypes of polyene chromophores that are important biologically for vision and energy production. Hudson, Kohler, and Schulten[131] have reviewed many different experimental and theoretical facets of linear polyenes, including some of the puzzling electron-correlation effects in excited states.

Here we shall focus only on symmetry properties of the PPP model. These are derived from subgroup chains of the unitary group U_{2n} which partition the energy according to exact spin and alternancy particle-hole symmetries. Different hierarchies of subgroups are found by symmetrizing Lie algebra generators in several different ways. In a spin-free picture the specific subgroup chains of interest are $U_n \supset R_n \supset R_3 \supset R_2$ and $U_n \supset Sp_n \supset R_3 \supset R_2$, where R_n (Sp_n) denotes the rotation (symplectic) group in n dimensions. The particular representations of interest are very similar to the usual seniority classifications of atomic and nuclear shell models. The important result derived here for polyenes is that pairing operators for seniority groups adapted to different types of particle–hole symmetry are exact constants of the motion for electron correlation in two model problems: (1) the Hubbard model, which has short-range (single atom) pairing symmetry, and (2) a harmonic coupling model which has long-range (between atoms at opposite ends of the molecule) pairing symmetry. The groups $R_3 \supset R_2$ describe the approximate pseudorotation symmetry of linear polyenes. Since many of the pseudorotation ideas are contained in the recent original work,[50] we shall focus here on their derivation from the more general procedure which uncovers the seniority subgroup chains. It is interesting to note that Wybourne[132] found similar types of seniority chains when degenerate states of an atomic orbital are split by a crystal field. Seniority was not considered physically realistic for correlation in molecules, however, until the pseudorotation symmetry of linear systems was found. Most other unitary-group studies of molecules have dealt with the problem of state labeling in the configuration-interaction matrix itself,[19-38] using complete sets of labels that are related to a canonical subgroup chain $U_n \supset U_{n-1} \supset \cdots \supset U_1$. Bincer[133] discusses the problem of constructing canonical representations of the rotation and symplectic groups in n dimensions.

The remainder of this section will describe the PPP model, its representation with Lie-algebra generators and tensors, our criteria for identifying seniority structure, and related properties of quasispin groups for pairing symmetry. Only key results from the mathematics are presented, in favor of interpretation of their physical relevance to the problem.

B. Unitary Structure of PPP Model

The goal here is to present a brief description of the PPP model, and its representation with Lie-algebra generators of U_{2n} and U_n that are linked to subsequent subgroup chains for approximate symmetries. The basic system is N electrons (spin-$\frac{1}{2}$ fermions) occupying an orthonormal basis of n atomic π-orbitals, with one orbital per atom. The most general form of the PPP hamiltonian in a second-quantization representation is

$$H_{PPP} = E_{nuc} + \sum_{q,r} \sum_{\sigma} f_{qr} X_{q\sigma}^\dagger X_{r\sigma} + \frac{1}{2} \sum_{q,r} \sum_{\sigma,\mu} \gamma_{qr} X_{q\sigma}^\dagger X_{r\mu}^\dagger X_{r\mu} X_{q\sigma} \quad (5.1)$$

Here $X_{q\sigma}^\dagger$ and $X_{q\sigma}$ are creation and annihilation operators for an electron with spin component $\sigma = \pm\frac{1}{2}$ at atom q; E_{nuc} is the nuclear repulsion, which is constant for fixed geometries; $f_{qr} = f_{rq}$ is the one-electron matrix element between orbitals on centers q and r; and $\gamma_{qr} = \gamma_{rq}$ is the effective two-electron Coulomb repulsion integral. The off-diagonal matrix element f_{qr} is represented as

$$f_{qr}(q \neq r) \equiv \beta_{qr} \tag{5.2}$$

where β_{qr} is a resonance parameter representing the chemical bonding attraction of an electron on atom r to atom q. The diagonal matrix element f_{qq} for the net binding of a valence electron at center q is approximated as

$$f_{qq} = -I_q - \sum_{r(\neq q)} Z_r \gamma_{qr} \tag{5.3}$$

where I_q is the effective ionization potential for the atomic orbital, and Z_r is the charge at center r. The nuclear repulsion term is also approximated with Coulomb repulsion integrals as

$$E_{\text{nuc}} = \sum_{q<r} Z_q Z_r \gamma_{qr} \tag{5.4}$$

More specific forms of the PPP parameters that give reasonable approximations of valence-shell spectra and potential energy surfaces are described in some of the recent configuration-interaction studies.[125-130] The symmetries of interest here are derived from a simplified version of (5.1) in which all one-center parameters are the same for each site:

$$\alpha \equiv -I_q, \qquad \gamma \equiv \gamma_{qq}, \qquad Z_q \equiv 1 \tag{5.5}$$

After taking into account the fermion anticommutation relations in Eq. (2.31), the PPP hamiltonian can be written in a variety of different ways, of which the following two are especially useful:

$$\hat{H}_{\text{PPP}} = \left[E_{\text{nuc}} + (\alpha + \gamma)\hat{N} \right] + \hat{\beta} + \hat{\gamma} \tag{5.6}$$
$$= \left[\alpha\hat{N} + \tfrac{1}{2}\gamma(\hat{N} - n) \right] + \hat{\beta} + \hat{V} \tag{5.7}$$

with

$$\hat{\beta} = \sum_{q,r} \beta_{qr} E_{qr} \tag{5.8}$$

$$\hat{\gamma} = \frac{1}{2} \sum_{q,r} \sum_{\sigma,\mu} \gamma_{qr} \left(X_{q\sigma}^\dagger X_{r\mu}^\dagger X_{r\mu} X_{q\sigma} - X_{q\sigma}^\dagger X_{q\sigma} \right)$$

$$= -\frac{1}{2} \sum_{q,r} \sum_{\sigma,\mu} \gamma_{qr} E_{q\sigma r\mu} E_{r\mu q\sigma} \tag{5.9}$$

and

$$\hat{V} = \frac{1}{2} \sum_{q,r} \gamma_{qr} (E_{qq} - 1)(E_{rr} - 1) \tag{5.10}$$

$\hat{N}$ is the operator for the total number of electrons,

$$\hat{N} = \sum_{q,\sigma} E_{q\sigma q\sigma} = \sum_{q} E_{qq} \tag{5.11}$$

while E_{qq} represents the number of electrons ($E_{qq} = 0, 1, 2$) at site q. The operator $\hat{\beta}$ is the one-electron resonance interaction. The operators $\hat{\gamma}$ and $\hat{V}$ represent slightly different versions of "correlation" effects originating in two-electron repulsion *and* one-electron core attraction. In a molecule like benzene, where f_{qq} is the same for each atom, all of the correlation originates in the two-electron repulsion. In less symmetrical systems such as linear polyenes f_{qq} is different for each atom, with the net effect being an attraction of the electron toward a lower-energy well at the center of the molecule. The overall correlation then represents a balance between this favorable core attraction near the middle of the molecule and two-electron repulsion, which drives two electrons toward opposite ends of the molecule. Their cancellation is responsible, in part, for the degree of approximate symmetry for correlation in linear polyenes.

Although the two correlation operators $\hat{\gamma}$ and $\hat{V}$ differ only by a constant (when $\hat{N}$ is constant),

$$\hat{V} = \hat{\gamma} + E_{\text{nuc}} + \frac{1}{2}\gamma(\hat{N} + n) \tag{5.12}$$

they represent two fundamentally different views of the problem. In the case of $\hat{V}$ the correlation is described with the *spin-free* Lie-algebra generators E_{qr} of the group U_n. Only sites that are vacant ($E_{qq} - 1 = -1$) or doubly occupied ($E_{qq} - 1 = +1$) contribute to the summation in Eq. (5.10), so that the correlation part of the energy is independent of the spin alignment of the singly occupied sites. In the case of $\hat{\gamma}$, on the other hand, correlation is described in terms of a quadratic form involving *spin-dependent* Lie-algebra generators of the group U_{2n}. In other words, generators $E_{r\mu q\sigma}$ appearing in Eq. (5.9) break the spin symmetry of the atomic site q, though this is recovered in the product $E_{q\sigma r\mu}E_{r\mu q\sigma} \equiv (E_{r\mu q\sigma})^\dagger E_{r\mu q\sigma}$. The difference becomes important in the way one partitions the energy contributions according to subgroup chains starting with $U_n \supset \cdots$ or $U_{2n} \supset \cdots$, since different hierarchies of seniority labels can arise.

The conventional way of interpreting the correlation in polyenes is with the spin-free approach. It is entirely reasonable to expect, however, that some additional insight (not obvious in the spin-free picture) could be more transparent in a spin-dependent picture. We shall leave open the possibility, then, that both views are important.

C. Spin, Alternancy, Particle–Hole Symmetry

The essential resonance and correlation interactions in the PPP model are contained in the operator

$$\hat{H} = \hat{\beta} + \hat{V} \tag{5.13}$$

which is effectively the hamiltonian in Eq. (5.7) if we omit the trivial constants in the first term there. So far the only restrictions on the resonance and Coulomb repulsion parameters in $\hat{H}$ are

$$\begin{aligned} \beta_{qr} &= \beta_{rq}, & \beta_{qq} &\equiv 0 \\ \gamma_{qr} &= \gamma_{rq}, & \gamma_{qq} &\equiv \gamma \end{aligned} \tag{5.14}$$

This leaves $\frac{1}{2}n(n-1)$ resonance parameters and $\frac{1}{2}n(n-1)+1$ Coulomb repulsion parameters, or a total of $n^2 - n + 1$ parameters that control the spectrum of $\hat{H}$. Though subsequent additional restrictions will reduce this number, the point is that in its most general form the only exact symmetry of (5.13) is conservation of total spin. Lie-algebra generators of the spin angular-momentum group (which is called R_3^S in this section) have a second-quantization representation as

$$S_+ = \sum_q X_{q\uparrow}^\dagger X_{q\downarrow}$$

$$S_- = (S_+)^\dagger = \sum_q X_{q\downarrow}^\dagger X_{q\uparrow} \tag{5.15}$$

$$S_0 = \frac{1}{2} \sum_q \left(X_{q\uparrow}^\dagger X_{q\uparrow} - X_{q\downarrow}^\dagger X_{q\downarrow} \right)$$

where $\uparrow$ and $\downarrow$ denote up ($\sigma = +\frac{1}{2}$) and down ($\sigma = -\frac{1}{2}$) spin components.

There is a second exact symmetry in the case of so-called alternant systems, in which the atoms are divided into two classes of "starred" and "unstarred" atoms such that direct bonding occurs only between two atoms in different classes. The idea of alternancy symmetry stems from the Hückel model for resonance between neighbor atoms.[134] The result is that the bond order between two atoms in the same class vanishes, and the orbital energies are paired symmetrically about a mean value for the molecule.[124] McLachlan[135] extended the concept of alternancy and pairing to many-electron states of the PPP hamiltonian, including non-nearest-neighbor resonance interactions between starred and unstarred atoms. The consequence of the alternancy symmetry assumption is that each level can be assigned a definite "parity" on the basis of properties of the wave function under exchange of electrons and holes. This is physical only in the case of neutral systems, in which $N = n$, because particle–hole exchange mixes the states of positive ($N < n$) and negative ($N > n$) molecular ions. Recent approaches[26, 136, 137] have focused on the assignment of particle–hole symmetry labels from the point of view of the unitary groups. One of the difficulties, however, has been the adoption of a uniform convention for assigning the alternancy parity labels. Part of the difficulty originates in the arbitrary phase associated with any particle–hole symmetry operator. Another difficulty is that the natural "even" or "odd" parity in a particle–hole symmetry approach does not always coincide with the more common labels "plus" and "minus" proposed by Pariser[138] for singly excited states. As a result of all this, one has to proceed with caution when trying to assess what "plus," "minus," "even," and "odd" mean throughout the literature.

Our preferred method for alternancy symmetry uses what we call a pseudorotation labeling of the atoms in each molecule. The idea is to adopt a labeling scheme that allows the alternancy state of each atom to be quantized more efficiently. Each molecule is assigned a quantum number j that is related to the total number of atoms by the formula

$$n = 2j + 1 \qquad (5.16)$$

Individual atoms are then labeled with values of the index q defined as

$$q = j, j-1, \ldots, -j+1, -j \qquad (5.17)$$

This is called a pseudorotation representation because the atoms are treated like components of a spherical tensor of rank j, A_{jq} ($-j \leqslant q \leqslant j$), although no real spherical symmetry is involved. Operators for a pseudorotation Lie algebra are defined in the second-quantization representation with the usual conventions for angular momentum from Section II,

$$j_\pm = \sum_q [(j \mp q)(j \pm q + 1)]^{1/2} E_{q \pm 1, q}$$
$$j_0 = \sum_q q E_{qq} \qquad (5.18)$$

We shall refer to the pseudorotation group as R_3^J in this section, where J denotes the quantum number for the total pseudorotation angular momentum in a many-electron system, $\mathbf{j}^2 = J(J + 1)$.

Some examples of possible pseudorotation labeling schemes in alternant systems are structures [1]–[4]. The two classes of atoms have different values

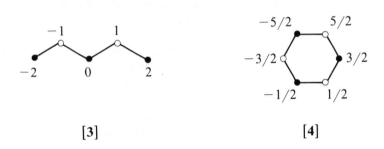

of the symmetry label T defined as

$$TA_{jq} = (-)^{j+q} A_{jq} \tag{5.19}$$

with $T = +1$ for starred atoms (●) and $T = -1$ for unstarred atoms (○). The PPP hamiltonian in Eq. (5.13) is then said to have alternancy symmetry when the resonance interaction vanishes between atoms of the same class,

$$\beta_{qr} \equiv 0 \quad \text{if} \quad q - r \text{ is } even \tag{5.20}$$

The precise number of independent PPP parameters left depends on whether the bonding is open, as in the case of linear polyenes, or has a closed-ring structure. In any case, the formula is generally different for *odd* and *even* values of n, and this is reflected in part in the different electronic-structure properties of odd-alternant and even-alternant hydrocarbons even at the Hückel level.[139] In pseudorotation notation the two classes of systems are distinguished by half-integer or integer values of j for irreducible representations of R_3^J,

$$\begin{aligned} even\ n\colon \quad & j = \tfrac{1}{2}, \tfrac{3}{2}, \tfrac{5}{2}, \ldots \\ odd\ n\colon \quad & j = 0, 1, 2, \ldots \end{aligned} \tag{5.21}$$

In the important case of neutral polyenes ($N = n$), the former corresponds to molecules and the latter corresponds to radicals.

The alternancy symmetry manifests itself naturally in the second-quantization picture via the transformation properties of operators under particle-hole conjugation. We have described this from the point of view of pseudorotations with a unitary conjugation operator P which transforms the Lie algebra generators of U_{2n} and U_n in the following way[50]:

$$PE_{q\sigma r\mu}P^{-1} = \delta_{qr}\delta_{\sigma\mu} - E_{r-\mu q-\sigma}(-)^{q-r}(-)^{\sigma-\mu}$$
$$PE_{qr}P^{-1} = 2\delta_{qr} - E_{rq}(-)^{q-r}. \tag{5.22}$$

The effect of particle–hole conjugation on the number of valence electrons is

$$P\hat{N}P^{-1} = 2n - \hat{N} \tag{5.23}$$

so that P mixes positive and negative ion states. In the case of neutral molecules and radicals $P^2 = 1$, and states can be classified with a parity label $P = +1$ or -1. The alternancy condition can be seen with the transformation properties of a general one-body operator $\hat{G}$ defined as

$$\hat{G} = \sum_{q,r} g_{qr}E_{qr} \tag{5.24}$$

with $g_{qr} = g_{rq}$:

$$P\hat{G}P^{-1} = 2\sum_{q} g_{qq} - \sum_{q,r} g_{qr}E_{qr}(-)^{q-r} \tag{5.25}$$

The form of $\hat{G}$ is left invariant by this particle–hole transformation when $g_{qr} \equiv 0$ for *even* values of $q - r$, and this is precisely the alternancy property for PPP resonance in (5.20). The operator $\hat{V}$ for correlation is also left unchanged, because the individual factors $E_{qq} - 1$ appearing in Eq. (5.10) are seen from (5.22) to be antisymmetric under particle–hole exchange. The net result is

$$P\hat{H}P^{-1} = \hat{H} \tag{5.26}$$

and the configuration-interaction matrix is blocked according to the parity of P.

Symmetric and antisymmetric components of the Lie-algebra generators of U_{2n} and U_n which satisfy

$$P\Omega_{q\sigma r\mu}P^{-1} = \Omega_{q\sigma r\mu}, \qquad P\Omega^{-}_{q\sigma r\mu}P^{-1} = -\Omega^{-}_{q\sigma r\mu}$$
$$P\Omega_{qr}P^{-1} = \Omega_{qr}, \qquad P\Omega^{-}_{qr}P^{-1} = -\Omega^{-}_{qr} \tag{5.27}$$

are

$$\Omega_{q\sigma r\mu} = \tfrac{1}{2}\left[E_{q\sigma r\mu} - (-)^{q-r}(-)^{\sigma-\mu}E_{r-\mu q-\sigma}\right]$$

$$\Omega^{-}_{q\sigma r\mu} = \tfrac{1}{2}\left[E_{q\sigma r\mu} + (-)^{q-r}(-)^{\sigma-\mu}E_{r-\mu q-\sigma} - \delta_{qr}\delta_{\sigma\mu}\right]$$

$$\Omega_{qr} = \tfrac{1}{2}\left[E_{qr} - (-)^{q-r}E_{rq}\right]$$

$$\Omega^{-}_{qr} = \tfrac{1}{2}\left[E_{qr} + (-)^{q-r}E_{rq} - 2\delta_{qr}\right]$$

$$(5.28)$$

The PPP resonance with alternancy symmetry can be expressed in terms of symmetric spin-free generators as

$$\hat{\beta} = \sum_{q,r} \beta_{qr}\Omega_{qr} \tag{5.29}$$

The correlation operator can be written in terms of spin-free generators as

$$\hat{V} = \frac{1}{2}\sum_{q,r} \gamma_{qr}\Omega^{-}_{qq}\Omega^{-}_{rr} \tag{5.30}$$

and in terms of spin-dependent generators as

$$\hat{V} = E_{\text{nuc}} + \tfrac{3}{4}\gamma n - \frac{1}{2}\sum_{q,r}\sum_{\sigma,\mu}\gamma_{qr}\Omega_{q\sigma r\mu}\Omega_{r\mu q\sigma}$$

$$-\frac{1}{2}\sum_{q,r}\sum_{\sigma,\mu}\gamma_{qr}\Omega^{-}_{q\sigma r\mu}\Omega^{-}_{r\mu q\sigma} \tag{5.31}$$

Note that both correlation operators include terms which break the effective one-electron alternancy parity, and in the case of the spin-free representation (5.30) all of the correlation is of this type. In the spin-dependent representation (5.31), on the other hand, a significant part of the correlation originates in terms that conserve the effective one-electron alternancy parity. This is an example of the different ways correlation is partitioned in the two approaches, though in the end the total energy is the same.

In addition to the spin and alternancy symmetries, most systems of physical interest have a center or plane of symmetry for the atoms. For the sake of generality we shall consider only the simplest case, in which the arrangement of atoms is left unchanged by a site-reflection operation R defined as

$$RA_{jq} = A_{j-q} \tag{5.32}$$

The actual physical meaning of R may be different in each case. Some examples are seen with the structures [1]–[4], where R corresponds to ordinary spatial reflection in [1], [3], and [4], and to a 180° rotation of the molecular skeleton in [2]. The site-reflection symmetry introduces the following additional restrictions on the PPP parameters:

$$\beta_{qr}=\beta_{-q-r}, \qquad \gamma_{qr}=\gamma_{-q-r} \qquad (5.33)$$

and the energy is blocked according to states with different symmetry labels $A\ (R=+1)$ and $B\ (R=-1)$.

The site reflection symmetry is implicit in a second unitary particle–hole conjugation operator C that has been defined in the pseudorotation picture.[50] The transformation of generators of U_{2n} and U_n is

$$CE_{q\sigma r\mu}C^{-1}=\delta_{qr}\delta_{\sigma\mu}-E_{-r-\mu-q-\sigma}(-)^{q-r}(-)^{\sigma-\mu}$$
$$CE_{qr}C^{-1}=2\delta_{qr}-E_{-r-q}(-)^{q-r} \qquad (5.34)$$

In neutral molecules and radicals $C^2=1$, and hence states have another parity label $C=+1$ or -1 that is related to the P-alternancy parity by

$$C=RP=PR \qquad (5.35)$$

C therefore commutes with the PPP Hamiltonian $\hat H$ only for systems which have both alternancy and site-reflection symmetries.

Symmetric and antisymmetric components of the generators which satisfy

$$C\Lambda_{q\sigma r\mu}C^{-1}=\Lambda_{q\sigma r\mu} \qquad C\Lambda^-_{q\sigma r\mu}C^{-1}=-\Lambda^-_{q\sigma r\mu}$$
$$C\Lambda_{qr}C^{-1}=\Lambda_{qr}, \qquad C\Lambda^-_{qr}C^{-1}=-\Lambda^-_{qr} \qquad (5.36)$$

are defined as

$$\Lambda_{q\sigma r\mu}=\tfrac12\big[E_{q\sigma r\mu}-(-)^{q-r}(-)^{\sigma-\mu}E_{-r-\mu-q-\sigma}\big]$$
$$\Lambda^-_{q\sigma r\mu}=\tfrac12\big[E_{q\sigma r\mu}+(-)^{q-r}(-)^{\sigma-\mu}E_{-r-\mu-q-\sigma}-\delta_{qr}\delta_{\sigma\mu}\big]$$
$$\Lambda_{qr}=\tfrac12\big[E_{qr}-(-)^{q-r}E_{-r-q}\big]$$
$$\Lambda^-_{qr}=\tfrac12\big[E_{qr}+(-)^{q-r}E_{-r-q}-2\delta_{qr}\big] \qquad (5.37)$$

When C commutes with the PPP Hamiltonian $\hat H$, the spin-independent form of the resonance and correlation energy operators is

$$\hat{\beta} = \sum_{q,r} \beta_{qr} \Lambda_{qr} \tag{5.38}$$

$$\hat{V} = \frac{1}{2} \sum_{q,r} \gamma_{qr} \Lambda_{qq} \Lambda_{rr} + \frac{1}{2} \sum_{q,r} \gamma_{qr} \Lambda^-_{qq} \Lambda^-_{rr} \tag{5.39}$$

The spin-dependent form of the correlation is

$$\hat{V} = E_{\text{nuc}} + \tfrac{3}{4} \gamma n - \frac{1}{2} \sum_{q,r} \sum_{\sigma,\mu} \gamma_{qr} \Lambda_{q\sigma r\mu} \Lambda_{r\mu q\sigma}$$

$$- \frac{1}{2} \sum_{q,r} \sum_{\sigma\mu} \gamma_{qr} \Lambda^-_{q\sigma r\mu} \Lambda^-_{r\mu q\sigma} \tag{5.40}$$

These forms of the energy symmetrized with respect to C are similar in appearance of the P-symmetrized operators in Eqs. (5.29)–(5.31). Note, however, that the spin-free correlation operator in Eq. (5.39) includes a part which conserves the effective one-electron C-alternancy symmetry, whereas $\hat{V}$ in Eq. (5.30) broke the effective one-electron P-alternancy symmetry. This is an indication of the different pictures implicit in the classification of energies with respect to P or C particle–hole exchange, though it is difficult to see just what these pictures are without more information about the specific form of the PPP parameters. The idea in an approximate-symmetry approach for correlation is to identify a particular zero-order form of the energy that contains the most of the correlation. We shall see subsequently that the P-alternancy scheme is more closely related to systems in which there is strong short-range electron correlation and pairing symmetry, while the C-alternancy scheme is more closely related to long-range correlation and pairing of electrons on opposite sides of the molecule. This is not obvious, however, without further investigation of the corresponding subgroup structure of U_{2n} or U_n for the different schemes.

Radiative selection rules[50,135–137] follow from the symmetry of the dipole operator. This is approximated as

$$\mathbf{D} = \sum_q \mathbf{d}_q E_{qq} \tag{5.41}$$

where $\mathbf{d}_q$ represents the position vector of atom q relative to a center of symmetry with

$$\sum_q \mathbf{d}_q = 0 \tag{5.42}$$

It is evident from the P-alternancy conjugation transformation of one-electron operators in Eq. (5.25) that P anticommutes with $\mathbf{D}$,

$$PDP^{-1} = -\mathbf{D} \qquad (5.43)$$

and can be represented with antisymmetric generators as

$$\mathbf{D} = \sum_q \mathbf{d}_q \Omega_{qq}^- \qquad (5.44)$$

In other words, the P-alternancy parity changes sign in a single-photon electric dipole transition. The selection rule for the C-alternancy parity depends on the symmetry of the transition moment axis under the site-reflection operation R. The transition operator generally includes symmetric and antisymmetric components,

$$\mathbf{D} = \frac{1}{2} \sum_q (\mathbf{d}_q - \mathbf{d}_{-q}) \Lambda_{qq} + \frac{1}{2} \sum_q (\mathbf{d}_q + \mathbf{d}_{-q}) \Lambda_{qq}^- \qquad (5.45)$$

If D_x is an axis with antisymmetric R-reflection symmetry, $x_q = -x_{-q}$, then C-alternancy parity is conserved in transitions. On the other hand, if D_y is an axis with symmetric R-reflection symmetry, $y_q = y_{-q}$, then the C-alternancy parity changes sign for these transitions. Molecular structure [2] is an example where both axes in the plane of the molecule are antisymmetric under R-reflection, and hence C-alternancy parity is conserved in both cases. Structure [3], on the other hand, has an antisymmetric horizontal axis and a symmetric vertical axis under R-reflection. The parity of C is conserved in the horizontal case, and broken for the transition moment component on the vertical axis. The P-alternancy parity is broken in each case.

D. Subgroup Chains, Quasispin Pairing

We have indicated how possible approximate correlation symmetries could be investigated by paritioning the energy with Lie-algebra generators that are symmetric and antisymmetric under particle–hole conjugation with the operators P or C. Similar partitioning could be carried out with other symmetry operations such as T (alternancy) or R (site reflection), or even other approximate symmetries that are usually assumed, such as double-bond alternation for localized charge in linear polyenes with the exciton picture. The interesting feature group-theoretically is that when the Lie-algebra basis is restricted to generators that are symmetric under a sequence $g_1, g_2, \ldots$ of these operations, the result is a subgroup chain $U_{2n} \supset G_1 \supset G_2 \supset \cdots$ that reflects the partitioning of the energy. The general idea is to investigate all realistic possibilities to see if quadratic invariants of the Lie algebras (operators that commute with all the Lie-algebra generators) are approximate constants of the motion for electron correlation. This is obviously a great task unless reasonable physical judgement is exercised.

Herrick and Liao[123] have carried out an analysis of subgroups derived from spin and particle–hole conjugation with C for polyenes, and found rotation subgroups $U_{2n} \supset R_{2n}$ and $U_n \supset R_n$, and symplectic subgroups $U_{2n} \supset Sp_{2n}$ and $U_n \supset Sp_n$ very similar to the groups for Racah seniority in atomic and nuclear spectroscopy. The goal here will be to outline the general procedure including both P and C conjugation, illustrating two very different types of seniority-pairing symmetry for polyenes. We present only the most elementary aspects of the Lie-algebra generators, invariants, and pairing operators with associated quasispin groups. The standard notation for irreducible representations of unitary groups, symplectic groups, and the generalized rotation groups for orbital representations of spin-$\frac{1}{2}$ fermions is a partition $(2_1 2_2 \ldots 2_\alpha 1_1 1_2 \ldots 1_{2\beta}) \equiv (2^\alpha 1^{2\beta})$ with the Young tableau:

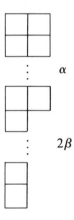

This contains $\alpha + 2\beta$ rows and at most two columns (for antisymmetry), with $2\alpha + 2\beta$ boxes in all.

We consider only Lie-algebra generators that are symmetrized with respect to spin and P conjugation, or spin and C conjugation. The following diagrams summarize the hierarchies of generators:

$$\begin{array}{ccc} & \Omega_{q\sigma r\mu} & \to S \oplus \Omega_{qr} \\ P \nearrow & & \nearrow \\ E_{q\sigma r\mu} & \to S \oplus E_{qr} & \\ C \searrow & & \searrow \\ & \Lambda_{q\sigma r\mu} & \to S \oplus \Lambda_{qr} \end{array} \qquad (5.46a)$$

the corresponding groups:

$$
\begin{array}{ccc}
\mathrm{Sp}_{2n}^{P} & \to & R_3^S \times R_n^P \\
{\scriptstyle P\nearrow} & & \nearrow \\
\mathrm{U}_{2n} \to R_3^S \times \mathrm{U}_n & & \\
{\scriptstyle C\searrow} & & \searrow \\
(R_{2n}^C, \mathrm{Sp}_{2n}^C) & \to & R_3^S \times (\mathrm{Sp}_n^C, R_n^C)
\end{array}
\tag{5.46b}
$$

and the irreducible representations for polyenes:

$$
\begin{array}{ccc}
\langle 1^w \rangle & \to & [S]\cdot[2^{\frac{1}{2}w-S}\,1^{2\beta}], \\
{\scriptstyle P\nearrow} & \nearrow & \beta \equiv \min(S,\kappa) \\
\{1^N\} \to [S]\cdot\{2^{\frac{1}{2}N-S}\,1^{2S}\} & & \\
{\scriptstyle C\searrow} & \searrow & \\
([1^N],\langle 1^v \rangle) & \to & [S]\cdot(\langle 2^{\frac{1}{2}v-t}\,1^{2t}\rangle, [2^{\frac{1}{2}v-S}\,1^{2\beta}]), \\
& & \beta \equiv \min(S,\lambda)
\end{array}
\tag{5.46c}
$$

Representations of spin angular momentum are labeled with the spin quantum number as $[S]$; N is the number of electrons; v and w label the "seniority" of the representations; t is called the "reduced spin" label; and so-called "quasispin" labels are defined as

$$
\begin{aligned}
\kappa &= \tfrac{1}{2}(n-w) & \text{for } R_n^P \\
\lambda &= \tfrac{1}{2}(n-v) & \text{for } R_n^C \\
\omega &= \tfrac{1}{2}(n-v) & \text{for } \mathrm{Sp}_n^C
\end{aligned}
\tag{5.47}
$$

The superscripts P or C are included in the labeling of groups to indicate that the Lie-algebra generators are left invariant under the respective type of particle–hole exchange conjugation. Some of the properties of each group are summarized in the following paragraphs.

1. U_{2n}. The complete configuration interaction basis is contained in the single irreducible representation $\{1^N\}$. The quadratic invariant is

$$
\Gamma_0 = \sum_{q,r}\sum_{\sigma,\mu} E_{q\sigma r\mu} E_{r\mu q\sigma} = N(2n+1-N)
\tag{5.48}
$$

and the eigenvalue depends only on the number of electrons and the number of atoms for this representation.

2. U_n. This is the spin-free unitary group that is the starting point for most other approaches based on canonical subgroup chains $U_n \supset U_{n-1} \supset U_{n-2} \supset \cdots \supset U_1$ in an arbitrary orbital bases. The quadratic invariant for U_n is

$$\Gamma = \sum_{q,r} E_{qr}E_{rq} = \tfrac{1}{2}N(2n+4-N) - 2S(S+1) \tag{5.49}$$

with a single irreducible representation for each value of the spin label $S = \tfrac{1}{2}N, \tfrac{1}{2}N - 1, \ldots, \tfrac{1}{2}$, or 0. These are exact symmetry blocks of the energy matrix.

3. Sp_{2n}^P. The hierarchy of seniority labels is $w = N, N-2, \ldots, 1$, or 0; and the eigenvalue of the quadratic invariant for each of these representations is

$$\Phi_0^P = \sum_{q,r} \sum_{\sigma,\mu} \Omega_{q\sigma r\mu}\Omega_{r\mu q\sigma} = \tfrac{1}{2}n(n+2) - 2\kappa(\kappa+1) \tag{5.50}$$

The form of the eigenvalue indicates why κ is called a quasispin label. Another way of describing this type of quasispin is with an angular-momentum group R_3^κ having Lie-algebra generators defined as

$$\kappa_+ = \sum_q (-)^{j+q} X_{q\uparrow}^\dagger X_{q\downarrow}^\dagger$$

$$\kappa_- = (\kappa_+)^\dagger = \sum_q (-)^{j+q} X_{q\downarrow} X_{q\uparrow} \tag{5.51}$$

$$\kappa_0 = \tfrac{1}{2}(\hat{N} - n)$$

These satisfy the usual angular-momentum commutation relations, and the square of the total angular momentum is $\kappa^2 = \kappa(\kappa+1)$. The operators κ_+ and κ_- describe collective creation and annihilation of pairs of electrons, and this clearly commutes with the ordinary electron spin, **S**. The pairing symmetry is expected to be most useful for short-range correlation effects, in the sense that the pairing occurs on single atoms in Eq. (5.51). The relationship between Sp_{2n}^P and R_3^κ is seen with a so-called "pairing operator" for the group, defined as

$$\mathscr{P}_0^P = \Gamma_0 - 2\Phi_0^P = \sum_{q,r} \sum_{\sigma,\mu} E_{q\sigma r\mu}E_{q-\sigma r-\mu}(-)^{q-r}(-)^{\sigma-\mu}$$

$$= 4\kappa_+\kappa_- - \hat{N} \tag{5.52}$$

4. R_n^P. The Lie algebra generators Ω_{qr} commute with the spin and the P-type particle–hole exchange. The diagonal quadratic invariant and the

spin-free pairing operators are

$$\Phi^P = \sum_{q,r} \Omega_{qr}\Omega_{rq} = \tfrac{1}{4}n(n+2) - \kappa(\kappa+1) - S(S+1) \qquad (5.53)$$

$$\mathcal{P}^P = \Gamma - 2\Phi^P = \sum_{q,r} (-)^{q-r} E_{qr}E_{qr}$$

$$= \hat{N} + 2\kappa_+\kappa_- \qquad (5.54)$$

Note that the eigenvalue of Φ^P is symmetric under exchange of spin and quasispin. This leads to an additional symmetry label for states when $S = \kappa$. The group-theoretical interpretation is that the spin-free subgroup chain is actually $U_n \supset O_n \supset R_n$ when n is *even*, and when $\kappa = S$ each irreducible representation of the full orthogonal group splits into two equal-dimension representations of the group of proper n-dimensional rotations,[140]

$$\left[2^{\frac{1}{2}n - 2\kappa}1^{2\kappa}\right] \rightarrow \left[2^{\frac{1}{2}n - 2\kappa}1^{2\kappa}\right]_+ + \left[2^{\frac{1}{2}n - 2\kappa}1^{2\kappa}\right]_- \qquad (5.55)$$

[An example of this was seen in Section IV.D, where intrashell states of doubly excited atoms were described with multiplets of O(4) that split into two representations of SO(4), or R_4.]

5. Sp_{2n}^C (*odd n*). Although the general form of the generators $\Lambda_{q\sigma r\mu}$ is the same for all values of n, there is a different group structure for *even* or *odd* n. The reason for this is the phase factor in Eq. (5.37), which causes generators $\Lambda_{q\sigma-q-\sigma}$ to vanish when n is *even* but not when n is *odd*. The result is $n(2n-1)$ generators for R_{2n}^C (*even n*), and $n(2n+1)$ generators for Sp_{2n}^C (*odd n*). Except for changes in notation, the structure of irreducible representations of Sp_{2n}^C is identical mathematically to that of the group Sp_{2n}^P. The corresponding diagonal quadratic invariant and pairing operator for Sp_{2n}^C are

$$\Phi_0^C(\textit{odd n}) = \sum_{q,r}\sum_{\sigma\mu}\Lambda_{q\sigma r\mu}\Lambda_{r\mu q\sigma} = \tfrac{1}{2}n(n+2) - 2\lambda(\lambda+1) \qquad (5.56)$$

$$\mathcal{P}_0^C = \Gamma_0 - 2\Phi_0^C = \sum_{q,r}\sum_{\sigma,\mu} E_{q\sigma r\mu}E_{-q-\sigma-r-\mu}(-)^{q-r}(-)^{\sigma-\mu}$$

$$= 4\lambda_+\lambda_- - \hat{N} \qquad (5.57)$$

Generators of the quasispin group R_3^λ are defined as

$$\lambda_+ = \sum_q (-)^{j-q}X_{q\uparrow}^\dagger X_{-q\downarrow}^\dagger$$

$$\lambda_- = (\lambda_+)^\dagger = \sum_q (-)^{j-q}X_{-q\downarrow}X_{q\uparrow} \qquad (5.58)$$

$$\lambda_0 = \tfrac{1}{2}(\hat{N} - n)$$

with $\lambda^2 = \lambda(\lambda + 1)$. Although the seniority labels and quasispin labels are identical for Sp_{2n}^C and Sp_{2n}^P, the two groups have very different effects on the configuration-interaction basis. The collective creation and annihilation of electron pairs in the operators λ_+ and λ_- is for two electrons at *different* atomic sites, q and $-q$, on opposite sides of the molecule with respect to the R-symmetry operation for site reflection. This type of pairing symmetry is expected to be most useful in radicals if there are long-range electron correlation effects. This contrasts sharply with the pairing symmetry of the group Sp_{2n}^P, which was short-range for both *even* and *odd n*.

6. R_{2n}^C (*even n*). The subgroup chain in this case is actually $U_{2n} \supset O_{2n} \supset R_{2n}$, with two representations of the group R_{2n} obtained as

$$\begin{aligned}
\{1^N\} &\rightarrow [1^N] \rightarrow [1^N] & (N \neq n) \\
\{1^N\} &\rightarrow [1^n] \rightarrow [1^n]_+ + [1^n]_- & (N = n)
\end{aligned} \tag{5.59}$$

The configuration-interaction matrix of neutral molecules therefore falls naturally into two classes of states according to the $\pm$ label for these representations. Since the group Sp_{2n}^C commutes with the particle–hole exchange operator C, all of the states in each class have the same C-alternancy parity. Herrick and Liao worked out tables of these representations, and found that the pseudorotation phase convention for particle–hole exchange has $C = \pm 1$ for $[1^n]_\pm$, respectively. Because each irreducible representation in Eq. (5.59) depends only on the number of electrons, N, there is no useful information contained in the Lie-algebra invariant and pairing operator for the polyene states for *even n*:

$$\Phi_0^C = \tfrac{1}{2}N(2n - N) \tag{5.60}$$

$$\mathscr{P}_0^C = \hat{N} \tag{5.61}$$

The value of the classification is that it provides a direct way of assigning particle–hole symmetry labels.

7. R_n^C (*odd n*). There is a similar distinction between the generators Λ_{qr} for *even* and *odd* values of n. In the former case there are $\tfrac{1}{2}n(n+1)$ generators for a representation of the Lie algebra Sp_n^C (*even n*), while generators Λ_{q-q} vanish in the latter case, leaving $\tfrac{1}{2}n(n-1)$ generators for a representation of the Lie algebra R_n^C (*odd n*). The mathematical similarity of Sp_{2n}^C and Sp_{2n}^P carries through to the respective subgroups R_n^C and R_n^P, with the following change in notation for the quadratic invariant and pairing operator:

$$\Phi^C = \sum_{q,r} \Lambda_{qr}\Lambda_{rq} = \tfrac{1}{4}n(n+2) - \lambda(\lambda+1) - S(S+1) \tag{5.62}$$

$$\mathscr{P}^C = \Gamma - 2\Phi^C = \sum_{q,r} (-)^{q-r} E_{qr}E_{-q-r}$$

$$= \hat{N} + 2\lambda_+\lambda_- \tag{5.63}$$

As we noted for Sp_{2n}^C, the real difference lies with the long-range nature of the pairing described by the quasispin group R_3^λ, versus the short-range pairing symmetry described by R_3^κ.

8. Sp_n^C (*even n*). We have saved this group for last because it involves a different type of quasispin group for pairing symmetry. The difference originates in the fact that λ_+ and λ_- describe a singlet spin coupling of two electrons when n is *odd*, but a triplet spin coupling when n is *even*. The result is that the quasispin generators no longer commute with the ordinary electron spin group R_3^S. Instead, the six generators λ and S are augmented by four additional generators

$$G_+ = \frac{1}{2}\sum_q (-)^{j-q} X_{q\uparrow}^\dagger X_{-q\uparrow}^\dagger$$

$$G_- = (G_+)^\dagger = \frac{1}{2}\sum_q (-)^{j-q} X_{-q\uparrow} X_{q\uparrow}$$

$$D_+ = \frac{1}{2}\sum_q (-)^{j-q} X_{q\downarrow}^\dagger X_{-q\downarrow}^\dagger \qquad (5.64)$$

$$D_- = (D_+)^\dagger = \frac{1}{2}\sum_q (-)^{j-q} X_{-q\downarrow} X_{q\downarrow}$$

and together the ten operators are generators of a five-dimensional quasispin group, $R_5^{\omega t}$. The operators defined in (5.64) vanish identically when n is *odd*. Physical states for *even n* have a classification with the subgroup $R_5^{\omega t} \supset R_3^S$, and λ is no longer a good quantum number. The relationship of quasispin to the diagonal quadratic invariant and pairing operator of the group Sp_{2n}^C is

$$\Phi^C = \sum_{q,r} \Lambda_{qr}\Lambda_{rq} = \tfrac{1}{4}n(n+6) - \omega(\omega+3) - t(t+1) \qquad (5.65)$$

$$\mathscr{P}^C = \Gamma - 2\Phi^C = \sum_{q,r} (-)^{q-r} E_{qr} E_{-q-r}$$

$$= 2\lambda_+\lambda_- + 4G_+G_- + 4D_+D_- - \hat{N} \qquad (5.66)$$

The label t is called the reduced spin, because it plays a role in Eq. (5.65) similar to that of the real electron spin S in Eq. (5.62) for Φ^C (*odd n*). The pairing symmetry involves long-range electron correlation in both cases.

Further discussion of quasispin seniority labels for polyenes can be found in References 50 and 123, where they were introduced. Additional information about rotation and symplectic groups from a general mathematical point of view is contained in Wybourne's book,[10] which has a useful appendix of

TABLE VIII
Irreducible Representations of $R_3^S \times R_n$ for Neutral Polyenes

Molecules (κ)					Radicals (κ or λ)				
S	0	1	2	3	S	0	1	2	3
$n=2$ 0	$[2]_\pm$	[0]			$n=3$ $1/2$	[2]	[1]		
1	[0]				$3/2$	[0]			
$n=4$ 0	$[2^2]_\pm$	[2]	[0]		$n=5$ $1/2$	$[2^2]$	[21]	[1]	
1	[2]	$[1^2]_\pm$			$3/2$	[2]	$[1^2]$		
2	[0]				$5/2$	[0]			
$n=6$ 0	$[2^3]_\pm$	$[2^2]$	[2]	[0]	$n=7$ $1/2$	$[2^3]$	$[2^21]$	[21]	[1]
1	$[2^2]$	$[21^2]_\pm$	$[1^2]$		$3/2$	$[2^2]$	$[21^2]$	$[1^3]$	
2	[2]	$[1^2]$			$5/2$	[2]	$[1^2]$		
3	[0]				$7/2$	[0]			

tables. Here we shall present some general results for neutral molecular or radical polyenes, which is the case of greatest experimental and theoretical interest.

The hierarchy of irreducible representations of the group $R_3^S \times R_n^P$ for both *even* and *odd* values of n is

$$S = \tfrac{1}{2}n, \tfrac{1}{2}n - 1, \ldots, \tfrac{1}{2}, \text{ or } 0$$
$$w = n, n-2, \ldots, 2S \tag{5.67}$$

The quasispin label κ is always an integer for neutral systems. Table VIII contains the irreducible representations for $n \leqslant 7$, which is enough to illustrate the general trend. Note that the table for each even value of n is symmetric with respect to exchange of κ and S. The polyene representations of the group $R_3^S \times R_n^C$ include only the *odd* values of n in Eq. (5.67), with seniority v instead of w; in Table VIII κ is replaced by λ. Although the same irreducible representations appear there for for both R_n^P and R_n^C, the actual configurations contained in identical representations of the two groups are, of course, different. As a concrete example, consider the atomic-orbital valence-bond configurations in which there is one electron at each atomic site. Computation of the short-range pairing symmetry with R_3^κ generators in Eq. (5.51) gives $\kappa = 0$ for all configurations, because there are no paired sites. The same valence-bond configurations (radicals only) have a range of values $0 \leqslant \lambda \leqslant \tfrac{1}{2}n - S$ when they are classified with the other quasispin group R_3^λ, which describes long-range pairing.

The representations of $R_3^S \times Sp_n^C$ for molecular polyenes are more difficult to describe because there are multiplicities. This means that the same irreducible representation (with fixed values of ω, t, S, and $C = \pm 1$) can occur more than once. Fortunately, the structure of the representations is broad enough so that the multiplicity problem occurs only very infrequently in systems of physical interest, and only when there are more than ten atomic sites. Tables in Reference 123 give the quasispin classification for molecules with fewer than twenty atoms. There are no multiplicities of irreducible representations for the important cases of singlet $(S = 0)$ and triplet $(S = 1)$ spin. The hierarchy of polyene representations $\langle 2^\alpha 1^{2t} \rangle$ of Sp_n^C for singlet states $(S = 0)$ is

$$\alpha = \nu, \nu - 2, \nu - 4, \dots, 1, \text{ or } 0$$
$$t = 0, 1, 2, \dots, \nu - \alpha \qquad (5.68)$$
$$C = (-)^t$$

with $\nu \equiv \frac{1}{2}n$. Seniority and quasispin labels for these singlet representations are $v = 2(\alpha + t)$ and $\omega = \nu - \alpha - t$. The number of configurations in each representation of Sp_n^C, neglecting spin multiplicities, is described by the formula

$$\dim \langle 2^{\frac{1}{2}v - t} 1^{2t} \rangle = \frac{(2t + 1)(2\omega + 3)(\omega - t + 1)(\omega + t + 2)}{(\nu + 1)(2\nu + 3)(\nu + 2)(\nu - \omega + t + 1)}$$
$$\times \left(\begin{array}{c} 2\nu + 3 \\ \nu - \omega + t \end{array} \right) \left(\begin{array}{c} 2\nu + 4 \\ \nu - \omega - t \end{array} \right) \qquad (5.69)$$

E. Exact Pairing Symmetries

Our investigation of polyene seniority groups has revealed so far two pairing symmetry groups that simplify the representation of the PPP Hamiltonian. What we know about the energy is

1. The resonance operator $\hat{\beta}$ is a linear combination of Lie-algebra generators of the groups, and hence has exact pairing symmetry. By this we mean it commutes with the quadratic invariant operator Φ, or equivalently with the pairing operator $\mathcal{P}$ and the invariant operator of the quasispin group.
2. The correlation operator $\hat{V}$ does not, in general, have exact pairing symmetry, though it can always be represented with sums of generator products and products of antisymmetric tensors.

A general procedure for investigating possible approximate symmetry for correlation would involve Clebsch–Gordan coupling of the antisymmetric tensors in $\hat{V}$ so that the operator is cast into a sum of two terms $\hat{V}_0 + \hat{V}_1$, of which $\hat{V}_0$ has pairing symmetry and $\hat{V}_1$ does not. One could then investigate the general class of Coulomb repulsion parameters γ_{qr} for which $\hat{V}_1 = 0$, or for which the symmetry-breaking effect of $\hat{V}_1$ is minimized in a multiconfiguration SCF approach with low-lying excited states, for example. Such a procedure has not yet been carried out.

What we shall offer instead is two different models which have exact pairing symmetries, and which illustrate nicely the difference between short-range pairing and long-range pairing. The Hubbard model[141] assumes only short-range (single center) repulsion, and some of its symmetry properties are already known.[142] The model is not very realistic for chemical systems, though it does reproduce many of the qualitative features of the PPP hamiltonian with Coulomb repulsion. A curious property of the PPP–Hubbard model is the occurrence of accidental degeneracies in spectra[38, 142] that have not been explained. A second model we have proposed[50] assumes the repulsion integral has a harmonic dependence on the distance between atoms. This does not mean the atoms are vibrating. The atoms are fixed and the *electrons* are interacting like coupled oscillators, though they are restricted to discrete atomic sites along the molecular skeleton in the second-quantization picture with the PPP hamiltonian. The result is that the harmonic model has exact long-range pairing symmetry, but only for *neutral* polyenes. In the special case of a linear polyene with equal spacing between atoms, the harmonic model reduces to the problem with exact pseudorotation symmetry that was found in our earlier approach.[50]

1. Hubbard Repulsion

The pairing symmetry can be derived in several ways, most conveniently in the spin-dependent generator approach or with quasispin. Since most of the work with unitary-group approaches these days is carried out in a spin-free picture, we shall outline the derivation from this point of view. The basic form of the Hubbard interaction is related to the PPP model with a constant one-center repulsion $\gamma_{qr} \equiv \gamma \delta_{qr}$. This reduces the correlation operator $\hat{V}$ to a sum of diagonal terms

$$\hat{V} = \tfrac{1}{2}\gamma \sum_q \left(E_{qq} - 1\right)^2 \tag{5.70}$$

The pairing symmetry is found after working out the commutator algebra in between the following steps:

$$
\begin{aligned}
[\mathcal{P}^P, \hat{H}] &= [\mathcal{P}^P, \hat{V}] \\
&= \gamma \sum_{q,r} (-)^{q-r}\left[\left(E_{rr} - E_{qq}\right)E_{qr}^2 + E_{qr}^2\left(E_{rr} - E_{qq}\right)\right] \\
&= 2\gamma \sum_{q,r} (-)^{q-r}\left(E_{rr}E_{qr}E_{qr} - E_{qr}E_{qr}E_{qq}\right) \tag{5.71}
\end{aligned}
$$

This is as far as we can reduce the commutator without breaking the spin-free picture. It is clear that terms on the right-hand side of (5.71) vanish when $q = r$. The remaining terms vanish individually because we are dealing with spin-$\frac{1}{2}$ particles, and no more than two particles can occupy a single atomic site. The term $E_{rr}E_{qr}E_{qr}$, for example, vanishes because the factor $E_{qr}E_{qr}$ removes two electrons from atom r, leaving $E_{rr} = 0$ for the remaining factor. In commutators with spin-dependent generators this type of cancellation is seen with products of creation operators $X_{q\sigma}^{\dagger} X_{q\mu}^{\dagger} X_{q\alpha}^{\dagger}$ which try to put three electrons on the same atom, or products $X_{q\sigma} X_{q\mu} X_{q\alpha}$ which try to remove three electrons from a single atom. The pairing symmetry of the Hubbard Hamiltonian for electrons is therefore a direct consequence of the Pauli exclusion principle, as opposed to commutation relations of the unitary groups themselves.

As a result of the pairing symmetry the configuration interaction matrix is blocked exactly according to the seniority label w and the spin label S for the group $R_3^S \times R_n^P$. Quasispin labels for the blocks were shown in Table VIII. Heilmann and Lieb[142] described a similar type of quasispin for the Hubbard model with nearest-neighbor resonance in the spectrum of benzene. It is interesting to see here how it falls out naturally in a more general approach based on seniority subgroups of the PPP hamiltonian described with unitary groups.

An important feature of the Hubbard model for chemical systems is that the correlation part of the energy is completely independent of the geometrical arrangement of atoms, since all repulsion is single-center and is independent of bond lengths or bond angles. What this means is that the seniority classification continues to block-diagonalize the energy even when motion of atoms over the PPP potential energy surface is taken into account. The degree to which seniority is conserved in actual chemical reactions would depend on many factors, such as the effect of longer-range correlation and the question of the applicability of the PPP model.

The pairing symmetry is also evident in the eigenspectrum of the correlation operator itself. This corresponds to the limit $|\beta/\gamma| = 0$ when the resonance and repulsion parameters are treated as continuous variables. Common lore has it that the Hubbard spectrum depends on the number of paired atomic sites. What is really important, however, is the number of singly occupied sites, in both neutral systems and ions. The site distribution of valence electrons is described with three numbers,

n_2 = number of atoms with two electrons

n_1 = number of atoms with one electron

n_0 = number of atoms with no electrons

which are related to the total number of atoms and electrons by the formulas

$$n = n_0 + n_1 + n_2$$
$$N = n_1 + 2n_2 \tag{5.72}$$

It is evident from Eqs. (5.70) and (5.72) that a general formula for the correlation energy is

$$\hat{V} = \tfrac{1}{2}\gamma(n_0 + n_2) = \tfrac{1}{2}\gamma(n - n_1) \equiv \gamma\rho \tag{5.73}$$

independent of the number of electrons in the system. It is clear from this result why there is pairing symmetry, because the quasispin generators of R_3^κ only increase or decrease the values of n_2 and n_0, and do not change the number of singly occupied sites. Evidently the quasispin group R_3^κ is only a subgroup of a larger group of operations which conserve the number of singly occupied sites. In this connection it is interesting to note that the index n_1 *itself* could be considered as a label for a different type of seniority with values

$$n_1 = N, N - 2, \ldots, 1, \text{ or } 0 \tag{5.74}$$

when $N \leqslant n$. The same values of n_1 are obtained when $N > n$ by particle–hole symmetry. The quantum number ρ that was defined in Eq. (5.73) would be the associated quasispin label. In neutral polyenes, for example, valence-bond configurations with all sites singly occupied would have $\rho = 0$.

2. Harmonic Repulsion

Harmonic repulsion is not a very good approximation of Coulomb integrals in actual molecules, but it has some of the qualitative features of electron correlation. The form of the repulsion integral is

$$\gamma_{qr} = \gamma - \mathfrak{f}(\mathbf{d}_q - \mathbf{d}_r)^2 \tag{5.75}$$

The symmetry results are independent of the sign and magnitude of the effective force constant $\mathfrak{f}$, and nothing is lost if the other constant is $\gamma \equiv 0$. The position of atom q is $\mathbf{d}_q$, and was used earlier in Eq. (5.41) for radiative transitions. The long-range pairing symmetry is found when (5.75) is inserted in Eq. (5.39), making use of the site-reflection symmetry properties

$$\Lambda_{qq}^- = \Lambda_{-q-q}^-$$
$$\sum_q \Lambda_{qq}^- = \hat{N} - n \tag{5.76}$$
$$\sum_q (f_q)^m \Lambda_{qq}^- = 0 \qquad (f_q = -f_{-q}, \quad m \text{ odd})$$

The result is a Hamiltonian $\hat{H} = \hat{H}_0 + \hat{H}_1$ with

$$\hat{H}_0 = \sum_{q,r} \beta_{qr} \Lambda_{qr} + \mathfrak{k} \left(\sum_q \mathbf{d}_q \Lambda_{qq} \right)^2$$

$$\hat{H}_1 = -\mathfrak{k}(\hat{N} - n) \sum_q (\mathbf{d}_q)^2 \Lambda_{qq}^- + \mathfrak{k} \sum_{q,r} (\mathbf{d}_q \Lambda_{qq}^-) \cdot (\mathbf{d}_r \Lambda_{rr}^-) \qquad (5.77)$$

$\hat{H}_0$ is a function of Lie algebra generators of R_n^C (*odd n*) or Sp_n^C (*even n*), and hence has exact pairing symmetry. $\hat{H}_1$ generally breaks the pairing symmetry, but vanishes identically in neutral ($N = n$) polyenes with atoms arranged so that

$$\mathbf{d}_q = -\mathbf{d}_{-q} \qquad (5.78)$$

in the plane of the molecule. This occurs in one-dimensional systems with reflection symmetry, or in linear polyenes which have a C_{2h} symmetry axis for 180° rotation (cf. structure [2]).

The quantization of $\hat{H}$ for these neutral polyenes involves exact symmetry blocks of the energy with respect to spin and seniority labels, because the Hamiltonian commutes with the pairing operator $\mathscr{P}^C$. The quantization is not complete, because there can be many configurations in each symmetry block. The number of configurations (neglecting spin multiplicity) is given by the dimension of the irreducible representation of Sp_n or R_n. The obvious physical picture of correlation in the model is collective motions of electrons like normal modes. The correlation interaction is a dipole–dipole coupling,

$$\hat{H}_0 = \hat{\beta} + \mathfrak{k} \mathbf{D} \cdot \mathbf{D} \qquad (5.79)$$

where $\mathbf{D}$ is the dipole operator in Equation (5.41). One-photon radiative transitions excite vibrationlike motions of an electron along the molecular backbone. The coupling of two electrons in (5.79) is thus similar to the types of correlation that would be excited in two-photon transitions.

3. Pseudorotations for Harmonic Model

There is an additional symmetry of the harmonic model for one-dimensional systems with equal bond lengths as in [5]:

$$-j \quad -j+1 \quad -j+2 \qquad \cdots \qquad j-2 \quad j-1 \quad j$$

$$x \rightarrow$$

[5]

In this case the position of each atom is simply $d_q = q\bar{R}$, where $\bar{R}$ is the bond distance between nearest-neighbor atoms. The result is that the dipole operator and the energy can be expressed in terms of the pseudorotation generator j_0 as

$$D_x = \bar{R}j_0, \qquad H_0 = \hat{\beta} + \mathfrak{k}\bar{R}^2 j_0^2 \qquad (5.80)$$

In other words, the configuration-interaction matrix for dipole transitions and electron correlation is blocked according to the selection rule $\Delta J = 0$ for the pseudorotation symmetry R_3^J. This is interesting because it extends the seniority chains further to $U_n \supset R_n^C \supset R_3^J \supset R_2^Q$ for *odd n* and $U_n \supset Sp_n^C \supset R_3^J \supset R_2^Q$ for *even n*. Here the label Q is for the eigenvalue $j_0 \equiv Q$ in the atomic basis. These groups for pseudorotation and seniority in polyenes are similar to the groups for the classifications of shell models in atoms and nuclei with exact spherical symmetry.

F. Linear Polyenes

1. Overview

The actual PPP correlation in low-lying states of linear polyenes investigated so far[50] is remarkably close to that predicted by long-range pairing and R_3^J pseudorotation symmetry. The reason for this agreement is not understood fully, because the calculations assumed a Coulomb repulsion integral approximated with the Ohno formula

$$\gamma_{qr} = \frac{14.397}{\left[(14.397/\gamma)^2 + R_{qr}^2\right]^{1/2}} \text{ eV} \qquad (5.81)$$

which has no obvious pseudorotation symmetry like the harmonic model. R_{qr} is the distance between atoms q and r in angstroms. A contributing factor to the symmetry was the discovery that the Hückel resonance for linear polyenes could be approximated with a pseudorotation generator $\hat{\beta} \simeq \frac{1}{2}\bar{\beta}(j_+ + j_-)$, where $\bar{\beta}$ is an effective resonance parameter. Diagonalization of this generator gives equally spaced levels, and corresponds to a pseudorotation molecular-orbital representation described as a second subgroup $R_3^J \supset R_2^M$. Here $M = J, J-1, \ldots, -J$ labels the eigenvalues of the diagonal generator, with $\hat{\beta} \simeq \bar{\beta}M$. Each pseudorotation molecular orbital is a linear combination of atomic orbitals with expansion coefficients that are Wigner d functions for an R_3^J-group rotation. Important features of the pseudorotation–Hückel spectrum are: (1) the ground level has the highest value of J and M, because $\bar{\beta}$ is negative; and (2) excited states with two or more pseudorotation quanta of excitation are highly degenerate, much like the degeneracies of N one-dimensional spin-$\frac{1}{2}$ fermion oscillators. The result is that a fair picture of the excitation spectrum is obtained by diagonalizing the correlation operator $\hat{V}$ within each manifold separately, this procedure giving the first-order splitting of the degeneracy. Interactions between the different manifolds shift the first-order spectra, but not substantially in low-n systems investigated so far. Assignments of spin–pseudorotation–quasispin term symbols $^{2S+1}(J)_\lambda$ for neutral radicals, or $^{2S+1}(J)_{\omega t}$ for neutral molecules, are made by comparing configuration-mixing structure in wave functions from the group theory and from diagonalization of the configuration-interaction energy. Only the spin and particle–hole conjugation are exact symmetries; pseudorotation and quasispin describe only approximate correlation symmetries, and their Lie-algebra invariants are approximate constants of the motion. Figures 18 and 19 illustrate the pseudorotation-splitting picture for the PPP spectra of C_3H_5 and C_4H_6 obtained[50] by diagonalizing the complete configuration-interaction matrix. The harmonic model predicts the correlation symmetry should weaken in the spectra of ions, and this type of behavior is found in the PPP spectra as well. The equal spacing of pseudorotation–Hückel resonance energies remains strong for systems with low degrees of ionization, however, and this leads to the possibility of correlation "shake-up" in PPP predictions of photoelectron spectra.[50]

More recent work[123] has investigated the pseudorotation tensor structure of the PPP Hamiltonian. This generally includes more terms, which break the pseudorotation symmetry, than the multipole expansions in atomic and nuclear problems with exact rotational symmetry. Nevertheless, two results

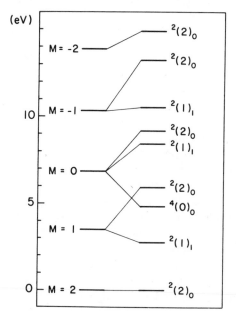

Fig. 18. Complete spectrum of PPP energy for allyl radical C_3H_5 with C_{2v} symmetry, illustrating correlation splitting of zero-order pseudorotation degeneracy manifolds. Term symbols are $^{2S+1}(J)_\lambda$, and other symmetry labels are $R = (-)^{M+1}$, $C = (-)^\lambda$, and $P = RC$.

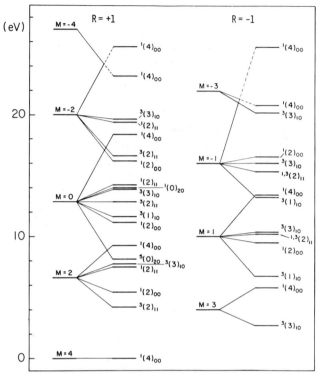

Fig. 19.   Complete spectrum of PPP energy for *trans*-butadiene (C_4H_6) with C_{2h} symmetry, illustrating correlation splitting of zero-order pseudorotation degeneracy manifolds. The term symbols are $^{2S+1}(J)_{\omega t}$, and the site-reflection symmetry is $R = +1$ (A_g) or $R = -1$ (B_u). Parity labels depend only on the values $^{2S+1}(\omega, t)$, and from Reference 123: $C = +1$ for $^{1,5}(2,0)$, $^{3}(1,1)$, and $^{1}(0,0)$: $C = -1$ for $^{1}(1,1)$ and $^{3}(1,0)$; and $P = RC$. In dipole transitions C is conserved, P changes sign, and the approximate pseudorotation selection rule is $\Delta J = 0$, $\Delta M = \pm 1$. Low-lying valence states of interest lie below the first ionization potential (at 9.1 eV); valence states above this are subject to autoionization.

were found that indicate the correlation in linear polyenes is something special: (1) contributions from coupled spin-dependent tensor vanish in symmetrical all-*trans* polyenes unless the two tensors have the same rank; and (2) there is an approximate cancellation of contributions from long- and short-range correlation effects, with the result that a large number of tensor expansion coefficients have approximately the same numerical value. The possible implications of this second result for explaining a high degree of pseudorotation symmetry are not understood yet. The first result is important because it shows how a lot of the symmetry breaking found in most systems goes away in linear polyenes, leaving a higher degree of pseudorotation symmetry overall. The fact that spin-dependent Lie-algebra tensors had to be used to find this result is one indication of the subtlety of the correlation if one sticks to the usual spin-free picture. A bonus feature of the spin-dependent tensor approach is that the expansion coefficients for symmetrical all-*trans* polyenes are independent of the number of carbon atoms in the chain. This could be important in considering how the spectra behave in the long-chain limit, which is not well understood.[131]

2. Tensors, Orbitals, Quasispin

The pseudorotation tensors that were proposed[123] for electron correlation in polyenes are represented here as

$$A_{d\alpha}^{ks} = \sum_{q,r} \sum_{\sigma,\mu} \langle jq, j-r|kd\rangle\langle \tfrac{1}{2}\sigma, \tfrac{1}{2}-\mu|s\alpha\rangle E_{q\sigma r\mu}(-)^{j-r+\frac{1}{2}-\mu} \quad (5.82)$$

There are $4n^2$ Lie-algebra generators of U_{2n} with $s=0,1$ and $k=0,1,\ldots,n-1$ for tensors $\mathbf{A}^{ks}$ with components $k \geq d \geq -k$ and $s \geq \alpha \geq -s$. Generators of subgroups are identified from the basic particle–hole symmetries. The spin-dependent generators are

$$\begin{aligned}
\text{Sp}_{2n}^P: &\quad A_{d\alpha}^{ks} - A_{-d-\alpha}^{ks} \\
R_{2n}^C: &\quad \mathbf{A}^{ks} \quad (k+s \text{ odd}, \ n \text{ even}) \quad\quad (5.83) \\
\text{Sp}_{2n}^C: &\quad \mathbf{A}^{ks} \quad (k+s \text{ odd}, \ n \text{ odd})
\end{aligned}$$

Generators of the spin-free Lie algebra U_n are $A_{d0}^{k0} \equiv A_d^k/\sqrt{2}$, with

$$A_d^k = \sum_{q,r} \langle jq, j-r|kd\rangle E_{qr}(-)^{j-r} \quad (5.84)$$

The number operator is $\hat{N} = \sqrt{2j+1}\,A_0^0$, and transformation properties of tensors with $k \neq 0$ are

$$\begin{aligned}
RA_d^k R^{-1} = (-)^k A_{-d}^k, &\quad TA_d^k T^{-1} = (-)^d A_d^k \\
CA_d^k C^{-1} = (-)^{k+1} A_d^k, &\quad PA_d^k P^{-1} = -A_{-d}^k
\end{aligned} \quad (5.85)$$

The Lie-algebra generators of the spin-free groups we have considered so far are recast in the tensor notation as

$$\begin{aligned}
R_n^P: &\quad A_d^k - A_{-d}^k \\[4pt]
\text{Sp}_n^C: &\quad \mathbf{A}^k \quad (k \text{ odd}, \ n \text{ even}) \\[4pt]
R_n^C: &\quad \mathbf{A}^k \quad (k \text{ odd}, \ n \text{ odd})
\end{aligned}$$

$$R_3^J: \qquad \mathbf{A}^1$$

$$R_2^Q: \qquad A_0^1$$

$$R_2^M: \qquad A_1^1 - A_{-1}^1 \tag{5.86}$$

Spin angular momentum itself is generated by $\mathbf{S} = \sqrt{j + \frac{1}{2}}\, \mathbf{A}^{01}$. The basic commutator for the spin-free tensors is

$$\left[A_\alpha^a, A_\beta^b \right] = \sum_{c,\gamma} C_{a\alpha,b\beta}^{c\gamma} A_\gamma^c$$

$$C_{a\alpha,b\beta}^{c\gamma} = (-)^{2j+1-\gamma} \begin{pmatrix} a & b & c \\ \alpha & \beta & -\gamma \end{pmatrix} \begin{Bmatrix} a & b & c \\ j & j & j \end{Bmatrix} \tag{5.87}$$

$$\times \left[(2a+1)(2b+1)(2c+1) \right]^{1/2} \left[1 - (-)^{a+b+c} \right]$$

An interesting result in Eq. (5.86) is that the molecular-orbital group R_2^M is a subgroup of the short-range pairing group R_n^P, while the atomic orbital group R_2^Q is not. Using G to denote possible intermediate groups in the subgroup chain $R_n^P \supset G \supset R_2^M$, the relationship between all the seniority groups and pseudorotation groups in the spin-free picture [cf. Eq. (5.46b)] is

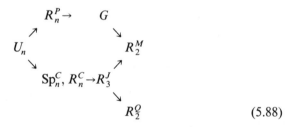

$$\tag{5.88}$$

An obvious possible candidate for the group G is another pseudorotation group we'll call R_3^L, though it has not been identified. Further investigation of this part of the short-range seniority subgroup chain could lead to additional symmetry properties of the Hubbard model. In the case of nonlinear polyenes, this might be related to accidental degeneracies found empirically in configuration interaction studies with the Hubbard model.[38, 142] Whether similar types of degeneracies are also part of the general Hubbard spectra of linear polyenes is not known.

The notation so far has been A_d^k for tensors in the atomic-orbital basis. The physical interpretation of the index d is an interaction between dth neighbors in a linear polyene. The Coulomb repulsion γ_{qr} between atoms q and r

has $d = q - r$, for example, and a Hückel nearest-neighbor resonance $\beta_{q\,q\pm1}$ has $d = \pm 1$. The related set of tensors for the pseudorotation molecular-orbital basis for R_2^M is denoted M_m^k. The general relationship between an arbitrary pseudorotation tensor $\mathcal{Q}_q^k$ in the atomic basis and $\mathfrak{M}_m^k$ in the molecular basis is

$$\mathfrak{M}_m^k = \sum_q d_{qm}^k \mathcal{Q}_q^k \tag{5.89}$$

where the Wigner d function is

$$d_{qm}^k = \langle km | \exp(-i\tfrac{1}{2}\pi j_y) | kq \rangle \tag{5.90}$$

Since the PPP model starts out in the atomic picture, Eq. (5.89) *defines* the pseudorotation molecular picture. In the special case when $\mathcal{Q}$ denotes an atomic orbital and $\mathfrak{M}$ denotes a molecular orbital, Eq. (5.89) represents the LCAO expansion with $k = j$. These pseudorotation orbitals are delocalized much like Hückel orbitals for low values of $n = 2j + 1$, and are the same as Hückel orbitals when $n \leqslant 3$. The two pictures are different in the long-chain limit, where Hückel orbitals for the outermost valence electrons remain delocalized, while the pseudorotation model freezes the outer electrons near the ends of the molecule. Another potentially useful way of looking at Eq. (5.89) is when a set of molecular orbitals are given to us (these could be *ab initio* SCF orbitals with either σ or π symmetry for bonds). In this case the inverse equation

$$\mathcal{Q}_q^k = \sum_m d_{qm}^k \mathfrak{M}_m^k \tag{5.91}$$

defines a set of localized pseudorotation atomic orbitals when $k = j$.

Quasispin is described in the molecular basis using creation and annihilation operators $X_{m\sigma}^\dagger$ and $X_{m\sigma}$. The notation is simplified if we now use the subscript m only for tensors in the molecular basis, and the subscript q only for the atomic basis. Generators in Eq. (5.51) for the short-range pairing symmetry R_3^κ have a molecular representation as

$$\kappa_+ = \sum_m X_{m\uparrow}^\dagger X_{-m\downarrow}^\dagger$$

$$\kappa_- = \sum_m X_{-m\downarrow} X_{m\uparrow} \tag{5.92}$$

$$\kappa_0 = \tfrac{1}{2}(\hat{N} - n)$$

The ground molecular-orbital configuration (highest J, M) has no pairing of this type, and hence has $\kappa = 0$. The highest-spin state $(S = \frac{1}{2}n)$ in a neutral polyene likewise has $\kappa = 0$. Comparisons like this with the pseudorotation phase convention for particle–hole exchange in linear polyenes[50] show that the parity of neutral states is

$$P = (-)^{\kappa} \qquad (5.93)$$

The quasispin groups R_3^{λ} and $R_5^{\omega t}$ for long-range correlation in the atomic basis commute with the pseudorotation group R_3^J. As a result of this their Lie-algebra generators in Eqs. (5.58) and (5.64) have the same identical form in the molecular basis after making the replacement $q \to m$. The ground molecular-orbital configuration has $\lambda = 0$ in neutral radicals and $\omega = t = 0$ in neutral molecules. The highest-spin state has $\lambda = 0$ in radicals, where the pairing is a singlet spin coupling of two electrons. In neutral molecules, on the other hand, the pair coupling has triplet spin, and hence the quasispin pairing has its highest value in the highest-molecular-spin state, $\omega = \frac{1}{2}n$ and $t = 0$. The pseudorotation phase convention for particle–hole exchange leads to the following results for neutral polyenes:

$$\text{Radicals}\left(R_n^C, \left[2^{\frac{1}{2}v - \beta}1^{2\beta}\right]\right): \qquad C = (-)^{\frac{1}{2}(v-1)}$$
$$\text{Molecules}\left(\text{Sp}_n^C, \langle 2^{\frac{1}{2}v - t}1^{2t}\rangle_{\pm}\right): \qquad C = \pm 1 \qquad (5.94)$$

The two types of parity are related by $P = RC$ in neutral polyenes, and in molecules this leads to the following result for singlet spin states:

$$P = (-)^{\frac{1}{2}n - t - M} \qquad (S = 0 \text{ only}) \qquad (5.95)$$

In addition to tensors for pseudorotations, there are also tensors for the quasispin groups. These generally provide information about the N dependence of matrix elements. The quasispin tensor structure of the PPP hamiltonian was worked out, but this was not particularly revealing about any cancellation of correlation contributions from one- and two-body parts of the energy.[50]

3. Resonance and Dipole Operators

Working in the pseudorotation molecular-orbital basis, the resonance tensor expansion for a system with alternancy and site-reflection symmetry is

$$\hat{\beta} = \sum_{k,m} \beta_m^k M_m^k \qquad (5.96)$$

with $1 \leqslant k < n$, and the coefficient $\beta_m^k = 0$ unless k is *odd* and m *even*. The usual assumption is nearest-neighbor resonance in the atomic basis, and evaluation of (5.96) for the Hückel model suggests that the leading-order term M_0^1 dominates for neutral linear polyenes. This is related to the fact that an expansion of Hückel orbital energies near the Fermi level gives equally spaced energies.[50] Components of the tensor $\mathbf{M}^1$ are proportional to a set of Lie-algebra generators $\mathbf{J}$ for the R_3^J angular momentum in the molecular basis. These are related to generators in the atomic basis by

$$J_x = -j_z, \qquad J_y = j_y, \qquad J_z = j_x \qquad (5.97)$$

The diagonal operators in the molecular basis are $\mathbf{J}^2 = J(J+1)$ and $J_z = M$, with $M = J, J-1, \ldots, -J$. The approximate symmetry of the Hückel resonance is a set of equally spaced levels described by

$$\hat{\beta} \simeq \bar{\beta} J_z \qquad (5.98)$$

where $\bar{\beta}$ is an effective pseudorotation resonance parameter. A similar analysis of the dipole operator for a system like [5], which has equal spacing $\bar{R}_x$ between atoms in the x direction, shows there is an exact pseudorotation symmetry

$$D_x = -\bar{R}_x J_x \qquad (5.99)$$

The overall picture is that $J_x = -Q$ in the atomic basis, and J_z describes nearest-neighbor transitions $\Delta Q = \pm 1$. In the molecular basis $J_z = M$, and J_x is the dipole interaction between neighbor levels $\Delta M = \pm 1$. We have $\Delta J = 0$ in both cases. The relationship is illustrated with the one-electron picture for four atoms (butadiene),

Atomic Sites (q) Molecular Levels (m)

The ground configuration and configurations in the first two excited resonance manifolds of the neutral system are

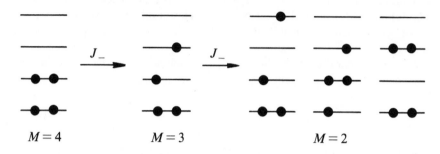

$$M = 4 \qquad\qquad M = 3 \qquad\qquad\qquad M = 2$$

where the operator $J_- = \frac{1}{2}(J_x - iJ_y)$ is the component of the dipole transition operator which lowers the value of M.

4. Correlation

The level $M = 2$ in butadiene is the simplest example of a degenerate collective excitation manifold in the pseudorotation picture, involving both one-
and two-electron jumps from the ground configuration. Most of the splitting
of the degeneracy in the full PPP spectrum shown in Fig. 19 originates in
the first-order contribution obtained by diagonalizing the correlation operator within the manifold. Higher-order contributions to the splitting involve
interactions $\Delta M \geqslant 2$ in both the resonance and correlation parts of the energy. Of particular experimental and theoretical interest[131] have been the first
two excited $^1A_g^-$ states, which have the same parity as the ground state. The
pseudorotation prediction of these is two terms $^1(2)_{00}$ and $^1(4)_{00}$ originating
in the zero-order manifold $M = 2$, with configuration mixings

$$^1(4)_{00} = \sqrt{\tfrac{3}{14}}\,(331-3) - \sqrt{\tfrac{3}{14}}\,(311-1) + \sqrt{\tfrac{8}{14}}\,(33-1-1)$$

$$^1(2)_{00} = -\sqrt{\tfrac{4}{14}}\,(331-3) + \sqrt{\tfrac{4}{14}}\,(311-1) + \sqrt{\tfrac{6}{14}}\,(33-1-1) \quad (5.100)$$

Here individual orbitals in each configuration are denoted with the value of
$2m$. The state $J = 4$ is the one obtained from the ground configuration $\psi_0 \equiv$
(3311) by the double excitation $J_-^2 \psi_0$. The first-order energy splitting predicted by the harmonic model has $^1(4)_{00} > {}^1(2)_{00}$, which agrees with the
ordering shown in Fig. 19.

Actual PPP wave functions computed with Ohno–Coulomb repulsion have
configuration mixings very similar to those in Eq. (5.100), indicating a very
high degree of pseudorotation symmetry for $\Delta M = 0$-type correlation. Wave
functions from the complete configuration interaction matrix suggest that $\mathbf{J}^2$

remains an approximate constant of the motion in low-lying levels, even when mixing ($\Delta M \neq 0$) between different pseudorotation manifolds is included. Because of the resonance energy spacing, the dominant interaction that mixes manifolds is $\Delta M = \pm 2$. This is the only off-diagonal correlation in the harmonic model, where the correlation is described by the operator J_x^2 in the molecular-orbital basis.

The reason for the approximate pseudorotation symmety of linear polyenes is certainly not obvious in an expansion of the correlation with spin-free tensors. First of all there is the fact that symmetry-breaking originates in both one- and two-body terms, and hence cancellation effects are contained in reduced matrix elements of tensors. Herrick and Liao proposed[123] that a way around this problem is an expansion in terms of spin-dependent one-body tensors, starting with the spin-dependent form of the correlation operator $\hat{\gamma}$ described here in Eq. (5.9). Since the total spin is conserved, the spin-dependent picture involves coupled tensors of the general form $(\mathbf{A}^{KS} \cdot \mathbf{A}^{K'S})_{00}^{k0}$, where $S = 0$ represents *singlet–singlet* spin coupling, and $S = 1$ represents *triplet–triplet* spin coupling of the one-body tensors. The pseudorotation ranks satisfy the usual triangle selection rule $|K - K'| \leq k \leq K + K'$ for coupled angular momentum, and $0 \leq K, K' \leq n - 1$.

The fact that the PPP correlation can be written in a spin-free picture means that the *singlet–singlet* and *triplet–triplet* tensors are not independent in the spin-free tensor expansion. The actual expansion is carried out in terms of a set of pseudorotation tensors $G(KK')_0^k$ defined as a linear combination

$$G(KK')_0^k = (-)^K (2K+1)^{1/2} \left[(\mathbf{A}^{K0} \cdot \mathbf{A}^{K'0})_{00}^{k0} - \sqrt{3} \, (\mathbf{A}^{K1} \cdot \mathbf{A}^{K'1})_{00}^{k0} \right]$$

$$(5.101)$$

This is essentially the mixing for the pairing state of coupled one-body singlet and triplet spin that is invariant under unitary transformations in four dimensions. To see this more clearly, the reader may wish to look back in Section IV.E and the pairing state for mixing of (s, s) and (p, p) atomic-orbital configurations in Eq. (4.44). In Eq. (5.101) the spin component $M_S = 0$ of $\mathbf{A}^{K0}$ is like the s orbital, and three components $M_S = 0, \pm 1$ of $\mathbf{A}^{K1}$ are like the atomic p orbital.

The general form of the tensor expansion of the correlation operator is

$$\hat{\gamma} = \sum_{k, K, K'} g_{KK'}^k G(KK')_0^k \qquad (5.102)$$

In systems of interest here, with alternancy and site-reflection symmetries, the only nonzero terms in the expansion are the ones with

$$
\begin{aligned}
&k = 0, 2, 4, \ldots, 2n - 2 \\
&|K - K'| \leqslant k \\
&K + K' \ even
\end{aligned}
$$
(5.103)

The general form of the expansion coefficient $g_{KK'}^k$ as a linear combination of Coulomb repulsion integrals γ_{qr} is contained in the original work of Herrick and Liao,[123] and involves nothing more than standard tensor-coupling algebra. This procedure could be carried out for any polyene with atoms labeled in a pseudorotation notation. Exact pseudorotation symmetry is associated with terms $k = 0$ in Eq. (5.102), which commute with generators of R_3^J, and the *singlet–singlet* coupling part of $G(11)_0^2$, which commutes with $\mathbf{J}^2$. Other tensors with $k > 0$ contain J-conserving parts ($\Delta J = 0$) and J-changing parts ($\Delta J \neq 0$), with the selection rule $|J - J'| \leqslant k$ derived from the Wigner–Eckart theorem for matrix elements of tensor operators. For a tensor $\mathcal{Q}_d^k$ in the atomic basis the general matrix element is

$$
\langle JQ|\mathcal{Q}_d^k|J'Q'\rangle = (-)^{J-Q}
\begin{pmatrix} J & k & J' \\ -Q & d & Q' \end{pmatrix}
\langle J\|\mathcal{Q}^k\|J'\rangle (2J+1)^{1/2}
$$
(5.104)

where $\langle J\|\mathcal{Q}^k\|J'\rangle$ is a reduced matrix element. Since the PPP correlation is diagonal in the atomic basis, all the tensors in Eq. (5.102) have $d = 0$, so that nonzero matrix elements of $\hat{\gamma}$ lie in blocks with $Q' = Q$. The approximate pseudorotation symmetry of linear polyenes found in the configuration-interaction studies suggests the contribution of $\Delta J \neq 0$ interactions for neutrals is relatively small in some cases, but this has not been unraveled completely yet.

Herrick and Liao focused on properties of the tensor expansion coefficients $g_{KK'}^k$ that could be related to the approximate symmetry as well. What they found was that all off-diagonal coefficients with $K' \neq K$ vanish identically for linear polyenes which have a sawtooth planar geometry with equal bond angles (θ), and equal bond lengths ($\bar{R}_{CC}$) between neighbor atoms, structure [6]

[6]

for an *even* number of atoms, and structure [7]

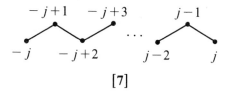

$$[7]$$

for an *odd* number of atoms. The bond angle in all-*trans* linear polyenes C_nH_{n+2} is $\theta \simeq 120°$. The actual bond lengths and angles in excited states are not known, but PPP potential-energy surfaces including some effects for the σ framework suggest[130] nearly equal bonds and angles that would be consistent with a zero-order sawtooth picture. The simplification of the pseudorotation tensor expansion is related to the corresponding simplification of the Coulomb repulsion, which is reduced to a description of repulsion between two electrons on d th-neighbor atoms. The third-neighbor interaction γ_{qq+3}, for example, is the same for each pair in the polyene, since we are assuming identical atomic orbitals at each atom. In formulas like the Ohno repulsion in Eq. (5.81) the repulsion depends on the distance between two atoms, R_{qr}. In the sawtooth geometry this is $R_{qr} = (X_{qr}^2 + Y_{qr}^2)^{1/2}$ with

$$X_{qr} = \bar{R}_{CC}\sin(\tfrac{1}{2}\theta)d \qquad (d = q - r)$$

$$Y_{qr} = \bar{R}_{CC}\cos(\tfrac{1}{2}\theta) \qquad (d\ odd)$$

$$= 0 \qquad (d\ even) \qquad\qquad (5.105)$$

The minimal set of repulsion parameters is then $\gamma_{qr} \equiv \gamma_d = \gamma_{-d}$ with $d = q - r$, and n independent values $\gamma_0, \gamma_1, \ldots, \gamma_{n-1}$.

The simplification of the correlation energy in the atomic basis is illustrated in Table IX, which gives the eigenvalue of $\hat{V}$ for each configuration of butadiene in which $Q \geqslant 0$. Configurations with $Q < 0$ are obtained from these by the site-reflection operation which takes each atomic site q into $-q$, and hence Q into $-Q$. The class of configurations with $Q = 0$ is symmetric under the spatial site-reflection operation, and includes the valence-bond configurations with each site singly occupied. An additional symmetry of pseudorotation states with $Q = 0$ and $\mathbf{J}^2$ diagonal is that the correlation energy remains blocked according to two classes with *even* or *odd* values of J. This symmetry is related to the fact that the 3-j symbol in Eq. (5.104) vanishes for $Q = 0$ unless $J + J' + k$ is an *even* integer. Since k is already *even*, interactions between pseudorotation states with $J + J'$ odd are zero. There is a

TABLE IX

Expansion of Eigenvalues of PPP Correlation Operator $\hat{V}^a$

Q	Atom (q)				$\hat{V}$
	$-\frac{3}{2}$	$-\frac{1}{2}$	$\frac{1}{2}$	$\frac{3}{2}$	
4	0	0	x^2	x^2	$2\gamma_0 + \gamma_1 - 2\gamma_2 - \gamma_3$
3	0	x	x	x^2	$\gamma_0 - \gamma_3$
2	0	x^2	0	x^2	$2\gamma_0 - 3\gamma_1 + 2\gamma_2 - \gamma_3$
	0	x	x^2	x $\left.\rule{0pt}{16pt}\right\}$	$\gamma_0 - \gamma_2$
	x	0	x	x^2	
1	0	x^2	x	x $\left.\rule{0pt}{24pt}\right\}$	
	x	0	x^2	x	$\gamma_0 - \gamma_1$
	x	x	0	x^2	
0	x^2	0	0	x^2 $\left.\rule{0pt}{16pt}\right\}$	$2\gamma_0 - \gamma_1 - 4\gamma_2 + 2\gamma_3$
	0	x^2	x^2	0	
	x	x	x	x	0

^{a}In terms of dth neighbor repulsion in atomic basis configurations of electrons (x) in structure [2], which is the approximate symmetry of *trans*-butadiene, C_4H_6.

similar selection rule (ΔJ *even*) for matrix elements in the pseudorotation molecular manifold with $M = 0$. The formula for the correlation energy when $\gamma_q = -d^2$ is $V = Q^2$, which is the result for the harmonic model (neutrals only). The Hubbard model has all $\gamma_d = 0$ for $d > 0$.

The fact that coefficients $g_{KK'}^k$ with $K' \neq K$ vanish is important because it cuts out a lot of symmetry-breaking tensors with $k > 0$ from Eq. (5.102). Furthermore, the remaining coefficients g_{KK}^k are independent of the number of atoms in the polyene, which suggests the approach may be useful for interpreting the n dependence of excitation spectra.

What is left in the expansion is written here as $\hat{\gamma} = \hat{\gamma}_0 + \hat{\gamma}_1$. The first term is the totally symmetric ($k = 0$) part,

$$\hat{\gamma}_0 = -\frac{1}{2} \sum_{K=0}^{n-1} g_K G(KK)_0^0$$

$$g_K = (2K+1)^{-1} \sum_{d=-K}^{K} \gamma_d$$

$$(5.106)$$

The term $\hat{\gamma}_1$ includes only tensors with $k > 0$. In the case of Ohno repulsion it is found that tensors $G(KK)_0^k$ with the same rank k contribute to $\hat{\gamma}_1$ *with nearly the same weight for each value of K*. The same approximate symmetry is also found with an exponential repulsion when the effective range is adjusted to values in the region of chemical interest. This is a relatively subtle effect whose precise link to approximate pseudorotation symmetry for neutral linear polyenes is not yet known. The implication is that[123]

$$\hat{\gamma}_1 \simeq \sum_{i=1}^{n-1} c_{2i} \Gamma_{2i} \qquad (5.107)$$

where c_{2i} represents an average tensor coefficient and the operator Γ_{2i} is defined as

$$\Gamma_{2i} = \sum_{K=i}^{n-1} G(KK)_0^{2i} \qquad (5.108)$$

This is a generalized tensor version of the quadratic invariant Γ_0 of the group $U(2n)$, defined in Eq. (5.48).

5. Tensor Model for Spectra

$\mathcal{C}_0^k$ and $\mathfrak{M}_m^k$ will now denote tensors for the PPP correlation operator $\hat{V}$ in the atomic and molecular bases. The specific form of the internal coupling for each tensor is not important here. The general form of the correlation expansion is

$$\hat{V} = \sum_k \mathcal{C}_0^k = \sum_{k,m} d_{0m}^k \mathfrak{M}_m^k \qquad (5.109)$$

with *even* values of k and m such that $0 \leqslant k \leqslant 2n-2$ and $0 < |m| < k$. The fact that only *even* values of m appear is related to the assumed site-reflection symmetry, which is $R = (-)^{Nj-M}$ in the molecular basis [each orbital has $R = (-)^{j-m}$]. Provided that we adhere to standard angular-momentum coupling rules, reduced matrix elements are the same in either basis,

$$\langle J \| \mathcal{C}^k \| J' \rangle = \langle J \| \mathfrak{M}^k \| J' \rangle \qquad (5.110)$$

and may be evaluated in whichever basis is more convenient. The point is that the general form of correlation matrix elements in the molecular pseudorotation basis is obtained from the Wigner–Eckart theorem as

$$\langle JM|\hat{V}|J'M'\rangle = \sum_{k,m} (-)^{J-M} (2J+1)^{1/2} \begin{pmatrix} J & k & J' \\ -M & m & M' \end{pmatrix}$$
$$\times \langle J \| \mathfrak{M}^k \| J' \rangle d_{0m}^k \qquad (5.111)$$

There is a similar result for the resonance energy expansion in Eq. (5.96) with *odd* values of k. Semiempirical values of reduced matrix elements could be assumed in systems where there is a high degree of pseudorotation symmetry, and these control the splitting of each degenerate manifold ($m = 0$). Coupling between different manifolds ($m \neq 0$) is controlled entirely by the 3-j symbol and the coefficient d_{0m}^k in Eq. (5.111).

The preceding result is interesting because it suggests model approximations of the PPP spectra if we assume each level has exact pseudorotation symmetry, with reduced matrix elements equal to zero unless $J' = J$. There would be similar selection rules for other quantum numbers we have left out of Eq. (5.111), like quasispin seniority and the exact spin symmetry. It is reasonable to expect that the dominant configuration mixing effects originate in coupling $\Delta M = \pm 2$, and what this leads to is a tridiagonal energy matrix with selection rules $\Delta J = 0$, $\Delta M = 0, \pm 2$. The harmonic model has this exact symmetry for matrix elements of correlation. The important result for the more general model is that spectra of tridiagonal matrices are easy to compute, and could even lead to analytical expressions for energies in the long-chain (large n) limit with suitable parametrization of the reduced matrix elements.

6. Coupling of Two Electrons

The degree of R_3^J pseudorotation symmetry for correlation weakens in higher ionic states, as suggested by the harmonic model. In the limit of just two electrons there is another type of pseudorotation coupling which describes an exact symmetry in the case of atoms with the sawtooth geometry.[50] Diagonalization of just the two-electron repulsion is probably more realistic in the case of doubly excited states in which two electrons occupy higher σ or π shells outside the valence shell. In this system it could be reasonable to assume that the motion of the valence electrons is much faster, so that the outer two electrons see an effective core charge that is uniform along the molecular backbone. If $\mathbf{J}_1$ and $\mathbf{J}_2$ denote the pseudorotation angular momenta for electrons 1 and 2, the operator $\mathbf{K}^2$ with

$$K_x = J_{1x} - J_{2x}, \qquad K_y = J_{1y} - J_{2y}, \qquad K_z = J_{1z} + J_{2z} \qquad (5.112)$$

is an exact constant of the motion for the electron repulsion in the PPP model. In the molecular basis $K_z = M$, and the energy is blocked according to eigenvalues $\mathbf{K}^2 = K(K+1)$.

States for this type of angular-momentum coupling were described in Eq. (2.8). They were also found to describe correlated wave functions for doubly excited states in the hypothetical two-dimensional helium atom in Eq. (4.60). In the three-dimensional helium atom the coupling of SO(4) Lie-algebra generators was $\mathbf{B}^2 = (\mathbf{b}_1 - \mathbf{b}_2)^2$. The point evidently is that correlation of just two electrons in excited states of both atoms and molecules involves unusual types of angular-momentum coupling, which are similar in both systems.

7. *Extension to Two Dimensions*

As another concrete example of how different subgroups of the unitary group can be useful for different arrangements of atoms, we consider briefly systems like [8]:

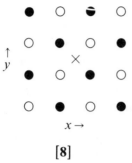

$$[8]$$

The filled and unfilled circles represent alternate atoms in the discrete, two-dimensional lattice; and $\times$ denotes an axis perpendicular to the plane, through the center of the structure. The obvious result is that now there are two pseudorotation groups, R_3^x and R_3^y, when the atomic sites are labeled with two indices as (q_x, q_y). The generalization is a finite lattice with $n_x \times n_y$ atoms in a rectangular $(n_x \neq n_y)$ or square $(n_x = n_y)$ array.

The interesting point of the model group-theoretically is that electron correlation in excited states could be described with representations of the group $R_4 \supset R_3 \supset R_2$, where the four-dimensional rotation group is defined as the product $R_4 = R_3^x \times R_3^y$. Properties of the group R_4, or SO(4), were described in Section II.B, and if [8] is rotated by 45° about the axis it looks like one of the Stark energy-level patterns of hydrogen illustrated in Fig. 2. The correlation described by the spherical representation would be a discrete lattice version of collective degrees of freedom for coupled two-dimensional oscillators. This would involve a quantization of states for clockwise and counterclockwise "rotations" around the center, and a radial breathing mode for "vibrations." It is possible that these types of correlation could be excited in the interaction of photons, electrons, atoms, or other molecules with a surface, when regions like [8] are treated as local subunits.

VI. CONCLUSION

The topics covered in this article have clarified further the relationship between Lie groups and collective degrees of freedom for electron correlation, but a general understanding of this problem is far from complete. The difficult task so far has been unraveling the spectra of these problems in new ways so that it is clear what approximate symmetries exist. Enough possible extensions of these ideas have been presented so that readers previously unfamiliar with this work can see where it is going. Another possibility for the electron–dipole coupling model is the application of noncompact groups like $SO(2, 1)$ or $SU(1, 1)$ for configuration mixing between the different thresholds. In the case of electron–molecule scattering these groups have Lie-algebra generators that could change the value of J for the molecule, and group rotations would lead to collective excitations of J that might block-diagonalize the coupled-channel equations even further. Similar types of collective excitation states could have useful applications in the problem of n and l mixing of Rydberg states in strong electric or magnetic fields. The idea would be to identify alternative bases or suitable group parameters for infinite-dimensional rotations that could approximately block-diagonalize the energy.

In two-electron atoms the important problem for the future is the identification of suitable collective coordinates that could explain the striking rotation–vibration structure that is found semiempirically with $SO(4)$ supermultiplets. We have already indicated one possible set of coordinates that incorporates both radial and angular degrees of freedom together, and these could be useful for resolving the behavior of correlated electrons near threshold for two-electron escape. In the group-theoretical picture the threshold behavior could be linked to representations of the noncompact group $SO(3, 1) \times SO(3, 1)$, whose Lie-algebra generators effect dipole transitions for all possible values of the one-electron angular momentum $l = 0, 1, \ldots$ via the analytic continuation $n \to in$ in matrix elements of the corresponding compact group $SO(4)$.

From a chemical point of view the ideas for approximate symmetry in molecules are most interesting, and certainly encompass the broadest range of topics. The real question is whether anything useful is gained by reformulating our understanding of molecular structure in terms of appropriate Lie group subgroup chains. An obvious place to start applying what we have already learned from polyenes is to correlation that could be important in triatomic molecules, for example ozone. Quasispin groups that change the number of electrons could shed new light on the Walsh rules[143] for the dependence of shapes of simple molecules on the number of electrons. In linear polyenes themselves an important question is the degree of coupling of electrons to vibrational modes of the atoms, for observing two-photon excitations.[131] One possibility is that propensity rules for matrix elements connecting the two degrees of freedom could be derived from the Lie-group structure for the picture of both problems as harmonic oscillators. Along a similar line, Kellman[144] has investigated recently an approximate symmetry for two coupled oscillators, and found spectra similar to the collective excitations we described for correlation in linear polyene with approximate pseudorotation symmetry.

While it is not reasonable to expect approximate symmetries with Lie groups in every atom or molecule, the possibility of other examples cannot be ignored. If the results described in the present article are typical, the greatest likelihood of approximate symmetry would occur in the class of systems in which two otherwise independent particles couple. The idea of approximate symmetry is that coupled representations of groups could mimic the correlation of the particles when Lie-algebra invariants are approximate constants of the motion. Of chemical interest are intermolecular forces ranging from coupling of just two atoms in a diatomic molecule, up to the pairing symmetry of two complementary polynucleotide strands in DNA.

Acknowledgments

Research for this article was supported in part by NSF grant CHE-79-09500, and by a Teacher–Scholar grant from the Camille and Henry Dreyfus Foundation.

References

1. V. Fock, Z. Phys. **98**, 145 (1935).
2. V. Bargmann, Z. Phys. **99**, 576 (1936).
3. E. Wigner, Phys. Rev. **51**, 106 (1937).
4. G. Racah, Phys. Rev. **76**, 1352 (1949).
5. B. H. Flowers, Proc. R. Soc. London A **212**, 248 (1952).
6. M. Hamermesh, Group Theory and Its Applications to Physical Problems, Addison-Wesley, Reading, Mass., 1962.
7. B. R. Judd, "Group theory in atomic spectroscopy," in E. M. Loebl, Ed., Group Theory and Its Applications, Academic, New York, 1968, p. 183.
8. B. R. Judd, Adv. Atom. Mol. Phys. **7**, 251 (1971).
9. B. G. Wybourne, Classical Groups for Physicists, Wiley-Interscience, New York, 1974.
10. B. G. Wybourne, Symmetry Principles and Atomic Spectroscopy (including an appendix of tables by P. H. Butler), Wiley-Interscience, New York, 1970.
11. K. T. Hecht, Annu. Rev. Nucl. Sci. **23**, 123 (1973).
12. R. Gilmore, Lie Groups, Lie Algebras, and Some of Their Applications, Wiley-Interscience, New York, 1974.
13. B. H. Flowers and S. Szpikowski, Proc. Phys. Soc. (Lond.) **84**, 673 (1964).
14. R. D. Lawson and M. H. MacFarlane, Nucl. Phys. **66**, 80 (1965).
15. K. Helmers, Nucl. Phys. **23**, 594 (1961).
16. K. T. Hecht, Nucl. Phys. **A102**, 11 (1967).
17. R. P. Hemenger and K. T. Hecht, Nucl. Phys. **A145**, 468 (1970).
18. M. Moshinsky, Group Theory and the Many-Body Problem, Gordon and Breach, New York, 1968.
19. W. G. Harter, Phys. Rev. A **8**, 2819 (1973).

20. J. Paldus, *J. Chem. Phys.* **61**, 5321 (1974).
21. F. A. Matsen, *Int. J. Quantum Chem.* **8S**, 379 (1974).
22. W. G. Harter and C. W. Patterson, *Phys. Rev. A* **13**, 1067 (1976).
23. J. Paldus, *Phys. Rev. A* **14**, 1620 (1976).
24. W. G. Harter and C. W. Patterson, *A Unitary Calculus for Electronic Orbitals*, Springer-Verlag, Berlin, 1976.
25. F. A. Matsen, *Int. J. Quantum Chem.* **10**, 525 (1976).
26. J. Paldus, "Many-electron correlation problem–group theoretical approach," in H. Eyring and D. J. Henderson, Eds., *Theoretical Chemistry: Advances and Perspectives*, Vol. 2, Academic, New York, 1976, p. 131.
27. C. W. Patterson and W. G. Harter, *J. Math. Phys.* **17**, 1125 (1976).
28. C. W. Patterson and W. G. Harter, *J. Math. Phys.* **17**, 1137 (1976).
29. I. Shavitt, *Int. J. Quantum Chem.* **11S**, 131 (1977).
30. C. W. Patterson and W. G. Harter, *Phys. Rev. A* **15**, 2372 (1977).
31. W. G. Harter and C. W. Patterson, *Int. J. Quantum Chem.* **11S**, 445 (1977).
32. J. Paldus, *Electrons in Finite and Infinite Structures*, Plenum, New York, 1977.
33. I. Shavitt, *Int. J. Quantum Chem.* **12S**, 5 (1978).
34. D. Braunschweig and K. T. Hecht, *J. Math. Phys.* **19**, 720 (1978).
35. F. A. Matsen, *Adv. Quantum Chem.* **11**, 223 (1978).
36. J. Paldus and M. J. Boyle, *Phys. Rev. A* **22**, 2299 (1980).
37. M. J. Boyle and J. Paldus, *Phys. Rev. A* **22**, 2316 (1980).
38. F. A. Matsen, "Lie groups, quantum mechanics, many-body theory and organic chemistry," in B. Gruber and R. S. Millman, Eds., *Symmetries in Science*, Plenum, New York, 1980.
39. J. Hinze, Ed., *The Permutation Group in Physics and Chemistry*, Springer-Verlag, Berlin, 1979.
40. J. P. Elliott, *Proc. R. Soc. London A* **245**, 128 (1958).
41. J. P. Elliott, *Proc. R. Soc. London A* **245**, 562 (1958).
42. A. Bohr and B. R. Mottelson, *Nuclear Structure*, Vol. 2, Benjamin, Reading, Mass., 1975.
43. A. Arima and F. Iachello, *Ann. Phys. (N.Y.)* **111**, 201 (1978).
44. A. Arima and F. Iachello, *Phys. Rev. Lett.* **40**, 385 (1978).
45. M. Gell-Mann and Y. Ne'eman, *The Eightfold Way*, Benjamin, New York, 1964.
46. D. R. Herrick and M. E. Kellman, *Phys. Rev. A* **21**, 418 (1980).
47. D. R. Herrick, M. E. Kellman, and R. D. Poliak, *Phys. Rev. A* **22**, 1517 (1980).
48. M. E. Kellman and D. R. Herrick, *J. Phys. B: Atom. Mol. Phys.* **11**, L755 (1978).
49. M. E. Kellman and D. R. Herrick, *Phys. Rev. A* **22**, 1536 (1980).
50. D. R. Herrick, *J. Chem. Phys.* **74**, 1239 (1981).
51. D. M. Brink and G. R. Satchler, *Angular Momentum*, Clarendon, Oxford, 1971.
52. E. U. Condon and G. H. Shortley, *Theory of Atomic Spectra*, London, Cambridge University Press, 1935.
53. L. C. Biedenharn, *J. Math. Phys.* **2**, 433 (1961).
54. M. Bander and C. Itzykson, *Rev. Mod. Phys.* **38**, 330 (1966).
55. M. J. Englefield, *Group Theory and the Coulomb Problem*, Wiley-Interscience, New York, 1972.

56. W. Pauli, *Z. Phys.* **36**, 336 (1926); English transl. in B. L. van der Waerden, Ed., *Sources of Quantum Mechanics*, North-Holland, Amsterdam, 1967.

57. L. D. Landau and E. M. Lifshitz, *Mechanics*, Addison-Wesley, Reading, Mass., 1960, p. 35.

58. M. Abromowitz and I. A. Stegun, *Handbook of Mathematical Functions*, Dover, New York, 1968.

59. D. Park, *Z. Phys.* **159**, 155 (1960).

60. H. A. Bethe and E. E. Salpeter, *Quantum Mechanics of One- and Two-Electron Atoms*, Springer-Verlag, Berlin, 1957, pp. 27, 238.

61. O. Sinanoğlu and D. R. Herrick, *J. Chem. Phys.* **62**, 886 (1975).

62. D. R. Herrick, *Phys. Rev. A* **12**, 413 (1975).

63. D. R. Herrick, *Phys. Rev. A* **12**, 1949 (1975).

64. J. R. Hiskes, C. B. Tarter, and D. A. Moody, *Phys. Rev.* **133**, A424 (1964).

65. J. R. Hiskes and C. B. Tarter, Lawrence Radiation Laboratory (Livermore) Report No. UCRL-7088, Rev. I (1964).

66. B. R. Judd, *Second Quantization and Atomic Spectroscopy*, Johns Hopkins, Baltimore, 1967.

67. N. F. Lane, *Rev. Mod. Phys.* **52**, 29 (1980).

68. R. B. Bernstein, Ed., *Atom–Molecule Collision Theory*, Plenum, New York, 1979.

69. P. C. Engelking, *Phys. Rev. A* **26**, 740 (1982).

70. D. R. Herrick and P. C. Engelking (unpublished).

71. M. J. Seaton, *Proc. Phys. Soc. London* **77**, 174 (1961).

72. M. Gailitis and R. Damburg, *Proc. Phys. Soc. London* **82**, 192 (1963).

73. P. G. Burke, *Adv. Phys.* **14**, 521 (1965).

74. J. Macek and P. G. Burke, *Proc. Phys. Soc. London* **92**, 351 (1967).

75. H. A. Hyman, V. L. Jacobs, and P. G. Burke, *J. Phys. B: Atom. Mol. Phys.* **5**, 2282 (1972).

76. D. R. Herrick, *Phys. Rev. A* **17**, 1 (1978).

77. D. R. Herrick and R. D. Poliak, *J. Phys. B: Atom. Mol. Phys.* **13**, 4533 (1980).

78. T. F. O'Malley, *Phys. Rev.* **137**, A1668 (1965).

79. S. I. Nikitin and V. N. Ostrovsky, *J. Phys. B: Atom. Mol. Phys.* **11**, 1681 (1978).

80. C. H. Greene, *Phys. Rev. Lett.* **44**, 869 (1980).

81. L. D. Landau and E. M. Lifshitz, *Quantum Mechanics*, Addison-Wesley, Reading, Mass., 1965, pp. 113, 123.

82. M. J. Seaton, *Proc. Phys. Soc. London* **79**, 1105 (1962).

83. R. M. Pengelly and M. J. Seaton, *Mon. Not. R. Astron. Soc.* **127**, 165 (1964).

84. U. Fano, "Doubly excited states of atoms," in V. W. Hughes, B. Bederson, V. W. Cohen, and F. M. J. Pichanick, Eds., *Atomic Physics*, Vol. 1, Plenum, New York, 1969, p. 209.

85. U. Fano, *Phys. Today* **29**:**9**, 32 (1976).

86. R. P. Madden and K. Codling, *Astrophys. J.* **141**, 364 (1965).

87. J. W. Cooper, U. Fano, and F. Prats, *Phys. Rev. Lett.* **10**, 518 (1963).

88. J. S. Alper and O. Sinanoğlu, *Phys. Rev.* **177**, 77 (1969).

89. C. E. Wulfman, *Phys. Lett. A* **26**, 397 (1968).

90. P. H. Butler and B. G. Wybourne, *J. Math. Phys.* **11**, 2519 (1970).

91. A. R. P. Rau, *Phys. Rev. A* **2**, 1600 (1970).

92. J. S. Alper, *Phys. Rev. A* **2**, 1603 (1970).
93. B. G. Wybourne, "Compact groups in atomic physics," in E. U. Condon and O. Sinanoğlu, Eds., *New Directions in Atomic Physics*, Vol. 1, Yale, New Haven, 1972, p. 63.
94. E. Chacon, M. Moshinsky, O. Novaro, and C. Wulfman, *Phys. Rev. A*. **3**, 166 (1971).
95. C. Wulfman, *Chem. Phys. Lett.* **23**, 370 (1973).
96. D. R. Herrick, *Phys. Rev. A* **22**, 1346 (1980).
97. D. R. Herrick, *J. Chem. Phys.* **67**, 5406 (1977).
98. D. R. Herrick and O. Sinanoğlu, *Phys. Rev. A* **11**, 97 (1975).
99. O. Sinanoğlu and D. R. Herrick, *Chem. Phys. Lett.* **31**, 373 (1975).
100. D. R. Herrick, *J. Math. Phys.* **16**, 1047 (1975).
101. D. R. Herrick and M. E. Kellman, *Phys. Rev. A* **18**, 1770 (1978).
102. S. I. Nikitin and V. N. Ostrovsky, *J. Phys. B: Atom. Mol. Phys.* **9**, 3141 (1976).
103. Y. K. Ho, *J. Phys. B: Atom. Mol. Phys.* **12**, 387 (1979).
104. Y. K. Ho., *Phys. Rev. A* **23**, 2137 (1981).
105. A. Temkin and J. F. Walker, *Phys. Rev.* **140**, A1520 (1965).
106. M. J. Seaton, *Mon. Not. R. Astron. Soc.* **118**, 504 (1958).
107. O. Bely, *Proc. Phys. Soc.* **88**, 833 (1966).
108. W. C. Martin, *J. Phys. Chem. Ref. Data* **2**, 257 (1973).
109. S. Bashkin and J. O. Stoner, Jr., *Atomic Energy Levels & Grotrian Diagrams*, Vol. 1, North-Holland, Amsterdam, 1975.
110. G. J. Schulz, *Rev. Mod. Phys.* **45**, 378 (1973).
111. F. H. Read, "Correlation effects in electron-atom scattering," in D. Kleppner and F. M. Pipkin, Eds., *Atomic Physics*, Vol. 7, Plenum, New York, 1981, p. 429.
112. O. Sinanoğlu, "Atomic structure, transition probabilities, and theory of electron correlation in ground and excited states," in V. W. Hughes, B. Bederson, V. W. Cohen, and F. M. J. Pichanick, Eds., *Atomic Physics*, vol. 1, Plenum, New York, 1969, p. 131.
113. P. Rehmus, M. E. Kellman, and R. S. Berry, *Chem. Phys.* **31**, 239 (1978).
114. P. Rehmus and R. S. Berry, *Chem. Phys.* **38**, 257 (1979).
115. H.-J. Yuh, G. Ezra, P. Rehmus, and R. S. Berry, *Phys. Rev. Lett.* **47**, 497 (1981).
116. G. Herzberg, *Molecular Spectra and Molecular Structure*, Vol. I, Van Nostrand, New York, 1950, p. 172.
117. L. Pauling, *J. Am. Chem. Soc.* **53**, 1367 (1931).
118. G. Ezra and R. S. Berry *Phys. Rev. A* **25**, 1513 (1982).
119. D. R. Herrick and F. H. Stillinger, *Phys. Rev. A* **11**, 42 (1975).
120. D. R Herrick, *J. Math. Phys.* **16**, 281 (1975).
121. L. D. Mlodinow and N. Papanicolaou, *Ann. Phys.* **131**, 1 (1981).
122. E. Witten, *Phys. Today* **33**:7, 38 (1980).
123. D. R. Herrick and C.-L. Liao, *J. Chem. Phys.* **75**, 4485 (1981).
124. R. G. Parr, *Quantum Theory of Molecular Structure*, Benjamin, New York, 1963.
125. K. Schulten, I. Ohmine, and M. Karplus, *J. Chem. Phys.* **64**, 4422 (1976).
126. I. Ohmine, M. Karplus, and K. Schulten, *J. Chem. Phys.* **68**, 2298 (1978).
127. P. Tavan and K. Schulten, *J. Chem. Phys.* **70**, 5407 (1979).
128. K. Balasubramanian and D. R. Yarkony, *Chem. Phys. Lett.* **70**, 374 (1980).

129. K. Schulten, U. Dinur, and B. Honig, *J. Chem. Phys.* **73**, 3927 (1980).

130. A. C. Lasaga, R. J. Aerni, and M. Karplus, *J. Chem. Phys.* **73**, 5230 (1980).

131. B. S. Hudson, B. E. Kohler, and K. Schulten, "Linear polyene electronic structure and potential surfaces," in E. C. Lim, Ed., *Excited States*, Vol. 5, Academic, New York, 1982, p. 1.

132. B. G. Wybourne, *Int. J. Quantum Chem.* **7**, 1117 (1973).

133. A. M. Bincer, *J. Math. Phys.* **21**, 671 (1980).

134. C. A. Coulson, B. O'Leary, and R. B. Mallion, *Hückel Theory for Organic Chemists*, Academic, New York, 1978.

135. A. D. McLachlan, *Mol. Phys.* **2**, 271 (1959).

136. F. A. Matsen, T. L. Welsher, and B. Yurke, *Int. J. Quantum Chem.* **12**, 985 (1977).

137. F. A. Matsen and T. L. Welsher, *Int. J. Quantum Chem.* **12**, 1001 (1977).

138. R. Pariser, *J. Chem. Phys.* **24**, 250 (1956).

139. K. Yates, *Hückel Molecular Orbital Theory*, Academic, New York, 1978, p. 78.

140. D. E. Littlewood, *The Theory of Group Characters*, Clarendon, Oxford, 1950.

141. J. Hubbard, *Proc. Roy. Soc. (London) A* **276**, 238 (1963).

142. O. J. Heilmann and E. H. Lieb, *Trans. N. Y. Acad. Sci.* **33**, 116 (1971).

143. A. D. Walsh, *J. Chem. Soc.* (1953), 2260.

144. M. E. Kellman, *J. Chem. Phys.* **76**, 4528 (1982).

VIBRONIC ENERGY LEVELS OF A LINEAR TRIATOMIC MOLECULE IN A DEGENERATE ELECTRONIC STATE: A UNIFIED TREATMENT OF THE RENNER–TELLER EFFECT

JOHN M. BROWN and FLEMMING JØRGENSEN[§]

Department of Chemistry
The University of Southampton
Southampton, SO9 5NH, England

CONTENTS

[§]Visitor from the Chemical Laboratory V, The H. C. Ørsted Institute, Universitetsparken 5, Copenhagen Ø, DK-2100, Denmark.

I. INTRODUCTION

When all three nuclei in a linear triatomic molecule are strictly on the axis, the component L_z of the electronic angular momentum around this axis must take one of the values $0, \pm 1, \pm 2, \ldots$, and the electronic states corresponding to the values $0, 1, 2, 3, \ldots$ of $\Lambda \equiv |L_z|$ are called $\Sigma, \Pi, \Delta, \Phi, \ldots$ states. When $\Lambda > 0$, the electronic state is doubly degenerate, although this is strictly true only in the linear configuration; as the molecule bends, the cylindrical symmetry is broken and the degeneracy is lifted. Hence, for $\Lambda > 0$, the movement of the nuclei is governed by two Born–Oppenheimer potentials which touch each other at the linear configuration. We call these potentials W_+ and W_-, where the index denotes the parity of the defining electronic wave function under a reflection in the molecular plane (a symmetry not broken by the bending). In terms of the normal coordinates Q_1 and Q_3 for the stretches and ρ for the bending, all defined relative to the average potential $(W_+ + W_-)/2$, the expansions of W_+ and W_- start as

$$W_\pm = \text{constant} + \tfrac{1}{2}\lambda_1 Q_1^2 + \tfrac{1}{2}\lambda_3 Q_3^2 + \tfrac{1}{2}\lambda_\pm \rho^2 + \cdots \qquad (1.1)$$

and the energies are, to a good first approximation, given by

$$\mathscr{E} = \omega_1\left(v_1 + \tfrac{1}{2}\right) + \omega_3\left(v_3 + \tfrac{1}{2}\right) + E + E_{\text{rot}} \qquad (1.2)$$

where v_k is the number of quanta in the kth stretching vibration with frequency ω_k, E_{rot} is the energy for end-over-end rotation, and the energy E can be determined within a model in which $Q_1 \equiv Q_3 = 0$ and the molecular axis $\mathbf{e}_z$ is fixed in space. From the latter assumption, the eigenvalue K of the angular momentum of electrons plus nuclei around the axis is a good quantum number. When the molecule is allowed to rotate, we get the familiar pattern of rotational levels

$$E_{\text{rot}} = B_e\{J(J+1) - K^2\}, \qquad J = K, K+1, \ldots \qquad (1.3)$$

where B_e is the rotational constant for the equilibrium configuration. The energy E was first discussed as early as 1934 by Renner,[1] a student of Teller's. He introduced a method in which a small "gap" between the potentials W_+ and W_- was considered as a perturbation and used second-order perturbation theory to calculate E_1 and E_2 in an expansion of the form

$$E = \omega(v+1) + \varepsilon E_1(v, K) + \varepsilon^2 E_2(v, K) + \cdots, \qquad \varepsilon = \frac{\lambda_+ - \lambda_-}{\lambda_+ + \lambda_-} \qquad (1.4)$$

where v is the bending vibration quantum number ($\equiv v_2$) and ω is the bending frequency corresponding to the force constant $\lambda = (\lambda_+ + \lambda_-)/2$. We note that ε, now called the Renner–Teller parameter, satisfies $|\varepsilon| < 1$ and $\lambda_\pm = (1 \pm \varepsilon)\lambda$. It can be shown [see Eq. (3.3) below] that $\varepsilon \equiv 0$ for all $\Lambda > 1$; Renner only considered the case $\Lambda = 1$. By use of numerical methods, Renner also estimated several energies $E(\varepsilon, v, K)$ for intermediate and large values of ε. Today E can be calculated to any desired accuracy for all relevant values of ε, but the formula (1.2) remains a first approximation because it ignores anharmonicities in the potential functions and the effects of nonrigidity on E_{rot}. The necessary corrections have been known for molecules in nondegenerate electronic states for almost 40 years[2] but a corresponding theory has not yet been given for degenerate states. The possible extension of Renner's work did not arouse any interest at the time because of the lack of experimental data. For many years afterwards, even with the advent of experimental observations, his treatment was considered to be too complicated to be extended with reasonable effort. At the present time however a large collection of high-quality data is available and points out an urgent need for a more sophisticated theory.

Let us briefly sketch the development of the theory of the Renner–Teller effect. The effect of electronic angular momentum about the axis was first detected experimentally, about 25 years after Renner's paper, by Dressler and Ramsay[3] in the electronic spectrum of NH_2, which is a molecule with a deep off-axis minimum in the lower potential (see Fig. 1b later). Pople and Longuet-Higgins gave the first theoretical discussion of this case.[4,5] Although the potential functions involved were of a shape not foreseen by Renner, Pople and Longuet-Higgins could have used his approach directly (see Appendix D below). However, they chose instead to introduce a new form of the effective hamiltonian which they formulated from qualitative symmetry considerations of a simplified one-electron model. Carrington, Fabris, Howard, and Lucas[6,7] found the same form by applying perturbation theory to a multipole expansion of the electrostatic interaction between electrons and nuclei. Pople and Longuet-Higgins's discussion of NH_2 was restricted to those bending vibrational states which were, in a first approximation, only sensitive to the upper potential. Very accurate data exist today for molecules in a wide range of vibrational states, and it has therefore also become necessary to devise a model that can cope with large bending vibrational amplitudes. Dixon, Duxbury, and coworkers[8,9] were the first to introduce these effects, but a complete treatment of the problem has only recently been published in a series of papers by Jungen, Merer, and Hallen.[10-14] The term "Renner–Teller effect" is now used in a broad sense to describe all the various effects of nonzero electronic angular momentum about the axis. For further references, the reader should consult the two recent reviews.[9,10]

In addition to these papers on the basic theory of the Renner–Teller effect, there have been several others that deal with more specific aspects. Pople[15] extended his and Longuet-Higgins's model to cover molecules in multiplet spin states by including the effect of spin–orbit coupling (most molecules studied so far have been in doublet states). Chang and Chiu[16] have included the magnetic interaction between vibrational, electronic, and spin angular momenta, an effect which is expected to be very small. Brown[17] has found an addition to Renner's original coupling term (see also Aarts[18]), and Gauyacq and Jungen[19] have given analytical expressions for the bending vibrational energy levels when Brown's "g_K correction" and the leading anharmonicities in the bending potentials are taken into account. Linear molecules in electronic Δ states were first discussed by Merer and Travis,[20] who explained deviations from harmonic vibrational behavior by the quartic anharmonicities in the bending potentials. The effect of anharmonicities for electronic Π states was first discussed by Hougen and Jesson,[21] who included the stretching vibrations and the spin–orbit interaction, but made some rather restrictive assumptions (see Section IX below). Hougen[22] has published a very detailed study of the rotational energies (including spin–orbit interaction) for a molecule in a Π state.

Renner's effective hamiltonian involves certain 2×2 matrices which should be considered as operators in their own right. However, he does not try to find the physical meaning of these, but rather plunges straight into the perturbation calculations. These are very heavy and cause the 2×2 matrices to be so mixed up with larger matrices that the fundamental simplicity of his approach is obscured. This is probably the reason why it has never, to our knowledge, been used in detail since. Part of the purpose of the present work is to resurrect Renner's treatment and to formulate it in such a way that the construction of the effective hamiltonian from the real one becomes obvious. We wish to show that it is simple to construct the effective hamiltonian in a consistent and logical fashion and to indicate how further corrections can be included when required by experiments with higher resolution. In the present article we do not go very far down this path, but confine ourselves to a model which allows only for small-amplitude bending and spin–orbit interaction. Within this model we reconsider, extend, and interrelate the various results that have been derived previously, and we build up the algebra of the operators and the wave functions introduced. The present results are used in Reference 23 to include the effects of the stretching vibrations and to derive the energy levels corrected for anharmonicity. The inclusion of end-over-end rotation is to be discussed in a later paper.

The scheme of the article is as follows. In Section II we introduce the kind of wave functions required to describe electronically degenerate states, and in Section III we derive Renner's effective hamiltonian in an approximate way which stresses the physical meaning of his 2×2 matrices. The systematic derivation is given in Section IV. In Section V the effective hamiltonian is written in a form suitable for perturbation calculations, a basis of zeroth-order wave functions is introduced, the matrix elements of the important operators are derived, and rules for their interrelation are pointed out. Section VI deals with the case considered originally by Renner where $\Lambda = 1$ and both potentials are assumed to be strictly harmonic. Applying modern standard perturbation theory, Eq. (1.3) can be quite easily extended to fourth order and the corresponding wave function derived to first order (second order when $|K| = v + 1$). In Section VII these results are extended to include the spin–orbit interaction. It is hoped that the fourth-order formulas will be of some practical value; more accurate results can be obtained by computer if required. The calculations demonstrate that the operators involved in Renner's approach are extraordinarily simple when treated appropriately. Very few degenerate perturbation problems of practical importance can be treated to fourth order. The present application may be of interest as a realistic test example for mathematicians studying perturbation theory as a subject in its own right.

In Section VIII we calculate the effects of small anharmonicities in the potentials on the levels found in Sections VI and VII. Electronic states with $\Lambda > 1$ are dealt with in Section IX. The formula of Merer and Travis[20] is found to be inadequate. The correct expression is closely related to that derived by Hougen and Jesson[21] for the case $\Lambda = 1$. In Section X we study various correlations caused by the gradual breaking of the $C_{\infty v}$ symmetry of the exact linear configuration as the molecule starts to bend. As we point out in Section XII, the previous discussions of this symmetry have been at a qualitative level only, subject to the same restrictions as imposed by Pople and Longuet-Higgins in their one-electron model. In Section XI we discuss the possibility of representing the two electronic wave functions by trigonometric functions $\cos \Lambda \theta$ and $\sin \Lambda \theta$ instead of by the column vectors

$$\begin{Bmatrix} 1 \\ 0 \end{Bmatrix} \quad \text{and} \quad \begin{Bmatrix} 0 \\ 1 \end{Bmatrix}$$

used in Renner's approach.

II. MOLECULAR STATES WITH ELECTRONIC ANGULAR MOMENTUM

The Schrödinger equation for a linear triatomic molecule can be written as

$$(\mathcal{H} - E)\Phi = 0, \qquad \mathcal{H} = T_e + T_n + C(r, q_n) \tag{2.1}$$

The Coulomb potential C depends upon the nuclear coordinates q_n and on the positions $r = (\dots, r_{ix}, r_{iy}, r_{iz}, \dots)$ of the electrons. The kinetic energy of the electrons T_e is represented by the well-known operator

$$2T_e = \sum_i \frac{p_{ix}^2 + p_{iy}^2 + p_{iz}^2}{m} \qquad (2.2)$$

If the stretching vibrations and end-over-end rotations are neglected, the kinetic energy T_n of the nuclei can be written as the kinetic energy of a two-dimensional harmonic oscillator in polar coordinates (ρ, χ), namely

$$2T_n = P_\rho^2 + \hbar^2 \big\{ (J_z - L_z)^2 - \tfrac{1}{4} \big\} \rho^{-2}, \qquad P_\rho = -i\hbar \frac{\partial}{\partial \rho}. \qquad (2.3)$$

Here the angular momentum of the nuclei around the molecular axis e_z is written as $J_z - L_z$, where J_z is the angular momentum of electrons + nuclei and L_z the angular momentum of the electrons alone, both in units of $\hbar$. Measuring the components of the electronic position vectors relative to the frame $\mathcal{K}$ of axes (e_x, e_y, e_z) which rotate with the molecule so that the $z - x$ plane coalesces with the molecular plane, one has

$$J_z = -i \frac{\partial}{\partial \chi} \quad \text{and} \quad L_z = \hbar^{-1} \sum_i (r_{ix} p_{iy} - r_{iy} p_{ix}) \qquad (2.4)$$

where χ is the azimuthal angle defining the orientation of the $z - x$ plane. The singularities of χ and r at $\rho = 0$ are discussed in Appendix B. The volume element for integration over nuclear coordinates is $dq_n = d\rho\, d\chi$. Many authors[1,5,24] use $dq_n = \rho\, d\rho\, d\chi$ instead $[F \equiv P_\rho^2 - (\hbar/2\rho)^2$ in Eq. (2.3) must then be replaced by $\rho^{-1/2} F \rho^{1/2} = \rho^{-2}(\rho P_\rho)^2 = P_\rho^2 - i\hbar\rho^{-1} P_\rho]$.

Within the adiabatic approximation[25-27] the wave function Φ can be factorized

$$\Phi(r, \rho, \chi) = \phi_s(r, \rho) \eta(\rho, \chi) \qquad (2.5)$$

where the electronic wave function ϕ_s is the real and normalized solution to the electronic eigenvalue problem

$$\{ H_e - W_s(\rho) \} \phi_s(r, \rho) = 0, \qquad H_e = T_e + C(r, \rho) \qquad (2.6)$$

which defines the Born–Oppenheimer potential W_s. Note that W_s, ϕ_s, and C are independent of χ. The index $s = \pm 1$ indicates the parity of ϕ_s under the reflection S of the electrons in the molecular plane,

$$S\phi_s = s\phi_s \qquad (2.7)$$

The expansions

$$\phi_s = \phi_s^0 + \phi_s'\rho + \phi_s''\rho^2 + \cdots$$
$$W_s = W_s^0 + W_s''\rho^2 + W_s''''\rho^4 + \cdots \qquad (2.8)$$

around $\rho = 0$ can be determined by applying perturbation theory to the problem

$$H_e = H_e^0 + C'\rho + C''\rho^2 + \cdots, \qquad H_e^0 = T_e + C^0 \qquad (2.9)$$

This is described in Section X and Appendix A. For reasons of symmetry, there are no odd powers of ρ in W_s. Here H_e^0 commutes not only with S, but also with L_z. Since

$$SL_z S^{-1} = -L_z \qquad (2.10)$$

L_z^2 also commutes with S and we have the relation

$$L_z^2 \phi_s^0 = \Lambda^2 \phi_s^0 \qquad (2.11)$$

which allows us to assign a value $\Lambda \in \{0, 1, 2, \dots\}$ to each solution of (2.6). When $\Lambda = 0, 1, 2, 3, \dots$, the molecule is said to be in an electronic Σ^s, Π, $\Delta, \Phi, \dots$, state.

The adiabatic approximation is only valid if the W_s in question is well separated from the other W's. This means that it is not valid if $\Lambda > 0$, because in that case the solutions to (2.6) come in pairs (ϕ_s, W_s), $s = \pm 1$, satisfying

$$W_+^0 = W_-^0 \qquad (2.12)$$

and

$$L_z \phi_s^0 = is\Lambda \phi_{-s}^0 \qquad (2.13)$$

[This is verified below. Equation (2.13) is only correct if the relative sign of the real functions ϕ_+ and ϕ_- is chosen appropriately.] As pointed out by Renner [1], such a pair of solutions gives rise to a molecular state which, to a good approximation, can be described by a wave function of the form

$$\Phi = \phi_+(r, \rho)\eta_+(\rho, \chi) + \phi_-(r, \rho)\eta_-(\rho, \chi) \qquad (2.14)$$

Such states are often called vibronic because they involve vibrational as well as electronic wave functions. When end-over-end rotation is also allowed for, the states are called rovibronic.

To verify the relations (2.12) and (2.13), note first that if $(H_e^0 - W_+^0)\phi_+^0 = 0$, then

1. $\Psi^0 \equiv L_z\phi_+^0$ is also in the level W_+^0 (because $[H_e^0, L_z] = 0$),
2. $S\Psi^0 = -L_z S\phi_+^0 = -\Psi^0$ ($\Rightarrow \langle \Psi^0 | \phi_+^0 \rangle = 0$),
3. $\langle \Psi^0 | \Psi^0 \rangle = \langle \phi_+^0 | L_z^2 | \phi_+^0 \rangle = \Lambda^2$.

Thus the level W_+^0 is at least doubly degenerate when $\Lambda > 0$. The degeneracy cannot be higher if all symmetry derives from the commutation of H_e^0 with L_z and S (neither of these operators can take a vector out of the space spanned by ϕ_+^0 and $\phi_-^0 \equiv -i\Lambda^{-1}\Psi^0$). If the nuclear equilibrium configuration is symmetric around the center nucleus, H_e^0 also commutes with the reflection S_h in the $x-y$ plane. However, since S_h commutes with S and L_z, this extra symmetry does not increase the twofold degeneracy already found. It only implies that an extra index is necessary if we want to indicate whether the two wave functions ϕ_+^0 and ϕ_-^0 are both even or both odd under S_h.

An Angle for the Collective Rotation of the Electrons.   In order to visualize the behavior of ϕ_s^0 under rotations one can change from the electronic coordinates r to $(|r|, \theta)$, where $|r|$ stands for some coordinates invariant under rotations around e_z and where θ is an angle defined so that the electrons undergo a collective rotation of angle $d\theta$ around e_z when $\theta \to \theta + d\theta$. One then has

$$L_z = -i\frac{\partial}{\partial\theta} \tag{2.15}$$

and the solution to (2.13) takes the form

$$\begin{aligned}
\phi_+^0(\theta) &= \phi_+^0(0)\cos\Lambda\theta - \phi_-^0(0)\sin\Lambda\theta \\
\phi_-^0(\theta) &= \phi_-^0(0)\cos\Lambda\theta + \phi_+^0(0)\sin\Lambda\theta
\end{aligned} \tag{2.16}$$

(the dependence on $|r|$ is not shown explicitly). Assuming that θ is measured from the x axis and is defined so that $S\theta S^{-1} = -\theta$, we get $S\phi_s^0(0) = s\phi_s^0(0)$. It is a common misconception that one can set either $\phi_-^0(0) = 0$ or $\phi_+^0(0) = 0$ in (2.16). This can actually be justified only within a one-electron model (where $|r|$ can be chosen invariant under S, ensuring that $\phi_-^0(0)$ vanishes). The misconception, which is unimportant for a qualitative discussion, goes back to Renner's original paper. See Appendix B.

III. THE RENNER–TELLER HAMILTONIAN

When the wave function Φ equals $\phi_+\eta_+ + \phi_-\eta_-$ as in (2.14), the eigenvalue problem (2.1) can also be written

$$(\bar{H} - E)\Phi = 0, \qquad \bar{H} \equiv T_n + W + \hat{W}S \tag{3.1}$$

where W and $\hat{W}$ are the two potentials

$$W = \tfrac{1}{2}(W_+ + W_-) = W^0 + W''\rho^2 + W''''\rho^4 + \cdots$$
$$\hat{W} = \tfrac{1}{2}(W_+ - W_-) = \qquad \hat{W}''\rho^2 + \hat{W}''''\rho^4 + \cdots \qquad (3.2)$$

[From the construction of ϕ_s one has $H_e\Phi = W_+\phi_+\eta_+ + W_-\phi_-\eta_- = (W + \hat{W}S)\Phi$.] We know from Appendix A [see also Eq. (10.7) and subsequent text] that

$$\hat{W} \text{ varies as } \rho^{2\Lambda} \qquad \text{at} \quad \rho = 0 \qquad (3.3)$$

Under the extra assumption (relaxed in Section IV) that the three relations

$$P_\rho\phi_s \simeq \phi_s P_\rho, \qquad L_z\phi_s \simeq is\Lambda\phi_{-s}, \qquad S\phi_s = s\phi_s \qquad (3.4)$$

all hold exactly (as they would do if ϕ_s did not deviate from ϕ_s^0), we can apparently solve the eigenvalue problem (3.1) without knowing anything more about ϕ_+, ϕ_-, L_z, and S. Hence, E, η_+, and η_- are not affected if we replace ϕ_+, ϕ_-, L_z, and S_z with other wave functions and operators which also satisfy (3.4). For instance, we could replace ϕ_s with ϕ_s^0 and leave L_z and S unchanged. Alternatively ϕ_+, ϕ_- can be replaced with $\cos\Lambda\theta$, $\sin\Lambda\theta$ as discussed in Section XI. Here we choose to introduce the Pauli matrices

$$\sigma_x = \begin{Bmatrix} 0 & 1 \\ 1 & 0 \end{Bmatrix}, \qquad \sigma_y = \begin{Bmatrix} 0 & -i \\ i & 0 \end{Bmatrix}, \qquad \sigma_z = \begin{Bmatrix} 1 & 0 \\ 0 & -1 \end{Bmatrix} \qquad (3.5)$$

and make the replacements

$$\phi_+ \rightarrow \begin{Bmatrix} 1 \\ 0 \end{Bmatrix}, \quad \phi_- \rightarrow \begin{Bmatrix} 0 \\ 1 \end{Bmatrix}, \quad L_z \rightarrow \Lambda\sigma_y, \quad S \rightarrow \sigma_z \qquad (3.6)$$

After the substitution, the problem (3.1) takes a form that is algebraically much more convenient:

$$(H - E)\Psi = 0, \qquad \Psi = \begin{Bmatrix} 1 \\ 0 \end{Bmatrix}\eta_+ + \begin{Bmatrix} 0 \\ 1 \end{Bmatrix}\eta_- = \begin{Bmatrix} \eta_+ \\ \eta_- \end{Bmatrix} \qquad (3.7)$$

where

$$H = T_\sigma + W + \hat{W}\sigma_z, \qquad 2T_\sigma = P_\rho^2 + \hbar^2\left\{(J_z - \Lambda\sigma_y)^2 - \tfrac{1}{4}\right\}\rho^{-2} \qquad (3.8)$$

This is the hamiltonian obtained by Renner[1] in his Eq. (16), except that he put $\Lambda = 1$ and assumed the potentials W_+ and W_- to be harmonic. In

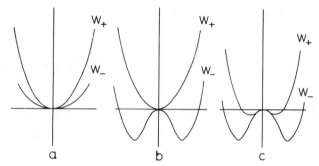

Fig. 1.   Sketch of various possible shapes of the potentials W_+ and W_-.

general the potentials can have other forms as illustrated in Fig. 1*b* and *c*. In these cases the first two of the assumptions in (3.4) become unreliable, but as we shall see in Section IV, we can still expect H to work well.

Evidently J_z in Renner's hamiltonian (3.8) still has the meaning of the total angular momentum around $\mathbf{e}_z$, and σ_z means the reflection of the electrons in the molecular plane. When ϕ_s only deviates a little from ϕ_s^0, we can think of $\Lambda\sigma_y$ as the operator L_z for the angular momentum of the electrons, but in the case of larger vibrational amplitudes this interpretation becomes less reliable.

In the figures we have assumed W_+ to be the upper potential. It is always permissible to start a particular theoretical study in this way. Should it turn out, at some later stage, that W_+ is the lower potential, it simply means that we have studied the unitary transform

$$\sigma_y H \sigma_y^{-1} = T_\sigma + W - \hat{W}\sigma_z \qquad (3.9)$$

instead of H itself.

We note that the unitary transformation

$$U = \exp\left(-i\chi\Lambda\sigma_y\right) = \cos(\Lambda\chi) - i\sin(\Lambda\chi)\,\sigma_y, \qquad (3.10)$$

transforms J_z into $J_z + \Lambda\sigma_y$ and therefore removes the σ_y term from T_σ:

$$U T_\sigma U^{-1} = T_\perp \equiv \tfrac{1}{2}P_\rho^2 + \tfrac{1}{2}\hbar^2\left\{J_z^2 - \tfrac{1}{4}\right\}\rho^{-2} \qquad (3.11)$$

Hence, if the two potentials W_+ and W_- coalesce, the energy levels of H are found as the eigenvalues of the Hamiltonian $T_\perp + W$ in accordance with expectation. If both potentials are close to the same parabola $\lambda\rho^2/2$, the energy levels deviate from those of the corresponding harmonic oscillator by small amounts which can be found by perturbation theory. The relevant perturbation problem is set up in Section V (see also Appendix B).

One can also write H as

$$H = h_+ \begin{Bmatrix} 1 & 0 \\ 0 & 0 \end{Bmatrix} + h_- \begin{Bmatrix} 0 & 0 \\ 0 & 1 \end{Bmatrix} - \left(\frac{\hbar}{\rho} \right)^2 J_z \Lambda \sigma_y$$

$$\equiv H_+ + H_- + H_{\text{coupl}} \tag{3.12}$$

$$h_s = \tfrac{1}{2} P_\rho^2 + \tfrac{1}{2} \hbar^2 \{ J_z^2 + \Lambda^2 - \tfrac{1}{4} \} \rho^{-2} + W_s$$

Since H commutes with J_z, one can replace the operator J_z with its eigenvalue $K = 0, \pm 1, \pm 2, \dots$ before diagonalization. When $J_z = K = 0$, the coupling term H_{coupl} in (3.12) vanishes and the eigenvalue problem $(H - E)\Psi = 0$ splits into two:

$$(h_+ - E_+)\eta_+ = 0 \quad \text{and} \quad (h_- - E_-)\eta_- = 0 \tag{3.13}$$

where the corresponding eigenvectors

$$\Psi = \eta_+ \begin{Bmatrix} 1 \\ 0 \end{Bmatrix} \quad \text{and} \quad \Psi = \eta_- \begin{Bmatrix} 0 \\ 1 \end{Bmatrix} \tag{3.14}$$

are eigenvectors of σ_z also (in agreement with $[H(K=0), \sigma_z] = 0$). The corresponding full wave functions $\Phi = \phi_+ \eta_+$ and $\Phi = \phi_- \eta_-$ also have the parities $s = +1$ and $s = -1$, respectively, under the reflection S in the molecular plane (in agreement with $[\mathcal{H}(K=0), S] = 0$).

Even when $K \neq 0$ it can sometimes be of interest to construct eigenvectors and eigenvalues of $H_+ + H_-$ by solving the two eigenvalue problems (3.13). If H_-, say, has some eigenstates far below the lowest H_+ state, these H_- states (strictly eigenstates of $H_+ + H_-$) are expected to be perturbed only a little by H_{coupl}, because the nearest state they can couple with is the remote, lowest H_+ state.

The Renner–Teller hamiltonian H also commutes with the hermitian and unitary operator

$$S_\sigma = S_\chi \sigma_z = \sigma_z S_\chi \tag{3.15}$$

where S_χ is the hermitian and unitary operator that changes the sign of χ and J_z. This symmetry of H under S_σ means that if Ψ_K is a normalized vector satisfying $H\Psi_K = E\Psi_K$ and $J_z \Psi_K = K\Psi_K$, then

$$\Psi_{-K} = S_\sigma \Psi_K \tag{3.16}$$

is a normalized vector satisfying

$$H\Psi_{-K} = E\Psi_{-K}, \quad J_z \Psi_{-K} = -K\Psi_{-K}, \quad \langle \Psi_K | \Psi_{-K} \rangle = \delta_{K, -K} \tag{3.17}$$

Thus a level characterized by a certain $K \neq 0$ is (at least) doubly degenerate. Also, if the problem $(H - E)\Psi = 0$ is solved for all $K > 0$, then the solutions for $K < 0$ follow trivially from (3.16). The trivial relationship between K and $-K$ could also have been established by use of the time-reversal operator. However, this is anti-unitary, and it is therefore easier to work with S_σ.

The coupled differential equations. Assuming the eigenfunction Ψ of H in (3.7) to be of the factorized form

$$\Psi = \left\{ \begin{matrix} \mu_+(\rho) \\ \mu_-(\rho) \end{matrix} \right\} e^{iKx} \tag{3.18}$$

one finds that E, μ_+, and μ_- can be determined by solving the pair of coupled differential equations

$$\left\{ h_K + \hat{W}(\rho)\sigma_z + \hbar^2 \Lambda \sigma_y \rho^{-2} - E \right\} \left\{ \begin{matrix} \mu_+ \\ \mu_- \end{matrix} \right\} = 0 \tag{3.19}$$

$$h_K \equiv \tfrac{1}{2} P_\rho^2 + \tfrac{1}{2}\hbar^2 \left\{ K^2 + \Lambda^2 - \tfrac{1}{4} \right\} \rho^{-2} + W(\rho)$$

This "radial equation" is important for numerical problems. Although such problems are not the subject of the present paper, we note that the term $\sigma_y \rho^{-2}$, which couples upper and lower components, becomes large at $\rho = 0$, implying that the equation is not well suited for the numerical treatment if the μ's have significant weight at $\rho = 0$. In such cases it has been found better to study the equivalent problem[8-11]

$$\left\{ h_K + \hat{W}(\rho)\sigma_x + \hbar^2 \Lambda \sigma_z \rho^{-2} - E \right\} \left\{ \begin{matrix} \tilde{\mu}_+ \\ \tilde{\mu}_- \end{matrix} \right\} = 0 \tag{3.20}$$

$$\left\{ \begin{matrix} \mu_+ \\ \mu_- \end{matrix} \right\} = u^{-1} \left\{ \begin{matrix} \tilde{\mu}_+ \\ \tilde{\mu}_- \end{matrix} \right\}, \qquad u = 2^{-1/2} \left\{ \begin{matrix} 1 & -i \\ 1 & i \end{matrix} \right\}$$

(To prove the equivalence, note that u is unitary and satisfies $u\sigma_r u^{-1} = \sigma_{r+1}$, $r = x, y, z, x, \ldots$.) Note that the matrix which couples upper and lower components (now σ_x) has been transferred to $\hat{W}$, which is small at $\rho = 0$. Note also that the $\tilde{\mu}$'s can be assumed real. Barrow et al.[8,9] have devised yet another form of Eq. (3.19) which is appropriate when the μ's have weight both at $\rho = 0$ and where $\hat{W}$ is large. In actual large-amplitude analysis, the bending coordinate ρ will usually not be defined as above. However, the essentials of Eqs. (3.19) and (3.20) are not changed.

IV. CORRECTIONS TO THE RENNER–TELLER HAMILTONIAN

In Section III we derived the Renner–Teller hamiltonian (3.8) assuming that (1) the form $\Phi = \phi_+ \eta_+ + \phi_- \eta_-$ for the wave function in (2.1) was exact and (2) the approximations of (3.7) were valid. We shall now investigate the corrections that arise when the second of these assumptions is dropped. Integrating $\phi_s(\mathcal{K} - E)\Phi = 0$ over the electronic coordinates, we find that E, η_+, and η_- are determined exactly by the eigenvalue problem (3.7) with

$$H = \begin{Bmatrix} H_{++} & H_{+-} \\ H_{-+} & H_{--} \end{Bmatrix}, \qquad H_{ss'} = \langle \phi_s | \mathcal{K} | \phi_{s'} \rangle. \tag{4.1}$$

Use of (2.6) gives

$$\begin{aligned} H_{ss} &= \langle \phi_s | T_n | \phi_s \rangle + W_s \\ H_{+-} &= -\hbar^2 J_z \langle \phi_+ | L_z | \phi_- \rangle \rho^{-2} = -H_{-+} \end{aligned} \tag{4.2}$$

and we can write H as

$$H = \tfrac{1}{2}(H_{++} + H_{--}) + \tfrac{1}{2}(H_{++} - H_{--})\sigma_z + iH_{+-}\sigma_y \tag{4.3}$$

A well-known calculation, first pointed out by Born[25,26] [see also for instance Messiah (Reference 27, pp. 790–791)] gives

$$\langle \phi_s | P_\rho^2 | \phi_s \rangle = P_\rho^2 + \langle P_\rho \phi_s | P_\rho \phi_s \rangle \tag{4.4}$$

$(\equiv P_\rho^2 + \hbar^2 \langle \partial \phi_s / \partial \rho | \partial \phi_s / \partial \rho \rangle)$. Hence, with h_s from (3.12),

$$H_{ss} = h_s + w_s(\rho),$$

where the *adiabatic correction potential* w_s is given by

$$\begin{aligned} 2w_s &= \langle P_\rho \phi_s | P_\rho \phi_s \rangle + \hbar^2 \langle \phi_s | (L_z^2 - \Lambda^2) | \phi_s \rangle \rho^{-2} \\ &\left[= \hbar^2 \langle \phi_s' | (L_z^2 - \Lambda^2 + 1) | \phi_s' \rangle \qquad \text{at} \quad \rho = 0 \right] \end{aligned} \tag{4.5}$$

The matrix element of L_z in H_{+-} can be written as

$$\langle \phi_+ | L_z | \phi_- \rangle = -i\Lambda \left(1 - \frac{g\rho^2}{\hbar^2} \right) \tag{4.6}$$

where g is the real function of ρ given by

$$g = \frac{\hbar^2}{\rho^2}(\Lambda - i\langle \phi_+ | L_z | \phi_- \rangle)\Lambda^{-1} = g^0 + g''\rho^2 + g''''\rho^4 + \cdots \tag{4.7}$$

$$g^0 = \hbar^2\{-i\Lambda^{-1}\langle \phi'_+ | L_z | \phi'_- \rangle + \tfrac{1}{2}(\langle \phi'_+ | \phi'_+ \rangle + \langle \phi'_- | \phi'_- \rangle)\}$$

The expression for g at $\rho = 0$ is obtained by substituting the expansion $\phi_s = \phi_s^0 + \phi'_s\rho + \phi''_s\rho^2 + \cdots$ in $1 = \langle \phi_s | \phi_s \rangle$ to give

$$\langle \phi_s^0 | \phi'_s \rangle = \langle \phi'_s | \phi_s^0 \rangle = 0$$

$$\langle \phi_s^0 | \phi''_s \rangle = \langle \phi''_s | \phi_s^0 \rangle = -\tfrac{1}{2}\langle \phi'_s | \phi'_s \rangle \tag{4.8}$$

$$\vdots$$

The first of these results implies that $\langle \phi_+ | L_z | \phi_- \rangle$ has no terms linear in ρ. The second result allows us to reduce $i\hbar^{-2}g^0\Lambda = \langle \phi_+^0 | L_z | \phi'' \rangle + \langle \phi'_+ | L_z | \phi'_- \rangle + \langle \phi''_+ | L_z | \phi_-^0 \rangle = -i\Lambda\langle \phi_-^0 | \phi'' \rangle + \langle \phi'_+ | L_z | \phi'_- \rangle - i\Lambda\langle \phi''_+ | \phi_+^0 \rangle$ as required. The lack of odd powers of ρ in the expansion of g and in the expansions

$$w \equiv \tfrac{1}{2}(w_+ + w_-) = w^0 + w''\rho^2 + w''''\rho^4 + \cdots \tag{4.9}$$

$$\hat{w} \equiv \tfrac{1}{2}(w_+ - w_-) = \hat{w}^0 + \hat{w}''\rho^2 + \hat{w}''''\rho^4 + \cdots$$

is justified in Appendix A, where it is also proven that

$$\hat{w} \text{ varies as } \rho^{2\Lambda} \quad \text{at} \quad \rho = 0 \tag{4.10}$$

[compare with (3.3), and see also Section X]. The results of (4.7) and (4.10) are more complete versions of results found previously in References 11 and 19, respectively.

We now see that H in (4.1) is obtained by adding the correction

$$\Delta H = w + \hat{w}\sigma_z + gJ_z\Lambda\sigma_y \tag{4.11}$$

to H in (3.8). At this point, the only approximation we have used to cast the Schrödinger equation (2.1) in the form $(H - E)\Psi = 0$ is the assumption that Φ lies entirely in the space of trial functions $\phi_+\eta_+ + \phi_-\eta_-$. Experience with molecules in nondegenerate electronic states suggests that deviations from this approximation are usually small compared with experimental accuracy. When the vibrational amplitude is small, within the Born-Oppenheimer ordering scheme[28,29] w_s and g have order of magnitude κ^4 ($\simeq \kappa^2 E_{\text{vib}} \simeq$ quartic potential anharmonicities), and we expect that it will be sufficient to approximate w_s and g by their equilibrium values. Then the constant w^0 can be omitted, the term $\hat{w}^0\sigma_z$ vanishes because $\hat{w}^0 = 0$ (though it would otherwise have been important), and we are left with

$$\Delta H = g_K J_z \Lambda \sigma_y, \qquad g_K \equiv g^0 \qquad (4.12)$$

The symbol g_K for g^0 is that used by Brown,[17] who first described this correction to the Renner–Teller hamiltonian. In Section X the expression for g^0 is discussed further. When the potentials are as in Fig. 1b and c, the corrections w and $\hat{w}\sigma_z$ can be accounted for by including w_s in the potential W_s. Experience with molecules in nondegenerate states suggests that this alters W_s only very little. Hence, under the reasonable assumption that $w + \hat{w}\sigma_z \sim g J_z \Lambda \sigma_y$, the correction ΔH does not affect the levels in any major way; it only gives rise to small corrections.

We note that g does not start to deviate significantly from g_K until ϕ_s starts to deviate from $\phi_s'\rho + \phi_s''\rho^2$, which happens when W_s starts to deviate from $W_s''\rho^2 + W_s''''\rho^4$ (because $W_s'',\ldots,W_s^{(2p)}$ only depend upon $\phi_s',\ldots,\phi_s^{(p)}$). The amplitude for this can be rather large. In NH_2, for instance, the potentials are as sketched in Fig. 1b, but they can still be described approximately without sextic terms[4,5] (see Appendix D). Jungen and Merer have included the g_K correction in their effective Hamiltonian for NH_2 and H_2O^+.[12] The term is found to improve the fit to the experimentally observed data.

The effects of stretching vibrations, end-over-end rotation, spin–orbit interaction, external magnetic or electric fields, etc., can be included in H simply by including them in $\mathcal{H}$ in (4.1) before working out the matrix elements $\langle\phi_{s'}|\mathcal{H}|\phi_s\rangle$. In Appendix E the result of including the spin–orbit interaction is given. We do not treat the other effects in the present article, because that would obscure the main purpose, which is to study the fundamental principles and develop the algebra for the operators and the wave functions involved in the approach. For the inclusion of the stretching vibrations, see Reference 23.

V. A BASIS FOR DIAGONALIZATION OF THE RENNER–TELLER HAMILTONIAN

The simple Renner-Teller Hamiltonian in (3.8) can be written as

$$H = \left(T_\sigma + \tfrac{1}{2}\lambda\rho^2\right) + \left(W - \tfrac{1}{2}\lambda\rho^2 + \hat{W}\sigma_z\right) \equiv H_\sigma + H'. \qquad (5.1)$$

The force constant λ is chosen so as to make H' as "small" as possible. From (3.11) we see that H_σ is the unitary transform of the twofold isotropic harmonic-oscillator hamiltonian

$$H_\perp = T_\perp + \tfrac{1}{2}\lambda\rho^2 = UH_oU^{-1} \tag{5.2}$$

with the well-known eigenvectors $|vl\rangle$:

$$
\begin{aligned}
|vl\rangle &= (2\pi)^{-1/2} R_{vl}(\rho)e^{ilx} \\
H_\perp|vl\rangle &= \omega(v+1)|vl\rangle, \qquad v = 0,1,2,\ldots \\
J_z|vl\rangle &= l|vl\rangle, \qquad v - |l| \in \{0,2,4,\ldots\}
\end{aligned} \tag{5.3}
$$

Note that we use ω ($=\lambda^{1/2}\hbar$) instead of $\hbar\omega$, to denote the quantum of energy. Introducing the dimensionless coordinate

$$q = \gamma^{1/2}\rho, \qquad \gamma = \lambda^{1/2}\hbar^{-1} = \omega\hbar^{-2}, \tag{5.4}$$

defined so that $\lambda\rho^2 = \omega q^2$, the radial function $R_{vl}(\rho)$ is

$$R_{vl}(\rho) = \left(\frac{\omega}{\hbar^2}\right)^{1/4} r_{vl}(q) = \left(\frac{\omega}{\hbar^2}\right)^{1/4} r_{vl}\left(\frac{\omega^{1/2}\rho}{\hbar}\right) \tag{5.5}$$

where $r_{vl}(q)$ is the real function solving

$$\left\{-\frac{d^2}{dq^2} + \left(l^2 - \tfrac{1}{4}\right)q^{-2} + q^2 - 2(v+1)\right\}r_{vl}(q) = 0, \tag{5.6a}$$

$$\int_0^\infty r_{v'l}(q)r_{vl}(q)\,dq = \delta_{v'v}. \tag{5.6b}$$

Thus R_{vl} is defined to within a sign. Choosing this as Di Lauro and Mills[24] do, one gets $R_{vl} = R_{v-l}$ and, in terms of $q_\pm = q\exp(\pm ix)$,

$$2^{1/2}q_\pm|vl\rangle = (v+2\pm l)^{1/2}|v+1,l\pm1\rangle + (v\mp l)^{1/2}|v-1,l\pm1\rangle. \tag{5.7}$$

This recursion relation is also valid for the $|vl\rangle$ basis which Messiah (Reference 27, p. 454) constructs in a very simple way by the use of creation and annihilation operators. Some results obtained by repeated application of (5.7) (and the relation $q^2 = q_+q_-$) are listed below.

To obtain the eigenvectors of H_σ, we first introduce the orthonormal eigenvectors $|+\rangle$ and $|-\rangle$ of σ_y,

$$|s\rangle = 2^{-1/2}\begin{Bmatrix} 1 \\ si \end{Bmatrix}, \qquad \begin{cases} \sigma_x|s\rangle = si|-s\rangle \\ \sigma_y|s\rangle = s|s\rangle \\ \sigma_z|s\rangle = |-s\rangle \end{cases} \tag{5.8}$$

and define

$$|vKs\rangle = U^{-1}|vl\rangle|s\rangle = e^{is\Lambda x}|vl\rangle|s\rangle \tag{5.9}$$

where $l = K - s\Lambda$. Then the set

$$|vKs\rangle = \tfrac{1}{2}\pi^{-1/2}R_{vl}(\rho)\begin{Bmatrix} 1 \\ si \end{Bmatrix}e^{iKx}$$
$$s = \pm 1, \qquad v = 0,1,2,\ldots, \qquad v - |K - s\Lambda| \in \{0,2,4,\ldots\} \tag{5.10}$$

is complete and orthonormal and one has

$$H_\sigma|vKs\rangle = \omega(v+1)|vKs\rangle \tag{5.11a}$$

$$\sigma_y|vKs\rangle = s|vKs\rangle \tag{5.11b}$$

$$J_z|vKs\rangle = K|vKs\rangle \tag{5.11c}$$

$$S_\sigma|vKs\rangle = |v,-K,-s\rangle \tag{5.11d}$$

where $S_\sigma = S_x\sigma_z$ is the unitary operator defined in (3.15). From the interpretation of $\Lambda\sigma_y$ we see that $|vKs\rangle$ can be thought of as a state with the value $\langle vKs|\Lambda\sigma_y|vKs\rangle = s\Lambda$ of L_z.

Actually Renner's expressions for the eigenvectors of H_σ are not quite correct [see his Eq. (25)]. The error amounts to having $\sigma_z|vKs\rangle$ instead of $|vKs\rangle$ and does not affect the final energies, because the perturbation H' in (5.1) commutes with the unitary operator σ_z, implying that H' has the same matrix elements with respect to the two bases $|vKs\rangle$ and $\sigma_z|vKs\rangle$.

For given K, the H_σ level $E = \omega(v+1)$ is spanned by

$$\begin{array}{ll} |vKs_K\rangle & \text{if } v \leqslant |K| \\ |vK+\rangle \text{ and } |vK-\rangle & \text{if } v > |K| \end{array} \tag{5.12}$$

Here s_K denotes the sign of K. When $v \leqslant |K|$, the level is nondegenerate and can be called *unique*.[10] When $v > |K|$, the level is doubly degenerate and so is *nonunique*.[10]

Consider now a hermitian operator F, which commutes with J_z and $S_\sigma = S_\chi \sigma_z$ and has only real matrix elements in the $|vKs\rangle$ basis. Then these matrix elements can be written as

$$\langle v'Ks|F|vKs\rangle = F^s_{v'v}$$
$$\langle v'K-|F|vK+\rangle = F_{v'v} \qquad\qquad (5.13)$$
$$\langle v'K+|F|vK-\rangle = F_{vv'}$$

where the F's are functions of K, satisfying

$$F^s_{vv'} = F^s_{v'v}, \qquad F^-_{vv'}(K) = F^+_{v'v}(-K)$$
$$F_{vv'}(K) = F_{v'v}(-K) \qquad\qquad (5.14)$$

The operators q^m and $q^m \sigma_z$, $m = 2, 4, \ldots$, serve as examples of F. One finds

$$(q^m)_{v'v} = (q^m \sigma_z)^s_{v'v} = 0$$
$$(q^m)^s_{v'v} = \langle v', l = K - s\Lambda | q^m | v, l = K - s\Lambda \rangle \qquad (5.15)$$
$$(q^m \sigma_z)_{v'v} = \langle v', l = K + \Lambda | q^{m-2\Lambda} q^{2\Lambda}_+ | v, l = K - \Lambda \rangle$$

where the matrix elements $\langle v'l' | \cdots | vl \rangle$ are found by repeated application of (5.7). (Note that $q^2 = q_+ q_- = q_- q_+$.) We give some examples for the case $\Lambda = 1$. For the matrix elements of the particularly important perturbation V, considered in Section VI, we introduce a special symbol and write the results in detail as follows:

$$V = \tfrac{1}{2}\lambda\rho^2\sigma_z = \tfrac{1}{2}\omega q^2\sigma_z$$
$$\langle v'Ks|V|vKs\rangle = 0$$
$$\langle v'K-|V|vK+\rangle = \omega a_{v'v}$$
$$\langle v'K+|V|vK-\rangle = \omega a_{vv'} \qquad\qquad (5.16)$$
$$a_{v\pm2,v} = \left(\tfrac{1}{2}q^2\sigma_z\right)_{v\pm2,v} = \tfrac{1}{4}\left[(\pm\tilde{v}+2+K)(\pm\tilde{v}+K)\right]^{1/2}$$
$$a_{vv} = \left(\tfrac{1}{2}q^2\sigma_z\right)_{vv} = \tfrac{1}{2}\left[\tilde{v}^2 - K^2\right]^{1/2}$$
$$\tilde{v} \equiv v+1, \qquad a_{vv'}(K) = a_{v'v}(-K)$$

Other examples are q^2, $q^4\sigma_z$, and q^4, which have all their matrix elements determined by

$$(q^2)^+_{v\pm2,v} = \tfrac{1}{2}[(\pm\tilde{v}+2-K)(\pm\tilde{v}+K)]^{1/2}$$

$$(q^2)^+_{vv} = \tilde{v}$$

$$(q^4\sigma_z)_{v\pm4,v} = \tfrac{1}{4}[(\pm\tilde{v}+4+K)(\pm\tilde{v}+2+K)(\pm\tilde{v}+K)(\pm\tilde{v}+2-K)]^{1/2}$$

$$(q^4\sigma_z)_{v\pm2,v} = \pm\tfrac{1}{2}(\pm2\tilde{v}+2-K)[(\pm\tilde{v}+2+K)(\pm\tilde{v}+K)]^{1/2}$$

$$(q^4\sigma_z)_{vv} = \tfrac{3}{2}\tilde{v}[\tilde{v}^2-K^2]^{1/2}$$

$$(q^4)^+_{v\pm4,v} = \tfrac{1}{4}[(\pm\tilde{v}+K)(\pm\tilde{v}+2-K)(\pm\tilde{v}+2+K)(\pm\tilde{v}+4-K)]^{1/2}$$

$$(q^4)^+_{v\pm2,v} = \pm(\pm\tilde{v}+1)[(\pm\tilde{v}+K)(\pm\tilde{v}+2-K)]^{1/2}$$

$$(q^4)^+_{vv} = \tfrac{1}{2}[3\tilde{v}^2-K(K-2)] \tag{5.17}$$

and the rules (5.14). We also give some results which are valid for arbitrary Λ. Here we include the operator p^2 which allows us to write H_σ as $\omega(p^2+q^2)/2$. Its nonvanishing matrix elements are fixed by the relations $\langle vKs|p^2-q^2|vKs\rangle = \langle v\pm2,K,s|p^2+q^2|vKs\rangle = 0$, and we find

$$(q^2)^+_{v\pm2,v} = -(p^2)^+_{v\pm2,v} = \tfrac{1}{2}[(\pm\tilde{v}+1)^2-(K-\Lambda)^2]^{1/2}$$

$$(q^2)^+_{vv} = (p^2)^+_{vv} = \tilde{v}$$

$$(q^4)^+_{v\pm4,v} = \tfrac{1}{4}[(\pm\tilde{v}+1)^2-(K-\Lambda)^2]^{1/2}[(\pm\tilde{v}+3)^2-(K-\Lambda)^2]^{1/2}$$

$$(q^4)^+_{v\pm2,v} = \pm(\pm\tilde{v}+1)[(\pm\tilde{v}+1)^2-(K-\Lambda)^2]^{1/2} \tag{5.18}$$

$$(q^4)^+_{vv} = \tfrac{1}{2}[3\tilde{v}^2-(K-\Lambda)^2+1]$$

It is not simple to determine $(q^2\sigma_z)_{v'v}$ when $\Lambda>1$ and $(q^4\sigma_z)_{v'v}$ when $\Lambda>2$. For $\Lambda=2$ one finds

$$(q^4\sigma_z)_{v\pm4,v} = \tfrac{1}{4}[(\pm\tilde{v}-1+K)$$

$$\times(\pm\tilde{v}+1+K)(\pm\tilde{v}+3+K)(\pm\tilde{v}+5+K)]^{1/2} \tag{5.19}$$

$$(q^4\sigma_z)_{v\pm2,v} = [(\pm\tilde{v}-1+K)(\pm\tilde{v}+3+K)]^{1/2}[(\pm\tilde{v}+1)^2-K^2]^{1/2}$$

$$(q^4\sigma_z)_{vv} = \tfrac{3}{2}[(\tilde{v}+1)^2-K^2]^{1/2}[(\tilde{v}-1)^2-K^2]^{1/2}$$

We finally note that if we drop the assumption that F in (5.13) is hermitian with real matrix elements, then $F_{vv'}$ in the last line of (5.13) must be replaced with $F_{v'v}(-K)$, and the first and the third result in (5.14) become invalid.

Time Reversal

The model of the molecule considered in the present paper ignores various movements and interactions. When these are allowed for, a more complicated situation arises and it can be helpful to make use of the time-reversal operation $\mathfrak{T}$. We therefore discuss this briefly.

Since the electronic wave functions are chosen real, the action of the time reversal operator $\mathfrak{T}$ on a trial function Φ is given by $\mathfrak{T}\Phi = \Phi^* = \phi_+\eta_+^* + \phi_-\eta_-^*$, where the asterisk denotes complex conjugation. This implies (we shall not go into details) that we must define $\mathfrak{T}$ on H's space by

$$\mathfrak{T}\Psi = \mathfrak{T}\left(\left\{\begin{matrix}1\\0\end{matrix}\right\}\eta_+ + \left\{\begin{matrix}0\\1\end{matrix}\right\}\eta_-\right) = \left\{\begin{matrix}1\\0\end{matrix}\right\}\eta_+^* + \left\{\begin{matrix}0\\1\end{matrix}\right\}\eta_-^* \tag{5.20}$$

Thus $\mathfrak{T}$ leaves the coordinates (ρ, χ) invariant, changes the sign of the conjugate momenta (P_ρ, J_z), and transforms the Pauli matrices as follows:

$$\mathfrak{T}(\sigma_x, \sigma_y, \sigma_z)\mathfrak{T}^{-1} = (\sigma_x, -\sigma_y, \sigma_z). \tag{5.21}$$

It is easy to verify that H is invariant under $\mathfrak{T}$. The action of $\mathfrak{T}$ on a wave function $|\Psi\rangle = \Sigma c_{vKs}|vKs\rangle$ is

$$\mathfrak{T}|\Psi\rangle = \sum c_{vKs}^*|v, -K, -s\rangle = S_\sigma \sum c_{vKs}^*|vKs\rangle \tag{5.22}$$

where the asterisk denotes complex conjugation. Thus $\mathfrak{T} = S_\sigma \mathcal{C} = \mathcal{C}S_\sigma$, where $\mathcal{C}$ is the complex-conjugation operator (Reference 27, p. 641) relative to the $|vKs\rangle$ basis. One sees that if an operator F commutes with S_σ then its $|vKs\rangle$ matrix elements are all real if, and only if, F is invariant under time reversal.

Matrix elements in $|vl\rangle$-basis. A nonvanishing matrix element $\langle v'l'|f_\pm|vl\rangle$ of an operator $f_\pm = q^m q_\pm^n$ can be written as $(f_\pm)_{v'v}^l$ (because we need not specify $l'=l\pm n$). All nonvanishing matrix elements of $q_\pm^2$, q^2, $q^2q_\pm^2$, q^4, and $q_\pm^4$ are then given by

$$\left(q_+^2\right)_{v\pm2,v}^l = \tfrac{1}{2}\left[(\pm\tilde{v}+1+l)(\pm\tilde{v}+3+l)\right]^{1/2}$$

$$\left(q_+^2\right)_{vv}^l = \left[\tilde{v}^2-(l+1)^2\right]^{1/2}$$

$$\left(q^2\right)_{v\pm2,v}^l = \tfrac{1}{2}\left[(\pm\tilde{v}+1)^2-l^2\right]^{1/2}$$

$$\left(q^2\right)^l_{vv} = \tilde{v}$$

$$\left(q^2 q_+^2\right)^l_{v\pm4,v} = \tfrac{1}{4}\left[(\pm\tilde{v}+1+l)(\pm\tilde{v}+3+l)(\pm\tilde{v}+5+l)(\pm\tilde{v}+1-l)\right]^{1/2}$$

$$\left(q^2 q_+^2\right)^l_{v\pm2,v} = \pm\tfrac{1}{2}(\pm2\tilde{v}+1-l)\left[(\pm\tilde{v}+1+l)(\pm\tilde{v}+3+l)\right]^{1/2}$$

$$\left(q^2 q_+^2\right)^l_{vv} = \tfrac{3}{2}\tilde{v}\left[\tilde{v}^2-(l+1)^2\right]^{1/2} \qquad\qquad (5.23)$$

$$\left(q^4\right)^l_{v\pm4,v} = \tfrac{1}{4}\left[(\pm\tilde{v}+1)^2-l^2\right]^{1/2}\left[(\pm\tilde{v}+3)^2-l^2\right]^{1/2}$$

$$\left(q^4\right)^l_{v\pm2,v} = \pm(\pm\tilde{v}+1)\left[(\pm\tilde{v}+1)^2-l^2\right]^{1/2}$$

$$\left(q^4\right)^l_{vv} = \tfrac{1}{2}(3\tilde{v}^2-l^2+1)$$

$$\left(q_+^4\right)^l_{v\pm4,v} = \tfrac{1}{4}\left[(\pm\tilde{v}+1+l)(\pm\tilde{v}+3+l)(\pm\tilde{v}+5+l)(\pm\tilde{v}+7+l)\right]^{1/2}$$

$$\left(q_+^4\right)^l_{v\pm2,v} = \left[(\pm\tilde{v}+1+l)(\pm\tilde{v}+3+l)(\pm\tilde{v}+5+l)(\pm\tilde{v}-1-l)\right]^{1/2}$$

$$\left(q_+^4\right)^l_{vv} = \tfrac{3}{2}\left[(\tilde{v}+1+l)(\tilde{v}+3+l)(\tilde{v}-3-l)(\tilde{v}-1-l)\right]^{1/2}$$

and the rule $(f_-)^l_{v'v} = (f_+)^{-l}_{v'v}$.

VI. THE HARMONIC APPROXIMATION WHEN $\Lambda = 1$

We shall consider the case where $\Lambda = 1$ and both potentials are strictly harmonic, $W_s = \tfrac{1}{2}\lambda_s\rho^2$. Introducing $\lambda = \tfrac{1}{2}(\lambda_+ + \lambda_-)$ and $\hat{\lambda} = \tfrac{1}{2}(\lambda_+ - \lambda_-)$, we can write the Renner–Teller hamiltonian as

$$H = H_\sigma + \varepsilon V, \qquad \begin{cases} V = \tfrac{1}{2}\lambda\rho^2\sigma_z = \tfrac{1}{2}\omega q^2\sigma_z \\[4pt] \varepsilon = \dfrac{\hat{\lambda}}{\lambda} = \dfrac{\lambda_+ - \lambda_-}{\lambda_+ + \lambda_-} \end{cases} \qquad (6.1)$$

$[H' = \varepsilon V$ in (5.1)]. The *Renner–Teller parameter* ε is seen to be in the interval $0 \leqslant |\varepsilon| < 1$, the limits $\varepsilon = \pm1$ corresponding to $\lambda_{\mp} = 0$ and the value $\varepsilon = 0$ corresponding to $\lambda_+ = \lambda_-$. It is clear from (3.12) that when $J_z = K = 0$, and $|\varepsilon|$ increases from zero, then the level $E_0 = \omega\tilde{v} \equiv \omega(v+1)$, spanned by $|v0+\rangle$ and $|v0-\rangle$ [see (5.12)], splits into two with energies and wave functions given by

$$E_s = \omega_s\tilde{v}, \qquad \omega_s \equiv \omega(1+s\varepsilon)^{1/2} \quad \text{and} \quad \tilde{v} \equiv v+1$$

$$|\Psi_+\rangle = 2^{-1/2}(|v0+\rangle^+ + |v0-\rangle^+) = (2\pi)^{-1/2}R^+_{v1}(\rho)\begin{Bmatrix}1\\0\end{Bmatrix} \qquad (6.2)$$

$$|\Psi_-\rangle = 2^{-1/2}(|v0+\rangle^- - |v0-\rangle^-) = (2\pi)^{-1/2}R^-_{v1}(\rho)\begin{Bmatrix}0\\1\end{Bmatrix}$$

where $|vK\pm\rangle^s$ is defined as $|vK\pm\rangle$ in (5.10) with

$$R_{vl}^s(\rho)\equiv(1+s\varepsilon)^{1/8}R_{vl}\big((1+s\varepsilon)^{1/4}\rho\big)\tag{6.3}$$

instead of $R_{vl}(\rho)$ [see also (5.5)].

In the general case we introduce

$$\tilde{\varepsilon}=\frac{\varepsilon}{4}\quad\text{and}\quad\Delta=\left\{1-\left(\frac{K}{\tilde{v}}\right)^2\right\}^{1/2}\tag{6.4}$$

and find, with $a_{n'n}$ from (5.16) (and the time-reversal operator $\mathfrak{T}$ described in Section V):

Unique level ($K=\pm\tilde{v}$):

$$E(-\varepsilon)=E(+\varepsilon),\qquad|\Psi(-\varepsilon)\rangle=\pm\sigma_y|\Psi(+\varepsilon)\rangle$$
$$|\Psi\rangle_{K=-\tilde{v}}=S_\sigma|\Psi\rangle_{K=+\tilde{v}}\ (=\mathfrak{T}\,|\Psi\rangle_{K=+\tilde{v}})\tag{6.5}$$

$$E=\omega\tilde{v}\{1-2\tilde{\varepsilon}^2(\tilde{v}+1)+2\tilde{\varepsilon}^4(\tilde{v}+1)\{\tilde{v}^2-3\tilde{v}-5\}+0(\tilde{\varepsilon}^6)\}\tag{6.6}$$

$$|\Psi\rangle=|vK\pm\rangle-\tilde{\varepsilon}|v+2,K,\mp\rangle\{\tilde{v}(\tilde{v}+1)\}^{1/2}$$
$$+\tfrac{1}{2}\tilde{\varepsilon}^2\{|v+4,K,\pm\rangle\{2\tilde{v}(\tilde{v}+1)\}^{1/2}+|v+2,K,\pm\rangle4\tilde{v}^{1/2}(\tilde{v}+1)$$
$$-|vK\pm\rangle\tilde{v}(\tilde{v}+1)\}+\cdots\tag{6.7}$$

Nonunique level ($|K|=\tilde{v}-2,\tilde{v}-4,\ldots;\ s=\pm1$):

$$E_s(-\varepsilon)=E_{-s}(+\varepsilon),\qquad|\Psi_s(-\varepsilon)\rangle=\sigma_y|\Psi_{-s}(+\varepsilon)\rangle$$
$$|\Psi_s\rangle_{-K}=sS_\sigma|\Psi_s\rangle_{+K}\ (=s\mathfrak{T}\,|\Psi_s\rangle_{+K})\tag{6.8}$$

$$E_s=\omega\tilde{v}\left\{1+s\tilde{\varepsilon}2\Delta-\tilde{\varepsilon}^22+s\tilde{\varepsilon}^3\Delta\left\{2(2-K^2)+\left(\frac{K}{\Delta}\right)^2\right\}\right.$$
$$\left.-\tilde{\varepsilon}^42(2K^2+5)+0(\tilde{\varepsilon}^5)\right\}\tag{6.9}$$

$$|\Psi_s\rangle=2^{-1/2}\left\{\left(1-s\tilde{\varepsilon}\frac{K}{2\Delta}\right)|vK+\rangle+\left(s+\tilde{\varepsilon}\frac{K}{2\Delta}\right)|vK-\rangle\right\}$$
$$+\tilde{\varepsilon}2^{1/2}\{|v-2,K,-\rangle a_{v-2,v}-|v+2,K,-\rangle a_{v+2,v}$$
$$+s|v-2,K,+\rangle a_{v,v-2}-s|v+2,K,+\rangle a_{v,v+2}\}+\cdots\tag{6.10}$$

To verify that $E_s(K=0)$ agrees with (6.2), consider $(1+s\varepsilon)^{1/2}=1+s\varepsilon2^{-1}-\varepsilon^22^{-3}+s\varepsilon^32^{-4}-\varepsilon^45\times2^{-7}+s\varepsilon^57\times2^{-8}+\cdots$ To check that the choice of phase factor of $|\Psi_s\rangle_{K=0}$ agrees with (6.2), note that the two expressions are identical at $\varepsilon = 0$.

The proof of the results (6.5)–(6.10) goes as follows. With the choice of phase factor made in (6.7) and (6.10), the results of (6.5) and (6.8) follow directly from (3.9) and (3.16). The expansions $E = E_0 + \varepsilon E_1 + \varepsilon^2 E_2 + \cdots$ and $|\Psi\rangle = |\Psi_0\rangle + \varepsilon|\Psi_1\rangle + \cdots$ in (6.6) and (6.7) are obtained from

$$|\Psi_1\rangle = \frac{Q_0}{a} V |\Psi_0\rangle$$

$$|\Psi_2\rangle = \frac{Q_0}{a} (V - E_1)|\Psi_1\rangle - \tfrac{1}{2}\langle\Psi_1|\Psi_1\rangle|\Psi_0\rangle \equiv |\Psi_2\rangle_\perp + |\Psi_2\rangle_0$$

$$E_1 = \langle\Psi_0|V|\Psi_0\rangle, \qquad E_2 = \langle\Psi_0|V|\Psi_1\rangle$$

$$E_3 = \langle\Psi_1|(V - E_1)|\Psi_1\rangle, \qquad E_4 = \langle\Psi_1|(V - E_1)|\Psi_2\rangle_\perp - \langle\Psi_1|\Psi_1\rangle E_2, \ldots$$

$$(6.11)$$

For a continuation of these recursion relations, see Messiah (Reference 27, p. 688). (Note that he normalizes $|\Psi\rangle$ by $\langle\Psi_0|\Psi\rangle = 1$, while we have $\langle\Psi|\Psi\rangle = 1$. Hence we do not have $|\Psi_2\rangle_0 \equiv |\Psi_0\rangle\langle\Psi_0|\Psi_2\rangle = 0$ as Messiah does.) The operator Q_0/a performs the sum over intermediate states. Writing E_s in (6.9) as ωe_s, we see that e_+ and e_- are the eigenvalues of the real, symmetric, and dimensionless 2×2 matrix[30-32]

$$e_{s's} = \frac{\langle vKs'|\varepsilon \mathcal{Q}_1 + \varepsilon^2 \mathcal{Q}_2 + \{\varepsilon^3 \mathcal{Q}_3 + \varepsilon^4 \mathcal{Q}_4\}_H + \cdots|vKs\rangle}{\omega} \qquad (6.12)$$

where $\{F\}_H \equiv \tfrac{1}{2}\{F + F^\dagger\}$ for arbitrary operator F and $\mathcal{Q} = \varepsilon\mathcal{Q}_1 + \varepsilon^2\mathcal{Q}_2 + \cdots$ is Bloch's effective hamiltonian,[33] with

$$\mathcal{Q}_1 = P_0 V P_0,$$

$$\mathcal{Q}_2 = P_0 V \frac{Q_0}{a} V P_0$$

$$\mathcal{Q}_3 = P_0 \left\{ V \frac{Q_0}{a} V \frac{Q_0}{a} V - V \frac{Q_0}{a^2} V P_0 V \right\} P_0$$

$$\mathcal{Q}_4 = P_0 \left\{ V \frac{Q_0}{a} V \frac{Q_0}{a} V \frac{Q_0}{a} V - V \frac{Q_0}{a} V \frac{Q_0}{a^2} V P_0 V \right.$$

$$- V \frac{Q_0}{a^2} V P_0 V \frac{Q_0}{a} V - V \frac{Q_0}{a^2} V \frac{Q_0}{a} V P_0 V$$

$$\left. + V \frac{Q_0}{a^3} V P_0 V P_0 V \right\} P_0,$$

$$\vdots$$

$$(6.13)$$

The formula for the eigenvalues is

$$e_s = c_+ + s[d^2 + c_-^2]^{1/2}$$

where $c_\pm = \tfrac{1}{2}(e_{++} \pm e_{--})$ and $d = e_{+-} = e_{-+}$ (6.14)

Since V and thus $\mathcal{Q}$ are invariant under S_σ from (5.11d), we deduce that $e_{--}(K) = e_{++}(-K)$ and $e_{-+}(K) = e_{+-}(-K)$. Hence,

$$c_\pm = \{c\}_K^\pm \quad (c \equiv e_{++}) \quad \text{and} \quad d(K) = d(-K) \qquad (6.15)$$

where $\{f\}_K^\pm$ means $\tfrac{1}{2}\{f(K) \pm f(-K)\}$ for arbitrary functions f of K. The matrix elements $\langle n'Ks'|V|nKs \rangle$ are given in (5.16). Since they vanish when $s' = s$, the expansions of $c_\pm$, d, and e_s simplify to

$$c_\pm = \varepsilon^2 c_{2\pm} + \varepsilon^4 c_{4\pm} + \cdots, \qquad d = \varepsilon d_1 + \varepsilon^3 d_3 + \cdots$$

$$e_s = s\varepsilon d_1 + \varepsilon^2 c_{2+} + s\varepsilon^3 \left\{ d_3 + \frac{1}{2} \frac{c_{2-}^2}{d_1} \right\} + \varepsilon^4 c_{4+} + \cdots \qquad (6.16)$$

Through first order in ε one has

$$|\Psi_s\rangle = \left(P_0 + \varepsilon \frac{Q_0}{a} V P_0 \right) (\alpha_{s+}|vK+\rangle + \alpha_{s-}|vK-\rangle) \qquad (6.17)$$

where

$$\left\{ \begin{matrix} \alpha_{s+} \\ \alpha_{s-} \end{matrix} \right\} = 2^{-1/2} \left(\left\{ \begin{matrix} 1 \\ s \end{matrix} \right\} - \tfrac{1}{2}\varepsilon\mu \left\{ \begin{matrix} s \\ -1 \end{matrix} \right\} \right), \qquad \mu \equiv \frac{K}{8\Delta} \qquad (6.18)$$

is the normalized eigenvector of

$$\left\{ \begin{matrix} e_{++} & e_{+-} \\ e_{-+} & e_{--} \end{matrix} \right\} = \varepsilon a_{vv} \left\{ \begin{matrix} 0 & 1 \\ 1 & 0 \end{matrix} \right\}$$

$$- \tfrac{1}{2}\varepsilon^2 \left\{ \begin{matrix} a_{v+2,v}^2 - a_{v-2,v}^2 & 0 \\ 0 & a_{v,v+2}^2 - a_{v,v-2}^2 \end{matrix} \right\} + \cdots$$

$$= \varepsilon a_{vv} \left\{ \begin{matrix} 0 & 1 \\ 1 & 0 \end{matrix} \right\} - \varepsilon^2 \tfrac{1}{8}\tilde{v} \left\{ \begin{matrix} 1+K & 0 \\ 0 & 1-K \end{matrix} \right\} + \cdots \qquad (6.19)$$

To check this, note that the eigenvectors of $e_{ss'}$ in (6.19) are identical to the eigenvectors of

$$(\varepsilon a_{vv})^{-1} \left[\{e_{ss'}\} + \varepsilon^2 \tfrac{1}{8}\tilde{v} \left\{ \begin{matrix} 1 & 0 \\ 0 & 1 \end{matrix} \right\} \right] = \left\{ \begin{matrix} 0 & 1 \\ 1 & 0 \end{matrix} \right\} - \varepsilon\mu \left\{ \begin{matrix} 1 & 0 \\ 0 & -1 \end{matrix} \right\} + O(\varepsilon^2)$$

because they are affected neither by the constant outside the square bracket, nor by the unit matrix inside the bracket. Now (6.10) is verified, and from (6.19) we also immediately verify (6.9) through second order. The observation $\langle vKs'|\mathcal{Q}_n^\dagger|vKs\rangle = \langle vKs|\mathcal{Q}_n|vKs'\rangle$ shows that the bracket $\{\ \}_H$ around $\mathcal{Q}_3$ and $\mathcal{Q}_4$ in (6.12) can be avoided by writing

$$d_3 = \{\langle vK-|\mathcal{Q}_3|vK+\rangle\}_K^+, \qquad c_4 = \langle vK+|\mathcal{Q}_4|vK+\rangle \qquad (6.20)$$

To calculate these we first notice the relations

$$a_{v,v\pm2}a_{v\pm2,v} = \tfrac{1}{4}a_{vv}a_{v\pm2,v\pm2} \qquad (6.21a)$$

$$a_{v\pm p,v\pm p} = \tfrac{1}{2}\left[(\pm\tilde{v}+p)^2 - K^2\right]^{1/2} \qquad (6.21b)$$

$$a_{v\pm4,v\pm2} = \tfrac{1}{4}\left[(\pm\tilde{v}+4+K)(\pm\tilde{v}+2+K)\right]^{1/2} \qquad (6.21c)$$

and we introduce $\{f\}_{\tilde{v}}^{\pm} = \tfrac{1}{2}\{f(\tilde{v})\pm f(-\tilde{v})\}$ for an arbitrary function f of $\tilde{v}$. If f depends on $\tilde{v}$ as well as K, one has

$$\{f\}_{K\tilde{v}}^{st} \equiv \{\{f\}_K^s\}_{\tilde{v}}^t = \{\{f\}_{\tilde{v}}^t\}_K^s \equiv \{f\}_{K\tilde{v}}^{ts} \qquad (6.22)$$

Noting results such as $f(\tilde{v}) = a_{v+2,v} \Rightarrow f(-\tilde{v}) = a_{v-2,v}$, we find

$$d_3 = \tfrac{1}{2}\{a_{v,v+2}a_{v+2,v+2}a_{v+2,v} - a_{v,v+2}^2 a_{vv}\}_{K\tilde{v}}^{++}$$

$$= \tfrac{1}{8}a_{vv}\{a_{v+2,v+2}^2 - 4a_{v+2,v}^2\}_{K\tilde{v}}^{++} = \tfrac{1}{16}a_{vv}(2-K^2)$$

$$c_4 = 2^{-3}2\{X\}_{\tilde{v}}^{-}, \qquad X = X_1 + X_2 + X_3 + X_4 + X_5$$

where

$$X_1 = -\tfrac{1}{2}a_{v+2,v}a_{v+2,v+4}a_{v+2,v+4}a_{v+2,v}$$

$$\qquad - a_{v+2,v}a_{v+2,v+2}a_{v+2,v+2}a_{v+2,v}$$

$$X_2 = X_4 = a_{v+2,v}a_{v+2,v+2}a_{v,v+2}a_{vv}$$

$$X_3 = a_{v+2,v}a_{v+2,v}a_{v+2,v}a_{v+2,v} \qquad\qquad (6.23)$$

$$X_5 = -a_{v+2,v}a_{v+2,v}a_{vv}a_{vv}$$

We have omitted the term $-a_{v+2,v}a_{v+2,v}a_{v-2,v}a_{v-2,v}$ from X_3 because it is even in $\tilde{v}$ and thus does not contribute to $\{X\}_{\tilde{v}}^{-}$. After use of (6.21a) in $X_2 = X_4$, we have

$$X_1 + X_3 = -\tfrac{1}{2}a_{v+2,v}^2\{a_{v+2,v+4}^2 + 2a_{v+2,v+2}^2 - 2a_{v+2,v}^2\}$$

$$X_2 + X_4 + X_5 = \tfrac{1}{2}a_{vv}^2\{a_{v+2,v+2}^2 - 2a_{v+2,v}^2\} \qquad (6.24)$$

that is,

$$X_1 + X_3 = -2^{-9}\{\tilde{v}^2 + (2K+2)\tilde{v} + (K^2 + 2K)\}$$
$$\times \{7\tilde{v}^2 + (-6K+34)\tilde{v} + (-9K^2 - 10K + 40)\}$$
$$X_2 + X_4 + X_5 = 2^{-7}\{\tilde{v}^2 - K^2\} \tag{6.25}$$
$$\times \{2\tilde{v}^2 + (-4K+12)\tilde{v} + (-6K^2 - 4K + 16)\}$$
$$c_4 = -2^{-8}\left[2\tilde{v}(2K^2 + 5) + K\tilde{v}(3\tilde{v}^2 - 5K^2 + 16)\right]$$

The result (6.9) is now verified. Note that E in (6.6) can also be found as

$$E = \omega\{\tilde{v} + e_{++}\}_{K=\tilde{v}} = \omega\{\tilde{v} + \varepsilon^2 c_2 + \varepsilon^4 c_4 + \cdots\}_{K=\tilde{v}} \tag{6.26}$$

A Numerical Example. In a recent paper,[19] Gauyacq and Jungen have calculated the exact values of the lowest E and E_s when $\varepsilon = -0.188$ and $\omega = 513$ cm^{-1}. The rows in Table I show how the expansions from (6.6) and (6.9) converge towards these exact values. All energies are in cm^{-1}. In comparison with the very good convergence in the other cases, that of $E_-(v=3, K=2)$ is surprisingly poor.

TABLE I
Convergence for E and $E_s{}^a$

v	K	s	0th order	1st order	2nd order	3rd order	4th order	Exact
0	1		513		508.47		508.40	508.40
1	2		1026		1012.40		1012.19	1012.19
	0	+	1026	929.56	925.02	924.60	924.55	924.54
	0	−	1026	1122.44	1117.91	1118.34	1118.29	1118.29
2	3		1539		1511.80		1511.50	1511.50
	1	+	1539	1402.61	1395.81	1395.34	1395.23	1395.23
	1	−	1539	1675.39	1668.59	1669.06	1668.96	1668.95
3	4		2052		2006.67		2006.57	2006.57
	2	+	2052	1884.95	1875.89	1875.64	1875.38	1875.42
	2	−	2052	2219.05	2209.98	2210.23	2209.97	2209.85
	0	+	2052	1859.11	1850.05	1849.19	1849.09	1849.08
	0	−	2052	2244.89	2235.82	2236.67	2236.57	2236.59

$^a \varepsilon = -0.188$ and $\omega = 513$ cm^{-1}.

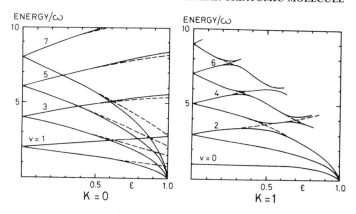

Fig. 2.   The bending-vibration energy levels of Π-electronic states as functions of ε for $K = 0$ and $K = 1$. The broken lines give the results of second-order perturbation theory [Eqs. (6.6) and (6.9)].

Larger Values of ε.   For given $K > 0$ and $\varepsilon > 0$, the lowest level of $H_\sigma + \varepsilon V$ in (6.1) is the unique level $E(\tilde{v} = K)$, and above that come the nonunique $E_s(\tilde{v}, K)$, $\tilde{v} = K + 2, K + 4, \ldots$, in the order

$$E(K) \equiv E_0 < E_-(K+2, K) \equiv E_1 < E_+(K+2, K) \equiv E_2$$

$$< E_-(K+4, K) \equiv E_3 < E_+(K+4, K) \equiv E_4 < \cdots \quad (6.27)$$

For small ε, this order follows directly from the explicit results (6.6) and (6.9), and as ε increases, the well-known noncrossing rule[34] prevents two different curves $E_i(\varepsilon)$ and $E_j(\varepsilon)$ from crossing (because they both correspond to the same eigenvalue K of J_z). When $K = 0$, $H_\sigma + \varepsilon V$ commutes also with σ_z, and an E_s curve is only prevented from crossing an $E_{s'}$ curve if $s = s'$. The lowest curves for $K = 0$ and $K = 1$ are sketched in Fig. 2.

While the curves for $K = 0$ are trivial to construct, those for $K = 1$ require more cumbersome numerical calculations. Figure 2 shows the results found by Renner in the original paper by the rather poor methods available at that time. In Herzberg's book[35] the partial curves are continued all the way across the interval $0 \leqslant \varepsilon \leqslant 1$ and the similar curves for $K = 2$ are drawn. Unfortunately they do not all end up at $E = 0$ for $\varepsilon = 1$ as intuition tells us they must do, and as Renner found in his formula for $E_i(\varepsilon)$ in the limit $\varepsilon \to 1$. Possibly too small a basis set was used in the computer diagonalization. Furthermore, the broken lines in Herzberg's figures are supposed to have the same meaning as in Fig. 2, but they all end up at $E = 0$, which they should not do.

VII. SPIN–ORBIT INTERACTION IN ELECTRONIC Π STATES

In this section we discuss how the spin–orbit interaction affects the levels found in Section VI. First we derive some general formulas.

On adding an extra perturbation h, assumed to commute with J_z, to $H_\sigma + \varepsilon V$ in (6.1), the energies E and E_s from (6.6) and (6.9) change to, say, $E^h = E + \Delta E$ and $E_s^h = E_s + \Delta E_s$ with corresponding wave functions $|\Psi^h\rangle$ and $|\Psi_s^h\rangle$. We can obtain these quantities by repeating the perturbation calculation from Section VI with $\varepsilon V + h$ instead of just εV. Thus we can determine E^h and $|\Psi^h\rangle$ by (6.11), and we can determine E_+^h and E_-^h as the eigenvalues of the 2×2 matrix

$$
\begin{aligned}
F_{s's} &= \langle vKs'| \left\{ (\varepsilon V + h) + (\varepsilon V + h)\frac{Q_0}{a}(\varepsilon V + h) + \cdots \right\} |vKs\rangle \\
&= \omega(e_{s's} + \Delta e_{s's})
\end{aligned}
\tag{7.1}
$$

$$
\text{where} \quad \Delta e_{s's} = \frac{1}{\omega}\langle vKs'| \left\{ h + \varepsilon\left(V\frac{Q_0}{a}h + h\frac{Q_0}{a}V \right) + \cdots \right\} |vKs\rangle
$$

and where $e_{s's}$ is the 2×2 matrix from (6.19). One finds

$$
\begin{aligned}
E_s^h &= \omega(\tilde v + c_+ + \Delta c_+ + se) \\
e &= s_\varepsilon \left\{ (c_- + \Delta c_-)^2 + |d + \Delta d|^2 \right\}^{1/2}
\end{aligned}
\tag{7.2}
$$

where s_ε is the sign of ε, $\Delta c_\pm = (\Delta e_{++} \pm \Delta e_{--})/2$, and $\Delta d = \Delta e_{+-}$ (the factor s_ε ensures that $E_s^h \to E_s$ in the limit $h \to 0$). The quantities $c_\pm = (e_{++} \pm e_{--})/2$ and $d = e_{+-} = e_{-+}$ are the same as in Section VI, i.e., to fourth order in $\tilde\varepsilon = \varepsilon/4$,

$$
\begin{aligned}
c_+ &= -2\tilde\varepsilon^2\tilde v\{1 + \tilde\varepsilon^2(2K^2 + 5)\} \\
c_- &= -2\tilde\varepsilon^2\tilde vK\{1 + \tfrac{1}{2}\tilde\varepsilon^2(3\tilde v^2 - 5K^2 + 16)\} \\
d &= 2\tilde\varepsilon\{\tilde v^2 - K^2\}^{1/2}\{1 + \tfrac{1}{4}\tilde\varepsilon^2(2 - K^2)\}
\end{aligned}
\tag{7.3}
$$

When the matrix elements $F_{s's}$ are real, F_{+-} equals F_{-+} and we can introduce an angle β such that

$$
\frac{F_{s's}}{\omega} = (c_+ + \Delta c_+)\begin{Bmatrix} 1 & 0 \\ 0 & 1 \end{Bmatrix} + e\cos 2\beta \begin{Bmatrix} 1 & 0 \\ 0 & -1 \end{Bmatrix} + e\sin 2\beta \begin{Bmatrix} 0 & 1 \\ 1 & 0 \end{Bmatrix}
$$

$$
(\cos 2\beta, \sin 2\beta) = \frac{(c_- + \Delta c_-, d + \Delta d)}{e}
\tag{7.4}
$$

The normalized eigenvector of $F_{s's}$, corresponding to the eigenvalue E_s^h, can then be written as

$$\begin{Bmatrix} \alpha_{s+} \\ \alpha_{s-} \end{Bmatrix} = \begin{Bmatrix} \cos\beta_s \\ \sin\beta_s \end{Bmatrix} \tag{7.5}$$

where the angle β_s is an arbitrary solution to

$$\cos 2\beta_s = s\cos 2\beta \quad \text{and} \quad \sin 2\beta_s = s\sin 2\beta \tag{7.6}$$

Choosing $0 \leqslant \beta < \pi$, $\beta_+ = \beta$, and $\beta_- = \beta - \pi/2$, one finds

$$|\Psi_s^h\rangle = \left\{ P_0 + \frac{Q_0}{a}(\varepsilon V + h)P_0 + \cdots \right\} \{\cos\beta_s |vK+\rangle + \sin\beta_s |vK-\rangle\} \tag{7.7}$$

and $|\Psi_s^h\rangle \to |\Psi_s\rangle$ as $h \to 0$. In the limit $h \to 0$, one has to first order in ε,

$$(\cos 2\beta, \sin 2\beta) = \left(-\varepsilon\frac{K}{4\Delta}, 1 \right) \tag{7.8}$$

(note that one must know c_- to *second* order in ε to establish this result), and hence $\beta \simeq \pi/4$ and

$$(\cos\beta_s, \sin\beta_s) = 2^{-1/2}\left(1 - s\varepsilon\frac{K}{8\Delta}, s + \varepsilon\frac{K}{8\Delta} \right) \tag{7.9}$$

in accordance with (6.17) [and thereby (6.10)].

In Eqs. (7.1) and (7.7), "$\cdots$" stands for terms of higher order in the perturbation. Most approaches to degenerate perturbation theory give the same expression for $F_{s's}$ to third order, but if fourth and higher orders are included, matrices $F_{s's}$ might result which differ by a unitary transformation, that is, by the value of the angle β in (7.4). Hence, to define β uniquely we must specify which perturbation method we use. The method used in (6.13) (called Soliverez's method in Reference 31) has become known as the "canonical method."[32]

Spin–Orbit Interaction. Most of the molecules subject to the Renner–Teller effect that have been studied so far appear in electronic spin doublet states and are therefore also subject to spin–orbit interaction. According to Pople[15] this can be taken into account by adding $h = H_{SO}$:

$$H_{SO} = \omega\xi\sigma_y, \quad \xi = \frac{AS_z'\Lambda}{\omega} \tag{7.10}$$

where the spin–orbit interaction constant A depends on the electronic Λ state in question and where S'_z is the eigenvalue (in units of $\hbar$) of the total electron spin along the z axis.

Pople just postulated that the effective spin–orbit operator should be $H_{SO} = AL_zS_z$, where L_z is the effective operator for the electronic angular momentum (i.e., $L_z = \Lambda\sigma_y$ in our approach). In Appendix E we derive this result from more fundamental premises. In Appendix F we discuss the spin-dependent wave function, which we avoid in the present section by replacing S_z with its eigenvalue S'_z.

When H_{SO} is included in the Renner–Teller hamiltonian, the eigenvalue K of J_z is still a good quantum number and it can be shown (by use of the symmetry operation R_σ described in Appendix F) that the energy levels must be invariant under simultaneous change of sign of K and ξ (i.e., K and S'_z). Levels with $|K| = 0, 1, 2, \ldots$ are called vibronic $\Sigma, \Pi, \Delta, \ldots$ levels and are denoted by $\Sigma_P, \Pi_P, \Delta_P, \ldots$, where the index P is the value of $|K + S'_z|$. Figure 3 shows Pople's sketch of how the spin–orbit interaction term affects the energy levels. We now use the theory developed above to extend his second-order formulas for E^{SO} and E_s^{SO} to order four (in $\varepsilon V + H_{SO}$) and to give the corresponding wavefunctions $|\Psi^{SO}\rangle$ and $|\Psi_s^{SO}\rangle$ to orders two and one, respectively. Pople did not derive the wave functions, and his calculations of the energies are different from those given here.

The perturbation calculations are greatly simplified by the fact that σ_y commutes with P_0 and Q_0/a (because σ_y commutes with H_a) and satisfies $\sigma_y V = -V\sigma_y$. Thus (6.11) gives very easily for the unique levels

$$E^{SO} = E + s_K \omega \xi \left\{ 1 - 2\tilde{\varepsilon}^2 \tilde{v}(\tilde{v} + 1)\{1 + s_K \xi\} \right\}$$

$$|\psi^{SO}\rangle = |\Psi\rangle - s_K \tilde{\varepsilon}\xi |v + 2, K, -s_K\rangle \{\tilde{v}(\tilde{v} + 1)\}^{1/2} \tag{7.11}$$

where E and $|\Psi\rangle$ are the same as in (6.6) and (6.7), and s_K is the sign of K ($= \pm \tilde{v}$). For a nonunique level one has (see below for proof) to fourth order

$$\Delta c_+ = -2\tilde{\varepsilon}^2 \xi \tilde{v} K - 2\tilde{\varepsilon}^2 \xi^2 \tilde{v}$$

$$\Delta c_- = \xi - 2\tilde{\varepsilon}^2 \xi \tilde{v} - 2\tilde{\varepsilon}^2 \xi^2 K \tilde{v} \tag{7.12}$$

$$\Delta d = -2\tilde{\varepsilon}^3 \xi \tilde{v} K \{\tilde{v}^2 - K^2\}^{1/2}$$

Now E^{SO}, correct to fourth order in $\varepsilon V + H_{SO}$, follows directly by insertion in (7.2). To second order (where $\Delta c_+ = \Delta d = 0$ and $\Delta c_- = \xi$) it is

$$E_s^{SO} = \omega\left(\tilde{v} - \tfrac{1}{8}\varepsilon^2 \tilde{v} + se\right)$$

$$e = s_\varepsilon \left\{\left(\xi - \tfrac{1}{8}\varepsilon^2 \tilde{v} K\right)^2 + \tfrac{1}{4}\varepsilon^2(\tilde{v}^2 - K^2)\right\}^{1/2} \tag{7.13}$$

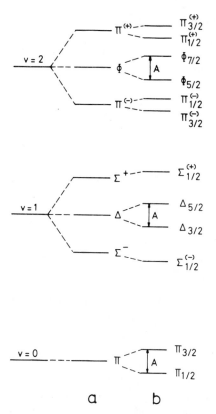

Fig. 3. Level splittings caused by (a) Renner–Teller effect alone, (b) Renner-Teller effect and spin-orbit coupling. Both ε and A are assumed positive. The index (s) shows that the energy and the wave function are given by (a) E_s and $|\psi_s\rangle$, (b) E_s^{SO} and $|\psi_s^{SO}\rangle$. In case (a), the parenthesis about s on Σ^s is omitted because in this special case s also means the parity of the vibronic wave function under a reflection in the plane defined by the molecule as it bends [see Eq. (3.14) and subsequent text]. All the levels under (b) are doubly degenerate, corresponding to the two values $\pm \frac{1}{2}$ of S_z'. We have indicated that the spin–orbit interaction splits the unique levels into two, separated by A. The independence of the quantum number v is only a first approximation; in the next approximation the separation is given by Reference 22, p. 520:

$$A^{eff} = A\left\{1 - \tfrac{1}{8}\varepsilon^2\tilde{v}(\tilde{v}+1)\right\}, \qquad \tilde{v} = v + 1$$

sometimes called the effective spin–orbit constant. Equation (7.11) shows that this expression remains unchanged in the subsequent approximation.

where s_ε is the sign of ε. Inserting this e into

$$(\cos 2\beta, \sin 2\beta) = \left(\frac{\xi - \tfrac{1}{8}\varepsilon^2 \tilde{v} K, \tfrac{1}{2}\varepsilon \tilde{v}\Delta}{e} \right) \quad (7.14)$$

and defining $\beta_+ = \beta$ and $\beta_- = \beta - \pi/2$, we get

$$|\Psi_s^{SO}\rangle = |vK+\rangle \cos\beta_s + |vK-\rangle \sin\beta_s$$
$$+ \tfrac{1}{2}\varepsilon\{|v-2, K, -\rangle a_{v-2,v} - |v+2, K, -\rangle a_{v+2,v}\} \cos\beta_s \quad (7.15)$$
$$+ \tfrac{1}{2}\varepsilon\{|v-2, K, +\rangle a_{v,v-2} - |v+2, K, +\rangle a_{v,v+2}\} \sin\beta_s$$

correct to first order in $\varepsilon V + H_{SO}$.

For such typical values as $\omega \simeq 500$ cm^{-1}, $A \simeq 100$ cm^{-1}, and $\varepsilon \simeq 0.2$ one gets $\xi = AS_z'/\omega \simeq 0.1$, $\varepsilon^2/4 \simeq 0.01$, $\xi\varepsilon^2/4 \simeq 0.001$ and can approximate

$$|e| \simeq Z^{1/2} - \tfrac{1}{8}\varepsilon^2\xi\tilde{v}KZ^{-1/2}, \quad Z \equiv \xi^2 + \tfrac{1}{4}\varepsilon^2(\tilde{v}^2 - K^2) \quad (7.16)$$

Insertion of this into (7.13) gives Pople's formula. It is not much simpler than that of (7.13), which has the advantage of remaining valid when ξ is small.

The formula (7.16) is in accordance with Hougen (Reference 22, p. 520) who corrected Pople's paper for some errors (and extended his second order result for a unique level to third order in accordance with the fourth order result (7.11) above). Hougen [22, Eq. (24)] also introduced a positive quantity r and an angle β which are identical to $\omega|e|$ and β in (7.13) and (7.14) if we replace (S_z', e) with $(+\tfrac{1}{2}, |e|)$ and $\xi - \varepsilon^2\tilde{v}K/8$ with just ξ in these formulas. Because of the way we use the angle, we cannot omit the ε^2 term. If we did so, $|\Psi_s^{SO}\rangle$ would not depend correctly on ε^1 in the limit $h \to 0$ (the K/Δ terms would drop out of $|\Psi_s\rangle$ in (6.10)).

The results of (7.12) are proven as follows. Substituting $\varepsilon V \to \varepsilon V + H_{SO}$ in the general formulas (6.12) and (6.13), one finds for $\Delta e_{s's}$ in (7.1)

$$\Delta e_{s's} = \langle vKs'| \left\{ \xi\sigma_y - 2\varepsilon^2\xi V \frac{Q_0}{a^2} V\sigma_y \right.$$
$$\left. + 2\varepsilon^3\xi \left\{ V\frac{Q_0}{a^2} V\frac{Q_0}{a} V + VP_0 V\frac{Q_0}{a^3} V \right\}_A \sigma_y \right\} |vKs\rangle \quad (7.17)$$

correct to fourth order [$\{F\}_A$ means $(F - F^\dagger)/2$ for arbitrary operator F]. The values of the two terms which vary quadratically with ε are read off directly from the term varying quadratically in (6.19), and the matrix $2\varepsilon^3\xi m_{s's}$ in (7.17) is hermitian with $m_{ss} = 0$. Only the second term in the bracket $\{\ \}_A$

contributes to m_{+-} ($=m_{-+}$), and with $\sigma_y|vK+\rangle=|vK+\rangle$ we find

$$
\begin{aligned}
m_{+-} &= \left\{\left\langle vK - |VP_0V\frac{Q_0}{a^3}V|vK+\right\rangle\right\}_K^- \\
&= (-\tfrac{1}{2})^3\left\{a_{vv}\left(a^2_{v+2,v} - a^2_{v-2,v}\right)\right\}_K^- \\
&= -2^{-2}a_{vv}\left\{a^2_{v+2,v}\right\}_{K\tilde{v}}^{--} = -2^{-6}\tilde{v}K\{\tilde{v}^2 - K^2\}^{-1/2} \quad (7.18)
\end{aligned}
$$

Hence, with $\tilde{\varepsilon} = \varepsilon/4$,

$$
\begin{aligned}
\Delta e_{s's} = \xi\begin{Bmatrix}1 & 0 \\ 0 & -1\end{Bmatrix} &- 2\tilde{\varepsilon}^2\tilde{v}\begin{Bmatrix}K+1 & 0 \\ 0 & K-1\end{Bmatrix} \\
- 2\tilde{\varepsilon}^2\xi^2\tilde{v}\begin{Bmatrix}1+K & 0 \\ 0 & 1-K\end{Bmatrix} &- 2\tilde{\varepsilon}^3\xi\tilde{v}K\{\tilde{v}^2 - K^2\}^{1/2}\begin{Bmatrix}0 & 1 \\ 1 & 0\end{Bmatrix}
\end{aligned}
$$

$$(7.19)$$

which proves the results of (7.12).

VIII.   THE EFFECT OF ANHARMONICITIES IN ELECTRONIC Π STATES

In Section VII we discussed how an extra perturbation h causes the energies E and E_s to change to $E^h = E + \Delta E$ and $E_s^h = E_s + \Delta E_s$, and we gave formulas for E^h and E_s^h. When h is much smaller than the Renner–Teller interaction εV, the perturbed energies can also be determined approximately by the simple relations

$$\Delta E = \langle\Psi|h|\Psi\rangle \quad\text{and}\quad \Delta E_s = \langle\Psi_s|h|\Psi_s\rangle \quad (8.1)$$

where the wave functions $|\Psi\rangle$ and $|\Psi_s\rangle$ for the levels E and E_s are given in (6.7) and (6.10). As an example we consider the perturbation

$$h = g_4 q^4 + \hat{g}_4 q^4\sigma_z + g_K J_z\Lambda\sigma_y \quad (8.2)$$

($\Lambda = 1$) arising when allowing for the g_K correction from (4.12) and for quartic anharmonicities in the two potentials W_+ and W_-. Using (C.5) we find

$$
\begin{aligned}
\Delta E &= g_4\tilde{v}(\tilde{v}+1)\{1 + \tilde{\varepsilon}^2 4(3\tilde{v}+4)\} \\
&\quad - 2\tilde{\xi}\hat{g}_4\tilde{v}(\tilde{v}+1)(\tilde{v}+2) + g_K\tilde{v}\{1 - 2\tilde{\varepsilon}^2\tilde{v}(\tilde{v}+1)\} \\
\Delta E_s &= g_4\left\{\tfrac{1}{2}(3\tilde{v}^2 - K^2) - s\tilde{\varepsilon}\left(6\tilde{v}^2\Delta + \frac{K^2}{\Delta}\right)\right\} \\
&\quad + \tfrac{3}{2}\hat{g}_4\tilde{v}^2\{s\Delta - 4\tilde{\varepsilon}\} - \frac{s\tilde{\varepsilon}g_K K^2}{\Delta}
\end{aligned}
$$

$$(8.3)$$

valid to second and first order, respectively, in $\tilde{\varepsilon} = \varepsilon/4$. If $K = 0$ one has [see (C.6) and (C.7)] for all values of ε

$$\Delta E_s = \tfrac{3}{2}(g_4 + s\hat{g}_4)(1 + 4s\tilde{\varepsilon})^{-1}\tilde{v}^2. \tag{8.4}$$

The formula (8.3) for ΔE_s is not reliable for levels with $0 < |K| < \tilde{v}$ if ε is so small that εV is comparable with h. In practical terms, we can expect the formula to be valid whenever Renner's original second-order formula works well (because then we know that the effect of h is small and that the ε^2 terms in the expansion of $|\Psi_s\rangle$ are ignorable). By use of (7.2) we can obtain a formula which remains valid as $\varepsilon \to 0$. If we include in $\Delta e_{s's}$ the terms shown explicitly in (7.1) and exclude terms with ε^n ($n > 2$) from $c_\pm$ and d in (7.3), the resulting formula is almost identical to that obtained recently by Gauyacq and Jungen [Reference 19, Eq. (23)] (the small difference is due to a difference in approach and is beyond the order of magnitude considered). Compared with Eq. (8.3), their result is rather complicated, and we do not reproduce it here. Gauyacq and Jungen also performed some numerical diagonalizations of $H_\sigma + \varepsilon V + h$, and in Table II we compare the results of these with the predictions of Eq. (8.3) and $\Delta E_{(s)}^{GJ} \equiv E_{(s)}^{GJ}(v, K, g_K, g_4, \hat{g}_4) - E_{(s)}^{GJ}(v, K, 0, 0, 0)$, where E^{GJ} and E_s^{GJ} denote Gauyacq and Jungen's energy expression. (Actually the numerical results are not exactly the same as in Reference 19, where sextic and higher anharmonicities are also taken into account. We thank the authors for repeating the calculations for us without these higher anharmonicities, which turn out to affect the energy levels as strongly as the ε^2 terms in ΔE.) It is seen that the simpler expressions in (8.3) reproduce the anharmonicity corrections just as well as $E_{(s)}^{GJ}$. We note that Gauyacq and Jungen choose the values of g_K, g_4, and $\hat{g}_4$ (g_K, g_{22}, and g_{22}' in their notation) from consideration of a certain model of the molecule according to which these parameters are interrelated.

Consider next the problem of determining the effect of anharmonicities and the g_K correction on the levels E^{SO} and E_s^{SO} (which already include the effect of spin–orbit interaction). Under the assumption that the deviations from the second-order results in (7.11) and (7.13) are small, we find as above

$$E_s = g_4 \langle q^4 \rangle_s^{SO} + \hat{g}_4 \langle q^4\sigma_z \rangle_s^{SO} + g_K \langle J_z\sigma_y \rangle_s^{SO} \tag{8.5}$$

(omit the index s when $|K| = \tilde{v}$) where the $\langle\ \rangle$'s are given in (C.8) and (C.9).

The model of the molecule considered in this section is unrealistic in that it neglects the stretching vibrations. In Reference 23, the present authors have shown how to include these and have derived the complete versions of the formulas (8.3) for molecules without spin–orbit interaction. The formula (8.5) has also been generalized. To obtain these results, a method is developed by which one can apply not only first-order perturbation theory to the levels of $H_\sigma + \varepsilon V + H_{SO}$ as above, but also second-order theory (for the cubic anharmonicities which couple stretching and bending).

TABLE II
Effect of Anharmonicities and the g_K Term in a Molecule[a]

g_K:		0.423			0.714			1.630			3.220		
$g_4 = \hat{g}_4$:		0.063			0.134			0.502			1.666		
Method:		a	b	c	a	b	c	a	b	c	a	b	c
v	K	s											
0	1		0.59	0.59	0.58	1.07	1.07	1.06	2.97	2.96	2.92	7.67	7.55	7.50
1	0	+	0.90	0.93		1.91	1.98		7.16	7.30		23.75	23.35	
	0	−	0.00	0.00		0.00	0.00		0.00	0.00		0.00	0.00	
	2		1.38	1.38	1.37	2.57	2.58	2.54	7.59	7.58	7.41	20.91	20.27	20.23
2	1	+	1.97	2.06	1.96	4.15	4.28	4.15	15.51	15.69	15.51	51.32	49.40	51.37
	1	−	0.00	0.00	0.00	0.01	−0.02	0.01	0.10	0.16	0.10	0.45	0.67	0.53
	3		2.40	2.42	2.38	4.58	4.62	4.51	14.19	14.13	13.76	40.83	38.84	39.12
3	0	+	3.60	3.72		7.64	7.82		28.65	28.82		94.99	89.91	
	0	−	0.00	0.00		0.00	0.00		0.00	0.00		0.00	0.00	
	2	+	3.34	3.44	3.35	7.05	7.24	7.08	26.19	26.27	26.27	86.36	81.33	86.49
	2	−	0.01	0.03	−0.03	0.06	0.12	0.03	0.45	0.74	0.37	1.97	2.92	2.06
	4		3.69	3.73	3.67	7.17	7.24	7.05	23.08	22.83	22.24	68.56	63.68	64.99

[a] $\omega = 513$ cm^{-1} and $\varepsilon = -0.188$, i.e., the effect on the levels in Table I. All numbers are in units of cm^{-1}. The increments ΔE (and ΔE_s) are worked out by three different methods, a, b and c: ΔE_a is the result of the formulas (8.3), ΔE_b is the result of an exact numerical diagonalization (see text) and ΔE_c is the result from the formula of Gauyacq and Jungen (see text). For $K = 0$ we have left the space for ΔE_c blank because the formula (and hence the number) for ΔE_c is identical to that for ΔE_a. For $K = \tilde{v} = v + 1$, ΔE_c is identical to ΔE_a, except that terms of second order in ε are not included in ΔE_c.

IX. ELECTRONIC STATES WITH $\Lambda > 1$

Consider again the Renner–Teller hamiltonian H from (3.8). When $\Lambda > 1$, (3.3) shows that $\hat{W}$ has no term quadratic in ρ. Hence, if we neglect sextic and higher anharmonicities in the potentials, include the g_K correction term from (4.12) and the spin–orbit interaction operator H_{SO} from (7.10), we can write

$$H = H_\sigma + H'$$
$$H' = g_4 q^4 + \hat{g}_4 q^4 \sigma_z + \{g_K J_z + A S_z'\} \Lambda \sigma_y \tag{9.1}$$

where H_σ and the dimensionless coordinate q for ρ are as in Section V, A is the spin–orbit interaction constant, and S_z' is the value (in units of $\hbar$) of the total electron spin along the molecular axis. When $\Lambda > 2$, $\hat{g}_4$ vanishes. We introduce the eigenvectors $|vKs\rangle$ of H_σ as in Section V, and, in terms of

$$w \equiv \langle vK - |\hat{g}_4 q^4 \sigma_z| vK + \rangle = 0 \quad \text{for} \quad \Lambda > 2$$
$$= \tfrac{3}{2} \hat{g}_4 \{ (\tilde{v}+1)^2 - K^2 \}^{1/2} \{ (\tilde{v}-1)^2 - K^2 \}^{1/2} \quad \text{for} \quad \Lambda = 2 \quad (9.2)$$

($\tilde{v} = v + 1$), first-order perturbation theory gives (see below) the following energy levels:

Unique level ($|K| \geq v$; $s_K = $ sign of K):

$$E = \omega\tilde{v} + \tfrac{3}{2} g_4 \tilde{v}^2 + g_{22} K^2 + \Lambda \{ (g_K - 2g_{22})|K| + s_K A S_z' \} \quad (9.3)$$

Nonunique level ($|K| < v$; $s = \pm 1$):

$$E_s = \omega\tilde{v} + \tfrac{3}{2} g_4 \tilde{v}^2 + g_{22} K^2 + s \{ w^2 + r^2 \}^{1/2}$$
$$r = \Lambda \{ (g_K - 2g_{22})K + A S_z' \} \quad (9.4)$$

where $g_{22} = -g_4/2$ [and a common constant, $g_4(1 - \Lambda^2)/2$, has been left out from all levels]. The leading corrections to these expressions can be ignored as long as the sextic anharmonicities are also, even if A is as large as ω. If $|A S_z'| \gg |w|$ (which is always the case when $\Lambda > 2$ because then $w = 0$), E_s simplifies to

$$E_s = \omega\tilde{v} + \tfrac{3}{2} g_4 \tilde{v}^2 + g_{22} K^2 + s|r| \quad (9.5)$$

To prove these results we first assume that A is as small as g_4 and $\hat{g}_4$. Then E equals $\langle vKs_K|H|vKs_K\rangle$, and E_+ and E_- are found as the eigenvalues of the real and symmetric 2×2 matrix $\langle vKs'|H|vKs\rangle$. The necessary matrix elements are all given in Section V, and the results of (9.3) and (9.4) follow immediately. As long as sextic anharmonicities are ignored, there is no reason to improve on these results by applying higher-order perturbation theory. The leading terms in such an improvement are of the order g_4^2/ω, i.e., of the same order as the sextic anharmonicities. The leading terms involving A are of the order $A g_4^2/\omega^2$, i.e., smaller than g_4^2/ω as long as $A \lesssim \omega$. For still larger values of $A S_z'$ we can consider $H_\sigma + A S_z' \Lambda \sigma_y$ as the zeroth-order hamiltonian and derive Eqs. (9.3) and (9.5) by applying first-order perturbation theory to the nondegenerate level $\omega\tilde{v} + s\Lambda|A S_z'|$.

When $\Lambda = 2$, Eq. (9.3) predicts the separation

$$\delta \equiv \{ E(K = v+2) - E(K = v) \}_{v=1} = 4g_K \quad (9.6)$$

Merer and Travis have found $\delta = 13.44 \text{ cm}^{-1}$ in an electronic $^2\Delta$ state of the

CCN radical.[20] This implies the value $g_K = 3.36$ cm^{-1}. However, at the time of Merer and Travis's paper, the g_K term had not yet been described and hence the two levels should not have been separated at all within the approximation of (9.3). Merer and Travis found an explanation by taking H' into account to second order, i.e., by adding the second-order term $\langle H', H' \rangle$ to E in (9.3) and correcting E_s in (9.4) similarly. For some reason they neglected the $g_4 q^4$ term. If we include it, then use of the terminology and the matrix elements from Section V gives

$$\langle H', H' \rangle = \frac{\hat{g}_4^2}{\omega} \left\{ -\tfrac{1}{4} \left(q^4 \sigma_z \right)_{v+4,v}^2 - \tfrac{1}{2} \left(q^4 \sigma_z \right)_{v+2,v}^2 \right\}$$

$$= -\frac{\hat{g}_4^2}{4\omega} \times \begin{cases} K(K+1)(K+2)(K+35) & \text{if} \quad v = K \\ (K-1)K(K+1)(K+2) & \text{if} \quad v = K-2 \end{cases}$$

$$(9.7)$$

and a similar correction to E_s, which we do not reproduce here. For the separation δ in particular, (9.7) gives

$$\delta = 24 \frac{\hat{g}_4^2}{\omega} \qquad (9.8)$$

This is a valid contribution to δ, but it is one order of magnitude smaller than the leading term $4g_K$ given in (9.6).

Merer and Travis's derivation of the second-order correction (9.7) is much lengthier than ours, and they obtain $4\hat{g}_4$ ($= 2\eta$ in their notation) where we have just $\hat{g}_4$. Thus their factor in the separation δ in (9.8) is $4^2 \times 24 = 384$, which is so large that the term could hardly be ignored had it been correct. However, the extra factor 4^2 arises because all the matrix elements in their Table III are missing a factor $\tfrac{1}{4}$. Jungen and Merer have argued (Reference 10, p. 140) that Merer and Travis's use of the parameter $\eta = 2\hat{g}_4$ is not "consistent": η is just an "effective" quantity. According to our analysis this is not so, but Merer and Travis leave out certain important interactions and include second-order interactions as small as the neglected sextic anharmonicities. Kroto[36] has estimated the separation δ for NCN in its $\tilde{a}\ ^1\Delta_g$ state as being negative, which should not be possible according to Eq. (9.8). The estimate should actually have been positive,[10] about 1.75 cm^{-1}, but as pointed out in Reference 23, it is not reliable, because it ignores comparable anharmonic effects in other levels. According to (9.6), the sign of δ is the same as the sign of g_K, which can be positive or negative.

In Reference 23, the present authors have shown that the effect of the stretching vibrations can be included in the formulas (9.3) and (9.4) by replacing $3g_4/2$ and g_{22} with x_{22} and g_{22} from the seven familiar anharmonicity parameters[2] $x_{11}, x_{12}, \ldots, g_{22}$ [relative to the potential $(W_+ + W_-)/2$] and adding $\omega_1 \tilde{v}_1 + \omega_3 \tilde{v}_3 + x_{11}\tilde{v}_1^2 + x_{13}\tilde{v}_1\tilde{v}_3 + x_{33}\tilde{v}_3^2 + x_{12}\tilde{v}_1\tilde{v}_2 + x_{32}\tilde{v}_3\tilde{v}_2$ ($\tilde{v}_k = v_k$ $+ \frac{1}{2}$, where v_k is the number of quanta in the kth stretching vibration and ω_k its frequency). We note that these modifications do not affect the result (9.6) for the separation δ (whatever the values of v_1 and v_3).

Hougen and Jesson have discussed molecules in special Π states ($\Lambda = 1$) in which the anharmonicities in $W_+ - W_-$ are ignorable and the Renner–Teller interaction εV (see Section VI) is as small as the anharmonicities in $W_+ + W_-$.[21] Apparently our discussion covers this case also if we simply replace w in (9.2) and (9.4) with $\langle vK - |\varepsilon V| vK + \rangle = \varepsilon\omega\{\tilde{v}^2 - K^2\}^{1/2}/2$ and put $\Lambda = 1$. Hougen and Jesson included the stretching vibrations, and their formulas are obtained by modifying Eqs. (9.3) and (9.4) as described above.[23]

X. IMPLICATIONS OF SYMMETRY

In this section we investigate some simple consequences of the invariance of the zeroth-order hamiltonian H_e^0 from (2.9) under the operators of the group $C_{\infty v}$, spanned by the reflection S and the rotations

$$U_\tau = \exp(-i\tau L_z) = 1 - i\tau L_z - \frac{1}{2}(\tau L_z)^2 + \cdots \qquad (10.1)$$

of the arbitrary angle τ. Repeated application of (2.13) gives

$$U_\tau \left\{ \begin{matrix} \phi_+^0 \\ \phi_-^0 \end{matrix} \right\} = \left\{ \begin{matrix} \cos\Lambda\tau & \sin\Lambda\tau \\ -\sin\Lambda\tau & \cos\Lambda\tau \end{matrix} \right\} \left\{ \begin{matrix} \phi_+^0 \\ \phi_-^0 \end{matrix} \right\} \qquad (10.2)$$

in particular,

$$U_{\pm\pi/2\Lambda}\phi_s^0 = \pm s\phi_{-s}^0. \qquad (10.3)$$

It can be shown (see Appendix A) that the terms in the expansion $C = C^0 + C'\rho + \cdots + C^{(p)}\rho^p + \cdots$ of the Coulomb potential can be written as

$$\begin{aligned} C^0 &= C_{0+}^0 \\ C' &= C_{1+}' \\ C'' &= C_{0+}'' + C_{2+}'' \\ C''' &= C_{1+}''' + C_{3+}''' \\ C'''' &= C_{0+}'''' + C_{2+}'''' + C_{4+}'''' \end{aligned} \qquad (10.4)$$

$$\vdots$$

where $C_{m+}^{(p)}$, when written as a function of coordinates $(|r|, \theta)$ defined as in

(2.16), is independent of θ when $m = 0$ and varies with θ as

$$C_{m+}^{(p)}(\theta) = C_{m+}^{(p)}(0)\cos m\theta + C_{m+}^{(p)}(\pi/m)\sin m\theta \tag{10.5}$$

when $m > 0$. One can also write $C_{m+}^{(p)}(\pi/m)$ as $C_{m-}^{(p)}(0)$, where

$$C_{m-}^{(p)} \equiv -U_{\pi/m}C_{m+}^{(p)}U_{\pi/m}^{-1}\left[= C_{m+}^{(p)}(\theta + \pi/m)\right] \tag{10.6}$$

The terms in the expansions (2.8) of wave functions and potentials can now be constructed by perturbation theory. For the first terms one finds

$$\phi_s' = \frac{Q_0}{a}C'|\phi_s^0\rangle$$

$$W_s'' = \langle\phi_s^0|\left\{C'\frac{Q_0}{a}C' + C''\right\}|\phi_s^0\rangle \tag{10.7}$$

where the operator Q_0/a performs the sum over intermediate states (Reference 27, p. 688) and is invariant under $C_{\infty v}$. The higher terms are considered in Appendix A; here we confine ourselves to a less systematic discussion of the lower terms. By use of (10.3) and (10.4) one finds immediately that $W_+'' = W_-''$ if $\Lambda \geq 2$. Hence, (3.3) is verified for $\Lambda = 1$ and 2. Straightforward use of the trigonometric addition formulas shows that ϕ_s' can be written as

$$\phi_s' = \phi_{\Lambda-1,s}' + \phi_{\Lambda+1,s}'$$

$$\text{where} \quad U_{\pm\pi/2m}\phi_{ms}' = \pm s\phi_{m,-s}' \quad \text{for} \quad m > 0 \tag{10.8}$$

(Hint: The operator Q_0/a is invariant under $C_{\infty v}$ and can therefore be left out of the argument.) It follows that if an operator A commutes with L_z, then

$$\langle\phi_{m+}'|A|\phi_{n+}'\rangle = \langle\phi_{m-}'|A|\phi_{n-}'\rangle\delta_{mn} \quad \text{for} \quad m > 0 \tag{10.9}$$

It is a trivial consequence of this result that $\hat{w}_0$ in (4.9) vanishes when $\Lambda = 1$. Hence, (4.10) is verified for $\Lambda = 1$. Another consequence is that $g_K = g^0$ from (4.12) and (4.7) can be written as

$$g^0 = \tfrac{1}{2}\hbar^2\{\|\phi_{0+}'\|^2 + \|\phi_{0-}'\|^2 - 2\|\phi_{2s}'\|^2\} \quad \text{for} \quad \Lambda = 1$$

$$g^0 = \hbar^2\Lambda^{-1}\{\|\phi_{\Lambda-1,s}'\|^2 - \|\phi_{\Lambda+1,s}'\|^2\} \quad \text{for} \quad \Lambda > 1 \tag{10.10}$$

where the right-hand sides are independent of s. One can consider $\|\phi_{ms}'\|^2$ as a measure of how strongly states with $|L_z| = m$ are mixed into the Λ state in question as the molecule starts to bend. Consider next W_s'' from (10.7), written as

$$W_s'' = \langle \phi_s' | C' | \phi_s^0 \rangle + \langle \phi_s^0 | C'' | \phi_s^0 \rangle \tag{10.11}$$

When $\Lambda = 1$ one finds

$$\langle \phi_+^0 | C_{0+}'' | \phi_+^0 \rangle = \langle \phi_-^0 | C_{0+}'' | \phi_-^0 \rangle$$

$$\langle \phi_+^0 | C_{2+}'' | \phi_+^0 \rangle = -\langle \phi_-^0 | C_{2+}'' | \phi_-^0 \rangle \tag{10.12}$$

$$\langle \phi_{2+}' | C' | \phi_+^0 \rangle = \langle \phi_{2-}' | C' | \phi_-^0 \rangle$$

(because C_{0+}'' is invariant under $U_{\pi/2}$, C_{2+}'' changes its sign, and $C'\phi_s^0$ can be written as $\psi_{0s} + \psi_{2s}$ where ψ_{ms} transforms as ϕ_{ms}' under $C_{\infty v}$). It follows that the force constant $\lambda = (W_+'' + W_-'')/4$ can be written as

$$\lambda = \frac{1}{4} \sum_s \langle \phi_s^0 | C' | \phi_s' \rangle + \tfrac{1}{2} \langle \phi_+^0 | C_{0+}^0 | \phi_+^0 \rangle \tag{10.13}$$

and that $\hat{\lambda} = (W_+'' - W_-'')/4$ can be written as $\hat{\lambda}_{HT} + \hat{\lambda}_{RT}$, where

$$\hat{\lambda}_{HT} = \frac{1}{4} \sum_s s \langle \phi_s^0 | C' | \phi_{0s}' \rangle$$

$$= \frac{1}{4} \sum_s s \langle \phi_s^0 | C' \frac{\Sigma^s}{a} C' | \phi_s^0 \rangle \tag{10.14}$$

$$\hat{\lambda}_{RT} = \tfrac{1}{2} \langle \phi_+^0 | C_{2+}'' | \phi_+^0 \rangle \left(= -\tfrac{1}{2} \langle \phi_-^0 | C_{2-}'' | \phi_-^0 \rangle \right)$$

Here the symbol Σ^s/a means Q_0/a, restricted to include only a sum over intermediate Σ^s states. The reason for the indices HT (Herzberg–Teller) and RT (Renner–Teller) is as follows. For qualitative purposes one can consider the deviation $C'\rho$ of the Coulomb potential from C^0 as being the dipole potential produced when "...neutralizing the positive charge of the [center] nucleus in the linear configuration by an equal negative charge, and by adding an equal positive charge in the non-linear configuration" (Reference 1, p. 178). This way of viewing the interaction between electrons and vibrations is due originally to Herzberg and Teller,[37] and the dipolar term $C'\rho$ from C's expansion is therefore now often called the Herzberg–Teller interaction term. The quadrupolar term $C_{2+}'' \rho^2$ is then called the Renner–Teller interaction term. One sees that $\hat{\lambda}_{HT}$ receives positive contributions from intermediate Σ^+ states, negative contributions from intermediate Σ^- states, and no contributions from intermediate $\Pi, \Delta, \Phi, \ldots$ states. Renner obtained a less complete, and purely qualitative, version of these results by considering a strongly simplified molecular model with only one electron, which interacts with the bending vibration via the Herzberg–Teller dipole. In such a model there is no quadrupolar contribution to $\hat{\lambda}$, and intermediate Σ^- states do not exist. Because of these qualitative considerations it has sometimes been thought (see e.g. Reference 6) that Renner's whole theory is restricted to small bending amplitudes and overlooks the quadrupolar contribution to $\hat{\lambda}$. However, Renner only used the primitive dipole model to discuss how intermediate states affect the force constants in the Born–Oppenheimer potentials W_+ and W_-; the other formulas in his paper do not depend on it.

XI. REPRESENTATION OF THE ELECTRONIC WAVE FUNCTIONS BY TRIGONOMETRIC FUNCTIONS. THE ONE-ELECTRON MODEL

In Section III we found Renner's form (3.7) for the eigenvalue problem for E, η_+, and η_- by use of the substitutions (3.6). We now discuss the alternative substitutions

$$\phi_+ \rightarrow \pi^{-1/2}\cos\Lambda\theta, \quad \phi_- \rightarrow \pi^{-1/2}\sin\Lambda\theta, \quad L_z \rightarrow L_\theta, \quad S \rightarrow S_\theta \quad (11.1)$$

where $L_\theta \equiv -i\partial/\partial\theta$ and S_θ is the unitary operator that changes the sign of θ (which is an angle without a physical meaning at this point). With these substitutions, the eigenvalue problem for E, η_+, and η_- takes the equivalent form $(H_\theta - E)\Psi_\theta = 0$, where

$$H_\theta = T_\theta + W + \hat{W}S_\theta, \qquad \Psi_\theta = \pi^{-1/2}\{\cos(\Lambda\theta)\,\eta_+ + \sin(\Lambda\theta)\,\eta_-\}$$

$$(11.2)$$

(T_θ is the result of substituting $L_z \rightarrow L_\theta$ in T_n). The hamiltonian obtained by replacing S_θ with $2\cos(2\Lambda\theta)$, that is,

$$H_\theta = T_\theta + W + 2\hat{W}\cos(2\Lambda\theta) \qquad (11.3)$$

is identical to H in (11.2) in the sense that each matrix element $\langle \Psi'_\theta | H_\theta | \Psi_\theta \rangle$ is independent of the form of H_θ used. However, as shown by the identity

$$\{2\cos(2\Lambda\theta) - S_\theta\}\Psi_\theta = \pi^{-1/2}\{\cos(3\Lambda\theta)\,\eta_+ + \sin(3\Lambda\theta)\,\eta_-\} \quad (11.4)$$

one does *not* have the simple operator identity $2\cos(2\Lambda\theta) \equiv S_\theta$; the right-hand side of (11.4) only vanishes when the inner product with $\langle \Psi'_\theta |$ is formed.

This means that algebraic manipulations cannot be carried out with un-thinking use of the fundamental operator rules if $2\cos(2\Lambda\theta)$ is used instead of S_θ. (Example: The insertion of $2\cos 2\theta$ for S_θ in $S_\theta[L_\theta, S_\theta]$ gives $4i\cos 2\theta \sin 2\theta = 2i\sin 4\theta = 0$ because $\langle\Psi_\theta'|\sin 4\theta|\Psi_\theta\rangle = 0$. The correct result is $S_\theta[L_\theta, S_\theta] = S_\theta L_\theta S_\theta - L_\theta = -2L_\theta$.) Since furthermore it is easier to act with S_θ than with the cosine, the S_θ version of H_θ seems preferable to the cosine version. The same is true for the unitary transformed hamiltonian

$$I_\theta = UH_\theta U^{-1}, \qquad U = e^{-i\chi L\theta} \tag{11.5}$$

for which one finds the two versions

$$I_\theta = T_\perp + W + \hat{W}S_{\theta-\chi} \tag{11.6a}$$

$$I_\theta = T_\perp + W + 2\hat{W}\cos 2\Lambda(\theta - \chi) \tag{11.6b}$$

($T_\perp$ is the result of omitting L_z from T_n), where $S_{\theta-\chi}$ changes the sign of $\theta - \chi$ by transforming θ to $2\chi - \theta$. When using H_θ, the value K of J_z is a good quantum number. When using I_θ, this quantum number is the eigenvalue of the transformed $UJ_zU^{-1} = J_z + L_\theta$. For given K, an eigenfunction of I_θ must therefore take the form

$$\tilde{\Psi}_\theta = (2\pi)^{-1}\sum_{s=\pm 1}e^{is\Lambda\theta}\tilde{\mu}_s(\rho)e^{i(K-s\Lambda)\chi} \tag{11.7}$$

and $\mu_+(\rho)$ and $\mu_-(\rho)$ in the corresponding eigenfunction $\Phi = (2\pi)^{-1/2}\{\phi_+\mu_+ + \phi_-\mu_-\}\exp(iK\chi)$ of $\mathcal{H}$ are related to $\tilde{\mu}_+$ and $\tilde{\mu}_-$ as in (3.20).

Let us consider a highly simplified model of the molecule which allows us to think of the operators H_θ and I_θ in a concrete way. It is assumed that the nuclei at the ends of the molecule are electrically neutral and that there is only *one* electron, constrained to move at a constant distance r_e from the z axis and in a plane through the center nucleus, parallel to the $x-y$ plane. The only electronic coordinate is then an angle θ. The angle from $\mathbf{e}_x$ to the electron is denoted by α, and we consider two possibilities for the origin of θ:

CASE I: θ is measured from $\mathbf{e}_x$ (i.e., $\alpha = \theta$).

CASE II: θ is measured from $\mathbf{e}_x^0 \equiv \mathbf{e}_x$ ($\chi = 0$) (i.e., $\alpha = \theta - \chi$).

In both cases we have $L_z = L_\theta$, but in case I we have $T_n = T_\theta$ and in case II we have $T_n = T_\perp$ (because $J_z \equiv -i\partial/\partial\chi$ in that case has the meaning of angular momentum of the nuclei alone). For the electronic hamiltonian H_e

in (2.6) we have

$$H_e = T_e + C, \qquad T_e = B_e L_z^2$$
$$C = C^0 \{1 - 2d\cos\alpha + d^2\}^{-1/2} = C^0 \{(1 - d_+)(1 - d_-)\}^{-1/2} \quad (11.8)$$

where B_e is the "rotational constant" $\hbar^2/2mr_e^2$, where $d = \rho/\rho_e$ is the distance from the z axis of the center nucleus in units of r_e (ρ_e is the appropriate constant), and where $d_\pm = d\exp(\pm i\alpha)$. The assumption of (2.14) takes the form

$$\Phi = \phi_+(\alpha,\rho)\eta_+(\rho,\chi) + \phi_-(\alpha,\rho)\eta_-(\rho,\chi) \quad (11.9)$$

and, assuming that the simplified H_e defines the same potentials as the real H_e, the eigenvalue problem (3.1), with the extra assumption that $\phi_s \simeq \phi_s^0$, takes the form

$$(\bar{H} - E)(\phi_+^0\eta_+ + \phi_-^0\eta_-) = 0 \qquad \begin{cases} \phi_+^0 = \pi^{-1/2}\cos\Lambda\alpha \\ \phi_-^0 = \pi^{-1/2}\sin\Lambda\alpha \end{cases} \quad (11.10)$$

where $\bar{H}$ stands for H_θ in case I and for I_θ in case II.

The simplified model is useful for a qualitative understanding of the problem. The content of multipoles in $C = C^0 + C'\rho + C''\rho^2 + \cdots$, for instance, is obtained easily from (11.8):

$$C' = C_1'\cos\alpha$$
$$C'' = C_0'' + C_2''\cos 2\alpha$$
$$C''' = C_1'''\cos\alpha + C_3'''\cos 3\alpha \quad (11.11)$$
$$\vdots$$

[compare with (10.4)] even though it is unlikely that a physical meaning can be attached to the constants $C_1', C_0'', C_2'', \ldots$. It is then straightforward to verify that $\hat{W}(\rho)$ varies as $\rho^{2\Lambda}$ at $\rho = 0$ as found in (3.3). We give a few more examples for the case $\Lambda = 1$. By first-order perturbation theory (and with the obvious meaning ΔE^{-1}) one finds $\phi_s' = \phi_{0s}' + \phi_{2s}'$, where

$$\phi_{0+}' = \frac{C_1'}{\Delta E_0} 2^{-1/2}(2\pi)^{-1/2}, \qquad \phi_{2+}' = \frac{C_1'}{\Delta E_2}\tfrac{1}{2}\pi^{-1/2}\cos 2\alpha$$
$$\phi_{0-}' = 0, \qquad \phi_{2-}' = \frac{C_1'}{\Delta E_2}\tfrac{1}{2}\pi^{-1/2}\sin 2\alpha \quad (11.12)$$

This result confirms that of (10.8). Unfortunately, the misleading result $\phi'_{0-} = 0$ also emerges. Σ^- states are perfectly well allowed for whenever there is more than one electron. As another example, consider the two force constants $\lambda = \frac{1}{4}(W''_+ + W''_-)$ and $\hat{\lambda} = \frac{1}{4}(W''_+ - W''_-)$, found by inserting (11.12) into (10.7):

$$\lambda = \left(\frac{1}{4} \frac{|C'_1|^2}{\Delta E_0} + \frac{1}{8} \frac{|C'_1|^2}{\Delta E_2} \right) + \frac{1}{2} C''_0$$

$$\hat{\lambda} = \frac{1}{4} \frac{|C'_1|^2}{\Delta E_0} + \frac{1}{4} C''_2$$

(11.13)

It is seen that the dipolar term C' from C only contributes in second order and that the contributions become particularly large when the Π state in question is close to a Σ^+ state (i.e. when ΔE_0 is small). Also it is seen that a nearby Δ state affects the value of λ (ΔE_2 small) but not the value of $\hat{\lambda}$. These conclusions were also drawn in Section X for the quantitatively correct equations (10.13) and (10.14).

If we insert $|\phi'_s\rangle$ from (11.12) in the corrections $\hat{w}^0$ and $g_K = g^0$ from (4.9) and (4.7), we find that $w^0 = 0$ [in accordance with (4.10)] and

$$g_K = \frac{\hbar^2}{4} \left\{ \frac{|C'_1|^2}{\Delta E_0^2} - \frac{|C'_1|^2}{\Delta E_2^2} \right\} = \hbar^2 \left\{ \frac{1}{2} \| \phi'_{0+} \|^2 - \| \phi'_{2s} \|^2 \right\}$$

(11.14)

where the last term is independent of s. This agrees with the correct result in (10.10), except that $\| \phi'_{0-} \|^2$ is missing.

Our simple model can be further simplified by truncating the expansion of C to

$$C = C^0 + C'\rho = C^0 + C'_1 \rho \cos \alpha$$

(11.15)

This is essentially the model used by Renner in his qualitative discussion of how the force constants in the two potentials W_+ and W_- depend on other electronic states (see Section X).

Discussion. Although the hamiltonian H_θ discussed in the present section is mathematically equivalent to Renner's H, there are several reasons for choosing the latter as is done in the present article. The algebraic rules for the Pauli matrices are much more familiar than those of the "angle variables" L_θ and S_θ, and the matrix notation allows us to formulate quite naturally several results which appear awkward in terms of L_θ and S_θ. The result (3.9), for instance, is nothing but the observation that σ_z changes its sign in the unitary transformation σ_y. The equivalent observation for the angle operators is that S_θ changes its sign in the unitary transformation $-i\Lambda^{-1}\partial/\partial\theta$ and is not directly obvious. Note also that it is rather awkward to formulate the equivalents of results such as (3.12) and $[\sigma_y, \sigma_z] = 2i\sigma_x$ and that the connection between the effective hamiltonian and the coupled differential equations discussed in Section III is more direct when 2×2 matrices are used in both. Furthermore the Pauli matrices lend themselves much more naturally than the angle operators to the construction of the complete effective hamiltonian from the four matrix elements $H_{s's}$ as discussed in Section IV. An advantage of the angle variables is that they can be given simple physical meanings within a one-electron model which is helpful for qualitative understanding.

The operator I_θ can be discussed in a similar way. It is closely analogous to $\tilde{H}$ in (B5).

XII. THE APPROACH BY POPLE AND LONGUET-HIGGINS

The coupling between degenerate electronic wave functions was first observed experimentally by Dressler and Ramsay[3] in 1957, about 25 years after the publication of Renner's paper. The molecule, NH_2, happened to be an example of the situation sketched in Fig. 1b, with potentials of such dimensions that there are several vibrational levels in W_-, below the lowest state in W_+. Pople and Longuet-Higgins, who gave the theoretical discussion, did not use Renner's formalism to calculate the energy levels (which can easily be done, as is shown in Appendix D). Instead they considered a simplified model with a single electron that was allowed only *one* degree of freedom, its azimuthal angle θ measured from the same laboratory-fixed axis as χ. They assert that "...this model incorporates the essential features of the situation...[such that] the resulting equations of motion are mathematically equivalent to those obtained by Renner from more sophisticated principles." Applying symmetry and plausibility considerations to their model, they reach the effective hamiltonian I_θ in our Eq. (11.6b):

$$I_\theta = \tfrac{1}{2}P_\rho^2 + \tfrac{1}{2}\hbar^2\{J_z^2 - \tfrac{1}{4}\}\rho^{-2} + W + 2\hat{W}\cos 2\Lambda(\theta - \chi) \qquad (12.1)$$

to be diagonalized on the space of the wave functions $\tilde{\Psi}_\theta$ from our Eq. (11.7). Pople and Longuet-Higgins's simplified model is similar to the one we considered in Section XI. Our purpose for introducing this model was to illustrate the operator I_θ which, by its construction in Section XI, is obviously equivalent to Renner's H. The mathematical equivalence also displays itself very clearly if we apply the formula $2\cos x = \exp(ix) + \exp(-ix)$ to the cosine in I_θ and compare with the unitary transform $\tilde{H}$ of H in (B.5). However, Pople and Longuet-Higgins work only within the framework of the simplified model and do not indicate any proof of the postulated equivalence to Renner's eigenvalue problem. Probably they have thought of the latter as the set of coupled differential equations from Renner's Eq. (14),

$$\left\{\begin{array}{cc} h_+ - E & i(\hbar/\rho)^2 \Lambda J_z \\ -i(\hbar/\rho)^2 \Lambda J_z & h_- - E \end{array}\right\}\left\{\begin{array}{c} \eta_+ \\ \eta_- \end{array}\right\} = 0 \qquad (12.2)$$

[see (3.12) for h_s], which is the form used by Longuet-Higgins in his review[5] from that time. It is straightforward to check the equivalence to $(I_\theta - E)\tilde{\Psi}_\theta = 0$. Subsequent authors have tried to derive the latter form directly, i.e. without reference to Renner's form, by expanding the complete hamiltonian

$$\mathcal{H} = T_n + T_e + C, \qquad C = C^0 + C'\rho + C''\rho^2 + \cdots \qquad (12.3)$$

and applying perturbation theory to the zeroth-order level spanned by wave functions of the form $\phi_+^0 \eta_+ + \phi_-^0 \eta_-$. The electrons have been described by coordinates $(|r|, \theta)$, where $|r|$ stands for some coordinates which are invariant under rotations around $\mathbf{e}_z$, and where θ is the azimuthal angle of the "first" electron. All perturbation calculations have been carried out under the assumption that one can write

$$\phi_+^0 = \cos \Lambda(\theta - \chi)f, \qquad \phi_-^0 = \sin \Lambda(\theta - \chi)f, \qquad (12.4)$$
$$C' = C_1'\cos(\theta - \chi), \qquad C'' = C_0^0 + C_2''\cos 2(\theta - \chi),\ldots$$

where $f, C_0^0, C_1', C_0'', C_2'',\ldots$ are functions of $|r|$ only. Unfortunately, it has not been fully realized that this amounts to the assumption that there is only one electron in the molecule (see Appendix B). Thus the approach is no more general than that founded on Pople and Longuet-Higgins's simplified model, but is in fact more complicated. Note in particular that the approach allows the existence of Σ^- states (which Pople and Longuet-Higgins's does not), but the assumptions of (12.4) exclude these from being mixed into ϕ_-. Carrington et al.[6] used the approach to formulate the force constants $\lambda = \frac{1}{2}W''$ and $\hat{\lambda} = \frac{1}{2}\hat{W}'''$, and found that the monopolar term C_0'' does not contribute to $\hat{\lambda}$, while the quadrupolar term $C_2''\cos 2(\theta - \chi)$ contributes (in first order) as much as does the dipolar term $C_1'\cos(\theta - \chi)$ (in second order). These results are also evident from (11.13). We note that when I_θ is constructed by perturbation theory founded on (12.4), the cosine term does not appear as the reflection operator in the molecular plane, but rather as an effective quadrupole which absorbs the effects of the multipole terms in C that couple the electronic wave functions. Several authors have taken this point of view.[6, 10, 17]

Brown[17] used the assumptions of (12.4) in a third-order calculation to find that a small correction

$$\Delta I_\theta = g_K(J_z + L_z)L_z, \qquad L_z = -i\frac{\partial}{\partial\theta} \tag{12.5}$$

should be added to I_θ in (12.1). His perturbation expression for the constant g_K is qualitatively correct in the same way as the expression in (11.14). Quantitatively, however, it is unlikely to be reliable.

APPENDIX A: FURTHER DISCUSSION OF SYMMETRY

In this appendix we extend the discussion of symmetry given in Section X. To do so in a systematic manner we introduce symbols for the irreducible representations of $C_{\infty v}$ as follows. If an electronic wave function ψ satisfies $(L_z^2 - m^2)\psi = 0$, $m = 0, 1, 2, \ldots$, and $S\psi = s\psi$, $s = \pm 1$, then we write $\psi \in \Gamma_{ms}$; and if a pair (ψ_+, ψ_-) of wave functions satisfy $L_z\psi_s = ism\psi_{-s}$, $m > 0$, and $S\psi_s = s\psi_s$, then we write $(\psi_+, \psi_-) \in \Gamma_m$ and have

$$U_\tau \left\{ \begin{matrix} \psi_+ \\ \psi_- \end{matrix} \right\} = \left\{ \begin{matrix} \cos m\tau & \sin m\tau \\ -\sin m\tau & \cos m\tau \end{matrix} \right\} \left\{ \begin{matrix} \psi_+ \\ \psi_- \end{matrix} \right\} \tag{A.1}$$

in particular,

$$\psi_{-s} = sU_{\pi/2m}\psi_s = -ism^{-1}L_z\psi_s \tag{A.2}$$

We use the same terminology for operators, i.e., if the operator A satisfies $[L_z, [L_z, A]] = m^2 A$ and $SAS^{-1} = sA$, we write $A \in \Gamma_{ms}$, and so on. Since $(\phi_+^0, \phi_-^0) \in \Gamma_\Lambda$, we have

$$\langle \phi_s^0 | A | \phi_s^0 \rangle = \langle \phi_s^0 | (A_{0+} + A_{2\Lambda+}) | \phi_s^0 \rangle$$
$$\langle \phi_+^0 | A_{0+} | \phi_+^0 \rangle = +\langle \phi_-^0 | A_{0+} | \phi_-^0 \rangle \tag{A.3}$$
$$\langle \phi_+^0 | A_{2\Lambda+} | \phi_+^0 \rangle = -\langle \phi_-^0 | A_{2\Lambda+} | \phi_-^0 \rangle$$

where A_{m+} is the part of A which transforms as Γ_{m+}.

The terms in the effective hamiltonian $G = G''\rho^2 + G''''\rho^4 + \cdots$, and the transformation operator $u = P_0 + u'\rho + \cdots$, which satisfy $W_s^{(p)} = \langle \phi_s^0 | G^{(p)} | \phi_s^0 \rangle$ and $\phi_s^{(p)} = u^{(p)}\phi_s^0$ can be constructed by perturbation theory as mentioned in (10.7). They will be products of the perturbations $C', C'', \ldots$, and the operators $P_0 = |\phi_+^0\rangle\langle\phi_+^0| + |\phi_-^0\rangle\langle\phi_-^0|$ and Q_0/a. Since the latter two are invariant under $C_{\infty v}$, $G^{(p)}$ and $u^{(p)}$ can both be decomposed exactly like

$C^{(p)}$ in (10.4), and it follows from (A.3) that $W_+ - W_-$ varies as $\rho^{2\Lambda}$ at $\rho = 0$ as postulated in (3.3). It also follows that the decomposition of $\phi_s^{(p)}$ is such that

$$\phi_s^0 = \phi_{\Lambda, s}^0$$
$$\phi_s' = \phi_{\Lambda-1, s}' + \phi_{\Lambda+1, s}'$$
$$\vdots$$
$$\phi_s^{(\Lambda)} = \phi_{0, s}^{(\Lambda)} + \phi_{2, s}^{(\Lambda)} + \cdots + \phi_{2\Lambda, s}^{(\Lambda)} \tag{A.4}$$
$$\phi_s^{(\Lambda+1)} = \phi_{1, s}^{(\Lambda+1)} + \phi_{3, s}^{(\Lambda+1)} + \cdots + \phi_{2\Lambda+1, s}^{(\Lambda+1)}$$
$$\phi_s^{(\Lambda+2)} = \phi_{0, s}^{(\Lambda+2)} + \phi_{2, s}^{(\Lambda+2)} + \cdots + \phi_{2\Lambda+2, s}^{(\Lambda+2)}$$
$$\vdots$$

where $\phi_{ms}^{(p)} \in \Gamma_{ms}$. That is, the part ϕ_{ms} of ϕ_s which transforms as Γ_{ms} is such that

$$\phi_s = \sum_{m=0}^{\infty} \phi_{ms}$$
$$\phi_{ms} = \rho^n \left(\phi_{ms}^{(n)} + \phi_{ms}^{(n+2)} \rho^2 + \cdots \right); \quad n \equiv |\Lambda - m| \tag{A.5}$$

It is proven below that the last Λ terms in the decomposition (A.4) of $\phi_+^{(p)}$ are related to the last Λ terms in the decomposition of $\phi_-^{(p)}$ by $L_z\phi_{ms}^{(p)} = ism\phi_{m,-s}^{(p)}$, that is,

$$\phi_{m,-s} \simeq sU_{\pi/2m}\phi_{ms} = -ism^{-1}L_z\phi_{ms} \quad \text{if} \quad m > 0 \tag{A.6}$$

where " $\simeq$ " means with the exception of terms of the order $\rho^{\Lambda+m}$.

We can now prove that $\hat{w}$ from (4.9) varies as $\rho^{2\Lambda}$ at $\rho = 0$: Inserting (A.5) into (4.5), we get $w_s = w_{0s} + w_{1s} + \cdots$, where

$$2w_{ms} = \hbar^2 \left\{ \left\| \frac{\partial \phi_{ms}}{\partial \rho} \right\|^2 + (m^2 - \Lambda^2) \left\| \frac{\phi_{ms}}{\rho} \right\|^2 \right\} \tag{A.7}$$

Noting that ϕ_{0s} varies as ρ^{Λ} at $\rho = 0$, and that $d\rho^{\Lambda}/d\rho = \Lambda\rho^{\Lambda}/\rho$, one sees that w_{0s} varies as $\rho^{2\Lambda}$ at $\rho = 0$. For $m > 0$, use of (A.6) shows that $w_{m+} - w_{m-}$ vanishes except for terms of the order $\rho^{2(\Lambda+m-1)}$. These results prove (4.10).

Allowing ρ to be negative, we evidently have $U_\pi H_e(\rho)U_\pi^{-1} = H_e(-\rho)$. From this result we can see that

$$U_\pi \phi_s(r, \rho) = (-1)^{\Lambda} \phi_s(r, -\rho) \tag{A.8}$$

Proof: The identity

$$U_\pi = \tfrac{1}{2}(U_\pi + U_{-\pi}) = \sum_{n=0}^{\infty} \frac{(-1)^n}{(2n)!} \left(\pi^2 L_z^2\right)^n \tag{A.9}$$

$[= \cos(\pi L_z)]$ shows that $U_\pi \phi_s$ is real and has the same parity under S as ϕ_s. Hence (A.8) must be correct except for a sign. This can be determined at $\rho = 0$. Q.E.D.

The result (A.8) implies that a function $f(\rho) \equiv \langle \phi_s | F | \phi_{s'} \rangle$ satisfies

$$f(-\rho) = \langle \phi_s | U_\pi^\dagger F(-\rho) U_\pi | \phi_{s'} \rangle \tag{A.10}$$

for an arbitrary ρ-dependent operator $F(\rho)$. It is now trivial to prove that w, $\hat{w}$, and g in (4.7) and (4.8) are even functions of ρ [note that since P_ρ commutes with U_π, we can replace $|\phi_s\rangle$ with $|P_\rho \phi_s\rangle$ in (A.10)].

Proof of the Results (A.6) and (10.4). As a preparation for the proof of (A.6), define for given $\psi_{m+} \in \Gamma_{m+}$, $m > 0$,

$$\tilde{\psi}_{ms} \equiv 2^{-1/2}(\psi_{m+} + is\psi_{m-}) \tag{A.11}$$

where ψ_{m-} is the wave function defined so that $(\psi_{m+}, \psi_{m-}) \in \Gamma_m$. Then we have

$$\psi_{m+} = 2^{-1/2}\left(\tilde{\psi}_{m+} + \tilde{\psi}_{m-}\right), \qquad \psi_{m-} = -i2^{-1/2}\left(\tilde{\psi}_{m+} - \tilde{\psi}_{m-}\right),$$
$$S\tilde{\psi}_{ms} = \tilde{\psi}_{m,-s} \quad \text{and} \quad L_z\tilde{\psi}_{ms} = sm\tilde{\psi}_{ms}. \tag{A.12}$$

Similar results are valid for operators. In particular we can decompose $u^{(p)}$ as described above (A.4) and thereby obtain operators $u_{m\pm}^{(p)}$ and $\tilde{u}_{m\pm}^{(p)}$ which satisfy relations analogous to (A.12). We then see that the two parts in the decomposition

$$u_{m+}^{(p)}\phi_s^0 = \psi_{\Lambda+m,s}^{(p)} + \psi_{\Lambda-m,s}^{(p)}$$

$$\psi_{\Lambda+m,+}^{(p)} = \frac{\tilde{u}_{m+}^{(p)}\tilde{\phi}_+^0 + \tilde{u}_{m-}^{(p)}\tilde{\phi}_-^0}{2}$$

$$\psi_{\Lambda+m,-}^{(p)} = -i\frac{\tilde{u}_{m+}^{(p)}\tilde{\phi}_+^0 - \tilde{u}_{m-}^{(p)}\tilde{\phi}_-^0}{2}$$

$$\psi_{\Lambda-m,+}^{(p)} = \frac{\tilde{u}_{m+}^{(p)}\tilde{\phi}_-^0 + \tilde{\phi}_{m-}^{(p)}\tilde{\phi}_+^0}{2} \tag{A.13}$$

$$\psi_{\Lambda-m,-}^{(p)} = -i\frac{-\tilde{u}_{m+}^{(p)}\tilde{\phi}_-^0 + \tilde{u}_{m-}^{(p)}\tilde{\phi}_+^0}{2},$$

$(m > 0)$ satisfy

$$L_z \psi^{(p)}_{\Lambda \pm m, s} = is(\Lambda \pm m) \psi^{(p)}_{\Lambda \pm m, -s} \tag{A.14}$$

When $\Lambda = 1$ we find

$$\phi^0_s = \psi^0_{1s}$$
$$\phi'_s = \psi'_{0s} + \psi'_{2s}$$
$$\phi''_s = \psi''_{-1,s} + \psi''_{1s} + \Delta\psi''_{1s} + \psi''_{3s}$$
$$\phi'''_s = \psi'''_{-2,s} + \psi'''_{0s} + \psi'''_{2s} + \psi'''_{4s} \tag{A.15}$$
$$\phi''''_s = \psi''''_{-3,s} + \psi''''_{-1,s} + \psi''''_{1s} + \Delta\psi''''_{1s} + \psi''''_{3s} + \psi''''_{5s}$$
$$\vdots$$

where $\Delta\psi^{(p)}_{1s} = u^{(p)}_{0+}\phi^0_s$. Similar results for $\Lambda = 2, 3, \ldots$ complete our proof.

Finally we prove the fundamental result of (10.4). Let $C(r, \rho)$ be the Coulomb potential as a function of (r, ρ), and let $\hat{C}(r, \rho\cos\chi, \rho\sin\chi)$ be the Coulomb potential as it would be if the components (r_{ix}, r_{iy}, r_{iz}) of the ith electron's position vector were defined relative to the laboratory-fixed frame $\mathcal{K}(\chi = 0)$. As also pointed out in (B.1), one then has

$$\hat{C}(r, \rho\cos\chi, \rho\sin\chi) = U_\chi C(r, \rho)U_\chi^{-1} \tag{A.16}$$

Writing the expansion of $\hat{C}$ in $(Q_x, Q_y) = (\rho\cos\chi, \rho\sin\chi)$ as

$$\hat{C} = C^0 + C_x Q_x + C_y Q_y + C_{xx} Q_x^2 + C_{xy} Q_x Q_y + C_{yy} Q_y^2 + \cdots \tag{A.17}$$

it follows from $\hat{C}(r, \rho, 0) = C(r, \rho)$ that

$$C' = C_x, \qquad C'' = C_{xx}, \qquad C''' = C_{xxx}, \ldots \tag{A.18}$$

The proof is now completed by showing that

$$C'_{1+} = C_x \quad \text{and} \quad C'_{1-} = C_y \tag{A.19}$$

are as required and that

$$C_{x\ldots, y\ldots} \text{ transforms as } C_x^{n_x} C_y^{n_y} \tag{A.20}$$

where n_x is the number of x's and n_y the number of y's. It goes as follows: According to (A.16), $U_\tau \hat{C} U_\tau^{-1}$ equals $U_{\chi+\tau} C(r, \rho) U_{\chi+\tau}^{-1} = \hat{C}(r, \rho\cos(\chi + \tau),$

$\rho \sin(\chi + \tau))$ and can therefore be found by substituting

$$\begin{Bmatrix} Q_x \\ Q_y \end{Bmatrix} \to \begin{Bmatrix} \rho\cos(\chi+\tau) \\ \rho\sin(\chi+\tau) \end{Bmatrix} = \begin{bmatrix} \cos\tau & -\sin\tau \\ \sin\tau & \cos\tau \end{bmatrix} \begin{Bmatrix} Q_x \\ Q_y \end{Bmatrix} \qquad (A.21)$$

in (A.17). Comparison with $U_\tau \hat{C} U_\tau^{-1} = C^0 + U_\tau C_x U_\tau^{-1} Q_x + \cdots$ shows that $C_{x\dots,\,y\dots}$ transforms as postulated under U_τ. In a similar way one finds

$$SC_{x\dots,\,y\dots}S^{-1} = (-1)^{n_y}C_{x\dots,\,y\dots} \qquad (A.22)$$

and the proof is complete. We note the relations

$$C_{0+}'' = \frac{C_{xx}+C_{yy}}{2}$$

$$C_{2+}'' = \frac{C_{xx}-C_{yy}}{2}, \qquad C_{2-}'' = \frac{C_{xy}}{2} \qquad (A.23)$$

APPENDIX B: SOME COMMENTS ON RENNER'S ORIGINAL PAPER

The coordinates χ and $r = (\dots, r_{ix}, r_{iy}, r_{iz}, \dots)$ defined in Section II are singular when all the nuclei are on the axis. Renner avoided the singularity in the coordinates r by defining r_{ix} and r_{iy} as the components of the position vector $\mathbf{r}_i$ along the axes $\mathbf{e}_x^0$ and $\mathbf{e}_y^0$, respectively, in the laboratory-fixed frame $\mathcal{K}_0 = \mathcal{K}(\chi = 0)$. Hence, his hamiltonian $\hat{\mathcal{K}}$ is obtained from our $\mathcal{K}$ by substituting $J_z - L_z \to J_z$ and $(r_{ix}, r_{iy}) \to (\cos\chi\, r_{ix} + \sin\chi\, r_{iy}, -\sin\chi\, r_{ix} + \cos\chi\, r_{iy})$. These substitutions can be carried out by a unitary transformation,

$$\hat{\mathcal{K}} = U_\chi \mathcal{K} U_\chi^{-1}, \qquad U_\chi = e^{-i\chi L_z} \qquad (B.1)$$

Hence, we expect the two approaches to be equally sound. Renner's electronic wave functions are related to ours by

$$\hat{\phi}_s = U_\chi \phi_s \qquad (B.2)$$

Thus, while our functions are independent of χ, those of Renner depend on χ in such a way that

$$J_z \hat{\phi}_s = \left(-i\frac{\partial U_\chi}{\partial\chi} \right)\phi_s = -L_z \hat{\phi}_s \qquad (B.3)$$

Hence, in accordance with Renner's Eq. (9),

$$J_z\hat{\phi}_s^0 = -is\Lambda\hat{\phi}_{-s}^0 \qquad (B.4)$$

Evidently Renner's 2×2 matrix $\langle\hat{\phi}_s|\hat{\mathfrak{K}}|\hat{\phi}_{s'}\rangle$ becomes strictly identical to our $\langle\phi_s|\mathfrak{K}|\phi_{s'}\rangle$ in (4.1). Hence, we must find exactly the same solutions $(\eta_+(\rho,\chi),\eta_-(\rho,\chi))$ as he does.

It might seem a little surprising that Renner's operator $J_z \equiv -i\,\partial/\partial\chi$ means the angular momentum of the nuclei alone when appearing in $\mathfrak{K}$, but means the total angular momentum when appearing in the effective hamiltonian. This is caused by the dependence of $\hat{\phi}_s$ on χ, and we can avoid the peculiarity, and obtain some other small advantages as well, by subjecting H first to the unitary transformation U from (3.11) and then to the unitary transformation u from (3.20). One finds

$$\begin{aligned}
\tilde{H} &\equiv uUHU^{-1}u^{-1} = T_\perp + W + \hat{W}\tilde{\sigma}_z \\
\tilde{\sigma}_z &\equiv uU\sigma_z U^{-1}u^{-1} = e^{2i\Lambda\chi}\sigma_- + e^{-2i\Lambda\chi}\sigma_+
\end{aligned} \qquad (B.5)$$

where $\sigma_\pm = (\sigma_x \pm i\sigma_y)/2$ are the familiar raising and lowering operators

$$\sigma_+ = \begin{Bmatrix} 0 & 1 \\ 0 & 0 \end{Bmatrix} \quad\text{and}\quad \sigma_- = \begin{Bmatrix} 0 & 0 \\ 1 & 0 \end{Bmatrix} \qquad (B.6)$$

In $\tilde{H}$ we must interpret $J_z[\,= uU(J_z - \Lambda\sigma_y)U^{-1}u^{-1}]$ as the angular momentum of the nuclei alone, $\Lambda\sigma_z[\,= uU(\Lambda\sigma_y)U^{-1}u^{-1}]$ as the angular momentum of the electrons, and $\tilde{\sigma}_z$ as the operator which reflects the electrons in the molecular plane. The eigenfunction ψ from (3.18) of H is transformed into the eigenfunction

$$\tilde{\Psi} = \begin{Bmatrix} 1 \\ 0 \end{Bmatrix}\tilde{\mu}_+(\rho)e^{i(K-\Lambda)} + \begin{Bmatrix} 0 \\ 1 \end{Bmatrix}\tilde{\mu}_-(\rho)e^{i(K+\Lambda)} \qquad (B.7)$$

$(= uU\Psi)$ of $\tilde{H}$. Here $\tilde{\mu}_+$ and $\tilde{\mu}_-$ are the same as in (3.20) (and the quantum number K is the eigenvalue of $J_z + \Lambda\sigma_z$). The transformed hamiltonian $\tilde{H}$ has some attractive properties in addition to J_z having the meaning expected from the definition of the coordinates (r,ρ,χ). In particular, the form of $\tilde{H}$ tells one immediately how to deal with Renner's original problem where W_+ and W_- are both strictly harmonic and $\hat{W} = \frac{1}{2}(W_+ - W_-)$ small: $T_\perp + W \equiv H_\perp$ is then the familiar hamiltonian for a two-dimensional harmonic oscillator, and its equidistant levels are perturbed by $\hat{W}\tilde{\sigma}_z$. The appearance of σ matrices in the perturbation is a well-known phenomenon from elementary spin problems.

To obtain the fundamental result of (B.4), Renner first asserted that ϕ_+^0 and ϕ_-^0 take the forms

$$\phi_+^0 = f(|r|)\cos \Lambda\theta, \qquad \phi_-^0 = f(|r|)\sin \Lambda\theta \qquad (B.8)$$

when formulated in coordinates $(|r|, \theta)$, defined as in Section II. He then observed that $\hat{\phi}_s^0(|r|, \theta, \chi) = \phi_s^0(|r|, \theta - \chi)$ satisfied (B.4). Comparing with (2.16), we see that Renner does not have the term $\sin \Lambda\theta$ in ϕ_+^0 and the term $\cos \Lambda\theta$ in ϕ_-^0. These terms do *not*, in fact, vanish with Renner's choice of θ as the azimuthal angle of the first electron (unless this is the only one in the molecule). This can be seen from a simple example with only two electrons, constrained to move on circles with constant distances from the z-axis. Note also that *if* θ can be defined in some other way such that the sine term in ϕ_+^0 and the cosine term in ϕ_-^0 *do* vanish, then an interesting new symmetry emerges: ϕ_s^0 becomes an eigenvector of the (unitary and hermitian) operator S_θ which changes the sign of θ and L_z. And since the level of ϕ_s^0 is chosen arbitrarily, this means that H_e^0 must commute with S_θ and thus be without any terms linear in L_z. This amounts to saying that there is an overall rotational motion (described by θ) and an internal motion (described by $|r|$) which can be time-reversed separately without changing the energy. We do not believe that such a symmetry exists, but many authors have it build into their formalism by the acceptance of (B.8) [see also Eq. (12.4) and subsequent text].

Renner, in his assertion of (B.6), cites as reference the paper by Born and Flügge.[38] These authors describe the electrons in a diatomic molecule in coordinates $(|r|, \theta)$ like those introduced in Section II, and they observe that since L_z commutes with the electronic hamiltonian H_e, the two electronic wave functions in a certain level can be chosen in the factorized form

$$\phi_s^0 = f_s(|r|)e^{is\Lambda\theta}, \qquad s = \pm 1 \qquad (B.9)$$

where f_s is an eigenfunction of $H_e(L_z \equiv s\Lambda)$. Since H_e is such that if ϕ is an eigenfunction of H_e, so is the complex conjugate ϕ^* also, we can furthermore choose $f_+ = f_-^*$. Apparently Renner did not notice that it is *not* justifiable to choose f_s *real* [because $H_e(+\Lambda)$ does *not* equal $H_e(-\Lambda)$].

The Limit $\varepsilon \to 1$. To draw the curves in Fig. 2 of the energies $E_i(\varepsilon)$ as functions of ε, Renner derived a formula [see Eq. (B.12) below] for their asymptotic behavior in the limit $\varepsilon \to 1$ (i.e. $\lambda_+ \gg \lambda_-$). In this limit we find that E_j and its eigenfunction

$$\eta_{j+}\begin{Bmatrix} 1 \\ 0 \end{Bmatrix} + \eta_{j-}\begin{Bmatrix} 0 \\ 1 \end{Bmatrix}$$

satisfy

$$E_j \simeq \omega(1-\varepsilon)^{1/2}\left\{(K^2+1)^{1/2}+2j+1\right\} \to 0$$

$$\|\eta_{j+}\|^2 \to 0 \tag{B.10}$$

or, inserting j as a function of $\tilde{v}$, K, and s,

$$E_s(\tilde{v}, K) \simeq \omega(1-\varepsilon)^{1/2}\left\{(K^2+1)^{1/2}+2(\tilde{v}-K)+s\right\} \to 0 \tag{B.11}$$

These results are established as follows: When $\lambda_+ \gg \lambda_-$, the lowest H_- levels [see (3.12)] will be far below the lowest H_+ level. Hence, they can be found approximately as $E = \omega_- e$, where e is the eigenvalue in the problem (5.6) with $l^2 = K^2+1$. One has (see below) $e = l+2N+1$, $N = 0,1,2,\ldots$, for arbitrary $l \geq 0$. [The quantum number v in (5.3) equals $l+2N$.] The number of states low enough to be determined in this way increases infinitely as $\varepsilon \to 0$. Hence, each E_j must tend to zero as asserted in (B.10).

The result $e = l+2N+1$ is established by the well-known polynomial method. Substituting the trial function $r_{vl} = w(q)f(q)$, $w \equiv q^{1/2}q^l\exp(-q^2/2)$, in (5.6), one sees (by multiplying on the left by w^{-1}) that

$$f'' + \left\{(2l+1)q^{-1}-2q\right\}f' + 2\{e-l-1\}f = 0$$

Inserting $f = \Sigma_i C_i q^i$, one gets $C_1 = 0$ and

$$(i+2)(i+2l+2)C_{i+2} = 2(i+l+1-e)C_i$$

Hence, C_i vanishes for *all* odd i. The result $e = l+2N+1$ now follows by requiring the series $f = C_0 + C_2q^2 + C_4q^4 + \cdots$ to terminate at $i = 2N$. (If the series does *not* terminate, then $C_{i+2}/C_i \to 2/i$, implying that $f(q) \to \exp(+q^2)$ and hence $r_{vl}(q) \to \infty$ as $q \to \infty$.)

For some reason, Renner disregarded the E_+ curves in his brief study of the limit $\varepsilon \to 1$. His result for the E_- curves is given in his Eq. (40) as

$$E_{2i+1, K} \simeq \omega(1-\varepsilon)^{1/2}\left\{(K^2+1)^{1/2}-K+2i+1\right\} \tag{B.12}$$

There appears to be something wrong with this result, at least if $i = 1,2,3,\ldots$, numbers the E_- curves in order of increasing energy as suggested by Renner's figures.

APPENDIX C: MATRIX ELEMENTS BETWEEN RENNER–TELLER WAVE FUNCTIONS ($\Lambda = 1$)

In this appendix we consider the construction of matrix elements $\langle F \rangle \equiv \langle \Psi | F | \Psi \rangle$, $\langle F \rangle_s \equiv \langle \Psi_s | F | \Psi_s \rangle$, $\langle F \rangle^{SO} \equiv \langle \Psi^{SO} | F | \Psi^{SO} \rangle$, and $\langle F \rangle_s^{SO} \equiv \langle \Psi_s^{SO} | F | \Psi_s^{SO} \rangle$, where the wave functions are those given in (6.7), (6.10), (7.12), and (7.16) and where F is an operator for which the terminology of (5.13) can be adopted. One finds $\langle F \rangle$ to second order in ε by inserting $K = \tilde{v}$ into

$$\langle F \rangle = F_{vv}^+ - 2\tilde{\varepsilon}F_{v+2,v}\{\tilde{v}(\tilde{v}+1)\}^{1/2}$$
$$+ \tilde{\varepsilon}^2\{F_{v+4,v}^+\{2\tilde{v}(\tilde{v}+1)\}^{1/2} + 4F_{v+2,v}^+\tilde{v}^{1/2}(\tilde{v}+1)$$
$$- (F_{vv}^+ - F_{v+2,v+2}^-)\tilde{v}(\tilde{v}+1)\} \tag{C.1}$$

($\tilde{\varepsilon} = \varepsilon/4$). To first order in ε one has

$$\langle F \rangle_s = \left\{ sF_{vv} + \left(1 - s\tilde{\varepsilon}\frac{K}{\Lambda}\right)F_{vv}^+ - 4\tilde{\varepsilon}(G + sG^+) \right\}_K^+$$
$$G = a_{v+2,v}F_{v+2,v} - a_{v-2,v}F_{v-2,v} \tag{C.2}$$
$$G^+ = a_{v,v+2}F_{v+2,v}^+ - a_{v,v-2}F_{v-2,v}^+$$

where $\{f\}_K^{\pm} = \{f(K) \pm f(-K)\}/2$ for arbitrary function f of K. To second order in $\varepsilon V + H_{SO}$ we have

$$\langle F \rangle^{SO} = \langle F \rangle - 2s_K\tilde{\varepsilon}\xi(F_{v+2,v})_{K=\tilde{v}}\{\tilde{v}(\tilde{v}+1)\}^{1/2} \tag{C.3}$$

(s_K = sign of K), and to first order

$$\langle F \rangle_s^{SO} = \{F_{vv}^+ - \varepsilon G\}_K^+ + s\{F_{vv}^+ - \varepsilon G\}_K^- \cos 2\beta + s\{F_{vv} - \varepsilon G^+\}_K^+ \sin 2\beta \tag{C.4}$$

where $(\cos 2\beta, \sin 2\beta)$ is given in (7.15) [and hence equals $(-\varepsilon K/4, 1)$ when $H_{SO} \to 0$; thus (C.4) reduces to (C.2) when $H_{SO} \to 0$].

By use of (C.1), (C.2) and the matrix elements in Section V we find (with $\tilde{\varepsilon} = \varepsilon/4$)

$$\langle J_z \sigma_y \rangle = \tilde{v}\{1 - 2\tilde{\varepsilon}^2 \tilde{v}(\tilde{v} + 1)\}$$

$$\langle J_z \sigma_y \rangle_s = \frac{-s\tilde{\varepsilon} K^2}{\Delta}$$

$$\langle q^2 \rangle = \tilde{v}\{1 + 6\tilde{\varepsilon}^2(\tilde{v} + 1)\}$$

$$\langle q^2 \rangle_s = \tilde{v}\{1 - 2s\tilde{\varepsilon}\tilde{v}\Delta\}$$

$$\langle q^2 \sigma_z \rangle = -2\tilde{\varepsilon}\tilde{v}(\tilde{v} + 1)$$

$$\langle q^2 \sigma_z \rangle_s = \tilde{v}\{s\Delta - 2\tilde{\varepsilon}\}$$

$$\langle q^4 \rangle = \tilde{v}(\tilde{v} + 1)\{1 + \tilde{\varepsilon}^2 4(3\tilde{v} + 4)\} \tag{C.5}$$

$$\langle q^4 \rangle_s = \tfrac{1}{2}(3\tilde{v}^2 - K^2) - s\tilde{\varepsilon}\left(6\tilde{v}^2\Delta + \frac{K^2}{\Delta}\right)$$

$$\langle q^4 \sigma_z \rangle = -2\tilde{\varepsilon}\tilde{v}(\tilde{v} + 1)(\tilde{v} + 2)$$

$$\langle q^4 \sigma_z \rangle_s = \tfrac{3}{2}\tilde{v}^2\{s\Delta - 4\tilde{\varepsilon}\}$$

$$\langle p^2 \rangle = \tilde{v}\{1 - 2\tilde{\varepsilon}^2(\tilde{v} + 1)\}$$

$$\langle p^2 \rangle_s = \tilde{v}\{1 + 2s\tilde{\varepsilon}\tilde{v}\Delta\}$$

For $K = 0$ we have the following exact relations:

$$\langle q^{2m}\sigma_z \rangle_s = s\langle q^{2m}\rangle_s$$

$$\langle q^{2m}\rangle_s = (1 + s\varepsilon)^{-m/2}\langle vl|q^{2m}|vl\rangle_{l=1} \tag{C.6}$$

$$\langle p^{2m}\rangle_s = (1 + s\varepsilon)^{+m/2}\langle vl|p^{2m}|vl\rangle_{l=1}$$

In particular, in agreement with (C.5),

$$\langle q^2 \rangle_s = (1 + s\varepsilon)^{-1/2}\tilde{v}$$

$$\langle q^4 \rangle_s = \tfrac{3}{2}(1 + s\varepsilon)^{-1}\tilde{v}^2 \tag{C.7}$$

$$\langle p^2 \rangle_s = (1 + s\varepsilon)^{1/2}\tilde{v}$$

The first result in (C.6) is trivial. To prove the second, first write the left-hand side as $^s\langle vl|q^{2m}|vl\rangle^s$, where $|vl\rangle^s$ is as in (6.2). Then write q^2 as $(\omega/\omega_s)(\gamma_s^{1/2}\rho)^2$, where $\gamma_s^{1/2}\rho$ is the dimensionless coordinate relating to the potential W_s. The last result is proven similarly.

When $H_{SO} \neq 0$, $\langle J_z \sigma_y \rangle^{SO}$ and $\langle q^{2m} \rangle^{SO}$ equal the corresponding quantities in (C.5) (to second order in $\varepsilon V + H_{SO}$), while (still to second order)

$$\langle q^m \sigma_z \rangle^{SO} = \langle q^m \sigma_z \rangle (1 + s_K \xi). \tag{C.8}$$

For $\langle \ \rangle_s^{SO}$ we find to first order in $\varepsilon V + H_{SO}$

$$\langle J_z \sigma_y \rangle_s^{SO} = sK \cos 2\beta$$

$$\langle q^2 \rangle_s^{SO} = \tilde{v} \{ 1 - 2s\tilde{\varepsilon}\Delta \sin 2\beta \}$$

$$\langle q^2 \sigma_z \rangle_s^{SO} = \tilde{v} \{ s\Delta \sin 2\beta - 2\tilde{\varepsilon}(1 + sK \cos 2\beta) \}$$

$$\langle q^4 \rangle_s^{SO} = \tfrac{1}{2}(3\tilde{v}^2 - K^2) + s(K \cos 2\beta - 6\tilde{\varepsilon}\tilde{v}^2 \Delta \sin 2\beta) \tag{C.9}$$

$$\langle q^4 \sigma_z \rangle_s^{SO} = s \{ \tfrac{3}{2}\tilde{v}^2 \Delta \sin 2\beta + (5\tilde{v}^2 - K^2 + 2)K \cos 2\beta \} - 6\tilde{\varepsilon}\tilde{v}^2$$

$$\langle p^2 \rangle^{SO} = \tilde{v} \{ 1 + 2s\tilde{\varepsilon}\Delta \sin 2\beta \}$$

For the nondiagonal matrix element

$$\langle F \rangle^{SO}_{+-} = \langle \Psi^{SO}_+ | F | \Psi^{SO}_- \rangle = \langle \Psi^{SO}_- | F | \Psi^{SO}_+ \rangle \tag{C.10}$$

one finds (to first order in $\varepsilon V + H_{SO}$),

$$\langle F \rangle^{SO}_{+-} = \{ F^+_{vv} - \varepsilon G \}^-_K \sin 2\beta - \{ F_{vv} - \varepsilon G^+ \}^+_K \cos 2\beta \tag{C.11}$$

APPENDIX D: THE TREATMENT OF NH₂ BY POPLE AND LONGUET-HIGGINS

In this appendix we investigate how the calculations of Pople and Longuet-Higgins[4] on NH_2 proceed when carried out with H from (5.1) instead of with their own I_θ (see the start of Section XII). Choosing λ as the force constant for W_+, and introducing the dimensionless coordinate q as in (5.4), the two potentials

$$W_+ = \omega \{ \tfrac{1}{2}q^2 + hq^4 \} \quad \text{and} \quad W_- = \omega \{ (\tfrac{1}{2} - f)q^2 + gq^4 \} \tag{D.1}$$

fit qualitatively with those in Fig. 1b if $f > \tfrac{1}{2}$ and $g > 0$. Insertion in H' in (5.1) gives

$$\frac{2H'}{\omega} = -fq^2 + (g+h)q^4 + fq^2 \sigma_z - (g-h)q^4 \sigma_z \tag{D.2}$$

When $K = 0$, H has a set of eigenstates which "feel" only the upper potential [see (3.13) and (3.14)]. In the hope that the same is approximately true for small K also, Pople and Longuet-Higgins guess at a set of eigenvectors

$$|\Psi_s\rangle \simeq 2^{-1/2}(|vK+\rangle + s|vK-\rangle), \qquad |K| \gg \tilde{v} \qquad \text{(D.3)}$$

which are those found in the limit $W_- \to W_+ \simeq \frac{1}{2}\omega q^2$ [i.e., the limit $\varepsilon \to 0$ in (6.10)]. The corresponding energies $E_s' = \langle \Psi_s|(H_\sigma + H')|\Psi_s\rangle = \omega\tilde{v} + \langle \Psi_s|H'|\Psi_s\rangle$ are found from (C.5) (with $\varepsilon = 0$). With $\Delta \simeq 1 - \frac{1}{2}(K/\tilde{v})^2$ we get

$$E_+' = \omega\{\tilde{v} + \frac{3}{2}h\tilde{v}^2 - \frac{1}{8}K^2(2f\tilde{v}^{-1} - g + 5h)\} \qquad \text{(D.4)}$$

We also find that $s = -1$ in (D.3) must be discarded because it leads to $E_-' < 0$. Pople and Longuet-Higgins estimated the values of ω, f, g, and h from the observed levels with $K = 0$, and found that (D.4) predicted several NH_2 levels very well. They stress in their conclusion that their "...theory is limited by a number of severe approximations." Some discussion of these has been given recently by Jungen, Hallin, and Merer[12] in the light of a numerical diagonalization of a more realistic hamiltonian. See also Dixon,[39] who first found that the upper potential in NH_2 is not quite like W_+ in (D.1) but has a little positive "bump" at $\rho = 0$.

APPENDIX E: THE SPIN–ORBIT INTERACTION IN SPIN DOUBLET STATES

Pople[15] postulated the form (7.10) of the effective spin–orbit interaction without a derivation from more fundamental principles. A satisfactory explanation of Pople's term must start out by adding to $\mathcal{H}$ in (2.1) the operator[40,41]

$$\mathcal{H}_{SO} = c\sum_i \mathbf{s}_i \cdot \mathbf{F}_i \qquad \text{(E.1)}$$

where c is a constant, $\mathbf{s}_i$ the spin (in units of $\hbar$) of the ith electron and

$$\mathbf{F}_i = \sum_\alpha Z_\alpha r_{i\alpha}^{-3}(\mathbf{r}_i - \mathbf{r}_\alpha) \times \mathbf{p}_i - \sum_{j \neq i} r_{ij}^{-3}(\mathbf{r}_i - \mathbf{r}_j) \times (\mathbf{p}_i - 2\mathbf{p}_j) \qquad \text{(E.2)}$$

Here Z_α is the charge number of the αth nucleus, $\mathbf{r}_\alpha$ is its position, $\mathbf{r}_i$ and $\mathbf{p}_i$ denote the position and momentum of the ith electron, and $r_{i\alpha}$ and r_{ij} are the distances $|\mathbf{r}_i - \mathbf{r}_\alpha|$ and $|\mathbf{r}_i - \mathbf{r}_j|$.

Following Pople, we confine ourselves to the case where the total spin is due to a single electron orbiting about a core with total spin zero, formed by the first $n-1$ electrons. More precisely we assume that ϕ_s can be factorized,

$$\phi_s(r, \rho) = \phi_{core}(r_1, \ldots, r_{n-1}, \rho)\chi_s(r_n, \rho) \qquad \text{(E.3)}$$

and that the trial functions are

$$\Phi = \sum_s \phi_s |0\rangle \eta_s, \qquad \eta_s = \sum_{S_z'} \eta_s(\rho, \chi, S_z') |S_z'\rangle \tag{E.4}$$

where $|0\rangle$ is a vector in the spin space of the first $n-1$ electrons such that $(\mathbf{s}_1 + \cdots + \mathbf{s}_{n-1})^2 |0\rangle = 0$, and where $|S_z'\rangle$, $S_z' = \pm \frac{1}{2}$, are orthonormal vectors in the spin space of the last electron such that $s_{nz} |S_z'\rangle = S_z' |S_z'\rangle$. For convenience we usually write $\mathbf{s}_n$ as $\hbar \mathbf{S}$ below. The component of $\mathbf{S}$ along $\mathbf{e}_r^0$ is called S_r^0, and the component along $\mathbf{e}_r$ is called S_r. The relations

$$\begin{Bmatrix} S_x \\ S_y \end{Bmatrix} = \begin{Bmatrix} \cos\chi & \sin\chi \\ -\sin\chi & \cos\chi \end{Bmatrix} \begin{Bmatrix} S_x^0 \\ S_y^0 \end{Bmatrix} \tag{E.5}$$

where χ is the angle from $\mathbf{e}_x^0$ to $\mathbf{e}_x$, show that only one of the sets (S_x, S_y) and (S_x^0, S_y^0) can be independent of χ. The same goes for all the sets (s_{ix}, s_{iy}) and (s_{ix}^0, s_{iy}^0), $i = 1, 2, \ldots$, and we choose the s_{ir}^0 components to be χ-independent, since then $J_z \equiv -i \partial/\partial \xi$ retains the meaning of total orbital angular momentum. (With the s_{ir} chosen χ-independent, J_z acquires the meaning of orbital plus spin angular momentum and we must make the replacement $J_z \to J_z - S_z$ in T_n.)

Proceeding as in Section IV, we see that the addition of $\mathcal{H}_{\mathrm{SO}}$ to $\mathcal{H}$ results in the addition of

$$H_{\mathrm{SO}} = \begin{Bmatrix} h_{++} & h_{+-} \\ h_{-+} & h_{--} \end{Bmatrix}, \qquad h_{ss'} = \langle \phi_s | \langle 0 | \mathcal{H}_{\mathrm{SO}} | 0 \rangle | \phi_{s'} \rangle, \tag{E.6}$$

to H from (4.1). With $\langle 0 | \mathbf{s}_i | 0 \rangle = 0$ for $i < n$ we get

$$\langle 0 | \mathcal{H}_{\mathrm{SO}} | 0 \rangle = c\hbar \left(S_x F_x + S_y F_y + S_z F_z \right) \tag{E.7}$$

Since $F_r \equiv \mathbf{e}_r \cdot \mathbf{F}$ is hermitian and $\langle \phi_s | F_r | \phi_{s'} \rangle$ imaginary, we have $h_{ss'} = -h_{s's}$, and since F_y is even under reflection in the molecular plane, we obtain

$$H_{\mathrm{SO}} = i h_{+-} \sigma_y = \left(A_x S_x + A_z S_z \right) \Lambda \sigma_y \tag{E.8}$$

where A_x and A_z are real functions of $\rho = \gamma^{-1/2} q$ given by

$$\begin{aligned} A_x &= i(\hbar\Lambda)^{-1} \langle \phi_+ | F_x | \phi_- \rangle = A'q + A'''q^3 + \cdots \\ A_z &= i(\hbar\Lambda)^{-1} \langle \phi_+ | F_z | \phi_- \rangle = A^0 + A''q^2 + A''''q^4 + \cdots \end{aligned} \tag{E.9}$$

$A^0, A', A'',\dots$, being constants. This result involves the straightforward use of some symmetry considerations [For example, $\langle \phi_+^0 | F_x | \phi_-^0 \rangle = 0$ because F_x can only add and subtract one unit of angular momentum. As to the missing even powers of ρ from A_x and odd powers of ρ from A_z, see (A.4) or (A.10). Note also that the term $r_{\alpha x} p_{ny}$ in F_z can transfer one unit of angular momentum, but is linear in ρ and thus gives no contribution to A^0.] Noting the rearrangement

$$qS_x = \frac{q_+ S_-^0 + q_- S_+^0}{2} \tag{E.10}$$

where $q_\pm = q\exp(\pm i\chi)$ and $S_\pm^0 = S_x^0 \pm iS_y^0$, it follows that for not too large vibrational amplitudes one has

$$H_{SO} = \left\{ A^0 S_z + \tfrac{1}{2}A'\left(q_+ S_-^0 + q_- S_+^0\right) + A''q^2 S_z \right\} \Lambda \sigma_y \tag{E.11}$$

where A^0, A', and A'' are constants. If, as a first approximation, we take only the A^0 term into account, $H + h$ commutes with S_z, which can therefore be replaced by its eigenvalue S_z'. The resulting H_{SO} is that of (7.10) (where the upper index on the constant A^0 is dropped). We may take the A' and the A'' terms into account as perturbations to the first approximation. The A' term gives no contribution to the energy in first order, and its second-order contribution is expected to be comparable with the first-order contribution from the A'' term.

We note that we would obtain the same H_{SO} as in (E.11) if we replace $\mathcal{H}_{SO}$ in (E.1) with

$$\mathcal{H}_{SO} = \left(A^0 + C^0 q^2\right)L_z S_z + B^0\left(L_x S_x + L_y S_y\right) \tag{E.12}$$

($= A^0 \mathbf{L}\cdot\mathbf{S}$ if $A^0 = B^0$ and $C^0 = 0$) and choose the constants B^0 and C^0 correctly. The connection between A'' and C^0 is seen to be

$$A'' = C^0 - g_K\frac{A^0}{\omega} \tag{E.13}$$

[approximate $g(\rho) \simeq g(0) \equiv g_K$ in (4.6) and insert $q = \gamma^{1/2}\rho$], where g_K is the constant from (4.12). Therefore, in a model that takes the spin–orbit interaction into account by $\mathcal{H}_{SO}$ from (E.12), the energy levels will depend slightly on the parameter g_K. Brown[17] and Gauyacq and Jungen[19] give some discussion of this dependence for the hypothetical case where $B^0 = C^0 = 0$.

Finally it should be mentioned that the above derivation of H_{SO} is not quite satisfactory, because the trial functions (E.4) have not been antisymmetrized in accordance with the Pauli principle. We do not attempt to correct for this in the present chapter.

APPENDIX F: INCLUSION OF THE SPIN VARIABLE IN THE BASIS VECTORS

The spin–orbit operator H_{SO} in (7.10) was obtained by replacing the operator S_z for the electron spin with its eigenvalue $S_z' = \pm\frac{1}{2}$. By doing so we avoided the introduction of the spin operator and the extension of the space of wave functions from that spanned by the $|vKs\rangle$ to that spanned by $|vKsS_z'\rangle = |vKs\rangle|S_z'\rangle$, where

$$|S_z' = +\tfrac{1}{2}\rangle = \begin{Bmatrix} 1 \\ 0 \end{Bmatrix}_S \quad \text{and} \quad |S_z' = -\tfrac{1}{2}\rangle = \begin{Bmatrix} 0 \\ 1 \end{Bmatrix}_S \qquad \text{(F.1)}$$

are the two eigenvectors of the spin operator

$$S_z = \frac{1}{2}\begin{Bmatrix} 1 & 0 \\ 0 & -1 \end{Bmatrix}_S \qquad \text{(F.2)}$$

(The index S indicates that the operator only acts on columns marked S.) In other contexts these quantities cannot be avoided, and we therefore discuss them briefly. The unitary (and hermitian) operator $S_\sigma = S_\chi\sigma_z$ from (3.15) and (5.11d) no longer commutes with H when H_{SO} is included. However, the unitary (and antihermitian) operator

$$R_\sigma = S_\sigma\begin{Bmatrix} 0 & -1 \\ 1 & 0 \end{Bmatrix}_S = \begin{Bmatrix} 0 & -1 \\ 1 & 0 \end{Bmatrix}_S S_\sigma \qquad \text{(F.3)}$$

does, and we have

$$R_\sigma|vKs,\pm\tfrac{1}{2}\rangle = \pm|v,-K,-s,\mp\tfrac{1}{2}\rangle \qquad \text{(F.4)}$$

[Note that the operator $\{\ \}_S$ in (F.3) commutes with σ_z because the two operators act on the different columns $\{\ \}_S$ and $\{\ \}$ in the wave functions.] It is now clear that each eigenvalue of $H + H_{SO}$ must be doubly degenerate, containing both the eigenvalues (K, S_z') and $(-K, -S_z')$ of (J_z, S_z). In fact this degeneracy remains even when the complete operator H_{SO} from (E.8) is used (R_σ commutes with the complete H_{SO} because it changes the sign of $S_x = \cos\chi\, S_x^0 + \sin\chi\, S_y^0$ as well as $S_z = S_z^0$). Although the complete H_{SO} does not commute with J_z and S_z, it does commute with the sum, implying that $P = K + S_z'$ is a good quantum number. [This fact also follows as a direct consequence of the cylindrical symmetry. The fact that each level contains a state with $+P$ as well as $-P$ can also be seen as a consequence of time-reversal symmetry. Actually the double degeneracy is an example of Kramer's degeneracy (Reference 27, p. 675).]

When end-over-end rotation of the molecule is allowed for, it becomes awkward to use a representation in which the molecule-fixed components S_x and S_y of the spin operators depend on the orientation of the rotating frame.

It is more natural to use S_z as in (F.2),

$$S_x = \frac{1}{2}\left\{\begin{matrix}0 & 1\\ 1 & 0\end{matrix}\right\}_s \quad \text{and} \quad S_y = \frac{1}{2}\left\{\begin{matrix}0 & -i\\ i & 0\end{matrix}\right\}_s \qquad (F.5)$$

When this change is made, $J_z = -i\partial/\partial\chi$ aquires the meaning of orbital plus spin angular momentum and we must therefore substitute $J_z \to J_z - S_z$ and replace the vector $|vKsS_z'\rangle$ with

$$|vKsS_z'\rangle = \tfrac{1}{2}\pi^{-1/2}R_{vl}(\rho)\left\{\begin{matrix}1\\ si\end{matrix}\right\}|S_z'\rangle e^{i(K+S_z')\chi} \qquad (F.6)$$

where $|S_z'\rangle$ is as in (F.1). The hamiltonian (5.1), including the full spin–orbit operator from (E.8) and the correction ΔH from (4.11), then takes the form

$$H = H_\sigma + H' + H_{SO}$$
$$H_\sigma = \tfrac{1}{2}\left\{P_\rho^2 + \left\{(J_z - \Lambda\sigma_y - S_z)^2 - \tfrac{1}{4}\right\}\rho^{-2} + \lambda\rho^2\right\}$$
$$H' = W - \tfrac{1}{2}\lambda\rho^2 + \hat{W}\sigma_z + w + \hat{w}\sigma_z + g(J_z - S_z)\Lambda\sigma_y$$
$$H_{SO} = (A_x S_x + A_z S_z)\Lambda\sigma_y$$
$$(= AS_z\Lambda\sigma_y \quad \text{at} \quad \rho = 0) \qquad (F.7)$$

where the real function $A_x(\rho)$ is odd $[= -A_x(-\rho)]$ and the real function $A_z(\rho)$ even $[= +A_z(-\rho)]$.

We note that if we subject H and $|vKsS_z'\rangle$ in the new representation to the unitary transformation $\exp(-i\chi S_z)$, we recover H and $|vKsS_z'\rangle$ in the old representation. Thus $J_z - S_z$ is transformed into just J_z, and S_x is transformed into

$$\left\{\begin{matrix}0 & 1\\ 1 & 0\end{matrix}\right\}_s \cos\chi + \left\{\begin{matrix}0 & -i\\ i & 0\end{matrix}\right\}_s \sin\chi$$

which is the operator for S_x in the old representation. Hence the matrix elements $\langle v'K's'S_z'|H|v''K''s''S_z''\rangle$ are the same in both representations (as they should be).

The wave functions in (F.6) satisfy

$$H_\sigma|vKsS_z'\rangle = \omega(v+1)|vKsS_z'\rangle$$
$$\sigma_y|vKsS_z'\rangle = s|vKsS_z'\rangle$$
$$J_z|vKsS_z'\rangle = (K+S_z')|vKsS_z'\rangle$$
$$S_z|vKsS_z'\rangle = S_z'|vKsS_z'\rangle$$
$$S_\mp|vKs\pm\tfrac{1}{2}\rangle = e^{\pm i\chi}|vKs\mp\tfrac{1}{2}\rangle$$
$$R_\sigma|vKs\pm\tfrac{1}{2}\rangle = \pm|v,-K,-s,\mp\tfrac{1}{2}\rangle$$

$$\text{(F.8)}$$

$(S_{\pm} = S_x \pm iS_y)$, and for any operator F that is independent of J_z, S_x, S_y, and S_z we have

$$\langle v'K's'S_z'| F | vKsS_z'\rangle = \langle v'K's'| F | vKs\rangle \qquad (F.9)$$

where $|vKs\rangle$ and $|v'K's'\rangle$ are as in Section V. To each function

$$|\Psi\rangle = \sum c_{vKsS_z'} | vKsS_z'\rangle \qquad (F.10)$$

in the space of the effective hamiltonian H, there corresponds, via the substitutions of (3.6), a function Φ in the space of the real hamiltonian $\mathcal{H}$, and vice versa. In particular, an eigenfunction $|\Psi\rangle$ of H corresponds to an eigenfunction Φ of $\mathcal{H}$.

The action of the time-reversal operator $\mathcal{T}$ on $|\Psi\rangle$ in (F.10) must be defined in accordance with this correspondence. Hence, from the known action of $\mathcal{T}$ on Φ (Reference 27, p. 670) and from the discussion of time reversal in Section V, we find that

$$\mathcal{T} |\Psi\rangle = R_\sigma \sum c^*_{vKsS_z'} | vKsS_z'\rangle \qquad (F.11)$$

i.e., $\mathcal{T} = R_\sigma \mathcal{C} = \mathcal{C} R_\sigma$, where $\mathcal{C}$ is the complex-conjugation operator relative to the $|vKsS_z'\rangle$ basis. For comparison with Reference 27, p. 670, note that the $\{\ \}_S$ matrix in R_σ can be written as

$$\begin{Bmatrix} 0 & -1 \\ 1 & 0 \end{Bmatrix}_S = -i \begin{Bmatrix} 0 & -i \\ i & 0 \end{Bmatrix}_S = e^{-i\pi S_y} \qquad (F.12)$$

We see that $\mathcal{T}$ changes the sign of S_x, S_y, and S_z, and transforms the remaining variables as described in Section V. Hence the hamiltonian H in (F.7) is invariant under time reversal. We note that if an operator F commutes with R_σ, its matrix elements are all real (pure imaginary) if, and only if, F is even (odd) under $\mathcal{T}$. If $FR_\sigma = -R_\sigma F$, the matrix elements are pure imaginary (real) if, and only if, F is even (odd).

Acknowledgments

We are grateful to Drs. Christian Jungen and Jim Watson for valuable comments on an earlier draft of the manuscript.

References

1. R. Renner, Z. Phys. **92**, 172 (1934).
2. H. H. Nielsen, Rev. Mod. Phys. **23**, 90 (1951).
3. K. Dressler and D. A. Ramsay, J. Chem. Phys. **27**, 971 (1957).
4. J. A. Pople and H. C. Longuet-Higgins, Mol. Phys. **1**, 372 (1958).

5. H. C. Longuet-Higgins, *Adv. Spectrosc.* **2**, 429 (1961).

6. A. Carrington, A. R. Fabris, B. J. Howard, and N. J. D. Lucas, *Mol. Phys.* **20**, 1961 (1971).

7. B. J. Howard, Ph.D. Thesis, 1970, University of Southampton.

8. T. Barrow, R. N. Dixon, and G. Duxbury, *Mol. Phys.* **27**, 1217 (1974).

9. G. Duxbury, *Molecular Spectroscopy*, Chemical Society Specialist Periodical Report, Vol. 3, 1975, p. 497.

10. A. J. Merer and Ch. Jungen, in K. Narahari Rao, Ed., *Molecular Spectroscopy: Modern Research*, Vol. II, Academic Press, New York, 1976, Chapter 3.

11. Ch. Jungen and A. J. Merer, *Mol. Phys.* **40**, 1 (1980).

12. Ch. Jungen, J. K-E. Hallin, and A. J. Merer, *Mol. Phys.*, **40**, 25 (1980).

13. Ch. Jungen, J. K-E. Hallin, and A. J. Merer, *Mol. Phys.* **40**, 65 (1980).

14. Ch. Jungen and A. J. Merer, *Mol. Phys.* **40**, 95 (1980).

15. A. J. Pople, *Mol. Phys.* **3**, 16 (1960).

16. C. F. Chang, and Y. N. Chiu, *J. Chem. Phys.* **53**, 2186 (1970).

17. J. M. Brown, *J. Mol. Spectrosc.* **68**, 712 (1977).

18. J. F. M. Aarts, *Mol. Phys.* **35**, 1785 (1978).

19. D. Gauyacq and Ch. Jungen, *Mol. Phys.* **41**, 383 (1981).

20. A. J. Merer and D. N. Travis, *Can. J. Phys.* **43**, 1795 (1965).

21. J. T. Hougen and J. P. Jesson, *J. Chem. Phys.* **38**, 1524 (1963).

22. J. T. Hougen, *J. Chem. Phys.* **36**, 519 (1962).

23. J. M. Brown and F. Jørgensen, *Mol. Phys.* **47** (to be published).

24. C. Di Lauro and I. M. Mills, *J. Mol. Spectrosc.* **21**, 386 (1966).

25. M. Born, *Gött. Nachr. Mat. Phys.* **6**, 1 (1951).

26. M. Born and K. Huang, *Dynamical Theory of Crystal Lattices*, Clarendon Press, Oxford, 1954.

27. A. Messiah, *Quantum Mechanics*, North-Holland, 1962.

28. M. Born, and J. R. Oppenheimer, *Ann. Physik* **84**, 457 (1927).

29. I. M. Mills, Vibration–rotation structure in asymmetric and symmetric top molecules, review article in K. N. Rao and C. W. Mathews, Eds., *Molecular Spectroscopy in Modern Research*, Academic, 1972.

30. C. E. Soliverez, *J. Phys.* **C2**, 2161 (1969).

31. F. Jørgensen, *Mol. Phys.* **29**, 1137. (1975).

32. B. Brandow, *Int. J. Quantum Chem.* **XV**, 207 (1979).

33. C. Bloch, *Nucl. Phys.* **6**, 329 (1958).

34. J. von Neumann and E. P. Wigner, *Z. Phys.* **30**, 467 (1929).

35. G. Herzberg, *Electronic Spectra of Polyatomic Molecules*, Van Nostrand—Reinhold, Princeton, N.J., 1966.

36. H. W. Kroto, *Can. J. Phys.* **45**, 1439 (1967).

37. G. Herzberg and E. Teller, *Z. Phys. Chem. (B)* **21**, 410 (1933).

38. M. Born and S. Flügge, *Ann. Phys.* **16**, 768 (1933).

39. R. N. Dixon, *Mol. Phys.* **9**, 357 (1965).

40. B. J. Howard and R. E. Moss, *Mol. Phys* **20**, 147 (1971).

41. J. H. Van Vleck, *Rev. Mod. Phys.* **23**, 213 (1951).

COUPLED-CLUSTER APPROACH IN THE ELECTRONIC-STRUCTURE THEORY OF MOLECULES

VLADIMÍR KVASNIČKA, VILIAM LAURINC, and STANISLAV BISKUPIČ

Faculty of Chemistry
Slovak Technical University
812 37 Bratislava, Czechoslovakia

MIROSLAV HARING

Computing Centre
Slovak Academy of Sciences
842 35 Bratislava, Czechoslovakia

CONTENTS

At present, the diagrammatic perturbation theory[1-3] is one of the most promising techniques for accounting for the correlation effects of small molecular systems. Its effectiveness is considerably increased when it is reformulated by making use of the coupled-cluster approach.[4-15] The individual perturbation terms are determined by a string of coupled nonlinear equations. Solving this string, we may simultaneously construct and evaluate the perturbation terms of arbitrarily high order and prescribed form of topology. That is, the coupled-cluster approach offers a relatively simple theory to evaluate infinite summations of diagrammatic perturbation terms. The resulting correlation energies are fully competitive with their counterparts calculated by the CI method. In this connection it should be emphasized that the coupled-cluster approach realization of many-body diagrammatic perturbation theory may be considered as a diagrammatic visualization of the so-called direct CI approach.[16-17] In particular, if the formal Rayleigh–Schrödinger perturbation theory is used for the diagonalization of the CI eigenvalue problem, then between CI and the coupled cluster approach there is a close one-to-one correspondence. The diagrammatic perturbation theory and the CI method represent, in fact, alternative realizations of a general method based upon a diagonalization of the total hamiltonian in a subspace spanned by preselected (in general, all possible) unperturbed state vectors (configurations).

II. NONDEGENERATE CCA

The nondegenerate version of CCA was initially elaborated in the framework of microscopic nuclear theory by Coester and Kümmel.[4-8] In the original pioneering article of Coester[4] the CCA was presented as an alternative proof of the fact that the nondegenerate ground-state energy of a many-nucleon system is determined by connected diagrammatic terms. The essential advance was achieved by Kümmel[7] when he derived a string of coupled nonlinear equations which determines the matrix elements of cluster operators. As will be shown below, this feature of the CCA allows one to formulate very effective procedures for the evaluation of infinite summations of diagrammatic terms with prescribed form of topology and level of excitations of intermediate virtual unperturbed states. This aspect of the CCA was further developed by Čížek and Paldus[9-15] in a form highly convenient for quantum-chemistry studies of correlation effects in molecular systems.

A. Synopsis of Nondegenerate Diagrammatic RSPT

Let us study a many-electron molecular system. Its *total hamiltonian* in the second-quantization form is[18-20]

$$H = \sum_{ij} \langle i|h|j \rangle X_i^\dagger X_j + \frac{1}{2} \sum_{ijkl} \langle ij|v|kl \rangle X_i^\dagger X_j^\dagger X_l X_k \qquad (2.1)$$

where $X_i^\dagger$ and X_j are annihilation and creation operators, respectively, obeying the usual Fermi anticommutation rules, and these operators are defined over an orthonormal set of one-particle functions. The matrix elements $\langle i|h|j \rangle$ and $\langle ij|v|kl \rangle$ correspond to the one- and the two-particle part of the total hamiltonian, respectively; the operator h describes the kinetic energy of an electron plus its attractive Coulomb interaction with the fixed lattice of nuclei; the operator v is the simple two-particle repulsive Coulomb interaction between electrons.

The hole–particle formalism[19-20] is introduced with respect to an unperturbed ground-state vector $|\Phi_0\rangle$; the one-particle states that are occupied (unoccupied) in $|\Phi_0\rangle$ are called the *hole* (*particle*) states. By using the state vector $|\Phi_0\rangle$ as a new reference vacuum state, the total hamiltonian (2.1) may be rewritten in a *normally ordered form*,

$$H = \langle \Phi_0|H|\Phi_0 \rangle + \sum_{ij} \langle i|f|j \rangle N \left[X_i^\dagger X_j \right]$$

$$+ \frac{1}{2} \sum_{ijkl} \langle ij|v|kl \rangle N \left[X_i^\dagger X_j^\dagger X_l X_k \right] \qquad (2.2)$$

where $N[\cdots]$ is the normal product of operators, and f is the Hartree–Fock (HF) operator defined with respect to the state $|\Phi_0\rangle$. In the framework of HF theory, the matrix elements $\langle i|f|j \rangle$ are diagonal, that is, $\langle i|f|j \rangle = \varepsilon_i \delta_{ij}$, where the ε's are the HF one-particle (orbital) energies. Introducing this assumption into (2.2), we get

$$H = \langle \Phi_0|H|\Phi_0 \rangle + H_0 + H_1 \qquad (2.3a)$$

$$H_0 = \sum_i \varepsilon_i N \left[X_i^\dagger X_i \right] \qquad (2.3b)$$

$$H_1 = \frac{1}{2} \sum_{ijkl} \langle ij|v|kl \rangle N \left[X_i^\dagger X_j^\dagger X_l X_k \right] \qquad (2.3c)$$

The *unperturbed hamiltonian* H_0 is a diagonal operator in the entire Hilbert space spanned by unperturbed states,

$$H_0|\Phi_0\rangle = 0 \qquad (2.4a)$$

$$H_0 X_{p_1}^\dagger \ldots X_{p_m}^\dagger X_{h_1} \ldots X_{h_n}|\Phi_0\rangle$$

$$= \left(\varepsilon_{p_1} + \cdots + \varepsilon_{p_m} - \varepsilon_{h_1} - \cdots - \varepsilon_{h_n}\right) X_{p_1}^\dagger \ldots X_{p_m}^\dagger X_{h_1} \ldots X_{h_n}|\Phi_0\rangle \qquad (2.4b)$$

for $p_1,\ldots,p_m \notin \Phi_0$ and $h_1,\ldots,h_n \in \Phi_0$. The *perturbation* H_1 corresponds to the net correlation effects; it gives rise to a lowering of the total energy with respect to the HF ground-state energy $\langle\Phi_0|H|\Phi_0\rangle$. The total hamiltonian H can be written in the form (2.3) for both closed-shell and open-shell systems, but for open-shell systems the HF scheme used should be of the unrestricted type,[21] (UHF) in contrast to closed-shell systems, for which the restricted (RHF) theory[22] is easily applicable.

A second alternative form of the normally ordered total hamiltonian (2.2) is

$$H = \langle\Phi_0|H|\Phi_0\rangle + H_0 + H_1 \qquad (2.5a)$$

$$H_0 = \sum_i \varepsilon_i N\left[X_i^\dagger X_i\right] \qquad (2.5b)$$

$$H_1 = \frac{1}{2}\sum_{ijkl}\langle ij|v|kl\rangle N\left[X_i^\dagger X_j^\dagger X_l X_k\right]$$

$$-\sum_{ij}\langle i|w|j\rangle N\left[X_i^\dagger X_j\right] \qquad (2.5c)$$

This form of the H differs, formally, from the previous (2.3) only in the perturbation H_1: now the operator H_1 is composed of one- as well as two-particle terms. A hermitian one-particle operator w in (2.5) determines an actual form of the orthonormal set of one-particle functions used: they are determined by $\langle i|(f+w)|j\rangle = \varepsilon_i\delta_{ij}$. The presented second alternative form of the total hamiltonian (2.5) is appropriate if other than the HF one-particle functions are used, or in the framework of RHF theory for open-shell molecular systems.[23]

In order to sketch the nondegenerate diagrammatic Rayleigh-Schrödinger perturbation theory (RSPT), we have, first of all, to introduce a diagrammatic interpretation of the perturbation H_1 (see Fig. 1). In our considerations we shall use the Hugenholtz diagrammatic terminology.[2] Following

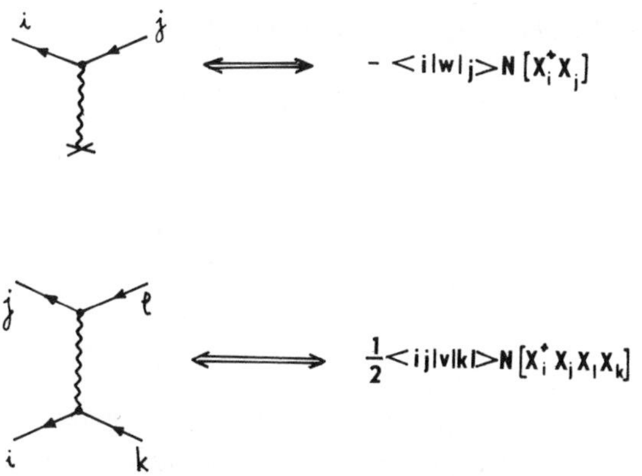

Fig. 1.   The diagrammatic interpretation of one- and two-particle terms of the perturbation H_1.

Goldstone,[1] the perturbed ground-state energy is diagrammatically determined by

$$E_0 = \langle \Phi_0 | H | \Phi_0 \rangle + \sum_{n \geqslant 1} \langle \Phi_0 | H_1 \left(\frac{1}{-H_0} H_1 \right)^n | \Phi_0 \rangle_C \qquad (2.6)$$

where the subscript C means that only the *connected* diagrammatic terms are considered. The corresponding perturbed ground-state vector (with intermediate normalization, $\langle \Phi_0 | \Psi_0 \rangle = 1$) is

$$|\Psi_0\rangle = |\Phi_0\rangle + \sum_{\alpha \neq 0} c_\alpha | \Phi_\alpha \rangle \qquad (2.7a)$$

$$c_\alpha = \frac{1}{D_\alpha} \sum_{n \geqslant 0} \langle \Phi_\alpha | H_1 \left(\frac{1}{-H_0} H_1 \right)^n | \Phi_0 \rangle_L \qquad (2.7b)$$

where $|\Phi_\alpha\rangle$ is an *l*-excited $(l = 1, 2, \ldots)$ unperturbed state vector, $|\Phi_\alpha\rangle = X^\dagger_{p_1} \ldots X^\dagger_{p_l} X_{h_1} \ldots X_{h_l} |\Phi_0\rangle$, and the denominator D_α is determined as a negative eigenenergy of $|\Phi_\alpha\rangle$, $H_0 | \Phi_\alpha \rangle = (-D_\alpha) | \Phi_\alpha \rangle$, $D_\alpha = \varepsilon_{h_1} + \cdots + \varepsilon_{h_l} - \varepsilon_{p_1} - \cdots - \varepsilon_{p_l}$. The subscript L in (2.7b) means that we take into account only the *linked* diagrams.

Hugenholtz[2] and Hubbard[3] have demonstrated that the perturbed state vector (2.7) may be written in an exponential form,

$$|\Psi_0\rangle = e^T | \Phi_0 \rangle \qquad (2.8a)$$

where the so-called *cluster operator* T is separated into various terms

$$T = \sum_{l \geq 1} T_l = T_1 + T_2 + T_3 + \cdots \qquad (2.8b)$$

The *l*-excited cluster operator T_l is determined as

$$T_l = \frac{1}{l!} \sum_{\substack{h_1,\ldots,h_l \in \Phi_0 \\ p_1,\ldots,p_l \notin \Phi_0}} \langle p_1 \ldots p_l | | h_1 \ldots h_l \rangle N\left[X_{p_1}^\dagger \ldots X_{p_l}^\dagger X_{h_l} \ldots X_{h_1} \right] \qquad (2.8c)$$

The matrix elements $\langle p_1 \ldots p_l | | h_1 \ldots h_l \rangle$ are diagrammatically determined by all *linked* and *connected* diagrammatic terms with l left hole–particle pairs of external lines,

$$\langle p_1 \ldots p_l | | h_1 \ldots h_l \rangle$$

$$= \frac{1}{D} \sum_{n \geq 0} \langle \Phi_0 | X_{h_l}^\dagger X_{p_l} \cdots X_{h_1}^\dagger X_{p_1} H_1 \left(\frac{1}{-H_0} H_1 \right)^n | \Phi_0 \rangle_{LC} \qquad (2.8d)$$

The energy denominator $D = \varepsilon_{h_1} + \cdots + \varepsilon_{h_l} - \varepsilon_{p_1} - \cdots - \varepsilon_{p_l}$ is a negative energy of the *l*-excited unperturbed state vector $X_{p_1}^\dagger X_{h_1} \cdots X_{p_l}^\dagger X_{h_l} | \Phi_0 \rangle$.

The algebraic interpretation of diagrams appearing in the diagrammatic RSPT is easily carried out by making use of the following overall numerical factor[3,20]:

$$(-1)^{h+l+n_1} w_\Gamma \qquad (2.9)$$

where h, l, n_1 denote the numbers of internal hole lines, closed loops, and one-particle vertices, respectively, and w_Γ is the weight factor. All the terms are assigned to a given diagram Γ. The resulting algebraic interpretation is summed over all internal one-particle states; the so-called exclusion-principle-violating (EPV) diagrams should be considered. Assuming that a studied molecular system is of the closed-shell type and that the one-particle functions were constructed by the RHF theory [that is, the total hamiltonian is determined by (2.3)], the algebraic interpretation of diagrams is essentially simplified into a spin-free form. The overall numerical factor is[20]

$$(-1)^{h+l} (2)^l w_\Gamma \qquad (2.10)$$

with the same meaning of its individual terms as in (2.9). But now, the resulting algebraic form is summed over only *spatial* parts of all internal one-particle states.

During the last decade the nondegenerate diagrammatic RSPT has been extensively used for the calculation of correlation energies of small molecules in limited basis sets of one-particle functions.[24] It was definitely demonstrated that this many-body approach offers very an effective technique to account for the correlation effects of molecular systems; the results obtained are in simple one-to-one correspondence with the correlation energies calculated by other many-body techniques (e.g., the CI method). The main obstacle to a wider exploitation of the diagrammatic RSPT is that the number of diagrams increases rapidly for higher-order perturbation terms. For instance, the third-order correlation energy is determined by 12 Feynman-Goldstone diagrams, whereas its fourth-order value is determined by 222 diagrams.[25] Therefore, it seems hopeless to apply the diagrammatic RSPT directly for the exact calculation of the correlation energy to a high order of perturbation theory. As a first step in the exploitation of the diagrammatic RSPT in quantum chemistry, this problem has been solved partially by the so-called denominator-shift technique.[26] This consists of an infinite summation of some preselected diagrammatic insertions via the geometric power series. Usually, such a simple approach led to an overestimation[27] of correlation energies. Therefore another, more general theory including the summation of diagonal as well as nondiagonal diagrammatic insertions appears to be highly desirable.

B. Exponential Wave Operator

The principal starting *Ansatz* of any version of the CCA is that the perturbed ground-state vector $|\Psi_0\rangle$ is expressed via an *exponential wave operator*,

$$|\Psi_0\rangle = U|\Phi_0\rangle \tag{2.11a}$$

$$U = e^T \tag{2.11b}$$

which is the intrinsic result of the diagrammatic RSPT; see Eq. (2.8a). Limiting ourselves to the case of a nondegenerate ground state, the cluster operator T is determined in the second-quantization formalism by (2.8b) and (2.8d). The CI form of the perturbed state vector $|\Psi_0\rangle$ constructed with respect to the unperturbed state vector $|\Phi_0\rangle$ may be written (we use the intermediate normalization, $\langle\Phi_0|\Psi_0\rangle = 1$) as

$$|\Psi_0\rangle = \left(1 + \sum_{l \geq 1} C_l\right)|\Phi_0\rangle \tag{2.12}$$

where C_l is an operator creating a fixed linear combination of all *l*-excited states (configurations). Introducing (2.8b) into (2.11), we get

$$|\Psi_0\rangle = \sum_{n_1, n_2, \ldots \geq 0} \frac{(T_1)^{n_1}(T_2)^{n_2}\cdots}{n_1! n_2! \cdots}|\Phi_0\rangle$$

$$= |\Phi_0\rangle + T_1|\Phi_0\rangle + \left(T_2 + \frac{1}{2!}T_1^2\right)|\Phi_0\rangle$$

$$+ \left(T_3 + T_1T_2 + \frac{1}{3!}T_1^3\right)|\Phi_0\rangle + \cdots \qquad (2.13)$$

Hence, by relating the terms on the right sides of (2.12) and (2.13), we arrive at

$$C_1 = T_1 \qquad (2.14a)$$

$$C_2 = T_2 + \frac{1}{2!}T_1^2 \qquad (2.14b)$$

$$C_3 = T_3 + T_1T_2 + \frac{1}{3!}T_1^3 \qquad (2.14c)$$

and so on. The following terminology is used[13]: The single terms are called *connected*, whereas the terms expressed as products of connected operators are called *disconnected*. The usual terminology is "linked" and "unlinked" instead of "connected" and "disconnected," respectively. This original terminology is slightly misleading from the standpoint of Hugenholtz's nomenclature of diagrammatic terms[2], where the terms linked and unlinked are reserved for a special type of diagrams which also may be connected and disconnected, respectively. A "connected term" ("disconnected term") is there interpreted as a term that is composed of a one (two or more) component(s).

The system of equations (2.14) may be simply solved for cluster operators

$$T_1 = C_1 \qquad (2.15a)$$

$$T_2 = C_2 - \tfrac{1}{2}C_1^2 \qquad (2.15b)$$

$$T_3 = C_3 - C_1C_2 - \tfrac{1}{3}C_1^3 \qquad (2.15c)$$

and so on. This means that the formulas (2.11) and (2.12) are fully equivalent: if we know the CI form (2.12) of $|\Psi_0\rangle$, then its exponential form (2.11) is unambiguously constructed by (2.15), and conversely.

Following Hugenholtz[2] and Primas,[28] an arbitrary many-body technique used for the calculation of correlation effects must fulfil the so-called *separability condition* (see also References 29 and 30). If a quantal system is composed of a number of noninteracting subsystems, then the total energy and the corresponding state vector should be additive and multiplicative, respectively. This principal property may be simply demonstrated when the exponential form (2.11) is used. For simplicity, let us assume that a given molecular system is separated into two noninteracting subsystems A and B; then the cluster operator T is split into two mutually commuting parts assigned to A and B, respectively:

$$T = T_A + T_B \qquad (2.16)$$

The total wave operator is equal to the simple product of the wave operators corresponding to A and B:

$$U = e^T = e^{T_A + T_B} = e^{T_A} e^{T_B} = U_A U_B \qquad (2.17)$$

The total hamiltonian H is easily separated into components of individual subsystems, $H = H_A + H_B$. Hence, the total energy is additive with respect to its constituents, $E = E_A + E_B$, which was to be demonstrated. Hugenholtz[2] was first (to our knowledge) to relate this separability condition to the diagrammatic RSPT. He has used the separability condition as the most general proof of the "linked-cluster" theorem (2.6) for the ground-state energy, which states that this quantity is determined by connected diagrammatic terms. Recently, in the quantum-chemistry literature, different many-body techniques have been reexamined to see whether they are *size-consistent* or not[24,31-33]; this is fully equivalent to our separability condition. For instance, let us study a system of N noninteracting H_2 molecules. Its total energy is determined by $E(NH_2) = NE(H_2)$. A truncated CI method involving only mono- and diexcited configurations does not have this property: for $N \to \infty$ the CI(0+1+2) energy varies as[31,32] $\sqrt{N}$. A molecule may be treated formally in a zeroth approximation as a system composed of noninteracting electron-pair bonds as in H_2, but it is obvious that a general many-body technique to be used for larger and larger molecules must fulfill the separability condition, or in other words, it must be size-consistent. We emphasize that the diagrammatic RSPT and the CCA are, *a priori*, of a size-consistent type, since both these techniques determine the correlation energy via connected diagrammatic terms.

C. "Linked-Cluster" Theorem

The nondegenerate version of the CCA can be formulated in the most concise way by the so-called "linked-cluster" theorem (LCT) of Coester and Kümmel[3-8] (for details see Reference 34). This theorem deals with an evaluation of the operator product $e^{-T}Ae^{T}$, where A is an object (see below), in terms of contracted algebraic expressions that are diagrammatically interpreted as connected. Its proof is a simple consequence of the generalized Wick theorem.[18,19]

Following Mukherjee et al.,[35] let us introduce the concept of an *object*, that is, an arbitrary normally ordered operator containing the same number of creation and annihilation operators. [For example, the cluster operator (2.8c) and the hamiltonian (2.3) are objects.] The LCT states[8,13,34]

$$e^{-T}Ae^{T} = \{Ae^{T}\}_{C} \tag{2.18}$$

that is, the operator product $e^{-T}Ae^{T}$ is determined by normally ordered terms that are composed of, reading from left to right, one vertex of A and one or more vertices of T, and the whole diagrammatic complex is connected. Since both T and A are normally ordered operators, internal hole lines simultaneously starting and terminating at the same vertex are strictly forbidden. Algebraically, the LCT may be expressed as follows:

$$e^{-T}Ae^{T} = A + N[\overline{AT}] + \frac{1}{2!}N[\overline{ATT}] + \cdots \tag{2.19}$$

where we have displayed in an explicit way that the contractions corresponding to internal hole or particle lines are formed only between object A and cluster operators T. The third alternative form of the LCT is[13,38]

$$e^{-T}Ae^{T} = \sum_{n\geq 1}\sum_{l_1\geq l_2\geq\cdots\geq l_n\geq 1} N(l_1,l_2,\ldots,l_n)\{AT_{l_1}T_{l_2}\cdots T_{l_n}\}_C \tag{2.20}$$

where $N(l_1,l_2,\ldots,l_n)$ is a numerical factor uniquely determined by (2.13), and $\{AT_{l_1}T_{l_2}\cdots T_{l_n}\}_C$ means that there is at least one internal line between vertices of A and T_{l_1}, A and T_{l_2}, and so on. The lines between cluster operators are omitted [see comment below (2.18)].

Assuming that an object from the LCT is equal to the unperturbed hamiltonian H_0 (this is a diagonal one-particle operator), then[4]

$$e^{-T}H_0e^{T} = \{H_0e^{T}\}_C = H_0 + [H_0,T]_-$$
$$= H_0 + N[\overline{H_0T}] \tag{2.21}$$

This relation is a direct consequence of the fact that the H_0 is a diagonal one-particle operator, so that there exists only one contraction (internal line) between H_0 and T. In similar way, setting $A = H_1$, we arrive at

$$e^{-T}H_1e^T = \{H_1e^T\}_C \qquad (2.22)$$

These relations (2.21) and (2.22) will be used in our forthcoming construction of the nondegenerate version of the CCA.

D. Construction of the CCA

We shall try to solve the *Schrödinger equation*

$$H|\Psi_0\rangle = E_0|\Psi_0\rangle \qquad (2.23)$$

for a nondegenerate ground state described by a perturbed state vector $|\Psi_0\rangle$. The total hamiltonian H is specified by (2.3) or (2.5); for the present construction of the CCA it is irrelevant whether the perturbation H_1 is composed of only two-particle or both one- and two-particle terms. Let us assume that the perturbed state vector $|\Psi_0\rangle$ is expressed in the exponential form (2.8a). Introducing these assumptions into the Schrödinger equation (2.23) and multiplying it from left by e^{-T}, we get

$$e^{-T}(H_0 + H_1)e^T|\Phi_0\rangle = E_{corr}|\Phi_0\rangle \qquad (2.24a)$$

$$E_{corr} = E_0 - \langle\Phi_0|H|\Phi_0\rangle \qquad (2.24b)$$

The relation (2.24a) is essentially simplified if the LCT results (2.21) and (2.22) are applied,

$$\left([H_0, T]_- + \{H_1e^T\}_C\right)|\Phi_0\rangle = E_{corr}|\Phi_0\rangle \qquad (2.25)$$

The unperturbed hamiltonian H_0 on the left in this relation has been omitted, since its action on $|\Phi_0\rangle$ is zero [see Eq. (2.4a)]. The diagrammatic form (2.25) of the original Schrödinger equation (2.23) serves for the determination of both the *correlation energy* E_{corr} and the cluster operators T. Let us multiply (2.25) on the left successively by $|\Phi_0\rangle$ and an *l*-excited unperturbed state vector $X_{p_1}^\dagger X_{h_1}\ldots X_{p_l}^\dagger X_{h_l}|\Phi_0\rangle$. We obtain two formulas, of which the first determines the correlation energy E_{corr}, while the second can be used for the evaluation of cluster-operator matrix elements[34,36-38]:

$$E_{corr} = \langle\Phi_0|H_1e^T|\Phi_0\rangle_C \qquad (2.26a)$$

$$\langle\Phi_\alpha|T_l|\Phi_0\rangle = \frac{1}{D_\alpha}\langle\Phi_\alpha|H_1e^T|\Phi_0\rangle_C \qquad (2.26b)$$

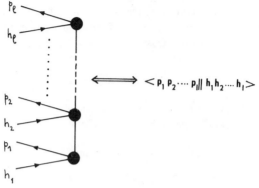

Fig. 2. The diagrammatic interpretation of "1-particle" terms of the cluster operator T.

The denominator D_α is determined as a negative eigenenergy of $|\Phi_\alpha\rangle$,

$$D_\alpha = \varepsilon_{h_1} + \cdots + \varepsilon_{h_l} - \varepsilon_{p_1} - \cdots - \varepsilon_{p_l} \qquad (2.26c)$$

and the subscript C means that in the evaluation of the matrix elements $\langle \Phi_0 | H_1 e^T | \Phi_0 \rangle$ and $\langle \Phi_\alpha | H_1 e^T | \Phi_0 \rangle$ we consider only connected terms.

In order to use the Feynman–Goldstone diagrammatic technique for the evaluation of the right sides of (2.26a) and (2.26b), we have to introduce a diagrammatic interpretation of cluster operators T_l. A matrix element of T_l is represented by l heavy dots connected by vertical solid line, as in Fig. 2. The algebraic interpretation of diagrams appearing in (2.26) is carried out by using a slightly modified Feynman–Goldstone diagrammatic technique[9,10,20]; the overall numerical factor is specified by (2.9). As in the diagrammatic RSPT, for closed-shell molecules with RHF one-particle functions the CCA may be realized in a spin-free form. The overall numerical factor is determined by (2.10), and the resulting algebraic expressions are summed over spatial parts of all internal lines.

Since the perturbation H_1 is composed of (at most) two-particle terms, the expression (2.26) determining the CCA correlation energy may be essentially simplified as follows:

$$E_{\text{corr}} = \langle \Phi_0 | H_1 \left(T_1 + T_2 + \frac{1}{2!} T_1^2 \right) | \Phi_0 \rangle_C \qquad (2.27)$$

the next connected as well as disconnected cluster operators cannot contribute. The diagrammatic interpretation of (2.27) is

$$E_{\text{corr}} = \quad (2.28)$$

or algebraically,

$$E_{\text{corr}} = - \sum_{ph} \langle h | w | p \rangle \langle p | | h \rangle$$

$$+ \frac{1}{2} \sum_{\substack{p_1 p_2 \\ h_1 h_2}} (\langle h_1 h_2 | v | p_1 p_2 \rangle - \langle h_1 h_2 | v | p_2 p_1 \rangle)$$

$$\times (\langle p_1 p_2 | | h_1 h_2 \rangle + \langle p_1 | | h_1 \rangle \langle p_2 | | h_2 \rangle) \tag{2.29}$$

We see that the CCA correlation energy is fully determined by mono- and di-excited cluster operators. The higher-excited cluster operators contribute to E_{corr} indirectly through their coupling with T_1 and T_2 (see below).

Now we turn our attention to (2.26b), which determines the cluster matrix elements. For an actual numerical realization it is advantageous to rewrite (2.26b) as[34,36-38]

$$\langle p_1 \ldots p_l | | h_1 \ldots h_l \rangle = \frac{1}{D} \sum A(p_{\alpha_1} \ldots p_{\alpha_l}, h_{\alpha_1} \ldots h_{\alpha_l}) \tag{2.30}$$

The denominator D is determined by (2.26c) and the summation runs over all $l!$ permutations

$$\begin{pmatrix} 1 & 2 & \cdots & l \\ \alpha_1 & \alpha_2 & \cdots & \alpha_l \end{pmatrix}$$

The matrix elements $A(p_{\alpha_1} \ldots p_{\alpha_l}, h_{\alpha_1} \ldots h_{\alpha_l})$ correspond to $\langle \Phi_\alpha | H_1 e^T | \Phi_0 \rangle_C$ with fixed labeling of external lines, and a pair of indices $(p_{\alpha_i}, h_{\alpha_i})$, for $i = 1, 2, \ldots, l$, is assigned to external lines of the same left excitation path:

$$A(p_{\alpha_1} \ldots p_{\alpha_l}, h_{\alpha_1} \ldots h_{\alpha_l}) = \tag{2.31}$$

The algebraic interpretation of diagrams in the rectangular block is carried out by using the overall numerical factor (2.9) or (2.10).

Formally, the relation (2.26b) can be written for $l = 1, 2$ as

$$
\begin{aligned}
T_1 = &[H_1]^{(1)} + [H_1 T_1]^{(1)} + [H_1 T_2]^{(1)} + [H_1 T_1 T_1]^{(1)} \\
&+ [H_1 T_3]^{(1)} + [H_1 T_1 T_2]^{(1)} + [H_1 T_1 T_1 T_1]^{(1)}
\end{aligned}
\tag{2.32a}
$$

$$
\begin{aligned}
T_2 = &[H_1]^{(2)} + [H_1 T_1]^{(2)} + [H_1 T_2]^{(2)} + [H_1 T_1 T_1]^{(2)} \\
&+ [H_1 T_3]^{(2)} + [H_1 T_1 T_2]^{(2)} + [H_1 T_1 T_1 T_1]^{(2)} \\
&+ [H_1 T_4]^{(2)} + [H_1 T_2 T_2]^{(2)} + [H_1 T_1 T_1 T_2]^{(2)} \\
&+ [H_1 T_1 T_3]^{(2)} + [H_1 T_1 T_1 T_1 T_1]^{(2)}
\end{aligned}
\tag{2.32b}
$$

and in similar way for "higher" cluster operators T_l. The term $[H_1 T_{l_1} \ldots T_{l_p}]^{(n)}$ represents those connected diagrammatic contributions to the cluster operator T_n that are composed of $p + 1$ vertices of $H_1, T_{l_1}, T_{l_2}, \ldots, T_{l_p}$. Hence the cluster operators T_l (or more precisely, their matrix elements) are determined by a *string of coupled nonlinear equations*. To outline this very interesting feature of the CCA, let us abbreviate the relations (2.32) and (2.26) as

$$
E_{corr} = f_0(T_1, T_2)
\tag{2.33a}
$$

$$
T_l = A_l + f_l(T_1, T_2, \ldots)
\tag{2.33b}
$$

for $l = 1, 2, \ldots$. The first relation (2.33a) determines the correlation energy as a scalar function of T_1 and T_2 [see Eq. (2.27)]. The second relation (2.33b) represents a string of coupled nonlinear equations. The operators A_l are constant parts of T_l; for the total hamiltonian with at most two-particle terms we have $A_l = 0$ for $l \geq 3$. The simplest and most transparent numerical procedure to solve the string (2.32b) is the method of successive iterations,

$$
T_l^{(k+1)} = A_l + f_l(T_1^{(k)}, T_2^{(k)}, \ldots)
\tag{2.34a}
$$

$$
T_l^{(0)} = A_l
\tag{2.34b}
$$

where $T_l^{(k+1)}$ is the $(k+1)$th-step value of T_l evaluated from the previous kth-step values of $T_1, T_2, \ldots$. Assuming a convergence of this simple iterative procedure, we interrupt it when, starting from $k \geq K$, we have achieved

$$
|T_l^{(k+1)} - T_l^{(k)}| < \varepsilon
\tag{2.35}
$$

for all $l \geq 1$. Finally, introducing the resulting cluster operators T_1 and T_2 into

(2.33a), the CCA correlation energy is obtained. As Paldus[13] points out, this iterative procedure is, in fact, nothing else than a recursive construction of all possible connected diagrams in (2.6). This is much simpler than determining the higher-order diagrammatic terms and then evaluating them, as is usually done in the standard realization of the diagrammatic RSPT. *The CCA presents very convenient recurrent formulas for simultaneous construction and evaluation of diagrammatic perturbation terms of arbitrarily high order and prescribed topology.*

E. Connection with Nondegenerate Diagrammatic RSPT

In the original article of Coester[4] the CCA method was presented as an alternative proof that the correlation energy is determined by the connected diagrams. This immediately follows from the structure of principal CCA relations (2.26a) and (2.26b). The cluster matrix elements are determined by connected terms; therefore if we introduce the cluster operators T_1 and T_2 into the relation (2.26a), then the correlation energy is expressed by connected terms only. A second problem, to say something more about the possible topological structure of these connected terms, is much more complex than the previous one. To prove an equivalence between nondegenerate versions of diagrammatic RSPT and CCA, we shall start from the formal RSPT; we demonstrate that it provides the CCA diagrammatic form of the Schrödinger equation (2.25), which in turn provides the CCA principal relations (2.26a) and (2.26b).

The formal RSPT is based upon the following commutator equation[29,39-41]:

$$[U, H_0]_- = H_1 U - E_{corr} U \qquad (2.36)$$

where U is a wave operator mapping the unperturbed state vector $|\Phi_0\rangle$ onto its perturbed counterpart, $|\Psi_0\rangle = U|\Phi_0\rangle$, and $\langle \Phi_0 | \Psi_0 \rangle = 1$. Solving the commutator relation (2.36), we arrive at

$$|\Psi_0\rangle = |\Phi_0\rangle + \frac{Q_0}{(-H_0)}(H_1 - E_{corr})|\Psi_0\rangle \qquad (2.37a)$$

$$E_{corr} = \langle \Phi_0 | H_1 | \Psi_0 \rangle \qquad (2.37b)$$

where Q_0 is a projector into a subspace spanned by all unperturbed states that are orthogonal to $|\Phi_0\rangle$. Now, let us assume that the wave operator in (2.36) is expressed in the exponential form (2.8). This is a plausible assumption in our derivation, since we know that the diagrammatic RSPT provides the wave operator in the exponential form (2.8). Then

$$[e^T, H_0]_- = H_1 e^T - E_{corr} e^T \qquad (2.38)$$

Its individual terms are transformed to a normally ordered form by making use of the generalized Wick theorem[18,19] (for details see Section III.A):

$$[e^T, H_0]_- = N[(TH_0 - H_0T)e^T] \tag{2.39a}$$

$$H_1 e^T = N[\{H_1 e^T\}_c e^T] \tag{2.39b}$$

where

$$\{H_1 e^T\}_C = N\left[H_1 + \overline{H_1 T} + \frac{1}{2!}\overline{\overline{H_1 TT}} + \cdots\right] \tag{2.40}$$

is identical with (2.22); it represents the LCT result for H_1. Finally, we introduce (2.39) into (2.38); after simple algebraic manipulations we obtain

$$N[(TH_0 - H_0T - \{H_1 e^T\}_C + E_{corr})e^T] = 0 \tag{2.41}$$

This relation is satisfied if the cluster operator T fulfils the following commutator equation:

$$[T, H_0]_- = \{H_1 e^T\}_C - E_{corr} \tag{2.42}$$

Multiplying it on the right by $|\Phi_0\rangle$, we obtain the CCA relation (2.25). Hence, we have demonstrated that postulating the exponential form of the wave operator (an intrinsic result of diagrammatic RSPT), the RSPT commutator relation (2.36) provides the CCA version. All intermediate steps in deriving (2.42) from (2.36) are fully reversible; that is, the nondegenerate version of diagrammatic RSPT and CCA are equivalent, which was to be demonstrated.

F. Connection with CI

For simplicity, let us assume that a perturbation H_1 is composed of two-particle terms—that is, our considerations are valid for both closed-shell and open-shell systems with RHF and UHF one-particle functions, respectively. For systems with a one-particle part of H_1 our theoretical considerations are simply generalized.

To relate the CI correlation energies to their counterparts calculated by the CCA, it is appropriate to split CI correlation energies into individual increments that are assigned to the terms with simultaneous occurrence of a given set of excitations (through tetraexcited states)[42]:

$$E_{CI,corr}(0+2) = \omega(2) \tag{2.43a}$$

$$E_{CI,corr}(0+1+2) = \omega(2) + \omega(1,2) \tag{2.43b}$$

$$E_{CI,corr}(0+2+3) = \omega(2) + \omega(2,3) \tag{2.43c}$$

$$E_{CI,corr}(0+1+2+3) = \omega(2) + \omega(1,2) + \omega(2,3) + \omega(1,2,3) \tag{2.43d}$$

$$E_{CI,corr}(0+2+4) = \omega(2) + \omega(2,4) \tag{2.43e}$$

$$E_{CI,corr}(0+1+2+4) = \omega(2) + \omega(1,2) + \omega(2,4) + \omega(1,2,4) \tag{2.43f}$$

$$E_{CI,corr}(0+2+3+4) = \omega(2) + \omega(2,3) + \omega(2,4) + \omega(2,3,4) \tag{2.43g}$$

$$E_{CI,corr}(0+1+2+3+4) = \omega(2) + \omega(1,2) + \omega(2,3)$$
$$+ \omega(2,4) + \omega(1,2,3) + \omega(2,3,4) + \omega(1,2,4)$$
$$+ \omega(1,2,3,4) \tag{2.43h}$$

where $\omega(i_1, i_2, \ldots)$ describes those terms of a given CI correlation energy that are involving simultaneously i_1-, i_2-,... excited states. Going successively from (2.43a) to (2.43h), we can calculate all the increments appearing in the CI calculations through tetraexcited states. By using the diagrammatic terminology, an arbitrary increment is unambiguously separated into two parts; the first is composed of connected terms, while the second contains disconnected terms. For example, the increment $\omega(2)$ corresponding to diexcited states is

$$\omega(2) = \omega_C(2) + \omega_{DC}(2) \tag{2.44}$$

The $\omega_C(2)$ is exactly equal to the correlation energy determined by the diagrammatic RSPT involving only the diexcited intermediate states. The disconnected part $\omega_{DC}(2)$ causes the so-called size inconsistency[31-33] of the CI correlation energy [see comment below (2.10)]. Since $\omega_{DC}(2) > 0$ (at least through fourth-order perturbation theory), this CI correlation energy is slightly positive than $\omega_C(2)$:

$$0 > E_{CI,corr}(0+2) > \omega_C(2) \tag{2.45}$$

The lowest order of nonzero terms of $\omega(1,2)$, $\omega(2,3)$, and $\omega(2,4)$ is the fourth order. Of these, $\omega^{(4)}(1,2)$ and $\omega^{(4)}(2,3)$ are determined by connected diagrammatic terms, but $\omega^{(4)}(2,4)$ is composed of both connected and disconnected terms:

$$\omega^{(4)}(1,2) = \omega_C^{(4)}(1,2) \tag{2.46a}$$

$$\omega^{(4)}(2,3) = \omega_C^{(4)}(2,3) \tag{2.46b}$$

$$\omega^{(4)}(2,4) = \omega_C^{(4)}(2,4) + \omega_{CD}^{(4)}(2,4) \tag{2.46c}$$

This disconnected part is partially canceled by disconnected terms of $\omega_{DC}(2)$; to achieve a complete cancellation we have to use the EPV diagrammatic terms:

$$-\omega_{C,\text{EPV}}^{(4)}(2,4) + \omega_{DC}^{(4)}(2,4) + \omega_{DC}^{(4)}(2) = 0 \tag{2.47}$$

Then, up to the fourth order, the CI correlation energy involving tetraexcited states can be diagrammatically classified as follows:

$$E_{\text{CI,corr}}^{(2)} = \omega_C^{(2)}(2) \tag{2.48a}$$

$$E_{\text{CI,corr}}^{(3)} = \omega_C^{(3)}(2) \tag{2.48b}$$

$$E_{\text{CI,corr}}^{(4)} = \omega_C^{(4)}(2) + \omega_C^{(4)}(1,2) + \omega_C^{(4)}(2,3)$$
$$+ \left[\omega_C^{(4)}(2,4) + \omega_{C,\text{EPV}}^{(4)}(2,4)\right] \tag{2.48c}$$

The disconnected terms (i.e., size inconsistency) appearing in correlation energies calculated by the CI method involving tetraexcited states begin to emerge at the sixth order of perturbation theory; within the framework of fifth order an analogous cancellation to (2.47) is satisfied.[20]

Summarizing, the dominant part of disconnected terms in CI results is canceled if tetraexcited states are taken into account. The CCA correlation energies can be related to the CI ones only when the tetraexcited states in both methods are considered.

G. CCA Realization of Nesbet's Hierarchy of Bethe–Goldstone Equations

The Nesbet method[43,44] of evaluation of correlation energies (often called the Nesbet's hierarchy of Bethe–Goldstone equations) is based upon successive CI calculations with increasing level of excitation. Let $E_{\text{CI,corr}}(h_1 h_2 \ldots h_p)$ be a CI correlation energy obtained by a diagonalization of the hamiltonian H in a space spanned by all configurations that are formed from a subset $\{h_1 h_2 \ldots h_p\}$ of occupied one-particle states. The CI correlation energies are used for a recursive definition of the so-called one-, two-,... body *net correlation increments*,

$$E_{CI,corr}(h) = \epsilon(h) \tag{2.49a}$$

$$E_{CI,corr}(h_1 h_2) = \epsilon(h_1) + \epsilon(h_2) + \epsilon(h_1 h_2) \tag{2.49b}$$

$$E_{CI,corr}(h_1 h_2 h_3) = \epsilon(h_1) + \epsilon(h_2) + \epsilon(h_3) + \epsilon(h_1 h_2)$$
$$+ \epsilon(h_2 h_3) + \epsilon(h_1 h_3) + \epsilon(h_1 h_2 h_3) \tag{2.49c}$$

and so on. Then the total correlation energy is equal to the sum of all net correlation increments,

$$E_{corr} = \sum_h \epsilon(h) + \sum_{h_1 < h_2} \epsilon(h_1 h_2) + \sum_{h_1 < h_2 < h_3} \epsilon(h_1 h_2 h_3) + \cdots \tag{2.50}$$

The main advantage of Nesbet's method is that the net correlation increments have the following simple physical interpretation: An *n*-body net correlation increment represents a "correlation" of *n* particles situated on *n* different one-particle functions, its probability tends rapidly to zero (1) with increasing number of mutually correlating particles and/or (2) if the given one-particle functions are localized in different parts of molecule. This second feature can be much more dominant than the first if the localized one-particle functions are used. Since the net correlation increments are determined via the CI technique, they are not size-consistent. A natural way to remove this shortcoming of Nesbet's method is to realize it in the framework of the CCA method[34,37]; then the CCA net correlation increments will be automatically of a size-consistent type.

Let $M = \{h_1 h_2 ... h_p\}$, for $1 \leqslant p \leqslant N$, be a subset of occupied one-particle states. A cluster operator T related to this preselected subset is

$$T^{(M)} = \sum_{l \geqslant 1} T_l^{(M)} \tag{2.51a}$$

$$T_l^{(M)} = \frac{1}{l!} \sum_{\substack{h_1, h_2, \ldots, h_l \in M \\ p_1, p_2, \ldots, p_l \notin \Phi_0}} \langle p_1 p_2 \cdots p_l | | h_1 h_2 \ldots h_l \rangle^{(M)}$$

$$\times N\left[X_{p_1}^\dagger \cdots X_{p_l}^\dagger X_{h_l} \cdots X_{h_1} \right] \tag{2.51b}$$

where the hole indices $h_1, h_2, \ldots, h_p$ belong to the subset M. A perturbed ground-state vector related to M is determined by

$$|\Psi_0^{(M)}\rangle = e^{T^{(M)}} |\Phi_0\rangle \tag{2.52}$$

We seek a solution of the Schrödinger equation

$$H|\Psi_0^{(M)}\rangle = E_0^{(M)}|\Psi_0^{(M)}\rangle \qquad (2.53)$$

in a subspace spanned by all unperturbed states from (2.52). The principal CCA relations (2.26a) and (2.26b) are still valid,

$$E_{\text{corr}}^{(M)} = \langle \Phi_0|H_1 e^{T^{(M)}}|\Phi_0\rangle_C \qquad (2.54a)$$

$$\langle \Phi_\alpha|T_l^{(M)}|\Phi_0\rangle = \frac{1}{D_\alpha}\langle \Phi_\alpha|H_1 e^{T^{(M)}}|\Phi_0\rangle_C \qquad (2.54b)$$

where $E_{\text{corr}}^{(M)}$ plays, formally, the same role as the CI correlation energies in (2.49), but now they are fully determined by connected diagrammatic terms. Introducing successively the correlation energies $E_{\text{corr}}^{(M)}$ into (2.49a), (2.49b), and so on, we get CCA correlation increments that are, *a priori*, size-consistent.

H. Approximate Versions of the CCA

Until now, we have mainly discussed the formal properties of the CCA and its relations to other many-body techniques. The aim of this section is to turn our attention to the physical content of the CCA. An exact solution of the string of coupled nonlinear equations (2.26b) is equivalent to the full CI method and/or diagrammatic RSPT carried out through infinite orders of perturbation theory. Therefore it seems impossible to realize the CCA numerically without a considerable reduction and simplification of its principal relations. We shall demonstrate, using a simple physical interpretation and some easily grasped perturbation-theory arguments that the algebraic structure of the CCA string of coupled equations may be essentially simplified while still accounting for dominant parts of correlation energy. Most probably, according to recently completed fourth-order diagrammatic RSPT calculations of correlation energies,[45-46] the role of individual connected as well as disconnected cluster operators may change if the number of electrons exceeds a critical threshold value of 15-20. Therefore, the physical arguments presented below in justification of approximate versions of the CCA are valid only for small molecular systems.

By using an analogy to an imperfect gas, Sinanoğlu[47,48] has interpreted a connected cluster operator T_l as an l-particle correlation event (collision) taking place at the same time and region of molecule, whereas a disconnected cluster operator, e.g. $T_l T_{l'}$, corresponds to different l- and l'-particle correlation events taking place at the same time but in different regions of the molecule. The monoexcited cluster operator T_1 adjusts the one-particle functions to the field of all other electrons to an extent beyond the HF theory. If the Brueckner (maximum-overlap) one-particle functions are used, then these correlation effects modifying a form of one-particle functions are exactly zero. Sinanoğlu has supported the above-presented interpretation of individual cluster operators by a few full CI results that have been available since the early sixties.[49,50] The CI calculations including different levels of excitations with respect to a single ground-state Slater determinant demonstrated that even though the effect of diexcited configurations is always dominant, the second most important contribution is usually the tetraexcited configurations, which are more important than mono- and triexcited configurations. Similarly, the hexaexcited configurations should be more important than pentaexcited configurations. Sinanoğlu[47,48] has observed that two types of tetraexcited configurations to the correlation energy are possible: the simultaneous interaction of four electrons (corresponding to T_4) and the simultaneous interaction of two pairs of electrons (corresponding to T_2T_2). The latter disconnected term is much more important to the CI correlation energy than its connected counterpart. Since these disconnected terms are simply related to the diexcited CI coefficients as their product, this part of the "tetraexcited" correlation effects can be treated in a direct manner. Including these two types of terms in a correlation-energy model, initiated by the works of Sinanoğlu[47,48] and Nesbet,[43,44] provides the bases for various "pair" theories, the most general form is the so-called *coupled-cluster many-electron theory* (CPMET) of Čížek et al.[10-11]

Unfortunately, the relative importance of different cluster operators T_2 and T_2^2 versus (in particular) T_1 and T_3 is not so simple. To simplify our forthcoming considerations we restrict ourselves to molecular systems with a perturbation composed of two-particle terms (i.e., mainly to closed-shell systems with RHF one-particle functions). We can see from Table I that the

TABLE I
The Lowest Order of Different Connected and
Disconnected Cluster Operators in Which
They Contribute to $|\Psi_0\rangle$ and E_{corr}

Cluster operator	$|\Psi_0\rangle$	E_{corr}
T_2	1	2
T_1	2	4
T_3	2	4
T_2T_2	2	4
T_1T_2	3	5
T_4	3	5
T_1T_1	4	5

TABLE II
Perturbation Breakdown of Correlation Energy[a]

Perturbation terms	Cluster operator	Energy (mH)		
		LiH	BH	FH
$E_{corr}^{(2)}$	T_2	−20.90	−64.43	−184.48
$E_{corr}^{(3)}$	T_2	−4.58	−16.02	−1.67
$E_{corr}^{(4)}$(mono)	T_1	−0.16	−0.36	−1.06
$E_{corr}^{(4)}$(di)	T_2	−1.43	−6.01	−2.28
$E_{corr}^{(4)}$(tri)	T_3	−0.01	−0.68	−2.31
$E_{corr}^{(4)}$(tetra)	$T_2 T_2$	+0.23	+1.12	+0.70

[a] Through the fourth order, for the molecules LiH ($R = 1.60$ Å), BH ($R = 1.20$ Å), and FH ($R = 0.90$ Å) in the 6-31G^{++} basis set of gaussian AOs.[46]

cluster operators T_1, T_3, and T_2^2 are of the same lowest order in which these operators contribute to E_{corr} and $|\Psi_0\rangle$. An actual role of individual cluster operators T_1, T_2, T_3, and T_2^2 may be well estimated by their fourth-order contributions to the E_{corr} (see Table II). Although the cluster operator T_2 is still the most important term, other cluster operators are roughly of the same importance. The contribution of the "triexcited" cluster operator, T_3, increases with the number of electrons, and the fourth-order correlation energy $E_{corr}^{(4)}$(tetra) produced by the disconnected cluster operator $T_2 T_2$ is usually positive. Summarizing these observations, we have no particular reason to consider the CPMET as a suitable method for the calculation of correlation energy. The inclusion of disconnected cluster T_2^2 is not properly balanced by a simultaneous inclusion of connected cluster operators T_1 and T_3 that are both of the same lowest order (see Table I) as T_2^2.

A logical step from the CPMET toward a more general approximate version of the CCA is the *extended* CPMET (ECPMET) of Paldus, Čížek, and Shavitt[12] (see also References 34 and 37). It yields the correlation energy exact up to the fourth order of perturbation theory. The cluster operators T_1 and T_3 are treated on an equal footing with the disconnected cluster operator T_2^2 already accounted for by CPMET. In the initial stage of justification of the ECPMET, one will be concerned about the role of the disconnected terms $T_1 T_1$ and $T_1 T_2$. According to Table I, these clusters should be of smaller importance than their connected counterparts T_2 and T_3, respectively. They lead to fifth- and higher-order perturbation terms, and we can expect that their role will be negligible in comparison with the connected clusters T_2 and T_3. This was initially checked numerically by Paldus et al.[12] for the BH_3 molecule. In the forthcoming subsections we present principal three approximate versions of the CCA that are appropriate for the calculation of correlation energy of molecules with a perturbation H_1 composed of only two-particle terms.

1. Linearized CCA (LCCA)[9,51-53]

The exponential wave operator e^T is approximated by

$$e^T \simeq 1 + T_2 \tag{2.55}$$

Then the expressions (2.26a) and (2.26b) are reduced to

$$E_{corr} = \langle \Phi_0 | H_1 T_2 | \Phi_0 \rangle_C \tag{2.56a}$$

$$\langle \Phi_\alpha | T_2 | \Phi_0 \rangle = 1/D_\alpha \langle \Phi_\alpha | H_1 (1 + T_2) | \Phi_0 \rangle_C \tag{2.56b}$$

where $|\Phi_\alpha\rangle = X_{p_1}^\dagger X_{h_1} X_{p_2}^\dagger X_{h_2} |\Phi_0\rangle$ is a diexcited unperturbed state. For better understanding of main features of the LCCA we rewrite (2.56) by introducing between the operators H_1 and T_2 the projector $\sum_{(\beta)} |\Phi_\beta\rangle\langle\Phi_\beta|$ into a subspace spanned by all diexcited states,

$$E_{corr} = \sum_\beta v_{0\beta} t_\beta^{(2)} \tag{2.57a}$$

$$t_\alpha^{(2)} = \frac{1}{D_\alpha} \left(v_{\alpha 0} + \sum_\beta v_{\alpha\beta} t_\beta^{(2)} \right) \tag{2.57b}$$

where $v_{\mu\nu} = \langle \Phi_\mu | H_1 | \Phi_\nu \rangle$ and $t_\mu^{(2)} = \langle \Phi_\mu | T_2 | \Phi_0 \rangle$. The formal diagrammatic counterparts of these relations are

$$(2.58a)$$

(a)

$$(2.58b)$$

(b)

The rectangular (round or oval) shaded block represents the matrix elements $t_\mu^{(2)}$ ($v_{\mu\nu}$), the vertical cut line represents the denominator $1/D_\alpha$, the oriented double (n-tuple) line indexed by μ corresponds to a μth diexcited (n-excited) state, and finally, the indices assigned to internal double (n-tuple) lines are summed over all diexcited (n-excited) states. Solving (2.58b) by iteration, the rectangular block is determined by infinite summation of diagrams:

$$(2.59)$$

Introducing this result into (2.58b), we get a formal diagrammatic expression for the correlation energy determined by the LCCA,

$$E_{corr} = \quad \overset{\beta}{\text{⬤┼◁⬤}} \quad + \quad \overset{\beta \quad \beta'}{\text{⬤┼◁⬤┼◁⬤}} \quad + \cdots \quad (2.60)$$

The resulting LCCA correlation energy is equal to the sum of all possible diagrammatic terms involving the diexcited intermediate states. This result is exact up to the third order; starting from the fourth order, the mono-, tri-, and tetraexcited states appear too. After the kth iteration we can calculate the correlation energy as the sum of all its components through the $(k + 1)$th ' order of perturbation theory.

Following Bartlett et al.,[51,52] the LCCA can be simply modified in such a way that to calculate the correlation energy through $2k$th order it is necessary to carry out only k iterations instead of $2k - 1$. The equation (2.58b) is modified as follows:

(a)
$$(2.61a)$$

(b)
$$(2.61b)$$

(c)
$$(2.61c)$$

where the integer k denotes the number of iterations (k is simply equal to the order of a given oval block). Then, the correlation energy (2.58a) is

$$E_{corr} = \sum_{k=1}^{\infty} \left(E_{corr}^{(2k-1)} + E_{corr}^{(2k)} \right) \quad (2.62a)$$

(b)
$$(2.62b)$$

(c)
$$(2.62c)$$

This is just an analogue of Wigner's $2k + 1$ rule for the diagrammatic RSPT.[36,42]

2. Coupled-Pair Many-Electron Theory (CPMET)

The CPMET of Čížek[9-11] is based on

$$e^T \simeq e^{T_2} = 1 + T_2 + \frac{1}{2!}T_2^2 + \cdots \qquad (2.63)$$

This approximation of e^T was initially suggested by Sinanoğlu,[47,48] who observed that the tetraexcited states in the CI expansion are well approximated by $(1/2!)T_2^2$, whereas the mono-, tri-, and pentaexcited states are negligible. Introducing (2.63) into (2.26a) and (2.26b), we get

$$E_{\text{corr}} = \langle \Phi_0 | H_1 T_2 | \Phi_0 \rangle_C \qquad (2.64a)$$

$$\langle \Phi_\alpha | T_2 | \Phi_0 \rangle = \frac{1}{D_\alpha} \langle \Phi_\alpha | H_1 \left(1 + T_2 + \frac{1}{2!}T_2^2 \right) | \Phi_0 \rangle_C \qquad (2.64b)$$

We see that the correlation energy is determined formally by the same expression (2.56a) as in the LCCA, but now the cluster operator T_2 is determined by a nonlinear relation (2.64b) containing on its right side both "linear" T_2 and "nonlinear" $T_2 T_2$ terms. Application of the same formal technique as was used just above in the LCCA gives[38]

$$t_\alpha^{(2)} = \frac{1}{D_\alpha} \left(v_{\alpha 0} + \sum_\beta v_{\alpha\beta} t_\beta^{(2)} + \sum_{\beta\beta'} v_{\alpha,\beta\beta'} t_\beta^{(2)} t_{\beta'}^{(2)} \right) \qquad (2.65)$$

where the summations run over all diexcited states. The formal diagrammatic interpretation of (2.65) is

$$(2.66)$$

V. KVASNIČKA ET AL.

Solving this system by iteration, we arrive at

$$(2.67)$$

That is, in the CPMET approximation the correlation energy (exact up to third order) is determined among others by special "treelike" many-body diagrams; as one goes from left to right, two-, four-, six-,... excited intermediate states appear. For finite many-electron systems these diagrams may contain higher-level intermediate excitations that cannot obviously be physically realized. Fortunately, they are of the EPV type and therefore should be considered in a corresponding many-body diagrammatic perturbation theory to achieve exact cancellation of disconnected terms with three or more components.

We demonstrate that the CPMET equation (2.65) may be rewritten[38] in a form analogous to the linearized CCA. An alternative form of the last term in (2.65) with the double summation is

$$\sum_{\beta\beta'} v_{\alpha,\beta\beta'} t_\beta^{(2)} t_{\beta'}^{(2)} = \sum_\beta \left\{ \sum_{\beta'} v_{\alpha,\beta\beta'} t_{\beta'}^{(2)} \right\} t_\beta^{(2)} = \sum_\beta g_{\alpha\beta} t_\beta^{(2)} \qquad (2.68)$$

Introducing this new notion into (2.65), we get the so-called *quasilinearized*

CPMET (QL-CPMET) equation

$$t_\beta^{(2)} = \frac{1}{D_\alpha}\left(v_{\alpha 0} + \sum_\beta \tilde{v}_{\alpha\beta} t_\beta^{(2)} \right) \tag{2.69a}$$

$$\tilde{v}_{\alpha\beta} = v_{\alpha\beta} + g_{\alpha\beta} \tag{2.69b}$$

with the same formal structure as (2.57b) of the LCCA. The resulting correlation energy calculated by CPMET and by QL-CPMET must be identical. The main advantage of (2.69) over (2.65) consists in its simpler form, allowing simultaneous tractability of the terms T_2 and T_2^2 with considerable reduction of numerical effort.

The QL-CPMET form of CCA serves[38] as a theoretical background for the derivation of CEPA (coupled electron-pair approximation) methods.[54-56] Let us introduce the following crucial approximation:

$$g_{\alpha\beta} \simeq \delta_{\alpha\beta} g_{\alpha\alpha} \tag{2.70}$$

that is, we neglect in (2.69) the off-diagonal matrix elements $g_{\alpha\beta}$. Owing to this approximation, the QL-CPMET relation (2.69) is essentially simplified:

$$t_\alpha^{(2)} = \frac{1}{D_\alpha - g_{\alpha\alpha}}\left(v_{\alpha 0} + \sum_\beta v_{\alpha\beta} t_\beta^{(2)} \right) \tag{2.71}$$

This relation can be used as a theoretical base of all known CEPA methods. First, neglecting the term $g_{\alpha\alpha}$, then (2.71) reduces to the LCCA method. Therefore, in the framework of the CCA, the CEPA methods are classified as many-body techniques ranged between the LCCA and CPMET. Second, diagonal matrix elements $g_{\alpha\alpha}$ are approximated by more or less justified formulas. Mainly, they are determined by EPV diagrammatic terms. The placing of $g_{\alpha\alpha}$ in the denominator corresponds to procedures of infinite summations of diagonal and fully factorizable diagrammatic insertions. The correlation energy calculated by a CEPA method may be related to a correlation energy calculated by the diagrammatic RSPT accounting for (1) all the terms involving biexcited intermediate states, and plus (2) all the diagrams that are successively constructed from the previous ones by introducing diagonal insertions of a special type. More details about these aspects of CEPA methods can be found in the literature.[57,58]

The QL-CPMET relation (2.69) is simply solved by the iterative method of successive approximations (2.34). To accelerate this simple iterative scheme, the Newton–Raphson method may be advantageously used.[37,38] The matrix form of (2.69) is

$$t = D^{-1}\{v + [V + G(t)]t\} \tag{2.72}$$

where t and v are column vectors with elements $t_\alpha^{(2)}$ and $v_{\alpha 0}$, respectively. D^{-1}, V, and $G(t)$ are square matrices with elements $\delta_{\alpha\beta}/D_\alpha$, $v_{\alpha\beta}$ and $g_{\alpha\beta}$, respectively. The iterative method of successive approximations is realized by

$$t^{(k+1)} = D^{-1}\{v + [V + G(t^{(k)})]t^{(k)}\} \tag{2.73}$$

and $t^{(0)} = D^{-1}v$. A quadratic estimation of t evaluated from the last two iterative steps $t^{(k+1)}$ and $t^{(k)}$ is determined by the so-called Newton–Raphson form of the CPMET equations,[37,38]

$$t \simeq t^{(k+1)} + D^{-1}A^{(k)}(t^{(k+1)} - t^{(k)}) \tag{2.74}$$

where the matrix elements of $A^{(k)}$ are

$$(A^{(k)})_{\alpha\beta} = v_{\alpha\beta} + \sum_{\beta'} (v_{\alpha,\beta\beta'} + v_{\alpha,\beta'\beta})t_{\beta'}^{(2)} \tag{2.75}$$

3. Extended CPMET (ECPMET)

At present, the most general approximate version of the CCA is the ECPMET of Paldus, Čížek, and Shavitt[12] (cf. also References 34, 37, and 59). The exponential wave operator is approximated by

$$e^T \simeq 1 + T_1 + T_2 + T_3 + \frac{1}{2!}T_2^2 \tag{2.76}$$

it provides the correlation energy exact up to the fourth order of perturbation theory. The ECPMET actual form of (2.26) is

$$E_{corr} = \langle \Phi_0 | H_1 T_2 | \Phi_0 \rangle_C \tag{2.77a}$$

$$\langle \Phi_\alpha | T_l | \Phi_0 \rangle = \frac{1}{D_\alpha} \langle \Phi_\alpha | H_1 \left(1 + T_1 + T_2 + T_3 + \frac{1}{2!}T_2^2\right) | \Phi_0 \rangle_C \tag{2.77b}$$

for $l = 1, 2, 3$. The unperturbed states $|\Phi_\alpha\rangle$ are, in dependence upon the index l, mono-, di-, and triexcited, respectively. The relation (2.77b) can be formally rewritten in a string of three coupled nonlinear equations,

$$t^{(1)}_{\alpha_1} = \frac{1}{D_{\alpha_1}} \left(\sum_{\beta_1} v_{\alpha_1\beta_1} t^{(1)}_{\beta_1} + \sum_{\beta_2} v_{\alpha_1\beta_2} t^{(2)}_{\beta_2} + \sum_{\beta_3} v_{\alpha_1\beta_3} t^{(3)}_{\beta_3} \right) \qquad (2.78a)$$

$$t^{(2)}_{\alpha_2} = \frac{1}{D_{\alpha_2}} \left(v_{\alpha_2 0} + \sum_{\beta_1} v_{\alpha_2\beta_1} t^{(1)}_{\beta_1} + \sum_{\beta_2} v_{\alpha_2\beta_2} t^{(2)}_{\beta_2} \right.$$

$$\left. + \sum_{\beta_3} v_{\alpha_2\beta_3} t^{(3)}_{\beta_3} + \sum_{\beta_2\beta_2'} v_{\alpha_2, \beta_2\beta_2'} t^{(2)}_{\beta_2} t^{(2)}_{\beta_2'} \right) \qquad (2.78b)$$

$$t^{(3)}_{\alpha_3} = \frac{1}{D_{\alpha_3}} \left(\sum_{\beta_2} v_{\alpha_3\beta_2} t^{(2)}_{\beta_2} + \sum_{\beta_3} v_{\alpha_3\beta_3} t^{(3)}_{\beta_3} \right.$$

$$\left. + \sum_{\beta_2\beta_2'} v_{\alpha_3, \beta_2\beta_2'} t^{(2)}_{\beta_2} t^{(2)}_{\beta_2'} \right) \qquad (2.78c)$$

where the indices α_i, β_i (for $i = 1, 2, 3$) denote the i-fold excited states. The formal diagrammatic representation of (2.78) is

(a)   (2.79a)

(b)   (2.79b)

(c)   (2.79c)

Solving these three coupled equations iteratively, we can see what kind of diagrams appear in the diagrammatic determination of the correlation energy in the framework of the ECPMET. The presence of the disconnected cluster operator T_2^2 in both equations (2.79b) and (2.79c) causes an existence of the abovementioned "treelike" diagrammatic terms [see comment below (2.67)].

A quasilinearized approach similar to the QL-CPMET presented in the previous subsection is applicable also for the ECPMET equations (2.78a) to (2.78c); we arrive at the so-called QL-ECPMET. Let us introduce the following two types of "effective" matrix elements:

$$g_{\alpha_2\beta_2} = \sum_{\beta_2'} v_{\alpha_2,\beta_2\beta_2'} t_{\beta_2'}^{(2)} \tag{2.80a}$$

$$g_{\alpha_3\beta_2} = \sum_{\beta_2'} v_{\alpha_3,\beta_2\beta_2'} t_{\beta_2'}^{(2)} \tag{2.80b}$$

Introducing these new entities into (2.78b) and (2.78c), we get

$$t_{\alpha_1}^{(1)} = \frac{1}{D_{\alpha_1}} \left(\sum_{\beta_1} v_{\alpha_1\beta_1} t_{\beta_1}^{(1)} + \sum_{\beta_2} v_{\alpha_1\beta_2} t_{\beta_2}^{(2)} + \sum_{\beta_3} v_{\alpha_1\beta_3} t_{\beta_3}^{(3)} \right) \tag{2.81a}$$

$$t_{\alpha_2}^{(2)} = \frac{1}{D_{\alpha_2}} \left(v_{\alpha_2 0} + \sum_{\beta_1} v_{\alpha_2\beta_1} t_{\beta_1}^{(1)} + \sum_{\beta_2} \tilde{v}_{\alpha_2\beta_2} t_{\beta_2}^{(2)} + \sum_{\beta_3} v_{\alpha_2\beta_3} t_{\beta_3}^{(3)} \right) \tag{2.81b}$$

$$t_{\alpha_3}^{(3)} = \frac{1}{D_{\alpha_3}} \left(\sum_{\beta_2} \tilde{v}_{\alpha_3\beta_2} t_{\beta_2}^{(2)} + \sum_{\beta_3} v_{\alpha_3\beta_3} t_{\beta_3}^{(3)} \right) \tag{2.81c}$$

where the matrix elements $\tilde{v}$ are determined by

$$\tilde{v}_{\alpha_2\beta_2} = v_{\alpha_2\beta_2} + g_{\alpha_2\beta_2} \tag{2.81d}$$

$$\tilde{v}_{\alpha_3\beta_2} = v_{\alpha_3\beta_2} + g_{\alpha_3\beta_2} \tag{2.81e}$$

We have obtained a string of three coupled equations that are nonlinear only in the second equation, which determines the cluster matrix elements $t_{\alpha_2}^{(2)}$ of T_2. The first and third equations are, in fact, linear with respect to the cluster matrix elements $t_{\alpha_1}^{(1)}$ and $t_{\alpha_3}^{(3)}$, respectively. Hence, an iterative solution of the QL-ECPMET may be accelerated by the same kind of Newton–Raphson procedure as was used for the QL-CPMET.

I. Feymman–Goldstone Diagrammatic Construction of the ECPMET

The aim of this section is to construct the ECPMET equations for closed-shell molecules with RHF one-particle functions, that is, the perturbation H_1 is composed of two-particle terms and their matrix elements are spin-independent. The algebraic interpretation of diagrams is simply carried out by making use of the overall numerical factor (2.10), and the summations run only over spatial part of all internal lines (denoted by capital letters). The spin-free ECPMET correlation energy is [see Eqs. (2.29) and (2.27a)]

$$E_{corr} = \sum_{H_1 H_2} \sum_{P_1 P_2} \langle H_1 H_2 | v | P_1 P_2 \rangle_A \langle P_1 P_2 | | H_1 H_2 \rangle \qquad (2.82)$$

where

$$\langle IJ | v | KL \rangle_A = 2\langle IJ | v | KL \rangle - \langle IJ | v | LK \rangle \qquad (2.83)$$

are "antisymmetrized" two-particle matrix elements of original Coulomb repulsion between electrons. The spin-free cluster matrix elements of T_1, T_2, and T_3 are determined by [see Eq. (2.30)]

$$\langle P | | H \rangle = \frac{1}{D(H, P)} A(P, H) \qquad (2.84a)$$

$$\langle P_1 P_2 | | H_1 H_2 \rangle = \frac{1}{D(H_1 H_2, P_1 P_2)}$$
$$\times \left[A(P_1 P_2, H_1 H_2) + A(P_2 P_1, H_2 H_1) \right] \qquad (2.84b)$$

$$\langle P_1 P_2 P_3 | | H_1 H_2 H_3 \rangle = \frac{1}{D(H_1 H_2 H_3, P_1 P_2 P_3)}$$
$$\times \left[A(P_1 P_2 P_3, H_1 H_2 H_3) + A(P_1 P_3 P_2, H_1 H_3 H_2) \right.$$
$$+ A(P_2 P_1 P_3, H_2 H_1 H_3) + A(P_2 P_3 P_1, H_2 H_3 H_1)$$
$$\left. + A(P_3 P_1 P_2, H_3 H_1 H_2) + A(P_3 P_2 P_1, H_3 H_2 H_1) \right] \quad (2.84c)$$

where $D(H_1..., P_1...) = \varepsilon_{H_1} + \cdots - \varepsilon_{P_1} - \cdots$. The entities $A(..., ...)$ correspond to all possible linked and connected topologically nonequivalent diagrammatic terms produced by matrix elements $\langle \Phi_\alpha | H_1 e^T | \Phi_0 \rangle_C$ with fixed labeling of pairs of external incoming and outgoing lines; see Eq. (2.31). Expanding the exponential wave operator e^T in power series (2.13), then the entities $A(..., ...)$ can be split into different terms that are attached to fixed products of cluster operators [cf. Eqs. (2.32a) and (2.32b)]:

$$A(P_1..., H_1...) = \sum_X A_X(P_1..., H_1...) \qquad (2.85)$$

for $X = T_1, T_2, ..., T_1 T_2, ..., T_{l_1} T_{l_2}, ...$. In particular, the ECPMET form of

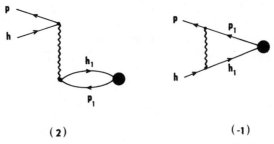

(2)   (-1)

Fig. 3. Diagrammatic terms contributing to $A_{T_1}(P, H)$.

$A(P_1..., H_1...)$ is

$$A(P, H) = A_{T_1}(P, H) + A_{T_2}(P, H) + A_{T_3}(P, H) \qquad (2.86a)$$

$$A(P_1 P_2, H_1 H_2) = A_{T_1}(P_1 P_2, H_1 H_2) + A_{T_2}(P_1 P_2, H_1 H_2)$$
$$+ A_{T_3}(P_1 P_2, H_1 H_2) + A_{T_2 T_2}(P_1 P_2, H_1 H_2) \qquad (2.86b)$$

$$A(P_1 P_2 P_3, H_1 H_2 H_3) = A_{T_2}(P_1 P_2 P_3, H_1 H_2 H_3) + A_{T_3}(P_1 P_2 P_3, H_1 H_2 H_3)$$
$$+ A_{T_2 T_2}(P_1 P_2 P_3, H_1 H_2 H_3). \qquad (2.86c)$$

The individual terms in (2.86) are determined as follows:

1. The term $A_{T_1}(P, H)$ (see Fig. 3):

$$A_{T_1}(P, H) = \sum_{H_1 P_1} \langle H_1 P | v | P_1 H \rangle_A \langle P_1 | | H_1 \rangle \qquad (2.87)$$

2. The term $A_{T_2}(P, H)$ (see Fig. 4):

$$A_{T_2}(P, H) = - \sum_{H_1 H_2} \sum_{P_1} \langle H_1 H_2 | v | P_1 H \rangle_A \langle P_1 P | | H_1 H_2 \rangle$$
$$+ \sum_{P_1 P_2} \sum_{H_1} \langle H_1 P | v | P_1 P_2 \rangle_A \langle P_1 P_2 | | H_1 H \rangle \qquad (2.88)$$

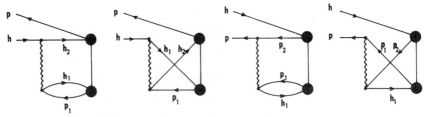

Fig. 4. Diagrammatic terms contributing to $A_{T_2}(P, H)$.

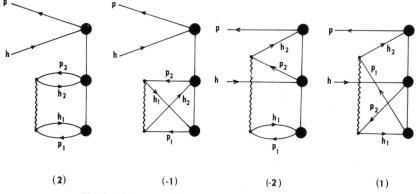

(2)   **(-1)**   **(-2)**   **(1)**

Fig. 5.   Diagrammatic terms contributing to $A_{T_3}(P, H)$.

3. The term $A_{T_3}(P, H)$ (see Fig. 5):

$$A_{T_3}(P, H) = \sum_{H_1 H_2} \sum_{P_1 P_2} \langle H_1 H_2 | v | P_1 P_2 \rangle_A$$

$$\times [\langle P_1 P_2 P | | H_1 H_2 H \rangle - \langle P_1 P_2 P | | H_1 H H_2 \rangle] \quad (2.89)$$

4. The term $A_{T_1}(P_1 P_2, H_1 H_2)$ (see Fig. 6):

$$A_{T_1}(P_1 P_2, H_1 H_2) = - \sum_{H_1'} \langle P_1 H_1' | v | H_1 H_2 \rangle \langle P_2 | | H_1' \rangle$$

$$+ \sum_{P_1'} \langle P_1 P_2 | v | H_1 P_1' \rangle \langle P_1' | | H_2 \rangle \quad (2.90)$$

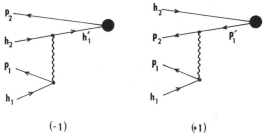

(-1)   **(+1)**

Fig. 6.   Diagrammatic terms contributing to $A_{T_1}(P_1 P_2, H_1 H_2)$.

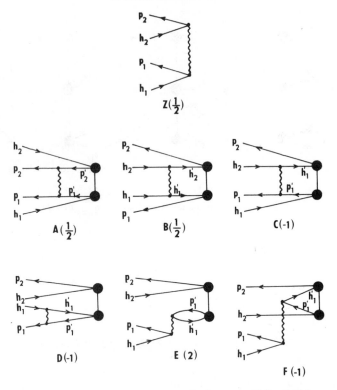

Fig. 7. Diagrammatic terms contributing to $A_{T_2}(P_1P_2, H_1H_2)$.

5. The term $A_{T_2}(P_1P_2, H_1H_2)$ (see Fig. 7):

$$A_{T_2}(P_1P_2, H_1H_2) = \tfrac{1}{2}\langle P_1P_2|v|H_1H_2\rangle$$

$$+ \frac{1}{2}\sum_{P_1'P_2'} \langle P_1P_2|v|P_1'P_2'\rangle\langle P_1'P_2'||H_1H_2\rangle$$

$$+ \frac{1}{2}\sum_{H_1'H_2'} \langle H_1'H_2'|v|H_1H_2\rangle\langle P_1P_2||H_1'H_2'\rangle$$

$$- \sum_{P_1'H_1'} [\langle P_1H_1'|v|P_1'H_2\rangle\langle P_1'P_2||H_1H_1'\rangle$$

$$- \langle P_1H_1'|v|H_1P_1'\rangle_A\langle P_1'P_2||H_1'H_2\rangle$$

$$+ \langle P_1H_1'|v|H_1P_1'\rangle\langle P_1'P_2||H_2H_1'\rangle] \qquad (2.91)$$

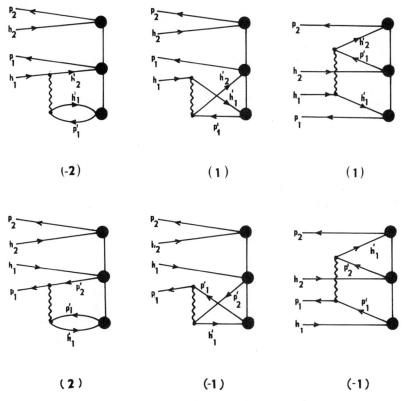

Fig. 8.   Diagrammatic terms contributing to $A_{T_3}(P_1 P_2, H_1 H_2)$.

6. The term $A_{T_3}(P_1 P_2, H_1 H_2)$ (see Fig. 8):

$$A_{T_3}(P_1 P_2, H_1 H_2) = - \sum_{H_1' H_2'} \sum_{P_1'} [\langle H_1' H_2' | v | P_1' H_1 \rangle_A \langle P_1' P_1 P_2 | | H_1' H_2' H_2 \rangle$$

$$- \langle H_1' H_2' | v | H_1 P_1' \rangle \langle P_1 P_1' P_2 | | H_1' H_2 H_2' \rangle]$$

$$+ \sum_{P_1' P_2'} \sum_{H_1'} [\langle H_1' P_1 | v | P_1' P_2' \rangle_A \langle P_1' P_2' P_2 | | H_1' H_1 H_2 \rangle$$

$$- \langle P_1 H_1' | v | P_1' P_2' \rangle \langle P_1' P_2' P_2 | | H_1 H_2 H_1' \rangle] \qquad (2.92)$$

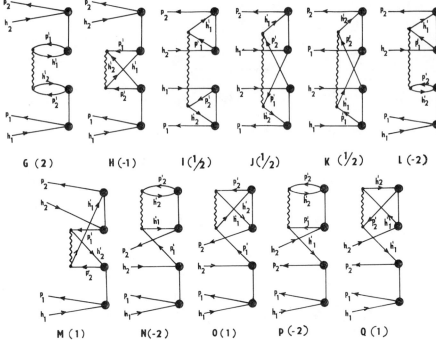

Fig. 9.   Diagrammatic terms contributing to $A_{T_2T_2}(P_1P_2, H_1H_2)$.

7. The term $A_{T_2T_2}(P_1P_2, H_1H_2)$ (see Fig. 9):

$$A_{T_2T_2}(P_1P_2, H_1H_2)$$

$$= \sum_{H_1H_2} \sum_{P_1P_2} \left[\langle H_2'H_1'|v|P_2'P_1'\rangle_A \langle P_1P_2'||H_1H_2'\rangle \langle P_1'P_2||H_1'H_2\rangle \right.$$

$$+ \tfrac{1}{2}\langle H_2'H_1'|v|P_2'P_1'\rangle \langle P_1P_2'||H_2'H_1\rangle \langle P_1'P_2||H_2H_1'\rangle$$

$$+ \tfrac{1}{2}\langle H_2'H_1'|v|P_1'P_2'\rangle \langle P_1P_2'||H_2'H_2\rangle \langle P_1'P_2||H_1H_1'\rangle$$

$$+ \tfrac{1}{2}\langle H_1'H_2'|v|P_1'P_2'\rangle \langle P_1'P_2'||H_1H_2\rangle \langle P_1P_2||H_1'H_2'\rangle$$

$$- \langle H_2'H_1'|v|P_2'P_1'\rangle_A \langle P_1P_2'||H_1H_2'\rangle \langle P_1'P_2||H_2H_1'\rangle$$

$$- \langle H_1'H_2'|v|P_1'P_2'\rangle_A \langle P_1P_1'||H_1H_2\rangle \langle P_2P_2'||H_1'H_2'\rangle$$

$$- \left. \langle H_1'H_2'|v|P_1'P_2'\rangle_A \langle P_1P_2||H_1H_1'\rangle \langle P_1'P_2'||H_2H_2'\rangle \right]. \qquad (2.93)$$

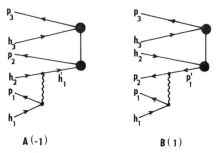

Fig. 10.   Diagrammatic terms contributing to $A_{T_2}(P_1 P_2 P_3, H_1 H_2 H_3)$.

8. The term $A_{T_2}(P_1 P_2 P_3, H_1 H_2 H_3)$ (see Fig. 10):

$$A_{T_2}(P_1 P_2 P_3, H_1 H_2 H_3) = -\sum_{H_1'} \langle P_1 H_1' | v | H_1 H_2 \rangle \langle P_2 P_3 | | H_1' H_3 \rangle$$

$$+ \sum_{P_1'} \langle P_1 P_2 | v | H_1 P_1' \rangle \langle P_1' P_3 | | H_2 H_3 \rangle \quad (2.94)$$

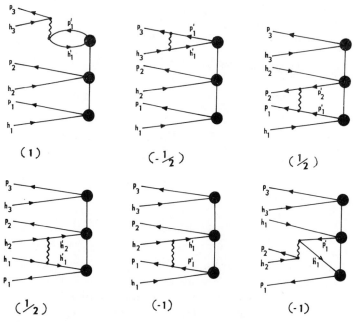

Fig. 11.   Diagrammatic terms contributing to $A_{T_3}(P_1 P_2 P_3, H_1 H_2 H_3)$.

9. The term $A_{T_3}(P_1P_2P_3, H_1H_2H_3)$ (see Fig. 11):

$$
\begin{aligned}
A_{T_3}(P_1P_2P_3, H_1H_2H_3) = & \frac{1}{2} \sum_{H_1'} \sum_{P_1'} \langle H_1'P_3|v|P_1'H_3\rangle_A \langle P_1P_2P_1'||H_1H_2H_1'\rangle \\
& + \frac{1}{2} \sum_{P_1'P_2'} \langle P_1P_2|v|P_1'P_2'\rangle \langle P_1'P_2'P_3||H_1H_2H_3\rangle \\
& + \frac{1}{2} \sum_{H_1'H_2'} \langle H_1'H_2'|v|H_1H_2\rangle \langle P_1P_2P_3||H_1'H_2'H_3\rangle \\
& - \sum_{H_1'} \sum_{P_1'} [\langle P_1H_1'|v|P_1'H_2\rangle \langle P_1'P_2P_3||H_1H_1'H_3\rangle \\
& + \langle P_2H_1'|v|H_2P_1'\rangle \langle P_1P_1'P_3||H_1'H_1H_3\rangle] \quad (2.95)
\end{aligned}
$$

10. The term $A_{T_2T_2}(P_1P_2P_3, H_1H_2H_3)$ (see Fig. 12):

$$
\begin{aligned}
& A_{T_2T_2}(P_1P_2P_3, H_1H_2H_3) \\
& = \sum_{H_1'H_2'} \sum_{P_1'} [-\langle H_2'H_1'|v|P_1'H_2\rangle_A \langle P_1P_1'||H_1H_2'\rangle \langle P_2P_3||H_1'H_3\rangle \\
& \quad + \langle H_2'H_1'|v|H_1P_1'\rangle \langle P_1P_1'||H_2'H_2\rangle \langle P_2P_3||H_1'H_3\rangle \\
& \quad + \langle H_2'H_1'|v|P_1'H_2\rangle \langle P_1'P_1||H_1H_2'\rangle \langle P_2P_3||H_1'H_3\rangle] \\
& \quad - \sum_{P_1'P_2'} \sum_{H_1'} \langle P_1H_1'|v|P_2'P_1'\rangle \langle P_2'P_1'||H_1H_2\rangle \langle P_2P_3||H_1'H_3\rangle \\
& \quad + \sum_{P_1'P_2'} \sum_{H_1'} [\langle H_1'P_2|v|P_2'P_1'\rangle_A \langle P_1P_2'||H_1H_1'\rangle \langle P_1'P_3||H_2H_3\rangle \\
& \quad - \langle P_1H_1'|v|P_2'P_1'\rangle \langle P_2'P_2||H_1H_1'\rangle \langle P_1'P_3||H_2H_3\rangle \\
& \quad - \langle H_1'P_2|v|P_2'P_1'\rangle \langle P_2'P_1||H_1H_1'\rangle \langle P_1'P_3||H_2H_3\rangle] \\
& \quad + \sum_{H_1'H_2'} \sum_{P_1'} \langle H_2'H_1'|v|H_1P_1'\rangle \langle P_1P_2||H_2'H_1'\rangle \langle P_1'P_3||H_2H_3\rangle \quad (2.96)
\end{aligned}
$$

1. Effective Interactions in ECPMET

As has been mentioned, the ECPMET equations may be reformulated [see Eqs. (2.80) and (2.81)] in such a way that a special type of effective interactions is introduced, which represents the so-called *quasilinearized* ECPMET. In the framework of this formalism we attach the entities $A(P_1\ldots, H_1\ldots)$ that are originated by the disconnected cluster operator T_2T_2 to their proper counterparts originated by the connected cluster operator T_2. In particular, the term $A_{T_2T_2}(P_1P_2, H_1H_2)$ is evaluated simultaneously with $A_{T_2}(P_1P_2, H_1H_2)$, and similarly, the terms $A_{T_2T_2}(P_1P_2P_3, H_1H_2H_3)$ with $A_{T_2}(P_1P_2P_3,$

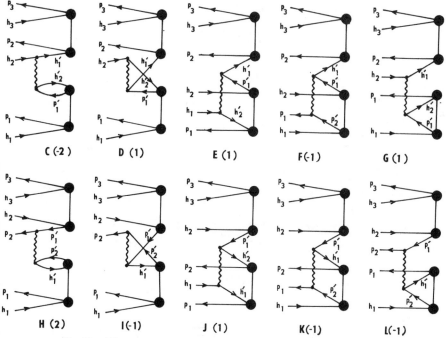

Fig. 12.   Diagrammatic terms contributing to $A_{T_2 T_2}(P_1 P_2 P_3, H_1 H_2 H_3)$.

$H_1 H_2 H_3$). Such a possibility is not only of theoretical interest, but also represents a considerable reduction of numerical effort in an actual realization of the ECPMET method.

The terms $A_{T_2}(P_1 P_2, H_1 H_2)$ and $A_{T_2 T_2}(P_1 P_2, H_1 H_2)$ determined by (2.91) and (2.93), respectively, are quasilinearized as follows[38] (see Fig. 13):

$$A_{T_2}(P_1 P_2, H_1 H_2) + A_{T_2 T_2}(P_1 P_2, H_1 H_2)$$

$$= \tfrac{1}{2}\langle P_1 P_2 | v | H_1 H_2 \rangle + \frac{1}{2} \sum_{P_1' P_2'} \langle P_1 P_2 | v | P_1' P_2' \rangle \langle P_1' P_2' | | H_1 H_2 \rangle$$

$$+ \frac{1}{2} \sum_{H_1' H_2'} \langle H_1' H_2' | \tilde{v} | H_1 H_2 \rangle \langle P_1 P_2 | | H_1' H_2' \rangle$$

$$- \sum_{H_1' P_1'} [\langle P_1 H_1' | \tilde{v} | P_1' H_2 \rangle \langle P_1' P_2 | | H_1 H_1' \rangle$$

$$- \langle P_1 H_1' | \tilde{v} | H_1 P_1' \rangle_A \langle P_1' P_2 | | H_1' H_2 \rangle + \langle P_1 H_1' | \tilde{v} | H_1 P_1' \rangle \langle P_1' P_2 | | H_2 H_1' \rangle]$$

$$+ \sum_{P_1'} \langle P_2 | \tilde{v} | P_1' \rangle \langle P_1 P_1' | | H_1 H_2 \rangle - \sum_{H_1'} \langle H_1' | \tilde{v} | H_2 \rangle \langle P_1 P_2 | | H_1 H_1' \rangle$$

$$(2.97)$$

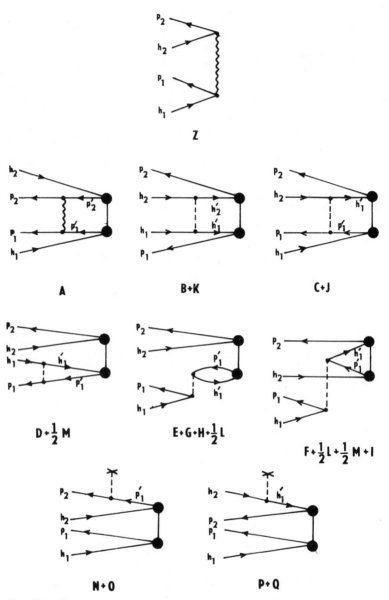

Fig. 13. The diagrammatic interpretation of the sum of $A_{T_2}(P_1P_2, H_1H_2)$ and $A_{T_2T_2}(P_1P_2, H_1H_2)$ with one- and two-particle effective interactions (vertical dashed lines).

Comparing (2.97) with (2.91), we see that the right side of (2.97) is formally almost identical with $A_{T_2}(P_1 P_2, H_1 H_2)$; they are different only in the last two terms in (2.97), corresponding to particle–particle and hole–hole effective interactions. The individual effective matrix elements $\langle \cdots | \tilde{v} | \cdots \rangle$ are diagrammatically defined in Figs. 14 to 20; their algebraic interpretations are[38]:

1. The effective matrix element $\langle H_1' | \tilde{v} | H_2 \rangle$ (see Fig. 14):

$$
\begin{aligned}
\langle H_1' | \tilde{v} | H_2 \rangle &= g(H_1', H_2) \\
&= \sum_{H_2'} \sum_{P_1' P_2'} \langle H_1' H_2' | v | P_1' P_2' \rangle_A \langle P_1' P_2' | | H_2 H_2' \rangle
\end{aligned}
\tag{2.98}
$$

2. The effective matrix element $\langle P_2 | \tilde{v} | P_1' \rangle$ (see Fig. 14):

$$
\begin{aligned}
\langle P_2 | \tilde{v} | P_1' \rangle &= g(P_2, P_1') \\
&= - \sum_{H_1' H_2'} \sum_{P_2'} \langle H_1' H_2' | v | P_1' P_2' \rangle_A \langle P_2 P_2' | | H_1' H_2' \rangle
\end{aligned}
\tag{2.99}
$$

3. The effective matrix element $\langle H_1' H_2' | \tilde{v} | H_1 H_2 \rangle$ (see Fig. 15):

$$
\langle H_1' H_2' | \tilde{v} | H_1 H_2 \rangle = \langle H_1' H_2' | v | H_1 H_2 \rangle + g(H_1' H_2', H_1 H_2)
\tag{2.100a}
$$

$$
g(H_1' H_2', H_1 H_2) = \sum_{P_1' P_2'} \langle H_1' H_2' | v | P_1' P_2' \rangle \langle P_1' P_2' | | H_1 H_2 \rangle
\tag{2.100b}
$$

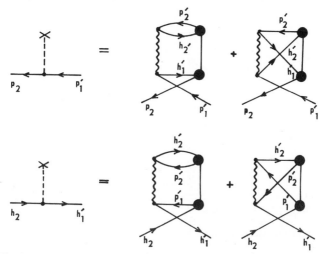

Fig. 14. The diagrammatic interpretation of the 1-particle–1-particle and 1-hole–1-hole effective interactions.

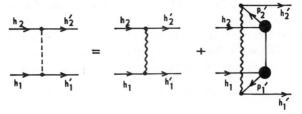

Fig. 15.   The diagrammatic interpretation of 2-hole–2-hole effective interactions.

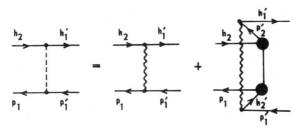

Fig. 16.   The diagrammatic interpretation of 2-particle–2-hole effective interactions.

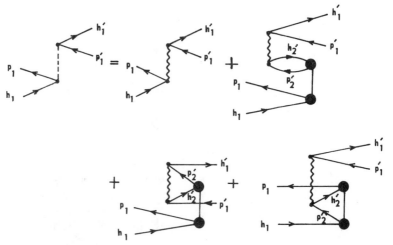

Fig. 17.   The diagrammatic interpretation of particle-hole–particle-hole effective interactions.

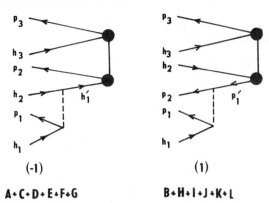

Fig. 18. The diagrammatic interpretation of the sum of $A_{T_2}(P_1P_2P_3, H_1H_2H_3)$ and $A_{T_2T_2}(P_1P_2P_3, H_1H_2H_3)$. with 2-hole–hole-particle and 2-particle–hole-particle effective interactions (vertical dashed lines).

4. The effective matrix element $\langle P_1 H_1' | \tilde{v} | P_1' H_2 \rangle$ (see Fig. 16):

$$\langle P_1 H_1' | \tilde{v} | P_1' H_2 \rangle = \langle P_1 H_1' | v | P_1' H_2 \rangle + g(P_1 H_1', P_1' H_2) \qquad (2.101a)$$

$$g(P_1 H_1', P_1' H_2) = -\frac{1}{2} \sum_{H_2' P_2'} \langle H_2' H_1' | v | P_1' P_2' \rangle \langle P_1 P_2' | | H_2' H_2 \rangle \qquad (2.101b)$$

5. The effective matrix element $\langle P_1 H_1' | \tilde{v} | H_1 P_1' \rangle$ (see Fig. 17):

$$\langle P_1 H_1' | \tilde{v} | H_1 P_1' \rangle = \langle P_1 H_1' | v | H_1 P_1' \rangle + g(P_1 H_1', H_1 P_1') \qquad (2.102a)$$

$$g(P_1 H_1', H_1 P_1') = \frac{1}{2} \sum_{H_2' P_2'} [\langle H_2' H_1' | v | P_2' P_1' \rangle_A \langle P_1 P_2' | | H_1 H_2' \rangle$$

$$- \langle H_2' H_1' | v | P_2' P_1' \rangle \langle P_2' P_1 | | H_1 H_2' \rangle] \qquad (2.102b)$$

The same approach of "effective" interactions may be used for the simultaneous evaluation of $A_{T_2}(P_1P_2P_3, H_1H_2H_3)$ and $A_{T_2T_2}(P_1P_2P_3, H_1H_2H_3)$. We get (see Fig. 18)

$$A_{T_2}(P_1P_2P_3, H_1H_2H_3) + A_{T_2T_2}(P_1P_2P_3, H_1H_2H_3)$$

$$= -\sum_{H_1'} \langle P_1 H_1' | \tilde{v} | H_1 H_2 \rangle \langle P_2 P_3 | | H_1' H_3 \rangle$$

$$+ \sum_{P_1'} \langle P_1 P_2 | \tilde{v} | H_1 P_1' \rangle \langle P_1' P_3 | | H_2 H_3 \rangle \qquad (2.103)$$

where both the effective matrix elements $\langle P_1 H_1' | \tilde{v} | H_1 H_2 \rangle$ and $\langle P_1 P_2 | \tilde{v} | H_1 P_1' \rangle$ are determined as follows:

1. The effective matrix element $\langle P_1 H_1' | \tilde{v} | H_1 H_2 \rangle$ (see Fig. 19):

$$\langle P_1 H_1' | \tilde{v} | H_1 H_2 \rangle = \langle P_1 H_1' | v | H_1 H_2 \rangle + g(P_1 H_1', H_1 H_2) \tag{2.104a}$$

$$g(P_1 H_1', H_1 H_2) = \sum_{P_1' H_2'} [\langle H_2' H_1' | v | P_1' H_2 \rangle_A \langle P_1 P_1' | | H_1 H_2' \rangle$$

$$- \langle H_2' H_1' | v | H_1 P_1' \rangle \langle P_1 P_1' | | H_2' H_2 \rangle$$

$$- \langle H_2' H_1' | v | P_1' H_2 \rangle \langle P_1' P_1 | | H_1 H_2' \rangle]$$

$$+ \sum_{P_1' P_2'} \langle P_1 H_1' | v | P_2' P_1' \rangle \langle P_2' P_1' | | H_1 H_2 \rangle \tag{2.104b}$$

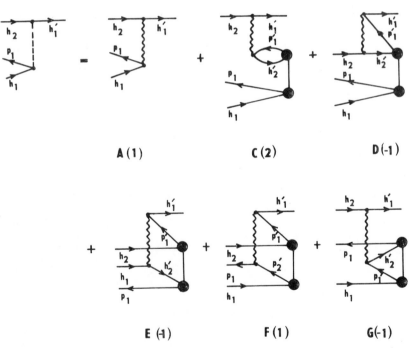

Fig. 19. The diagrammatic interpretation of the 2-hole–hole-particle effective interactions.

2. The effective matrix element $\langle P_1 P_2 | \tilde{v} | H_1 P_1' \rangle$ (see Fig. 20):

$$\langle P_1 P_2 | \tilde{v} | H_1 P_1' \rangle = \langle P_1 P_2 | v | H_1 P_1' \rangle + g(P_1 P_2, H_1 P_1')$$

$$(2.105a)$$

$$g(P_1 P_2, H_1 P_1') = \sum_{P_2' H_1'} [\langle H_1' P_2 | v | P_2' P_1' \rangle_A \langle P_1 P_2' | | H_1 H_1' \rangle$$

$$- \langle P_1 H_1' | v | P_2' P_1' \rangle \langle P_2' P_2 | | H_1 H_1' \rangle$$

$$- \langle H_1' P_2 | v | P_2' P_1' \rangle \langle P_2' P_1 | | H_1 H_1' \rangle]$$

$$+ \sum_{H_1' H_2'} \langle H_2' H_1' | v | H_1 P_1' \rangle \langle P_1 P_2 | | H_2' H_1' \rangle \quad (2.105b)$$

Introducing (2.104) and (2.105) into (2.103), we can see by inspection that a result is exactly identical with the sum of (2.94) and (2.96).

The main obstacle appearing in an implementation of these ECPMET formulas is an enormous number of "three-particle" matrix elements $\langle P_1 P_2 P_3 | | H_1 H_2 H_3 \rangle$ corresponding to T_3. This gives rise to some special difficulties in coding the ECPMET method, since all the matrix elements cannot be simultaneously stored in high-speed memory. In order to surmount this problem we have suggested a kind of two-level searching method,[42] it needs only a fraction of matrix elements to be stored at high speed memory.

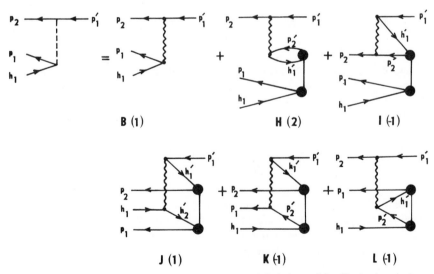

Fig. 20. The diagrammatic interpretation of the 2-particle–hole-particle effective interactions.

Following Paldus,[13,14] the total number of "three-particle" cluster matrix elements may be partially reduced by making use of his spin-adapted CPMET method.

2. Approximate Versions of the CCA

By using the ECPMET equations (2.84a) to (2.84c) we can simply specify both the LCCA and CPMET approximations. The LCCA approximate version of the CCA is determined by

$$A(P, H) = 0 \qquad (2.106a)$$

$$A(P_1 P_2, H_1 H_2) = A_{T_2}(P_1 P_2, H_1 H_2) \qquad (2.106b)$$

$$A(P_1 P_2 P_3, H_1 H_2 H_3) = 0 \qquad (2.106c)$$

where the term $A_{T_2}(P_1 P_2, H_1 H_2)$ is defined by (2.91). For an actual numerical implementation of the LCCA these relations are not very advantageous; a much better approach is represented by the relations (2.61) and (2.62) initially suggested by Bartlett et al.[51,52]

The CPMET approximate version of the CCA is specified by

$$A(P, H) = 0 \qquad (2.107a)$$

$$A(P_1 P_2, H_1 H_2) = A_{T_2}(P_1 P_2, H_1 H_2) + A_{T_2 T_2}(P_1 P_2, H_1 H_2)$$

$$(2.107b)$$

$$A(P_1 P_2 P_3, H_1 H_2 H_3) = 0 \qquad (2.107c)$$

where the term $A_{T_2 T_2}(P_1 P_2, H_1 H_2)$ is defined by (2.93). If we use the technique of effective interactions, the right side of (2.107b) in the QL-CPMET form is determined by (2.97) and (2.102). The CEPA approximation of the CPMET [see Eqs. (2.70) and (2.71)] is, in general, specified by the following modification of (2.84b):

$$\langle P_1 P_2 | | H_1 H_2 \rangle = \frac{A_{T_2}(P_1 P_2, H_1 H_2) + A_{T_2}(P_2 P_1, H_2 H_1)}{D(H_1 H_2, P_1 P_2) + \Delta(H_1 H_2, P_1 P_2)} \qquad (2.108)$$

where the Δ's play the same role as the diagonal matrix elements $g_{\alpha\alpha}$ in (2.71). Putting $\Delta(H_1 H_2, P_1 P_2) = 0$, we get from (2.108) the CEPA-0 method, identical with our LCCA approximation. If we approximate $\Delta(H_1 H_2, P_1 P_2)$ by formulas accounting for different linear combinations of the so-called pair-correlation energies, then we arrive at the CEPA-1, CEPA-2,... methods.[54–56,58] A precise determination of $\Delta(H_1 H_2, P_1 P_2)$ was given in our recent work.[38]

J. CCA for Systems with One-Particle Perturbation Terms

Until now, we have studied the approximate versions of CCA that are appropriate for molecular systems with a perturbation composed of two-particle terms (i.e., closed-shell molecules with RHF one-particle functions or open-shell molecules with UHF one-particle functions). We turn our attention to molecular systems with a perturbation composed of both one- and two-particle terms. In particular, the theory outlined will be useful for simple open-shell molecular systems well approximated by a single Slater determinant built up over an orthonormal set of the RHF one-particle functions.[60] In order to obtain an approximate version of the CCA which produces the correlation energy exact through the third order of perturbation theory, the exponential wave operator must be approximated by

$$ e^T \simeq 1 + T_1 + \frac{1}{2!} T_1^2 + T_2 \tag{2.109} $$

Then the CCA relations (2.26a) and (2.26b) are

$$ E_{\text{corr}} = \langle \Phi_0 | H_1 \left(1 + \frac{1}{2!} T_1^2 + T_2 \right) | \Phi_0 \rangle_C \tag{2.110a} $$

$$ \langle \Phi_\alpha | T_l | \Phi_0 \rangle = \frac{1}{D_\alpha} \langle \Phi_\alpha | H_1 \left(1 + T_1 + \frac{1}{2!} T_1^2 + T_2 \right) | \Phi_0 \rangle_C \tag{2.110b} $$

for $l = 1, 2$. We see that the correlation energy E_{corr} is formally determined by the same relation as in the general CCA; see Eqs. (2.27) to (2.29). The relation (2.110b) offers a formal string of two coupled nonlinear equations,

$$ t_{\alpha_1}^{(1)} = \frac{1}{D_{\alpha_1}} \left(v_{\alpha_1 0} + \sum_{\beta_1} v_{\alpha_1 \beta_1} t_{\beta_1}^{(1)} + \sum_{\beta_2} v_{\alpha_1 \beta_2} t_{\beta_2}^{(2)} + \sum_{\beta_1 \beta_1'} v_{\alpha_1, \beta_1 \beta_1'} t_{\beta_1}^{(1)} t_{\beta_1'}^{(1)} \right) \tag{2.111a} $$

$$ t_{\alpha_2}^{(2)} = \frac{1}{D_{\alpha_2}} \left(v_{\alpha_2 0} + \sum_{\beta_1} v_{\alpha_2 \beta_1} t_{\beta_1}^{(1)} + \sum_{\beta_2} v_{\alpha_2 \beta_2} t_{\beta_2}^{(2)} + \sum_{\beta_1 \beta_1'} v_{\alpha_2, \beta_1 \beta_1'} t_{\beta_1}^{(1)} t_{\beta_1'}^{(1)} \right) \tag{2.111b} $$

where the indices $\alpha_1, \beta_1, \beta_1'$ (β_2, α_2) denote the monoexcited (diexcited) unperturbed states. Their formal diagrammatic interpretation is

$$ \tag{2.112a} $$

(a)

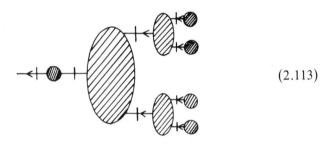

$$(2.112b)$$

$$(b)$$

One can tell by inspection after a few iterations, that this system of coupled equations produces the "treelike" diagrammatic terms, e.g.

$$(2.113)$$

In an analogous way to the QL-ECPMET, the relations (2.111a) and (2.111b) may be quasilinearized. Let us introduce the following two types of effective matrix elements:

$$g_{\alpha_1\beta_1} = \sum_{\beta_1'} v_{\alpha_1,\beta_1\beta_1'} t_{\beta_1'}^{(1)} \tag{2.114a}$$

$$g_{\alpha_2\beta_1} = \sum_{\beta_1'} v_{\alpha_2,\beta_1\beta_1'} t_{\beta_1'}^{(1)} \tag{2.114b}$$

Then the QL form of (2.111) is

$$t_{\alpha_1}^{(1)} = \frac{1}{D_{\alpha_1}} \left(v_{\alpha_1 0} + \sum_{\beta_1} \tilde{v}_{\alpha_1\beta_1} t_{\beta_1}^{(1)} + \sum_{\beta_2} v_{\alpha_1\beta_2} t_{\beta_2}^{(2)} \right) \tag{2.115a}$$

$$t_{\alpha_2}^{(2)} = \frac{1}{D_{\alpha_2}} \left(v_{\alpha_2 0} + \sum_{\beta_1} \tilde{v}_{\alpha_2\beta_1} t_{\beta_1}^{(1)} + \sum_{\beta_2} v_{\alpha_2\beta_2} t_{\beta_2}^{(2)} \right) \tag{2.115b}$$

where

$$\tilde{v}_{\alpha_1\beta_1} = v_{\alpha_1\beta_1} + g_{\alpha_1\beta_1} \tag{2.115c}$$

$$\tilde{v}_{\alpha_2\beta_2} = v_{\alpha_2\beta_2} + g_{\alpha_2\beta_1} \tag{2.115d}$$

Retaining in (2.115c) only the diagonal matrix elements $g_{\alpha_1\alpha_1}$, an analogue of the CEPA may be constructed. For molecular systems with one-particle terms in H_1 the outlined approximate version of CCA is, roughly speaking, of the same quality as the LCCA for systems with a perturbation containing only two-particle terms.

Feynman–Goldstone Diagrammatic Construction. Since the perturbation H_1 (in particular its one-particle part) is not spin-independent, the algebraic interpretation of diagrams appearing in this approximate version of the CCA cannot be simply carried out in a spin-free form. The cluster matrix elements of T_1 and T_2 are determined by [see Eq. (2.30)]

$$\langle p||h\rangle = \frac{A(p,h)}{D(h,p)} \tag{2.116a}$$

$$\langle p_1 p_2||h_1 h_2\rangle = \frac{A(p_1 p_2, h_1 h_2) + A(p_2 p_1, h_2 h_1)}{D(h_1 h_2, p_1 p_2)} \tag{2.116b}$$

where $D(h_1 \ldots, p_1 \ldots) = \varepsilon_{h_1} + \cdots - \varepsilon_{p_1} - \cdots$. The entities $A(\ldots, \ldots)$ may be split into components originating from different cluster operators:

$$A(p,h) = A_{T_1}(p,h) + A_{T_2}(p,h) + A_{T_1 T_1}(p,h) \tag{2.117a}$$

$$A(p_1 p_2, h_1 h_2) = A_{T_1}(p_1 p_2, h_1 h_2) + A_{T_2}(p_1 p_2, h_1 h_2)$$
$$+ A_{T_1 T_1}(p_1 p_2, h_1 h_2) \tag{2.117b}$$

These terms are diagrammatically interpreted in Figs. 21 to 26. The algebraic interpretation of diagrams is simply carried out by making use of the overall numerical factor (2.9):

1. The term $A_{T_1}(p,h)$ (see Fig. 21):

$$A_{T_1}(p,h) = -\langle p|w|h\rangle + \sum_{h_1'}\langle h_1'|w|h\rangle\langle p||h_1'\rangle$$
$$- \sum_{p_1'}\langle p|w|p_1'\rangle\langle p_1'||h\rangle$$
$$+ \sum_{p_1' h_1'}\langle h_1' p|v|p_1' h\rangle_A\langle p_1'||h\rangle \tag{2.118}$$

2. The term $A_{T_2}(p,h)$ (see Fig. 22):

$$A_{T_2}(p,h) = -\sum_{h_1 h_2}\sum_{p_1}\langle h_1 h_2|v|p_1 h\rangle_A\langle p_1 p||h_1 h_2\rangle$$
$$+ \sum_{p_1 p_2}\sum_{h_1}\langle h_1 p|v|p_1 p_2\rangle_A\langle p_1 p_2||h_1 h\rangle$$
$$- \sum_{h_1 p_1}\langle p_1|w|h_1\rangle(\langle p_1 p||h_1 h\rangle - \langle p_1 p||hh_1\rangle) \tag{2.119}$$

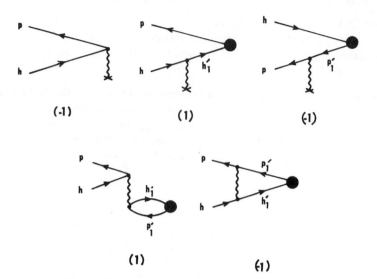

Fig. 21. Diagrammatic terms contributing to $A_{T_1}(p, h)$.

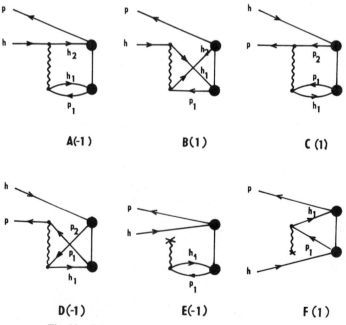

Fig. 22. Diagrammatic terms contributing to $A_{T_2}(p, h)$.

230

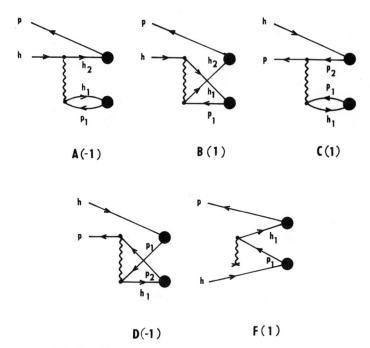

Fig. 23. Diagrammatic terms contributing to $A_{T_1 T_1}(p, h)$.

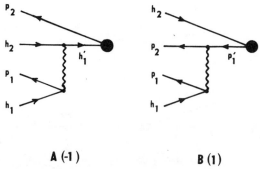

Fig. 24. Diagrammatic terms contributing to $A_{T_1}(p_1 p_2, h_1 H_2)$.

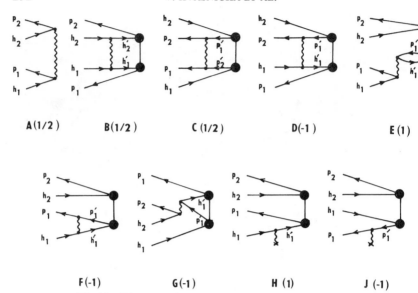

Fig. 25.   Diagrammatic terms contributing to $A_{T_2}(p_1p_2, h_1h_2)$.

3. The term $A_{T_1T_1}(p, h)$ (see Fig. 23):

$$
\begin{aligned}
A_{T_1T_1}(p, h) = &- \sum_{h_1h_2}\sum_{p_1} \langle h_1h_2|v|p_1h\rangle_A \langle p||h_2\rangle\langle p_1||h_1\rangle \\
&+ \sum_{p_1p_2}\sum_{h_1} \langle h_1p|v|p_1p_2\rangle_A \langle p_2||h\rangle\langle p_1||h_1\rangle \\
&+ \sum_{p_1h_1} \langle p_1|w|h_1\rangle\langle p||h_1\rangle\langle p_1||h\rangle
\end{aligned}
\tag{2.120}
$$

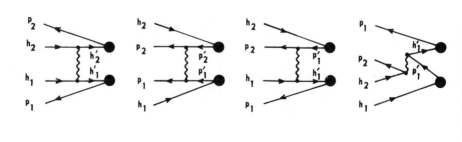

Fig. 26.   Diagrammatic terms contributing to $A_{T_1T_1}(p_1p_2, h_1h_2)$.

4. The term $A_{T_1}(p_1, p_2, h_1 h_2)$ (see Fig. 24):

$$A_{T_1}(p_1 p_2, h_1 h_2) = -\sum_{h_1'} \langle p_1 h_1' | v | h_1 h_2 \rangle \langle p_2 | | h_1' \rangle$$

$$+ \sum_{p_1'} \langle p_1 p_2 | v | h_1 p_1' \rangle \langle p_1' | | h_2 \rangle. \qquad (2.121)$$

5. The term $A_{T_2}(p_1 p_2, h_1 h_2)$ (see Fig. 25):

$$A_{T_2}(p_1 p_2, h_1 h_2) = \tfrac{1}{2} \langle p_1 p_2 | v | h_1 h_2 \rangle$$

$$+ \frac{1}{2} \sum_{h_1' h_2'} \langle h_1' h_2' | v | h_1 h_2 \rangle \langle p_1 p_2 | | h_1' h_2' \rangle$$

$$+ \frac{1}{2} \sum_{p_1' p_2'} \langle p_1 p_2 | v | p_1' p_2' \rangle \langle p_1' p_2' | | h_1 h_2 \rangle$$

$$- \sum_{h_1' p_1'} [\langle h_1' p_2 | v | h_1 p_1' \rangle \langle p_1 p_1' | | h_1' h_2 \rangle$$

$$- \langle p_1 h_1' | v | h_1 p_1' \rangle_A \langle p_1' p_2 | | h_1' h_2 \rangle$$

$$+ \langle p_2 h_1' | v | h_2 p_1' \rangle \langle p_1' p_1 | | h_1 h_1' \rangle]$$

$$+ \sum_{h_1'} \langle h_1' | w | h_1 \rangle \langle p_1 p_2 | | h_1' h_2 \rangle$$

$$- \sum_{p_1'} \langle p_1 | w | p_1' \rangle \langle p_1' p_2 | | h_1 h_2 \rangle \qquad (2.122)$$

6. The term $A_{T_1 T_1}(p_1 p_2, h_1 h_2)$ (see Fig. 26):

$$A_{T_1 T_1}(p_1 p_2, h_1 h_2) = \frac{1}{2} \sum_{h_1' h_2'} \langle h_1' h_2' | v | h_1 h_2 \rangle \langle p_1 | | h_1' \rangle \langle p_2 | | h_2' \rangle$$

$$+ \frac{1}{2} \sum_{p_1' p_2'} \langle p_1 p_2 | v | p_1' p_2' \rangle \langle p_1' | | h_1 \rangle \langle p_2' | | h_2 \rangle$$

$$- \sum_{h_1' p_1'} [\langle h_1' p_2 | v | h_1 p_1' \rangle \langle p_1 | | h_1' \rangle \langle p_1' | | h_2 \rangle$$

$$+ \langle p_2 h_1' | v | h_2 p_1' \rangle \langle p_1' | | h_1 \rangle \langle p_1 | | h_1' \rangle] \qquad (2.123)$$

Now, the approach of effective interactions (2.114) and (2.115) may be simply introduced, but in this special case of interest it is more advantageous to use another technique. In particular, the entities $A_{T_2}(\ldots, \ldots)$ and $A_{T_1 T_1}(\ldots, \ldots)$ may be, owing to their algebraic structure, simultaneously evaluated; the terms from $A_{T_1 T_1}(\ldots, \ldots)$ differ from their counterparts in $A_{T_2}(\ldots, \ldots)$ only in that the cluster matrix elements $\langle ij | | kl \rangle$ are replaced by the product $\langle i||k\rangle\langle j||l\rangle$. Hence, the term $A_{T_1T_1}(\ldots,\ldots)$ can be evaluated simultaneously with $A_{T_2}(\ldots,\ldots)$ if in the latter some pertinent cluster matrix elements $\langle ij||kl\rangle$ are replaced by $\langle ij||kl\rangle + \langle i||k\rangle\langle j||l\rangle$.

III. QUASIDEGENERATE CCA

The diagrammatic RSPT is easily generalized in the so-called degenerate[61-63] (or more generally, quasidegenerate[29,39,64,65]) versions, where a model hamiltonian is defined in a multidimensional model space spanned by a number of unperturbed state vectors which belong to a fixed eigenenergy (or to a few eigenenergies in the case of quasidegenerate perturbation theory). The matrix elements of the model hamiltonian are determined by linked and connected diagrammatic terms that can be further classified as folded and/or unfolded; their algebraic form is closely related to the formal model hamiltonian.[29] In an analogous way to the case of nondegenerate CCA, we would like to derive a CCA recursive construction of this diagrammatic model hamiltonian. To solve a problem like this we have, in particular, to generalize the CCA in a quasidegenerate form. The first extension of CCA to simple degenerate systems was initially reported by Coester.[66] By now it has been studied by many authors.[35,67-74] The most frequent shortcoming of these generalizations is that the exponential wave operator produces the so-called spurious terms,[75,76] which are represented by diagrams that have no proper counterpart within the formalism of the quasidegenerate diagrammatic RSPT. A simple and straightforward one-to-one correspondence between quasidegenerate versions of the CCA and the diagrammatic RSPT was achieved by Lindgren.[72] He solved the problem of spurious terms by placing the exponential wave operator inside of a normal product. The resulting CCA nonhermitian model hamiltonian is in a simple one-to-one correspondence with the model hamiltonian constructed by diagrammatic RSPT. Recently, Kvasnička[74] enlarged Lindgren's formalism in such a way that the resulting model hamiltonian is a hermitian operator.

Let us introduce the hole–particle formalism with respect to an unperturbed "core" state vector $|\Phi_0\rangle$. The states that are distributed near the Fermi level are called *active*.[63] The particle (hole) states that are above (below) these active states are called *passive*; see Fig. 27. A multi-dimensional *model space* D_0 with the projector P_0 is spanned by all possible n-particle and m-hole unperturbed state vectors generated from the "core" state $|\Phi_0\rangle$ by creating all possible n active particles and m active holes,

$$D_0 \equiv \left\{ X_{p_1}^{\dagger}\ldots X_{p_n}^{\dagger} X_{h_1}\ldots X_{h_m}|\Phi_0\rangle \right\} \qquad (3.1)$$

The projector P_0 into D_0 is

$$P_0 = \sum_{\alpha} P_0(\alpha) = \sum_{\alpha} |\Phi_\alpha\rangle\langle\Phi_\alpha| \qquad (3.2)$$

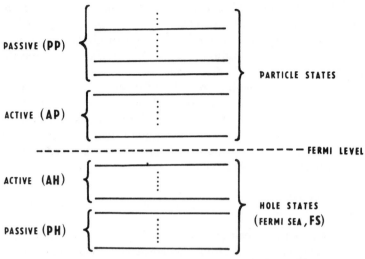

Fig. 27. The classification of one-particle states in the many-body quasidegenerate perturbation theory.

where the summation index α runs over all states from D_0. If N_p and N_h are the numbers of active particle and hole states, respectively, then the dimension of D_0 is

$$\dim D_0 = \text{Tr}(P_0) = d = \binom{N_p}{n}\binom{N_h}{m} \tag{3.3}$$

A wave operator Ω (i.e., $\Omega P_0 = \Omega$) maps the model space D_0 into a subspace spanned by a finite number d of perturbed state vectors $|\Psi_i\rangle = \Omega|\varphi_i\rangle$ for $|\varphi_i\rangle \in D_0$. A *model hamiltonian* $H_{\text{mod}} = \langle\Phi_0|H|\Phi_0\rangle + H_0 + G$, referred to the total hamiltonian $H = \langle\Phi_0|H|\Phi_0\rangle + H_0 + H_1$, is determined by $H\Omega = \Omega H_{\text{mod}}$, or in the form of the commutator equation[39-41]

$$\Omega H_0 - H_0\Omega = [\Omega, H_0]_- = H_1\Omega - \Omega G \tag{3.4}$$

where a *model interaction* G is defined in the model space D_0, $G = P_0 G P_0$.

Let us now specify the wave operator Ω. Following Lindgren,[72] we postulate this exponential wave operator in a normally ordered form[74]

$$\Omega = (N_0)^{-1/2} U = (N_0)^{-1/2} N[e^S] \tag{3.5a}$$

$$S = T + F \tag{3.5b}$$

The total cluster operator S is split into two parts T and F, corresponding to the so-called *correlation* and *relaxation* effects, respectively. The correlation cluster operator T corresponds to a cluster operator of nondegenerate "core" subsystem; the whole of Section II is devoted to the construction of this cluster operator. A normalization constant N_0 in (3.5a) will be specified below in accordance with the type of model hamiltonian (whether this operator is hermitian or nonhermitian). The relaxation cluster operator F describes the interaction processes between "core" and active particles as well as between active particles themselves. In order to specify this cluster operator more sharply, we use the following terminology of Lindgren[72]: An arbitrary operator A defined on the model space is formally split into the so-called *open* (op) and *closed* (cl) parts,

$$A = AP_0 = A_{op} + A_{cl}, \qquad A_{op} = (1 - P_0)A, \quad A_{cl} = P_0 A \qquad (3.6)$$

In the diagrammatic formalism this means that a closed operator A_{cl} is determined by diagrams with external lines labeled by the active states only (i.e., from subsets AP and/or AH; see Fig. 27), whereas an open operator A_{op} is determined by diagrams with left external lines labeled by at least one passive state (that is, from subsets PP and/or PH; see Fig. 27) and right external lines labeled by active states. Hence, the relaxation cluster operator can be written as

$$F = F_{op} + F_{cl} \qquad (3.7)$$

It will be demonstrated below that the open part of F is determined by a string of coupled nonlinear equations, while its closed part is fully determined by a type of normalization of perturbed state vectors. The second-quantization representation of F is

$$F = \sum_{l \geqslant 1} F_l \qquad (3.8a)$$

$$F_l = \frac{1}{l!} \sideset{}{'}\sum_{\substack{i_1,\ldots,i_l \\ j_1,\ldots,j_l}} \langle i_1 \ldots i_l | | j_1 \ldots j_l \rangle N\left[X_{i_1}^\dagger \ldots X_{i_l}^\dagger X_{jl} \ldots X_{j1} \right] \qquad (3.8b)$$

where the primed summation means that its indices are restricted by the following two conditions:

1. The right outgoing and incoming lines are indexed only by active states, and
2. The left outgoing and incoming lines may be indexed by both active and passive states.

A. Generalized "Linked-Cluster" Theorem (G-LCT)

The original form of the LCT presented in Section II.C is not applicable to the present case, since Ω^{-1} is not identical with $(N_0)^{1/2}N[e^{-S}]$. Therefore, the purpose of this section is to formulate a generalized LCT (G-LCT) which does not require us to know an explicit form of Ω^{-1}. It will be proved along a line initially suggested by K. O. Friedrichs[77] in the early sixties, in his lecture notes on the mathematical problems of perturbation theory. Almost the same considerations have been used by Lindgren[72] in his work on the quasidegenerate version of CCA with nonhermitian model hamiltonian; see also Reference 34.

Let us consider an expression $A\Omega$, where A is an object and Ω is the exponential wave operator (3.5). By using the generalized Wick's theorem[18,19] we get

$$A\Omega = N[A\Omega] + N[\overline{A\Omega}] \tag{3.9}$$

The term $N[\overline{A\Omega}]$ denotes all the contracted normally ordered products of A and Ω; it can formally be written as

$$N[\overline{A\Omega}] = (N_0)^{-1/2} \sum_{n=1}^{\infty} \frac{1}{n!} N[\overline{AS^n}] \tag{3.10}$$

where for a fixed integer n the term $N[\overline{AS^n}]$ is equal to the following sum:

$$
\begin{aligned}
N[\overline{AS^n}] &= \sum_{m=1}^{n} \binom{n}{m} N[\overline{A(S)}^m S^{n-m}] \\
&= \binom{n}{1} N[\overline{ASS^{n-1}}] + \binom{n}{2} N[\overline{ASSS^{n-2}}] + \cdots \\
&\quad + \binom{n}{n} N[\overline{AS...S}]
\end{aligned} \tag{3.11}
$$

Introducing (3.11) and (3.10) into (3.9), we arrive at

$$
\begin{aligned}
A\Omega &= (N_0)^{-1/2} \sum_{n=0}^{\infty} \frac{1}{n!} \sum_{m=0}^{n} \binom{n}{m} N[\overline{A(S)}^m S^{n-m}] \\
&= (N_0)^{-1/2} \sum_{n=0}^{\infty} \sum_{p=0}^{\infty} \frac{1}{n!p!} N[\overline{A(S)}^n S^p] \\
&= \sum_{n=0}^{\infty} \frac{1}{n!} N[\overline{A(S)}^n \Omega] = N[\{AU\}_C \Omega]
\end{aligned} \tag{3.12}
$$

where $\{AU\}_C$ is determined as the sum of all possible "connected" terms containing contractions (internal lines) only between A and S operators, and the contractions between S operators are strictly forbidden:

$$\{AU\}_C = N\left[A + \frac{1}{1!}\overline{AS} + \frac{1}{2!}\overline{ASS} + \cdots\right] \qquad (3.13a)$$

Analogous algebraic considerations are applicable if we have to normally order an inverse product ΩA; a difference is that instead of $\{AU\}_C$ there appears a term $\{UA\}_C$ determined by

$$\{UA\}_C = N\left[A + \frac{1}{1!}\overline{SA} + \frac{1}{2!}\overline{SSA} + \cdots\right] \qquad (3.13b)$$

Summarizing above considerations, we can formulate two alternative forms of the so-called *generalized LCT of Friedrichs*[77] (cf. Reference 34):

$$A\Omega = N[\{AU\}_C\Omega] \qquad (3.14a)$$

$$\Omega A = N[\{UA\}_C\Omega] \qquad (3.14b)$$

where the terms $\{\cdots\}_C$ are determined by (3.13).

The G-LCT will be applied to individual terms of the commutator equation (3.4)

$$H_1\Omega = N[\{H_1U\}_C\Omega] \qquad (3.15a)$$

$$\Omega G = N[\{UG\}_C\Omega] \qquad (3.15b)$$

$$\Omega H_0 - H_0\Omega = N[(SH_0 - H_0S)\Omega] \qquad (3.15c)$$

where in deriving the last expression we keep in mind that H_0 is a diagonal one-particle operator. Introducing (3.15) into the commutator equation (3.4) gives

$$N[(SH_0 - H_0S - \{H_1U\}_C - \{UG\}_C)\Omega] = 0 \qquad (3.16)$$

This condition is satisfied if the cluster operator S fulfils the following relation[72,74]:

$$[S, H_0]_- = \{H_1U - UG\}_C = \{H_1U\}_C - \{UG\}_C \qquad (3.17a)$$

where $\{H_1U\}_C$ and $\{UG\}_C$ are

$$\{H_1U\}_C = N\left[H_1 + \frac{1}{1!}\overline{H_1S} + \frac{1}{2!}\overline{H_1SS} + \cdots\right] \qquad (3.17b)$$

$$\{UG\}_C = N\left[G + \frac{1}{1!}\overline{SG} + \frac{1}{2!}\overline{SSG} + \cdots\right] \qquad (3.17c)$$

The commutator equation (3.17a) represents the principal relation of quasi-degenerate CCA; it allows us to construct both nonhermitian and hermitian model hamiltonians.

B. Nonhermitian Model Hamiltonian

This type of the model hamiltonian, initially derived by Lindgren,[72] is based upon the intermediate normalization of perturbed state vectors. Then the wave operator Ω is specified by

$$P_0\Omega = P_0U = P_0, \qquad N_0 = 1 \qquad (3.18)$$

It implies that the closed part of the relaxation cluster operator is zero:

$$F_{cl} = 0 \qquad (3.19)$$

that is, the action of the cluster operator S or its powers on an arbitrary unperturbed state vector from the model space D_0 produces the states that are orthogonal to D_0. Multiplying (3.17a) on both sides by P_0, we arrive at

$$G = \{P_0H_1UP_0\}_C = \{H_1U\}_{C,cl} \qquad (3.20a)$$

The nonhermitian model interaction is determined by all possible connected and closed diagrammatic terms attached to the product H_1U. The model interaction G can be split into the connected ground-state diagrams and linked–connected (LC) diagrams; the sum of the former terms is exactly equal to the correlation energy $E_{corr,core}$ of the "core" subsystem,

$$G = E_{corr,core} + \{H_1U\}_{LC,cl} \qquad (3.20b)$$

Then the model eigenproblem fully determined in the finite-dimensional model space D_0 is

$$\left(H_0 + \{H_1U\}_{LC,cl}\right)|\varphi_i\rangle = \Delta E_i|\varphi_i\rangle \qquad (3.21)$$

where $\Delta E_i = E_i - E_0$ is an "excitation" energy between the ground state of the "core" subsystem and an n-particle, m-hole perturbed state vector $|\Psi_i\rangle = \Omega|\varphi_i\rangle$.

By using the intermediate normalization (3.18), the commutator equation (3.17a) may be solved[29,30] for the cluster operator as follows:

$$S = \sum_\alpha \frac{1 - P_0}{E_\alpha^{(0)} - H_0} \{H_1U - UG\}_C P_0(\alpha) \qquad (3.22)$$

where the summation index α runs over all unperturbed states $|\Phi_\alpha\rangle$ belonging to the model space D_0, $H_0|\Phi_\alpha\rangle = E_\alpha^{(0)}|\Phi_\alpha\rangle$. Let $|\Phi_\beta\rangle \in D_0^\perp$ be an unperturbed state vector orthogonal to D_0; the second alternative form of (3.22) is

$$\langle\Phi_\beta|S|\Phi_\alpha\rangle = \frac{1}{E_\alpha^{(0)} - E_\beta^{(0)}}\langle\Phi_\beta|(H_1U - UG)|\Phi_\alpha\rangle_{\mathrm{LC,op}} \qquad (3.23)$$

Since the total cluster operator S is composed of two parts T and F, it is easy to demonstrate that (3.23) provides two relations which determine these operators separately. The first is identical with (2.26b) for the correlation cluster operator T, and the second determines the relaxation cluster operator F. The algebraic structure of this second relation is formally equivalent to the original relation (3.23), but now we consider only those linked–connected and open diagrams that contain at least one one hole (particle) path running from left (right) to right (left). The diagrammatic terms of T contain merely the left excitation paths.

The present form of the quasidegenerate version of CCA offers, as in the case of its nondegenerate closed-shell version, a recursive algorithm to construct the nonhermitian model interaction G. Since we have started in our construction of the CCA from the RSPT commutator equation (3.4), the resulting model interaction should be in a simple one-to-one correspondence with that one constructed by the quasidegenerate diagrammatic RSPT.[29,39] The so-called folded diagrams, representing one of principal concepts of diagrammatic RSPT, are generated in the CCA formalism by the product UG on the right side of (3.23).

Ionization potentials. The CCA theory of nonhermitian model interaction will be illustrated by an example of direct evaluation of vertical ionization potentials; a similar problem in the framework of the diagrammatic RSPT was studied by Kvasnička, Hubač, et al.[78-80] Let the model space D_0 be spanned by an unperturbed "1-hole" state vector

$$D_0 \equiv \{|\Phi_h\rangle\}, \qquad P_0 = |\Phi_h\rangle\langle\Phi_h|, \qquad d = 1 \qquad (3.24)$$

Now we specify the exponential wave operator $\Omega = U = N[e^S]$. In this case it can be significantly simplified as follows:

$$U = N[e^T(1 + F_h)] \qquad (3.25a)$$

$$F_h = F_{h,0} + \sum_{l \geqslant 1} F_{h,l} \qquad (3.25b)$$

$$F_{h,0} = \sum_{h' \neq h} \langle h||h'\rangle N[X_h^\dagger X_{h'}] \qquad (3.25c)$$

$$F_{h,l} = \frac{1}{l!} \sum_{\substack{h_1,\ldots,h_l\ h' \\ p_1,\ldots,p_l}} \sum \langle p_1\ldots p_l h||h_1\ldots h_l h'\rangle N[X_{p_1}^\dagger\ldots X_{p_l}^\dagger X_h^\dagger X_{h'} X_{h_l}\ldots X_{h_1}] \qquad (3.25d)$$

where the relaxation cluster operator F_h is explicitly referred to the state $|\Phi_h\rangle$; it describes a "relaxation" process when the hole state h is replaced by another one h' and this is accompanied by simultaneous l-fold excitations. For $l = 0$ the summation index h' runs over all occupied states except that one indexed by h; this restriction comes from the above requirement that the relaxation cluster operator F_h have only open components. The diagrammatic interpretation of the few first vertices of F_h is shown in Fig. 28.

By using the model eigenproblem (3.21), the vertical ionization potential $(IP)_h$ is

$$(IP)_h = -\varepsilon_h + \Delta_h \tag{3.26a}$$

$$\Delta_h = \langle \Phi_h | G | \Phi_h \rangle = \langle \Phi_h | H_1 N [e^T (1 + F_h)] | \Phi_h \rangle_{LC} \tag{3.26b}$$

In general, the relaxation cluster operator F_h is determined by (3.23); we get

$$\langle \Phi_{h'} | F_{h,0} | \Phi_h \rangle = \frac{1}{\varepsilon_{h'} - \varepsilon_h} \langle \Phi_{h'} | (H_1 U - UG) | \Phi_h \rangle_{LC} \tag{3.27a}$$

$$\langle \Phi_{\alpha} | F_{h,l} | \Phi_h \rangle = \frac{1}{D_{\alpha}} \langle \Phi_{\alpha} | (H_1 U - UG) | \Phi_h \rangle_{LC} \tag{3.27b}$$

where $|\Phi_{\alpha}\rangle = X_{p_1}^{\dagger} X_{h_1} \cdots X_{p_l}^{\dagger} X_{h_l} |\Phi_h\rangle$ is an "l-excited" unperturbed state vector orthogonal to $|\Phi_h\rangle$, $D_{\alpha} = \varepsilon_{h_1} + \cdots + \varepsilon_{h_l} - \varepsilon_{p_1} - \cdots - \varepsilon_{p_l} - \varepsilon_h + \varepsilon_{h'}$. Since the model space D_0 is one-dimensional, the model interaction G is, in fact, a scalar entity $\Delta_h = \langle \Phi_h | G | \Phi_h \rangle$. This means that the relations (3.27a) and (3.27b) may be simplified as follows:

$$\langle \Phi_{h'} | F_{h,0} | \Phi_h \rangle = \frac{1}{\Delta_h + \varepsilon_{h'} - \varepsilon_h} \langle \Phi_{h'} | H_1 U | \Phi_h \rangle_{LC} \tag{3.28a}$$

$$\langle \Phi_{\alpha} | F_{h,l} | \Phi_h \rangle = \frac{1}{D_{\alpha} + \Delta_h} \langle \Phi_{\alpha} | H_1 U | \Phi_h \rangle_{LC} \tag{3.28b}$$

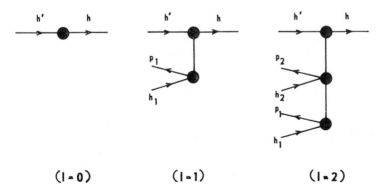

$(l=0)$ $(l=1)$ $(l=2)$

Fig. 28. The diagrammatic interpretation of "zero-," "one-," and "two-particle" terms of the relaxation cluster operator F_h.

The term UG on the right side of (3.27) produces the so-called folded diagrams of Morita and Brandow.[61,62] Their suppression in (3.28) is paid for by the occurrence of Δ-dependent energy denominators. The relations (3.26) and (3.28) may be considered as a CCA realization of the Bloch–Horowitz diagrammatic perturbation theory[81] for the evaluation of vertical ionization potentials. This is easy to understand when one recollects Brandow's original construction[62] of diagrammatic RSPT. He started from the Bloch–Horowitz diagrammatic PT, and his suppression of Δ-dependent energy denominators results in the occurrence of the folded diagrams.

The relations (3.28a) and (3.28b) represent the CCA string of coupled nonlinear equations which determine the relaxation cluster matrix elements; their second alternative form, more appropriate for an actual implementation, is [see eqs. (2.30) and (2.31)]

$$\langle h||h'\rangle = \frac{1}{\Delta_h + \varepsilon_{h'} - \varepsilon_h} A(h, h') \tag{3.29a}$$

$$\langle p_1 \ldots p_l h||h_1 \ldots h_l h'\rangle = \frac{1}{\Delta_h + D_\alpha} \sum A(p_{\alpha_1} \ldots p_{\alpha_l} h, h_{\alpha_1} \ldots h_{\alpha_l} h') \tag{3.29b}$$

where the summation runs over all $l!$ permutations

$$\begin{pmatrix} 1 & \cdots & l \\ \alpha_1 & \cdots & \alpha_l \end{pmatrix}.$$

The matrix elements $A(h, h')$ and $A(p_{\alpha_1} \ldots h, h_{\alpha_1} \ldots h')$ correspond to $\langle \Phi_{h'} | H_1 U | \Phi_h \rangle_{LC}$ and $\langle \Phi_\alpha | H_1 U | \Phi_h \rangle_{LC}$, respectively, with fixed labelling of external lines,

$$A(h, h') = \tag{3.30a}$$

(a)

$$A(p_{\alpha_1} \ldots p_{\alpha_l} h, h_{\alpha_1} \ldots h_{\alpha_l} h') = \tag{3.30b}$$

(b)

The algebraic interpretation of the diagrams in the above rectangular blocks is carried out by making use of the overall numerical factors (2.9) and (2.10), respectively.

Now, the general theory is illustrated by its approximate version for closed-shell molecular systems; the resulting ionization potentials will be exact through the third order of perturbation theory. The exponential wave operator U is approximated by

$$U = 1 + T_1 + T_2 + F_{h,1} + F_{h,2} \qquad (3.31)$$

where T_1 and T_2 are correlation cluster operators, already determined for the given closed-shell molecular system. The vertical ionization potential $(IP)_h$ is determined by (3.26); the diagrammatic terms of Δ_h are presented in Fig. 29.

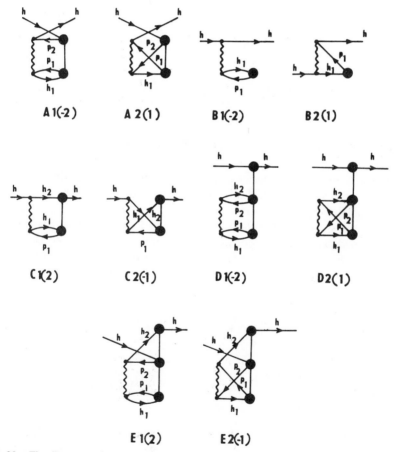

Fig. 29.   The diagrammatic terms for the vertical ionization potential $(IP)_h$, exact through the third order of perturbation theory.

Their spin-free algebraic interpretation is simply carried out by making use of slightly modified overall numerical factor (2.10),

$$(-1)^{h+l+1}(2)^l w_\Gamma \tag{3.32}$$

that is, we have introduced an additional minus owing to the existence of one hole path in these diagrams.[20,63] We get

$$\Delta_h = \Delta_{h,\text{corr}} + \Delta_{h,\text{relax}} \tag{3.33a}$$

$$\Delta_{h,\text{relax}} = \sum_{H_1 H_2} \sum_{P_1} \langle H_1 H_2 | v | P_1 H \rangle_A \langle P_1 H | | H_1 H_2 \rangle$$

$$+ \sum_{H_1 H_2} \sum_{P_1 P_2} \langle H_1 H_2 | v | P_1 P_2 \rangle_A$$

$$\times (-\langle P_1 P_2 H | | H_1 H_2 H \rangle + \langle P_1 P_2 H | | H_1 H H_2 \rangle)$$

$$\tag{3.33b}$$

$$\Delta_{h,\text{corr}} = -\sum_{P_1 P_2} \sum_{H_1} \langle H_1 H | v | P_1 P_2 \rangle_A \langle P_1 P_2 | | H_1 H \rangle$$

$$-\sum_{P_1} \sum_{H_1} \langle H_1 H | v | P_1 H \rangle_A \langle P_1 | | H_1 \rangle \tag{3.33c}$$

where $\langle IJ | v | KL \rangle_A = 2\langle IJ | v | KL \rangle - \langle IJ | v | LK \rangle$. The many-body terms from the entity Δ_h may be separated into the so-called correlation and relaxation contributions. Perhaps such a partitioning of Δ_h might be of value for deeper physical understanding of processes occurring in the ionization of closed-shell molecular systems.

Let us assume that the matrix elements of T_1 and T_2 have been already determined in an independent CCA calculation of correlation energy involving both these operators. What remains to be determined is the relaxation cluster matrix elements of $F_{h,1}$ and $F_{h,2}$. In general, they are determined by (3.29b):

$$\langle P_1 H | | H_1 H' \rangle = \frac{A(P_1 H, H_1 H')}{\Delta_h + \varepsilon_{H_1} + \varepsilon_{H'} - \varepsilon_{P_1} - \varepsilon_H} \tag{3.34a}$$

$$\langle P_1 P_2 H | | H_1 H_2 H' \rangle = \frac{A(P_1 P_2 H, H_1 H_2 H') + A(P_2 P_1 H, H_2 H_1 H')}{\Delta_h + \varepsilon_{H_1} + \varepsilon_{H_2} + \varepsilon_{H'} - \varepsilon_{P_1} - \varepsilon_{P_2} - \varepsilon_H}$$

$$\tag{3.34b}$$

The entities $A(\ldots, \ldots)$ are split into terms originating from different cluster operators [cf. Eqs. (2.85) and (2.86)]. In the framework of our approximate scheme providing ionization potentials exact through third order of perturbation theory, we have [see comment below (3.30b)]

$$A(P_1H, H_1H') = A_{T_2}(P_1H, H_1H') + A_{F_{h,1}}(P_1H, H_1H') \tag{3.35a}$$

$$A(P_1P_2H, H_1H_2H') = A_{T_2}(P_1P_2H, H_1H_2H') + A_{F_{h,1}}(P_1P_2H, H_1H_2H') \tag{3.35b}$$

The individual terms on the right side of (3.35) are determined as follows:

1. The term $A_{T_2}(P_1H, H_1H')$ (see Fig. 30):

$$
\begin{aligned}
A_{T_2}(P_1H, H_1H') = {} & \langle P_1 H | v | H_1 H' \rangle \\
& + \sum_{H_1'P_1'} \big(\langle H_1'H | v | P_1'H' \rangle_A \langle P_1 P_1' | | H_1 H_1' \rangle \\
& - \langle H_1'H | v | P_1'H' \rangle \langle P_1' P_1 | | H_1 H_1' \rangle \\
& - \langle H_1'H | v | H_1 P_1' \rangle \langle P_1 P_1' | | H_1' H' \rangle \big) \\
& + \sum_{P_1'P_2'} \langle P_1 H | v | P_1' P_2' \rangle \langle P_1' P_2' | | H_1 H' \rangle \tag{3.36}
\end{aligned}
$$

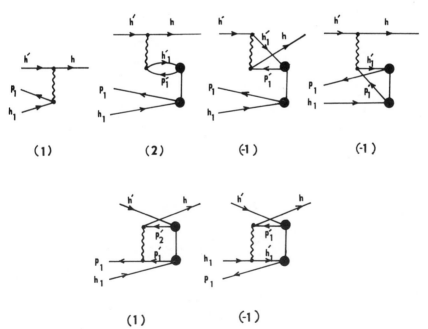

(1)   (2)   (-1)   (-1)

(1)   (-1)

Fig. 30.   Diagrammatic terms contributing to $A_{T_2}(P_1H, H_1H')$.

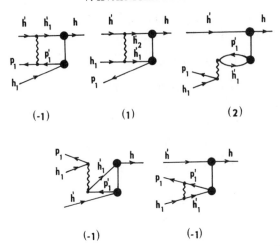

Fig. 31. Diagrammatic terms contributing to $A_{F_{h,1}}(P_1H, H_1H')$.

2. The term $A_{F_{h,1}}(P_1H, H_1H')$ (see Fig. 31):

$$A_{F_{h,1}}(P_1H, H_1H') = \sum_{P_1'H_1'} (-\langle P_1H_1'|v|P_1'H'\rangle \langle P_1'H||H_1H_1'\rangle$$

$$+ \langle H_1'P_1|v|P_1'H_1\rangle_A \langle P_1'H||H_1'H'\rangle$$

$$- \langle H_1'P_1|v|P_1'H_1\rangle \langle P_1'H||H'H_1'\rangle)$$

$$+ \sum_{H_1'H_2'} \langle H_1'H_2'|v|H_1H'\rangle \langle P_1H||H_1'H_2'\rangle \quad (3.37)$$

3. The term $A_{T_2}(P_1P_2H, H_1H_2H')$ (see Fig. 32):

$$A_{T_2}(P_1P_2H, H_1H_2H') = \sum_{P_1'} (\langle P_2H|v|P_1'H'\rangle \langle P_1P_1'||H_1H_2\rangle$$

$$+ \langle P_2H|v|H_2P_1'\rangle \langle P_1P_1'||H_1H'\rangle)$$

$$- \sum_{H_1'} \langle H_1'H|v|H_2H'\rangle \langle P_1P_2||H_1H_1'\rangle \quad (3.38)$$

4. The term $A_{F_{h,1}}(P_1P_2H, H_1H_2H')$ (see Fig. 33):

$$A_{F_{h,1}}(P_1P_2H, H_1H_2H') = -\sum_{H_1'} (\langle H_1'P_2|v|H'H_2\rangle \langle P_1H||H_1H_1'\rangle$$

$$+ \langle H_1'P_2|v|H_1H_2\rangle \langle P_1H||H_1'H'\rangle)$$

$$+ \sum_{P_1'} \langle P_1P_2|v|P_1'H_2\rangle \langle P_1'H||H_1H'\rangle \quad (3.39)$$

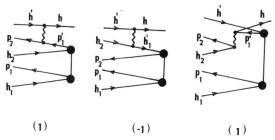

(1)   (-1)   (1)

Fig. 32.   Diagrammatic terms contributing to $A_{T_2}(P_1 P_2 H, H_1 H_2 H')$.

C. Hermitian Model Hamiltonian

The hermitian model hamiltonian[74] is constructed from the commutator equation (3.17a) under the assumption that the exponential wave operator is unitary, $\Omega^\dagger = \Omega^{-1}$. This immediately implies the so-called unit normalization of perturbed state vectors, $\langle \Psi_i | \Psi_j \rangle = \langle \varphi_i | \Omega^\dagger \Omega | \varphi_j \rangle = \delta_{ij}$. The normalization constant N_0 in (3.5a) is determined by

$$N_0 = \langle \Phi_0 | e^{T^\dagger} e^T | \Phi_0 \rangle = \exp\left(\langle \Phi_0 | e^{T^\dagger} e^T | \Phi_0 \rangle_C \right) \qquad (3.40)$$

That is, the normalized perturbed "core" ground-state vector is $|\Psi_0\rangle = (N_0)^{-1/2} e^T |\Phi_0\rangle$. In order to ensure the unitary property of Ω we must assume that the closed part of the relaxation operator F is nonzero and hermitian:

$$F_{cl} \neq 0, \qquad F_{cl}^\dagger = F_{cl} \qquad (3.41)$$

The hermiticity of F_{cl} is required to achieve the same property of the wave

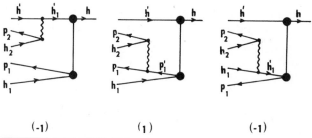

(-1)   (1)   (-1)

Fig. 33.   Diagrammatic terms contributing to $A_{F_{h,1}}(P_1 P_2 H, H_1 H_2 H')$.

operator Ω projected into the model space, $(P_0 \Omega P_0)^\dagger = (P_0 \Omega P_0)$. This requirement is a fundamental one in the formal quasidegenerate RSPT[41,82] with the hermitian model hamiltonian.

Multiplying (3.17a) on the left by the projector P_0, we get

$$[F_{cl}, H_0]_- = P_0 \{H_1 U - UG\}_C P_0$$
$$= \{H_1 U - UG\}_{C,cl} \qquad (3.42)$$

Let us introduce an operator

$$R = P_0 U P_0 - P_0 = N\left[F_{cl} + \frac{1}{2!} F_{cl} F_{cl} + \cdots\right] = R^\dagger \qquad (3.43)$$

which expresses the closed perturbation terms of Ω. Then the relation (3.42) may be written in a form which determines the model interaction,

$$G = \{H_1 U - RG\}_{C,cl} - [F_{cl}, H_0]_- \qquad (3.44)$$

Since the operators G and R are hermitian, a second alternative form is

$$G = \{U^\dagger H_1 - GR\}_{C,cl} + [F_{cl}, H_0]_- \qquad (3.45)$$

Adding (3.44) and (3.45), we get the model interaction in a manifestly hermitian form:

$$G = \tfrac{1}{2}(\hat{G} + \hat{G}^\dagger) \qquad (3.46a)$$

$$\hat{G} = \{H_1 U - RG\}_{C,cl} \qquad (3.46b)$$

The open part of the relaxation cluster operator, F_{op}, is fully determined by the commutator relation (3.17a),

$$[F_{op}, H_0]_- = \{H_1 U - UG\}_{LC,op} \qquad (3.47)$$

The wave operator U appearing in the second term on the right side of this relation can be reduced to the simpler form $N[e^{F_{op}}]$, since the correlation cluster operator T produces here only the disconnected terms, and the closed relaxation cluster operator F_{cl} cannot contribute to open diagrams. Formally, the solution of (3.47) is [see Eqs. (3.22) and (3.23)]

$$F_{op} = \sum_\alpha \frac{1 - P_0}{E_\alpha^{(0)} - H_0} \{H_1 U - UG\}_{LC,op} P_0(\alpha) \qquad (3.48)$$

The closed part F_{cl} of the relaxation cluster operator will be determined from the condition that Ω is a unitary operator, or

$$P_0 \Omega^\dagger \Omega P_0 = P_0 \tag{3.49}$$

Applying the standard diagrammatic analysis, one can demonstrate that the expression on the left of this relation is [cf. Eq. (3.40)]

$$(N_0)^{-1} \exp\left(\langle \Phi_0 | e^{T^\dagger} e^T | \Phi_0 \rangle_C\right) P_0 N [e^Z] P_0 = P_0 \tag{3.50a}$$

$$Z = \{ U^\dagger U \}_{LC,cl} \tag{3.50b}$$

where the operator Z contains only the linked–connected and closed diagrammatic terms. Owing to the condition (3.49), the operator Z must vanish, which directly determines the hermitian closed part of relaxation operator,

$$F_{cl} = -\tfrac{1}{2} N\left[\overline{S^\dagger S} + \frac{1}{2!} \left(\overline{S^\dagger S^\dagger S} + \overline{S^\dagger S S} \right) + \cdots \right]$$

$$= -\tfrac{1}{2} \{ \chi^\dagger \chi \}_{LC,cl} \tag{3.51}$$

for $\chi = \sum_{n \geq 1} (n!)^{-1} N[S^n]$. Thus, the relaxation cluster operator F is fully determined; its open part can be evaluated by using the relation (3.48), whereas the closed part is determined by (3.51). Finally, if we separate from the model interaction (3.46) those diagrammatic terms that are connected and ground-state (their sum is equal to the "core" correlation energy), then the model interaction $G' = G - E_{corr,core}$ is determined by the linked and connected diagrams. This means that the corresponding model eigenproblem is

$$(H_0 + G') | \varphi_i \rangle = \Delta E_i | \varphi_i \rangle \tag{3.52}$$

where

$$G' = \tfrac{1}{2} (\hat{G}' + \hat{G}'^\dagger) \tag{3.53a}$$

$$\hat{G}' = \{ H_1 U - RG' \}_{LC,cl} \tag{3.53b}$$

Multiconfiguration HF Theory. The main task remaining to be accomplished in the general diagrammatic perturbation method is to elaborate a theory that is potentially applicable to closed-shell as well as open-shell molecular systems in an arbitrary geometry on their energy hypersurfaces. Unfortunately, this could not be done in the framework of a one-determinantal approximation, even for the UHF type; therefore we turn our attention to a multideterminantal reference function. This is just the case of natural applicability of the above-elaborated quasidegenerate CCA with hermitian model hamiltonian. Here we have to deal with a proper preselection of the "core" state vector $|\Phi_0\rangle$ which is stable (not greatly varying) over the whole hypersurface. This property of $|\Phi_0\rangle$ is simply achieved if it contains only the core electrons (in the sense of quantum chemistry, the core electron are those ones that do not directly participate in chemical bonds). The particle states are classified [see comment below (3.1)] as passive and active. A trial ground-state vector $|\varphi_0\rangle$ is formed as a linear combination of *all possible* unperturbed states belonging to the model space D_0, $|\varphi_0\rangle = \Sigma_\alpha C_\alpha$ $|\Phi_\alpha\rangle$, where $|\Phi_\alpha\rangle = X_{p_1}^\dagger \dots X_{p_n}^\dagger |\Phi_0\rangle$. The model space D_0 is spanned by *all possible* n-particle state vectors $|\Phi_\alpha\rangle = X_{p_1}^\dagger \dots X_{p_n}^\dagger |\Phi_0\rangle$ that can be formed from $|\Phi_0|$ by creating n *active* particle states,

$$D_0 \equiv \left\{ |\Phi_\alpha\rangle = X_{p_1}^\dagger \dots X_{p_n}^\dagger |\Phi_0\rangle \right\} \tag{3.54}$$

Finally, applying a multiconfiguration HF (MC-HF) scheme (see e.g. Reference 83), we get an orthonormal set of one-particle functions which will be used in our forthcoming considerations as a background for the CCA theory. Hence, the quasidegenerate CCA allows us to replace the standard HF (restricted or unrestricted) one-particle functions with much more appropriate MC-HF ones, which correctly describe a given molecular system on the whole energy hypersurface. The total hamiltonian constructed over the set of MC-HF one-particle functions is

$$H = \langle \Phi_0 | H | \Phi_0 \rangle + \sum_{ij} \langle i | f_0 | j \rangle N \left[X_i^\dagger X_j \right]$$
$$+ \frac{1}{2} \sum_{ijkl} \langle ij | v | kl \rangle N \left[X_i^\dagger X_j^\dagger X_l X_k \right] \tag{3.55}$$

where f_0 and $N[\cdots]$ are the HF operator and the normal product, respectively, both determined with respect to the core state vector $|\Phi_0\rangle$. The diagonal terms in the one-particle part of H form an unperturbed hamiltonian H_0, while the remaining (one-particle nondiagonal and all two-particle) terms are set up as a perturbation H_1, that is, the total hamiltonian is split into $H = \langle \Phi_0 | H | \Phi_0 \rangle + H_0 + H_1$.

After this preliminary discussion of theoretical tools, we are ready to apply the quasidegenerate CCA elaborated in Section III.C. In the second-quantization formalism the resulting hermitian model interaction (3.53) is

$$G' = \sum_{l \geqslant 1} G^{(l)\prime} \tag{3.56a}$$

$$G^{(l)\prime} = \frac{1}{l!} \sum_{\substack{p_1 \dots p_l \\ p_1' \dots p_l'}} \langle p_1 \dots p_l | g' | p_1' \dots p_l' \rangle N \left[X_{p_1}^\dagger \dots X_{p_l}^\dagger X_{p_l'} \dots X_{p_1'} \right] \tag{3.56b}$$

where the summation runs over *active* particle states, and the matrix elements $\langle p_1 \cdots p_l | g' | p'_1 \cdots p'_l \rangle$ are symmetric with respect to an arbitrary permutation

$$\begin{pmatrix} 1 & \cdots & l \\ \alpha_1 & \cdots & \alpha_l \end{pmatrix}$$

of both left and right indices $p_1, \ldots, p_l$ and $p'_1, \ldots, p'_l$:

$$\langle p_1 \cdots p_l | g' | p'_1 \cdots p'_l \rangle = \langle p_{\alpha_1} \cdots p_{\alpha_l} | g' | p'_{\alpha_1} \cdots p'_{\alpha_l} \rangle \qquad (3.57)$$

Owing to the relation (3.53a), these matrix elements are manifestly hermitian,

$$\langle p_1 \cdots p_l | g' | p'_1 \cdots p'_l \rangle = \tfrac{1}{2} [\langle p_1 \cdots p_l | \hat{g}' | p'_1 \cdots p'_l \rangle + \langle p'_1 \cdots p'_l | \hat{g}' | p_1 \cdots p_l \rangle]$$
$$(3.58)$$

where the matrix elements $\langle p_1 \cdots p_l | \hat{g}' | p_1 \cdots p_l \rangle$ are diagrammatically determined by (3.53b). Diagonalizing the model hamiltonian $H_{\mathrm{mod}} = H_0 + G'$ in the entire model space D_0, we obtain the spectrum $\{\Delta E_\alpha\}$ of its eigenvalues. The total energy E_α corresponding to a state $|\Psi_\alpha\rangle$ whose dominant part is inside the model space is determined by

$$E_\alpha = \langle \Phi_0 | H | \Phi_0 \rangle + E_{\mathrm{corr,core}} + \Delta E_\alpha \qquad (3.59)$$

where $E_{\mathrm{corr,core}}$ is the correlation energy of the core subsystem. We believe that the outlined implementation of the quasidegenerate CCA employing the MC-HF one-particle functions provides, in general, a very effective many-body diagrammatic theory accounting properly for the correlation effects on whole molecular-energy hypersurfaces.

IV. ILLUSTRATIVE CALCULATIONS

The aim of this section is to present a few illustrative calculations to demonstrate for smaller molecular systems the efficiency of CCA methods carried out in different basis sets of gaussian AOs; the CCA correlation energies are compared with their counterparts calculated by the CI, diagrammatic RSPT, and CEPA methods.

A. ECPMET Correlation Energies

The ECPMET has been tested for the BH_3 molecule by Paldus et al.[12] Recently, we have published[37] other calculations of this type for the LiH, BeH_2, BH, and CH^+ molecules. In a most detailed form it is presented for the BH

TABLE III
Classification of Approximate Versions of the CCA

Version	Approximation of e^T
A (LCCA, Section II.H.1)	$1 + T_2$
B	$1 + T_1 + T_2$
C	$1 + T_2 + T_3$
D	$1 + T_1 + T_2 + T_3$
E (CPMET, Section II.H.2)	$1 + T_2 + \frac{1}{2} T_2 T_2$
F	$1 + T_1 + T_2 + \frac{1}{2} T_2 T_2$
G	$1 + T_2 + T_3 + \frac{1}{2} T_2 T_2$
H (ECPMET, Section II.H.3)	$1 + T_1 + T_2 + T_3 + \frac{1}{2} T_2 T_2$

molecule calculated in Mayer's basis set of lobe gaussian AOs.[84] The CI correlation-energy increments are listed in Table IV, by using these increments one can immediately set up an arbitrary CI correlation energy through tetra-excited configurations. The CCA correlation energies are presented in Table V, where we have listed all the possible approximate versions of the CCA, from the simplest LCCA through the ECPMET method (see Table III). We have to recall that the CCA correlation energies calculated by methods A to D cannot be directly compared with the corresponding CI correlation energies. If we correct these CCA correlation energies for about the diexcited disconnected term $\omega_{DC}(2) = E_{CI,corr}(0+2) - E_{LCCA,corr}$, then they may be simply related to CI correlation energies; such corrected CCA correlation energies are listed in Table V in parentheses. In Tables VI to IX are listed the correlation energies of the LiH, BeH_2, BH, and CH^+ molecules calculated by methods A and E to H (see Table III), together with the corre-

TABLE IV
CI Correlation-Energy Increments of the BH Molecule[a]

$\omega(2)$	-0.031424
$\omega(1,2)$	-0.000136
$\omega(2,3)$	-0.000159
$\omega(2,4)$	-0.000231
$\omega(1,2,3)$	-0.000004
$\omega(1,2,4)$	-0.000002
$\omega(2,3,4)$	-0.000001
$\omega(1,2,3,4)$	-0.000000
Sum	-0.031957
Full CI	-0.031958

[a] In au ($R = 2.35$ au); gaussian lobe AOs.[84]

TABLE V
CCA and CI Correlation Energies of the BH Molecule[a]

Version of CCA		Version of CI	
A (LCCA)	-0.032091 $(-0.031424)^{b}$	$(0+2)$	-0.031424
B	-0.032244 $(-0.031567)^{b}$	$(0+1+2)$	-0.031560
C	-0.032265 $(-0.031598)^{b}$	$(0+2+3)$	-0.031583
D	-0.032420 $(-0.031753)^{b}$	$(0+1+2+3)$	-0.031723
E (CPMET)	-0.031643	$()+2+4)$	-0.031655
F	-0.031788	$(0+1+2+4)$	-0.031793
G	-0.031805	$(0+2+3+4)$	-0.031815
H (ECPMET)	-0.031940	$(0+1+2+3+4)$	-0.031957
		Full CI	-0.031958

[a] From Reference 37; in au. $R_{B-H} = 2.35$ au; gaussian lobe AOs.[84]
[b] These CCA correlation energies are modified in the diexcited disconnected term $\omega_{DC}(2)$.

TABLE VI
CCA and CI Correlation Energies of the LiH Molecule[a]

Version of CCA		Version of CI	
A (LCCA)	-0.01876	$(0+2)$	—
E (CPMET)	-0.01840	$(0+1+2)$	-0.01840
F	-0.01938	$(0+1+2+4)$	-0.01938
G	-0.01841	$(0+2+3+4)$	-0.01841
H (ECPMET)	-0.01939	$(0+1+2+3+4)$	-0.01939

[a] From Reference 37. $R_{Li-H} = 1.6409$ Å; 6-31G basis set of gaussian AOs.[85]

TABLE VII
CCA and CI Correlation Energies of the BeH$_2$ Molecule[a]

Version of CCA		Version of CI	
A (LCCA)	-0.04093	$(0+2)$	—
E (CPMET)	-0.04023	$(0+2+4)$	-0.04026
F	-0.04059	$(0+1+2+4)$	-0.04060
G	-0.04062	$(0+2+3+4)$	-0.04065
H (ECPMET)	-0.04099	$(0+1+2+3+4)$	-0.04104

[a] From Reference 37. Linear $R_{Be-H} = 2.5217$ au, 6-31G basis set of gaussian AOs.[85]

TABLE VIII
CCA and CI Correlation Energies of the BH Molecule[a]

Version of CCA		Version of CI	
A (LCCA)	−0.06543	(0+2)	—
E (CPMET)	−0.06191	(0+2+4)	−0.06201
F	−0.06270	(0+1+2+4)	−0.06271
G	−0.06302	(0+2+3+4)	−0.06296
H (ECPMET)	−0.06383	(0+1+2+3+4)	−0.06368

[a]From Reference 37. $R_{B-H} = 1.2313$ Å; 6-31G basis set of gaussian AOs.[85]

sponding CI correlation energies. All these computations were carried out for the 6-31G basis set of gaussian AOs.[85]

Comparing the CCA and CI correlation energies presented in Tables V to IX, we may immediately conclude that the CCA approximate versions, which are ranged between CPMET and ECPMET, provide correlation energies that are in very close agreement with the CI. The ECPMET energies reproduce with high precision the CI energies involving all excitations through tetraexcited states. This observation might be of general validity for smaller molecules up to 15–20 electrons and a geometry not too different from equilibrium.

B. Potential Curves of the BH Molecule

In order to study the efficiency of CCA methods in geometries differing from equilibrium, we present illustrative calculations for the BH molecule. The basis set of gaussian AOs use is specified in Reference 55. Similar studies were recently published by Bartlett et al.[24] for the N_2 molecule and by Kvasnička for the LiH, BH, and FH molecules.[46] In Table X are listed the

TABLE IX
CCA and CI Correlation Energies of CH^+ Molecule[a]

Version of CCA		Version of CI	
A (LCCA)	−0.07361	(0+2)	—
E (CPMET)	−0.07031	(0+2+4)	−0.07048
F	−0.07099	(0+1+2+4)	−0.07117
G	−0.07159	(0+2+3+4)	−0.07139
H (ECPMET)	−0.07232	(0+1+2+3+4)	−0.07216

[a]From Reference 37. $R_{C-H} = 2.0967$ au; 6-31G basis set of gaussian AOs.[85]

TABLE X

Valence Correlation Energies of the BH Molecule[a]

Method	Correlation energy (au)	% of full CI
MBPT-2	−0.04769	65.6
MBPT-3	−0.06212	85.4
MBPT-4	−0.06773	93.2
LCCA (CEPA-0)	−0.07432	102.2
CPMET	−0.07069	97.2
CEPA-1[b]	−0.0720	99.0
CEPA-2[b]	−0.0731	100.5
CEPA-3[b]	−0.0706	97.1
CI $(0+2)$[b]	−0.0694	95.5
Full CI[b]	−0.0727	100.0

[a] Calculated by different many-body techniques.
$R = 2.336$ au; contracted gaussian AOs.[55]
[b] Reference 55.

TABLE XI

SCF and Valence Correlation Energies of the BH Molecule[a]

Inter-nuclear distance R (au)	SCF energy[b] (au)	Valence correlation energy[b] $(10^{-3}$ au)					
		MBPT-3	MBPT-4	LCCA	CPMET	CEPA-PNO[c]	Full CI
1.836	25.064362	61.943	67.170	73.807	70.170	70.800	72.06
2.086	25.097990	61.738	67.113	73.658	70.067	70.888	
2.236	25.104639	61.900	67.405	73.961	70.357	71.264	
2.336	25.105478	62.116	67.727	74.317	70.689	71.624	72.72
2.436	25.104230	62.413	68.143	74.786	71.110	72.059	
2.586	25.099533	62.996	68.931	75.698	71.925	72.885	
2.836	25.087030	64.293	70.632	77.715	73.689	74.847	76.09
3.336	25.055157	67.932	75.305	83.474	78.537	79.854	
4.336	24.992759	79.097	89.479	102.065	92.556	93.460	99.12
5.0	24.957414	89.374	102.704	120.704	104.054	104.517	
6.0	24.913670	109.931	130.415	156.650	122.601	122.621	

[a] Basis set of contracted gaussian AOs.[55]
[b] Energy is minus the tabulated value.
[c] Canonical CEPA-PNO, Karlsruher program.[93]

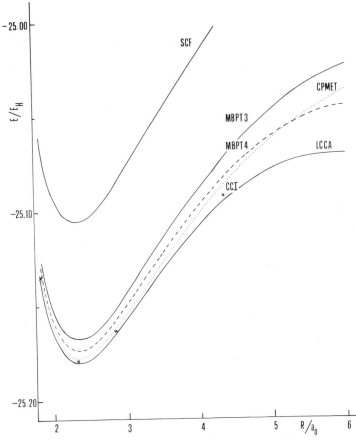

Fig. 34. The potential curves of the BH molecule calculated by different diagrammatic perturbation techniques.

valence correlation energies calculated for the fixed internuclear distance
$R_{B-H} = 2.336$ au. Since all the CEPA energies contain contributions due to
the connected cluster operator T_1, it is impossible to reach a definite conclusion about the comparison of the CEPA and CPMET methods. Nevertheless, it seems from the results of Taylor et al.[86-87] and of Zirz and Ahlrichs[56]
that the CEPA-1 method provides a correlation energy which is almost identical to that of the CPMET; if so, this represents a very considerable reduction of numerical effort in an approximate realization of the CPMET. The
RHF and correlation energies for different internuclear distances near equilibrium are listed in Table XI. The corresponding potential curves are pre-

TABLE XII
Spectroscopic Constants of the BH Molecule Calculated by Different Methods[a]

Method	R_e (Å)	ω_e (cm^{-1})	B_e (cm^{-1})	α_e (cm^{-1})	$\alpha_e x_e$ (cm^{-1})
SCF	1.229	2485	12.10	0.402	47.87
MBPT-2	1.232	2432	12.04	0.424	51.02
MBPT-3	1.235	2396	11.97	0.437	52.41
MBPT-4	1.238	2369	11.91	0.447	53.79
LCCA (CEPA-0)	1.239	2353	11.89	0.462	64.45
CPMET	1.238	2363	11.91	0.449	52.54
CEPA-PNO	1.238	2372	11.91	0.481	63.79
CI$(0+1+2)^b$	1.238	2368	11.91	0.446	54.53
Exp.^{c}	1.232	2367	12.02	0.412	49.39

[a] Basis set of contracted gaussian AOs.[55]
[b] E. L. Mehler, G. A. Van der Velde, and W. C. Niewpoort, *Int. J. Quantum Chem. Symp.* **9**, 245 (1975).
[c] J. W. C. Johns, F. A. Grimm, and R. F. Porter, *J. Mol. Spectrosc.* **22**, 435 (1967).

sented in Fig. 34. The spectroscopic constants of the BH molecule from these potential curves are listed in Table XII. The individual correlation energies are plotted versus the internuclear distance in Fig. 35. We see that if the internuclear distance is increasing, then the correlation energies are raised to nonphysical values. This is an indication of the well-known breakdown of the one-determinantal approximation to $|\Phi_0\rangle$.

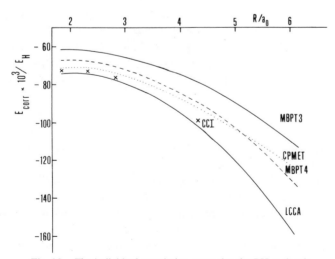

Fig. 35.   The individual correlation terms for the BH molecule.

TABLE XIII
Electronic Ground-State Energies of the BeH Molecule[a]

Basis set	RHF-LHP	Full CI[b]	MBPT-3	CCA
A	− 16.59639	− 16.60587	− 16.60489	− 16.60590
B	− 16.63111	− 16.66166	− 16.65965	− 16.66185
C	− 16.63302	− 16.66544	− 16.66354	− 16.66562
D	− 16.63303	− 16.66610	− 16.66418	− 16.66619
E	− 16.67441	− 16.69258	− 16.69052	− 16.69261
F	− 16.67732	− 16.70607	− 16.70442	− 16.70639

[a] For different basis sets of gaussian AOs[88]; $R_{Be-H} = 2.538$ au.
[b] Reference 88.

C. CCA Calculation for the BeH Molecule[60]

The CCA theory formulated in Section II.J has been tested for the BeH molecule. We used the six basis sets of gaussian AOs denoted A to F by Tiňo et al.,[88] for which the full CI results are known.[88] The one-particle functions were calculated by the Longuet-Higgins–Pople RHF scheme[89] (LHP-RHF). In Table XIII are listed the LHP-RHF, full CI, diagrammatic RSPT through third order, and CCA electronic energies. From these results we see that the full CI energy is tightly bounded by $E_{RSPT}^{(1-3)}$ and E_{CCA},

$$E_{CCA} < E_{CI} < E_{RSPT}^{(1-3)},$$

where the differences $|E_{CI} - E_{RSPT}^{(1-3)}|$ and $|E_{CI} - E_{CCA}|$ for all six bases are of the order $(1-2) \times 10^{-3}$ and 1×10^{-4} au, respectively. Here it is easy to show that the UHF energy[21] of a given molecule can be calculated within the CCA version used. The exponential operator should be approximated by $e^T \simeq e^{T_1}$, and only the one-particle terms of H_1 are considered; then $E_{CCA} = E_{UHF}$.

V. CONCLUSIONS

We have seen that the CCA allows us to formulate very effective algorithms for the calculation of infinite summations of the diagrammatic perturbation terms with preselected form of the topology as well as the level of excitations of intermediate states. *A priori*, the resulting correlation energies are automatically size-consistent, that is, the total energy is additive for a system composed of two or more noninteracting subsystems. The quaside-generate CCA with the hermitian model hamiltonian, combined with the MC-HF theory, provides the diagrammatic technique, which is applicable for whole energy hypersurfaces. At the present stage of the art of programming, the CCA is fully competitive with the up-to-date direct CI methods based upon the unitary-group approaches.[90-92]

References

1. J. Goldstone, *Proc. R. Soc. London, Ser. A* **239**, 267 (1957).

2. N. M. Hugenholtz, *Physica* **23**, 481 (1957).

3. J. Hubbard, *Proc. R. Soc. London, Ser. A* **240**, 539 (1957).

4. F. Coester, *Nucl. Phys.* **7**, 421 (1958).

5. F. Coester and H. Kümmel, *Nucl. Phys.* **17**, 477 (1960).

6. H. Kümmel, *Nucl. Phys.* **22**, 177 (1961).

7. H. Kümmel, in E. R. Caianiello, Ed., *Lectures on the Many-Body Problem*, Academic, New York, 1962, p. 265.

8. H. Kümmel, K. H. Lührmann, and J. G. Zabolitzky, *Phys. Rep.* **36**, 1 (1978).

9. J. Čížek, *J. Chem. Phys.* **45**, 4256 (1966).

10. J. Čížek, *Adv. Chem. Phys.* **14**, 35 (1969).

11. J. Čížek and J. Paldus, *Int. J. Quantum Chem.* **5**, 359 (1971).

12. J. Paldus, J. Čížek, and I. Shavitt, *Phys. Rev. A* **5**, 50 (1972).

13. J. Paldus, *J. Chem. Phys.* **67**, 303 (1977).

14. B. G. Adams and J. Paldus, *Phys. Rev. A* **20**, 1 (1979).

15. J. Čížek and J. Paldus, *Phys. Scr.* **21**, 251 (1980).

16. B. O. Roos and E. M. Siegbahn, in H. F. Schaefer III, Ed., *Methods of Electronic Structure Theory*, Plenum, New York, 1977, p. 277.

17. R. F. Hausman, Jr., and C. F. Bender, in H. F. Schaefer III, Ed., *Methods of Electronic Structure Theory*, Plenum, New York, 1977, p. 319.

18. S. S. Schweber, *An Introduction to Relativistic Quantum Field Theory*, Row-Peterson, Evanston, Ill., 1961, Chapter 4.

19. D. A. Kirzhnitz, *Field Theoretical Methods in Many-Body Systems*, Pergamon, Oxford, England, 1967, Chapter III, Sections 10–11.

20. J. Paldus and J. Čížek, *Adv. Quantum Chem.* **9**, 105 (1975).

21. J. A. Pople and R. K. Nesbet, *J. Chem. Phys.* **22**, 571 (1954).

22. C. C. J. Roothaan, *Rev. Mod. Phys.* **23**, 69 (1951).

23. C. C. J. Roothaan, *Rev. Mod. Phys.* **32**, 179 (1960).

24. R. J. Bartlett and G. D. Purvis III, *Phys. Scr.* **21**, 255 (1980) (and references therein).

25. U. Kaldor, *J. Comp. Phys.* **20**, 432 (1976).

26. K. F. Freed, *Annu. Rev. Phys. Chem.* **22**, 313 (1971).

27. R. J. Bartlett and I. Shavitt, *Chem. Phys. Lett.* **50**, 190 (1977); **57**, 157 (1978).

28. H. Primas, in O. Sinanoğlu, Ed., *Modern Quantum Chemistry, Part 2*, Academic, New York, 1965, p. 45.

29. V. Kvasnička, *Adv. Chem. Phys.* **36**, 345 (1977).

30. V. Kvasnička, *Czech. J. Phys. Ser. B* **27**, 599 (1977).

31. J. A. Pople, J. S. Binkley, and R. Seeger, *Int. J. Quantum Chem. Symp.* **10**, 1 (1976).

32. J. A. Pople, R. Seeger, and R. Krishnan, *Int. J. Quantum Chem. Symp.* **11**, 149 (1977).

33. R. J. Bartlett and G. D. Purvis, *Int. J. Quantum Chem.* **14**, 561 (1978).
34. V. Kvasnička, V. Laurinc, and S. Biskupič, *Phys. Rep.* (in press).
35. D. Mukherjee, R. K. Moitra, and A. Mukhopadhayay, *Mol. Phys.* **30**, 1861 (1975).
36. V. Kvasnička, V. Laurinc, and S. Biskupič, *Mol. Phys.* **39**, 143 (1980).
37. V. Kvasnička, *Phys. Rev. A* **25**, 671 (1982).
38. V. Kvasnička, *J. Chem. Phys.* **75**, 2471 (1982).
39. I. Lindgren, *J. Phys. B* **7**, 2441 (1974).
40. V. Kvasnička and A. Holubec, *Chem. Phys. Lett.* **32**, 489 (1975).
41. J. Jorgensen, *Mol. Phys.* **29**, 1137 (1975).
42. V. Kvasnička, V. Laurinc, and S. Biskupič, *Czech. J. Phys., Ser. B* **31**, 41 (1981).
43. R. K. Nesbet, *J. Chem. Phys.* **40**, 3619 (1964).
44. R. K. Nesbet, *Adv. Chem. Phys.* **9**, 321 (1965); **14**, 1 (1969).
45. V. Kvasnička, V. Laurinc, and S. Biskupič, *Chem. Phys. Lett.* **67**, 81 (1979).
46. V. Kvasnička, *Chem. Phys. Lett.* **78**, 98 (1981).
47. O. Sinanoğlu, *J. Chem. Phys.* **36**, 706 (1961).
48. O. Sinanoğlu, *Adv. Chem. Phys.* **6**, 315 (1964); **14**, 237 (1969).
49. R. E. Watson, *Phys. Rev.* **119**, 170 (1960).
50. D. D. Ebbing, Ph.D. thesis, 1960, Department of Chemistry, Indiana University, Bloomington.
51. R. J. Bartlett and D. M. Silver, in J.-L. Calais et al., Eds., *Quantum Science*, Plenum, New York, 1976.
52. R. J. Bartlett and I. Shavitt, *Chem. Phys. Lett.* **50**, 190 (1977).
53. V. Kvasnička and V. Laurinc, *Theor. Chim. Acta* **45**, 197 (1977).
54. W. Kutzelnigg, in H. F. Schaefer III, Ed., *Methods of Electronic Structure Theory*, Plenum, New York, 1977, p. 129.
55. R. Ahlrichs, *Compt. Phys. Comm.* **17**, 31 (1979).
56. C. Zirz and R. Ahlrichs, in M. F. Guest and S. Wilson, Eds., *Electron Correlation*, Proceedings of the Daresbury Study Weekend, 17–18 November 1979, p. 80.
57. M. A. Robb, in G. H. F. Diercksen et al., Eds., *Computational Techniques in Quantum Chemistry*, Reidel, Dordrecht, Holland, 1975, p. 435.
58. K. Jankowski and J. Paldus, *Int. J. Quantum Chem.* **18**, 1243 (1980).
59. A. C. Hurley, *Electron Correlation of Small Molecules*, Academic, New York, 1976.
60. V. Kvasnička, S. Biskupič, and V. Laurinc, *Mol. Phys.* **42**, 1345 (1981).
61. T. Morita, *Progr. Theor. Phys.* **29**, 351 (1963).
62. B. H. Brandow, *Rev. Mod. Phys.* **39**, 771 (1967).
63. M. Johnson and M. Baranger, *Ann. Phys. (N.Y.)* **62**, 172 (1972).
64. V. Kvasnička, *Czech. J. Phys. Ser. B* **24**, 605 (1974).
65. V. Kvasnička, *Czech. J. Phys. Ser. B* **25**, 371 (1975).
66. F. Coester, in E. T. Mahanthappa and W. E. Brittin, Eds., *Lectures in Theoretical Physics*, Vol. XIB, Gordon and Breach, New York, 1969, p. 157.
67. R. Offermann, W. Ey, and H. Kümmel, *Nucl. Phys. A* **273**, 349 (1976).
68. R. Offermann, *Nucl. Phys. A* **273**, 365 (1976).
69. F. Harris, *Int. J. Quantum Chem. Symp.* **11**, 403 (1977).

70. W. Ey, *Nucl. Phys. A* **296**, 189 (1978).

71. J. Paldus, J. Čížek, M. Saute, and L. Laforgue, *Phys. Rev. A* **17**, 805 (1978).

72. I. Lindgren, *Int. J. Quantum Chem. Symp.* **12**, 33 (1978).

73. H. Reitz and W. Kutzelnigg, *Chem. Phys. Lett.* **66**, 111 (1979).

74. V. Kvasnička, *Chem. Phys. Lett.* **79**, 89 (1981).

75. B. H. Brandow, *Adv. Quantum Chem.* **10**, 187 (1977).

76. B. H. Brandow, *Int. J. Quantum. Chem.* **15**, 207 (1979).

77. K. O. Friedrichs, *Perturbation Theory of Spectra in Hilbert Space*, American Mathematical Society, Providence, Rhode Island, 1965, Chapter III, Section 12.

78. V. Kvasnička and I. Hubač, *J. Chem. Phys.* **60**, 4483 (1974).

79. I. Hubač, V. Kvasnička, and A. Holubec, *Chem. Phys. Lett.* **23**, 381 (1973).

80. I. Hubač and M. Urban, *Theor. Chim. Acta* **45**, 185 (1977).

81. C. Bloch and J. Horowitz, *Nucl. Phys.* **8**, 91 (1958).

82. V. Kvasnička and A. Holubec, *Chem. Phys. Lett.* **32**, 489 (1975).

83. K. Ruedenberg, L. M. Cheung, and S. T. Elbert, *Int. J. Quantum Chem.* **16**, 1069 (1979).

84. I. Mayer, *Int. J. Quantum Chem.* **14**, 29 (1978).

85. W. J. Hehre, R. Dietchfield, and J. A. Pople, *J. Chem. Phys.* **56**, 2257 (1972).

86. P. R. Taylor, G. B. Bacskay, N. S. Hush, and A. C. Hurley, *Chem. Phys. Lett.* **41**, 444 (1976).

87. P. R. Taylor, G. B. Bacskay, N. S. Hush, and A. C. Hurley, *J. Chem. Phys.* **69**, 1971 (1978).

88. J. Tiňo, V. Klimo, T. A. Claxton, and B. Burton, *J. Chem. Soc. Faraday Trans. 2* **75**, 1307 (1979).

89. H. C. Longuett-Higgins and J. A. Pople, *Proc. Phys. Soc. London Sect. A* **68**, 591 (1955).

90. J. Paldus, in H. Eyring and D. Henderson, Eds., *Theoretical Chemistry, Advances and Perspectives*, Vol. 2, Academic, New York, 1976, p. 131.

91. I. Shavitt, *Int. J. Quant. Chem. Symp.* **11**, 131 (1977); **12**, 5 (1978).

92. B. R. Brooks and H. F. Schaefer III, *J. Chem. Phys.* **70**, 5092 (1979).

93. R. Ahlrichs, H. Lischka, V. Staemmler, and W. Kutzelnigg, *J. Chem. Phys.* **62**, 1225 (1975).

MOLECULAR BEAM PHOTOIONIZATION STUDIES OF MOLECULES AND CLUSTERS

C. Y. NG[§]

Ames Laboratory,[‖] U.S. Department of Energy
and
Department of Chemistry
Iowa State University
Ames, Iowa 50011

CONTENTS

[§]Alfred P. Sloan Research Fellow.
[‖]Operated for the U.S. Department of Energy by Iowa State University under contract No. W-7405-Eng-82. This work was supported by the Director for Energy Research, Office of Basic Energy Science.

I. INTRODUCTION

When a molecule absorbs a vacuum ultraviolet (VUV) photon, ionization and dissociation are usually the predominant processes. The measurement of the ionization cross sections for a particular ionizing channel with variable photon excitation energies is the subject of photoionization experiments. Under thin-target conditions, when only a very small fraction of photons is absorbed by the gas, and the density of the sample gas, as well as the length of the ionization region, is kept constant, relative photoionization cross sections can be obtained by measuring the photoionization efficiency (the ratio of the number of ions produced to the number of transmitted photons) as a function of photon energy. Photoionization-efficiency (PIE) data for positive and negative ions, taken at sufficiently high resolution, contain valuable information on photoionization, photofragmentation, ion-pair formation, predissociation, and autoionization processes which cannot be obtained by other methods. The high-resolution photoionization experiment is one of the most sensitive probes for investigating the coupling mechanisms between excited Rydberg states and ionic states. Experimental determinations of breakdown curves by photoionization have provided critical tests for statistical theories for unimolecular decompositions.[1,2] Many of the most accurate ionization potentials and vibrational frequencies of molecular ions known today were obtained by the photoionization method. From the appearance energies of fragment ions, reliable values for heats of formation and bond dissociation energies can be deduced. Combined with photoabsorption, photoelectron, and fluorescence spectroscopy, photoionization mass spectrometry has also played an important role in establishing multidimensional potential-energy diagrams which make possible correlations of ion–molecule processes.

Although ions can be detected with almost unit efficiency, the resolution achieved in photoionization mass-spectrometric experiments has been limited mainly by the low light intensity obtainable in the VUV region. In order to compensate for this difficulty, conventional photoionization studies usually employ the gas-cell method with fairly high pressure in the gas cell. The low light intensity is partly due to the low reflectivity of diffraction gratings in this region. Nevertheless, after more than two decades of intense activity, much advancement has been made in this direction. At the present techno- logical level, a windowless VUV monochromator operated by extensive dif- ferential pumping and intense laboratory continuum VUV light sources covering the region $\simeq 600$–2000 Å are available to enable instrumental reso- lution of 1–10 meV without suffering from low ion intensity.

The other factor which determines the ultimate resolution is rotational and low-frequency vibrational excitations of molecules at a given temperature. For diatomic molecules at room temperature, the half-width of the rota- tional envelope is approximately 25 meV, while that of polyatomic mole- cules is about 39 meV. These excitations not only sometimes prevent the identification and examination of the line shape of autoionization structure observed in photoionization efficiency curves, but they also introduce tailing structure and shift the ionization and appearance potentials to lower en- ergies. For gases with low freezing points, such as N_2, O_2, CO, H_2, and CH_4, the hot-band effects can be suppressed by cooling the sample gas to lower temperatures. In the case of hydrogen, which has a relatively large rotational energy separation, an energy resolution as high as 2 cm^{-1} has been achieved by Dehmer and Chupka[3] by cooling H_2 in a gas cell to liquid-nitrogen tem- perature.

Obviously, this method is not applicable to condensable gases with high freezing points. One way to circumvent this difficulty is to use the super- sonic molecular-beam method. In the expansion of polyatomic molecules, rotational and vibrational degrees of freedom in general are not in equi- librium either with the stagnation reservoir or with each other, and both will have characteristic temperatures well below the nozzle stagnation tempera- ture. Rotational and vibrational temperatures as low as 0.17 K[4] and 20–50 K, respectively, have been attained for polyatomic molecules by this method. The other serious problem encountered in gas-cell photoionization studies is the difficulty of distinguishing ions from the primary process and secondary processes. This can also be alleviated by the molecular-beam method, if the molecular beam is sampled downstream in a sufficiently low-density region. The progress of research in the application of supersonic jets in molecular optical spectroscopy has been reviewed recently by Smalley, Levy, Wharton, and Auerbach.[5-8] The first photoionization experiment using the super- sonic-beam method to demonstrate the cooling effect was reported by Parr and Taylor[9,10] in 1973.

Since rotational and vibrational relaxations are much faster than condensation, the abrupt supersonic expansion process in effect allows internal cooling before extensive condensation takes place. However, the concentrations of dimers and clusters are usually found to increase substantially in comparison with those observed under equilibrium conditions at the same nozzle stagnation temperature. The sampling and characterization of dimers and clusters synthesized in a supersonic beam expansion has broadened the scope of conventional photoionization experiments. In a series of molecular-beam photoionization studies,[9-32] it has been demonstrated that these dimers and clusters are excellent molecular precursors for the investigation of interaction energies between ions and neutral species. Photoionization of these precursors produces molecular ions which cannot be obtained from ordinary stable molecules, thus making it a useful means of determining thermochemical quantities for new molecular species. Photoionization studies of van der Waals molecules have shed light on the structures of these species. From the point of view of chemical dynamics, photoionization and photoexcitation in the VUV region of dimers and clusters offers a direct and general route for preparation of collision complexes and examination of the decomposition of these complexes as a function of internal excitation; hence it is a valuable tool for investigating ion–molecule reactions and high-Rydberg-state chemistry.

However, molecular-beam photoionization experiments are subject to an even more severe sensitivity problems than are encountered in conventional gas-cell experiments. In order to maintain a "collision-free environment" in the ionization region, differential pumping is required which limits the number density of target gas molecules at the collision center to a density of ~ 0.01 to 1 mTorr, as compared to ≥ 10 mTorr employed in gas-cell studies. Furthermore, the concentration of dimers and higher clusters formed in the beam is usually only a small fraction of that of the monomer. This and the low intensity of available VUV photons make photoionization studies of van der Waals molecules difficult. Not until the last few years has this sensitivity problem been improved to a level making the full potential of the molecular beam photoionization method realizable.

Excellent and extensive reviews on subjects related to conventional photoionization mass spectrometry using the gas-cell or effusive-source technique have been published by Berkowitz,[33] by Chupka,[34,35] and by Marr.[36] This review will be devoted mainly to the recent progress in this field, and it will emphasize the novel aspects of the application of the molecular-beam method to photoionization experiments. Moreover, since most of the papers published in this area used laboratory discharge lamps as the light sources, in this review we will limit the photon energy in the range from $\simeq 6$–20 eV (600–2000 Å). Because of the different excitation processes involved, work in the area of molecular-beam laser multiphoton ionization will be excluded.

II. EXPERIMENTAL CONSIDERATIONS

The apparatus used in molecular-beam photoionization mass-spectromet-ric studies involves a molecular-beam production system coupled to a pho-toionization mass spectrometer.

A. Photoionization Mass Spectrometer

Basically, a photoionization mass spectrometer consists of four essential components: light source, monochromator, light detector, and mass spectrometer. Since all these elements either have been discussed previously[33,35] or are available commercially, only a brief description will be given here. For details of the techniques and instrumentation for experimental work in the vacuum-ultraviolet region, readers are referred to the excellent monographs of Samson[37] and Berkowitz.[33]

1. Light Source

There are many laboratory light sources which have been designed and used for vacuum-ultraviolet studies. A line spectrum is often more intense than a continuum source. For measurements of photoionization cross sections as a function of photon energy, it is preferable to use a continuum source. True continuous sources are mandatory for very high-resolution studies of photoionization processes. In the wavelength region from $\simeq 600$ Å to longer than 2000 Å, the three most commonly used discharge light sources are the helium Hopfield continuum[38-40] ($\simeq 584-1100$ Å), the argon continuum ($\simeq 1050-1550$ Å),[41] and the many-lined hydrogen pseudocontinuum ($\simeq 850-2000$ Å). The many-lined hydrogen pseudocontinuum can be obtained simply by a dc discharge in hydrogen at low pressure ($\simeq 1-5$ Torr). Since the production mechanism of the rare-gas continua involves a long-duration afterglow,[42] repetitive high-power pulses are required. The Ar (as well as Kr and Xe) continua can also be produced less efficiently in a microwave discharge. In 1965, Huffman et al.[43] introduced a thyratron-triggered circuit pulsed at a repetition rate of ~ 10 kHz to generate the He and Ar continua. When a higher power and repetition rate (~ 100 kHz)[44] were used, light intensities obtainable from these continua were found to increase significantly. The Argonne group also found that light output increases with the length of the discharge lamp.[45] A high-power pulse generator which is similar in design to those used by the Argonne and Berkeley groups[46] is now available commercially.[47] The photon intensity obtainable at the maximum ($\simeq 800$ Å) of the helium Hopfield continuum is approximately 10^{10} photon/sec Å.[46] The average intensity at approximately 1200 Å provided by the H_2 pseudocontinuum is about one order of magnitude higher than this value.

For continuous emission below 600 Å, one must rely on radiation from electron synchrotrons and storage rings. The techniques and applications of synchrotron radiation have been described in a monograph edited by Kunz.[48] The spectral distribution and intensity of the emission from an electron storage ring depend on the ring radius and operating conditions, such as the electron-beam energy, current, and duty cycles. Usable continuous radiation extending from the visible into the x-ray region can be obtained. A comparison of spectral intensities from the high-power pulsed helium and argon continua with the effective light output of the 240-MeV storage ring (TANTALUS I) at Stoughton, Wisconsin and the 1.8-GeV ring (DORIS) in Germany was made by Radler and Berkowitz.[49] It was found that at $\simeq 800$ Å, the maximum-light-output region of the helium Hopfield continuum, obtainable light intensities from the storage rings are similar to that from the He Hopfield continuum. However, in the near future, when newly dedicated synchrotron rings will be operating at increased ring current, the intensity of synchrotron sources will certainly offer great promise for future photoionization studies. The major drawback of using synchrotron radiation as a light source is the difficulty of eliminating second and higher orders of radiation with a single dispersing element. Near the ionization onsets of clusters and the dissociative ionization onsets of fragment ions from clusters, where the ionization cross sections are usually low, ions produced by higher-order vacuum-ultraviolet light will make the identification of the true thresholds extremely difficult. Moreover, the broad spectral range emitted by synchrotron sources might also prevent a monochromator from being operated at second or higher order, which is often necessary for very high-resolution studies.

2. Vacuum-Ultraviolet Monochromator

For practical mechanical reasons, it is desirable for the entrance and exit slits of the vacuum-ultraviolet monochromator used for photoionization mass-spectrometric studies to be fixed and in focus at the exit slit while the wavelength is scanned. The light emanating from the exit slit should also remain in the same direction. In the wavelength region $\simeq 500-2000$ Å, the monochromators which satisfy these requirements are the Seya–Namoika type[50–52] and the near-normal-incidence type.[53–56] Since the Seya–Namoika type gives an astigmatic and curved image at the exit slit which causes a loss of intensity and limits the ultimate resolution, the near-normal-incidence monochromator is preferable for high-resolution photoionization studies. The light intensity at specific wavelengths can be optimized by selecting the blaze angle and surface coating of the diffraction grating. For the wavelength region $\simeq 1100-2000$ Å, aluminum plus a thin film of MgF_2 has the highest reflectivity, while in the region $\simeq 500-1100$ Å, an osmium coating should be used. For high-resolution experiments, it is advantageous to use a monochromator designed for a grating with a large radius of curvature. The loss in throughput due to the increase in the radius of curvature can usually be more than compensated for by using wider entrance and exit slits and a larger grating. The better collimation and wider separation between the entrance and exit slits thus obtained will result in a higher ion-collecting efficiency and will simplify the mechanical construction of the apparatus.

Although materials for windows are available for photon energies below $\simeq 11.9$ eV ($\simeq 1050$ Å), for best performance in the vacuum ultraviolet region the monochromator should be windowless. When laboratory gas-discharge light sources are used, this requires differential pumping to maintain sufficiently low pressure ($< 10^{-4}$ Torr) in the monochromator. In order to maximize the obtainable light intensity, it is important to have an extensive differential pumping system which will allow the monochromator to be operated with wide slits. If synchrotron radiation is used as the light source, the vacuum requirements are more stringent. To ensure cleanliness and high vacuum ($10^{-9}-10^{-10}$ Torr) in the storage ring, diffusion pumps are prohibited. Pumping must rely on other more expensive alternatives such as turbomolecular pumps, ion pumps, and cryogenic pumps.

3. VUV Photon Detector

The various modes of interaction between radiation and matter provide the underlying principles for all detectors. For vacuum-ultraviolet radiation, these interactions involve the photoionization of gases, the ejection of photoelectrons from metals, fluorescence, and others. The most commonly used vacuum-ultraviolet detectors in photoionization mass spectrometry include a metallic photoelectric cathode, an electron multiplier, or a fluorescence converter coupled with a photomultiplier. The latter detector involves coating a glass window or the glass surface of a phototube with a layer of scintillating material. Fluorescent signals produced by impinging vacuum-ultraviolet photons are amplified by the photomultiplier and monitored by photon counting or by measuring the direct output current. Sodium salicylate is the most attractive scintillating compound because it has a high fluorescence efficiency and a nearly constant quantum yield in the range $\simeq 300-2000$ Å.[57,58] The maximum intensity of the fluorescence of sodium salicylate, which is located at $\simeq 4200$ Å,[59] is also found to coincide with the maximum sensitivity of a photomultiplier. For high-resolution experiments, when the photon signal is low, it is necessary to cool the photomultiplier in order to obtain a favorable signal-to-noise level. Using a nude electron multiplier, such as a Channeltron, operated in the photon counting mode, is a more sensitive approach. However, in this case, the spectral response of the multiplier must be determined.

4. Mass Spectrometer

Both magnetic and quadrupole mass spectrometers have been widely used in photoionization studies. The transmission attained in a quadrupole mass filter is dependent upon the resolution and mass-to-charge ratio. The compactness and simple ion optics required are the advantages of a quadrupole mass spectrometer. Magnetic mass spectrometers can be used to detect metastable ions, which have proven to be a useful means of studying the unimolecular decomposition of molecular ions. However, in order to obtain high transmission through a magnetic sector mass spectrometer, it is necessary to use an ion-optical system such as an electrostatic quadrupole ion lens[60] to collect ions from a large area and focus them to the shape of the entrance slit.

B. Molecular-Beam Production System

1. Supersonic Molecular Beams

The study of supersonic beams was pioneered by Kantrowitz and Grey,[61] by Kistiakowsky and Slichter,[62] and by Becker and Bier.[63] After nearly three decades of development and due to the advances in vacuum technology, the supersonic molecular beam has become a popular tool in many research areas involving spectroscopy and molecular dynamics. Review articles[64-66] have been devoted to the discussions of the development and application of these techniques. The following sections will only briefly discuss some important properties of supersonic molecular beams which are relevant to photoionization mass spectroscopy.

A molecular beam is formed when a gas or the vapor of a liquid or solid flows from a reservoir at a pressure P_0 and a temperature T_0 through a small orifice into an evacuated chamber at a pressure P_1. In order to minimize the scattering of particles in the beam by background molecules which will result in attenuation of the beam, P_1 is usually maintained at $\leqslant 10^{-3}$ Torr. Molecular-beam sources can be characterized by a Knudsen number K_n which is defined to be the ratio λ_0/D, where D and λ_0 represent the diameter of the source orifice and the mean free path of the gas or vapor in the reservoir. In an effusive molecular-beam source ($K_n > 1$) the particles pass through the orifice without undergoing any collisions, whereas in a supersonic beam ($K_n < 1$) the particles undergo many collisions while flowing through the orifice. It is these collisions during the hydrodynamic expansion that give rise to some unique properties of a supersonic beam.

One of the most important properties of a supersonic molecular beam, which appears to be invaluable in photoionization mass spectrometry, is the cooling of the rotational and vibrational degrees of freedom. The primary effect of the expansion is the conversion of random thermal energy into directed mass flow, and this causes the decrease in translational temperature with concomitant increase in the mass-flow velocity u. Assuming the expansion to be isentropic, u is related to the change in enthalpy of the gas by

$$\tfrac{1}{2}mu^2 = \int_T^{T_0} C_p\, dT \qquad (2.1)$$

For an ideal gas, the temperature T of the jet can be found from the relation[67]

$$\frac{T}{T_0} = \left(\frac{n}{n_0}\right)^{\gamma-1} = \left[1 + \frac{\gamma-1}{2M^2}\right]^{-1} \qquad (2.2)$$

where n_0 and n are the density of the gas in the reservoir and in the isentropic part of the expansion, respectively; γ is the heat-capacity ratio C_p/C_v; and M is the Mach numbers which is defined to be the ratio of u to the local speed of sound. In the high-density region of the jet where the two-body collision rate is high, the Mach number can be estimated[68] according to

$$M = A\left(\frac{x}{D}\right)^{\gamma-1} \qquad (2.3)$$

where x is the distance from the nozzle, and A is a constant which only depends on γ and is equal[66] to 3.26 or 3.64 for $\gamma = \tfrac{5}{3}$ or $\tfrac{7}{5}$, respectively. Equations (2.2) and (2.3) predict that as the expansion proceeds, the Mach number increases, while the temperature and density decrease. However, as the flow passes over into the free-molecular-flow region where the mean free path in the jet becomes so great that collisions between molecules in the beam are negligible, M will approach a terminal value M_T. From experiments with argon jets, Anderson and Fenn[69] found

$$M_T = 1.17(K_n)^{-0.4} \propto (P_0 D)^{0.4} \qquad (2.4)$$

They also showed that the 0.4-power law can be explained by the application of simple kinetic theory. Thus, the product $P_0 D$ is a measure of the total number of collisions of molecules in the beam and determines the ultimate cooling and mass-flow velocity of the beam.

Energy transfer from internal to translational degrees of freedom occurs in molecular collisions during the expansion and is the prime mechanism for the lowering of the vibrational and rotational temperatures. The rapid expansion of a jet results in nonequilibrium between the translational, rotational, and vibrational degrees of freedom of diatomic or polyatomic molecules. The number of collisions required for relaxation between the various degrees of freedom is in general greatest for vibrational–translational energy exchanges and least for exchanges between the three translational degrees of freedom. As the density of the jet decreases, the vibrational temperature T_{vib} and rotational temperature T_{rot} will also approach asymptotic limits as the axial translational temperature T_{trans}. The ultimate temperatures achieved in a given expansion condition usually follow the order

$$T_0 > T_{vib} > T_{rot} > T_{trans}$$

The limits of these temperatures which can be attained in a supersonic expansion are also governed by the extent of condensation in the beam. Since the heat released from exothermic condensation processes may be absorbed by molecules in the beam and reheat the beam molecules, in order to achieve the lowest possible temperatures for a given expansion condition, the formation of clusters should be minimized. One of the best ways to reduce the clustering effect is by seeding a low concentration of the sample gas in a helium supersonic beam. Due to the nearly ideal-gas behavior of helium, extensive clustering of helium atoms is not anticipated. The helium jet essentially serves as a cold bath to absorb the heat released from finite cluster formations of the sample gas. For example, oxalyl fluoride has been cooled to a rotational temperature of 0.17 K using the helium seeded-beam method.[4] However, since the intensity of the molecules obtainable at the photoionization region is approximately proportional to its partial pressure in the gas reservoir, the seeded ratio and the stagnation pressure should be a compromise between the available pumping capacity and a satisfactory signal-to-noise ratio in a given experiment.

2. Cluster Beams

For experiments concerned with the photoionization of dimers and clusters, it is necessary to maximize the concentrations of these species by optimizing the nozzle expansion conditions. The extent of condensation depends upon the degree of cooling in an expansion; hence one would expect to obtain higher concentrations of dimers and clusters in a supersonic beam when a higher stagnation pressure, a larger nozzle diameter, and a lower nozzle stagnation temperature are used. As a consequence of the extreme translational cooling attainable in a helium supersonic beam, higher concentrations of cluster species can also be obtained by seeding a proper amount of sample gas in a helium jet, than in an expansion with the pure sample gas at the same stagnation pressure. The rotational and vibrational temperatures of dimers and clusters of the sample gas produced in a seeded helium supersonic jet can be quite low.

The time scale of expansion is the most important criterion in determining the extent of condensation. Wind-tunnel experiments have shown that a slow expansion favors condensation processess. The temperature of the gas attained in an isentropic nozzle expansion is usually well below its freezing point, so if the time scale of the expansion is long, it is more likely to lead to the collapse of the supersaturated state by the spontaneous formation and growth of clusters. By adopting a proper nozzle design, the time scale of an expansion can be lengthened and thus the concentrations of clusters can be increased substantially. Experimentally, it is found that conical supersonic nozzles[70] are more efficient for the formation of clusters than sonic nozzles,[71] namely, for the same throat diameter and supply conditions (P_0, T_0), conical nozzles give higher cluster beam intensity and larger clusters. A conical supersonic nozzle has a diverging conical opening and is characterized by its throat diameter, cone angle, and length of diverging section. The length of the diverging section is usually much longer than the throat diameter. A sonic nozzle is the simplest type of nozzle and has a diameter-to-wall thickness ratio of about one.

In a static reservoir of an unsaturated monoatomic gas where equilibrium exists between dimers and monomers, assuming a Lennard-Jones (6-12) potential, a statistical model predicts the concentration of bound dimers relative to that of monomers, X_2, as roughly proportional to $\sigma^3 P_0 \varepsilon^{3/2} T_0^{-5/2}$. Here ε is the maximum depth of the potential well, and σ is the lower limit of the internuclear distance for which the intermolecular potential is zero. Therefore, dimers are more readily formed for gases having large attractive potentials ε. In a supersonic expansion, X_2 is usually found to be proportional to P_0^n with a value of about 2 for n. Figure 1 shows that using a room-temperature nozzle with $D = 0.127$ mm, $X_2(Kr_2)$ is approximately proportional to P_0^3, whereas $X_2(Ar_2)$ is only proportional to $P_0^{1.5}$. The concentrations of heavier clusters are found to have an even greater pressure dependence than that of the dimer.

The concentration of the heterogeneous dimers formed in a supersonic expansion is quite sensitive to the mixing ratio of the corresponding species at a given stagnation pressure. The proper mixing ratio which gives rise to the highest concentration of a particular dimer can be determined experimentally. Figure 2 shows the result of such an exercise in optimizing the concentrations of the XeKr, XeAr, and KrAr molecules.[15]

The temperature dependence of cluster beams has not been studied in detail. The ratio of the concentration of a particular cluster to that of the monomer measured experimentally in a supersonic beam can be fitted with a function AT_0^{-n}, where A and n are constants. The value of n for X_2 is usually greater than that predicted by the statistical model for an unsaturated vapor in a static reservoir. As shown in Fig. 3,[72] $X_2((NO)_2)$ is found to be

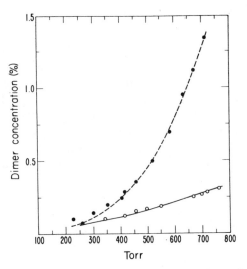

Fig. 1. Variation of the concentrations of the Kr_2 and Ar_2 rare-gas dimers, relative to that of the corresponding monomer, versus nozzle stagnation pressure at a nozzle temperature of 298 K. ● experimental point for Kr_2 obtained at 870 Å; – – – approximate fit, $X_2(Kr_2) = 1.81 \times 10^{-5}(P_0D)^3$; ○ experimental point for Ar_2 obtained at 780 Å; —— approximate fit, $X_2(Ar_2) = 1.01 \times 10^{-4}(P_0D)^{1.5}$, where D ($= 0.0127$ cm^{-1}) and P_0 (Torr) are the nozzle diameter and nozzle stagnation pressure, respectively.[13]

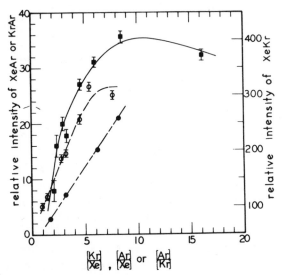

Fig. 2. ●, – – – : Relative intensity of XeKr versus [Kr]/[Xe] measured at 800 Å and a total nozzle stagnation pressure of $\simeq 400$ Torr. ○, — — : Relative intensity of KrAr versus [Ar]/[Kr] measured at 780 Å and a total nozzle stagnation pressure of $\simeq 700$ Torr. ■, — : Relative intensity of XeAr versus [Ar]/[Xe] measured at 780 Å and a total nozzle stagnation pressure of $\simeq 700$ Torr. The nozzle temperature is 298 K.[15]

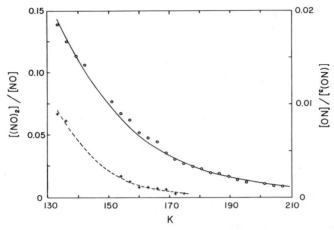

Fig. 3. Variation of the concentration of $(NO)_2$ and $(NO)_3$ relative to that of NO. The photoionization cross sections for $(NO)_2$ and $(NO)_3$ are assumed to be a factor of 2 and 3, respectively, larger than that for NO. O experimental points; — best fit, $[(NO)_2]/[NO] = kT_0^{-6}$. + experimental points; – – – best fit, $[(NO)_3]/[NO] = k'T_0^{-11}$, where k and k' are constants.[72]

proportional to T_0^{-6}, and the ratio of the concentration of $(NO)_3$ to that of NO is approximately proportional to T_0^{-11}. Since X_2 has a greater power dependence in temperature than in pressure, for a given pumping capacity it is more effective and advantageous to produce higher intensities of dimers and clusters by lowering the stagnation nozzle temperature than by increasing the stagnation pressure. In general this is also true for the production of larger clusters.

In a photoionization mass-spectrometric study of a particular cluster, it is sometimes desirable to minimize the formation of clusters heavier than that of interest in order to avoid the contamination of PIE curves by fragment ions of the heavier clusters. If one monitors the intensity of dimers in a supersonic beam as a function of stagnation pressure, one usually find the intensity of the dimers goes through a maximum and then decreases as the intensity of the trimers starts to increase. This indicates that dimers are the building blocks of trimers. In studies of the supersonic expansion of Ne, Ar, H_2, N_2, and O_2, van Deursen and Ruess[73] found that there is a limiting nozzle stagnation pressure, P_L, below which only dimers are formed with little contamination from heavier clusters. The value of P_L in Torr is given approximately by

$$P_L \simeq 5.25 \times 10^{-4} \frac{\varepsilon}{R_e^3} \left(\frac{kT_0}{\varepsilon} \right)^{(2-3\gamma)/(2-2\gamma)} \left(\frac{R_e}{D} \right)^{0.55} \quad (2.5)$$

where R_e and k are the equilibrium internuclear distance in cm and the

276   C. Y. NG

Boltzmann constant, respectively; ε, D, T_0 are in ergs, cm, and K, respectively. The intensity profiles for trimers and higher clusters are expected to be similar to those of the dimers except that these curves are shifted to higher pressure. Therefore, in principle, one can choose a particular stagnation pressure at which adequate intensity of the cluster of interest is formed while the intensities of heavier clusters are negligible. However, since the pressure-dependent curves for heavier clusters are usually quite congestive, this method can usually be applied to separate dimers and trimers from heavier clusters.

C. Description of a Molecular-Beam Photoionization Apparatus

Although different light sources, mass spectrometers, and VUV monochromators have been used in the construction of molecular-beam photoionization mass spectrometers, the basic designs of all these apparatuses are quite similar. To illustrate the basic experimental arrangement, we shall describe below the photoionization apparatus used in our laboratory.[24]

Figure 4 shows a detailed cross section of the differential pumping arrangement of the molecular-beam production system, ionization region, ion optics, and quadrupole mass spectrometer. The schematic diagram of a variable-temperature nozzle design is also shown in the figure. By changing the

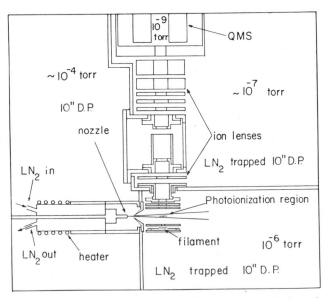

Fig. 4.   Cross-sectional view of the differential pumping arrangement of the molecular-beam production system, ionization region, ion optics, and quadrupole mass spectrometer.[24]

feeding rate of liquid nitrogen to cool the nozzle or the input power of the heater to heat it, the nozzle temperature can be varied from $\simeq$ 90–600 K. The beam expansion chamber is evacuated by a 10-in. diffusion pump having a pumping speed of $\simeq$ 4000 l/sec. Using a 0.07-mm-diameter nozzle and a stagnation pressure of $\simeq$ 1–2 atm, the beam expansion chamber can be mantained at a pressure of approximately 10^{-4} Torr. The high-intensity central portion of the supersonic jet is collimated into the ionization chamber by a 0.76-mm-diameter conical skimmer. This chamber, which is evacuated by a liquid-nitrogen-trapped 10-in. diffusion pump ($\simeq$ 4000 l/sec), is usually operated in the 10^{-5}–10^{-6}-Torr range. The skimmed molecular beam intersects at 90° with the dispersed light beam emitted from the exit slit of the VUV monochromator. The number density of target gas molecules at the collision center is estimated to be $\sim 10^{-3}$ Torr. The photoions generated at the collision center are deflected into a third chamber, which is also evacuated by a liquid-nitrogen-trapped 10-in. diffusion pump ($\simeq$ 4000 l/sec) which maintains a pressure of approximately 10^{-7} Torr. The ion lenses focus the photoions onto the quadrupole mass spectrometer, and the resulting mass-selected ions are detected with a Daly-type particle detector using pulse-counting techniques. The entire detector chamber is pumped differentially by a 120-l/sec ion pump and a liquid-nitrogen trap, which maintains a pressure of approximately 10^{-9} Torr in the chamber. This arrangement not only provides a higher molecular-beam intensity at the ionization center than that obtainable using a previous arrangement,[11] but it also maintains a low enough background pressure to prevent interferences due to secondary scattering processes.

A side view of the apparatus which shows the coupling of the 3-m monochromator (McPherson 2253 M) and the scattering chamber is shown in Fig. 5. Due to the length of the monochromator exit arm and the heavy weight of both the scattering chamber and the monochromator, the units are coupled together with flexible bellows ⑬ to avoid accidental distortion of the exit arm. Alignment of the photon and molecular beams is achieved by three X–Y translational benches, ②, on which the monochromator is supported. In order to prevent any change in relative positions of the units during pumping cycles, the monochromator is coupled rigidly to the monochromator stand, which is in turn attached securely to the stand of the scattering chamber.

The design of the discharge lamp is similar to that described previously.[3] In order to operate the light-source–monochromator in a windowless mode, the source is pumped differentially by a Roots blower ($\simeq$ 300 CFM) and an ejector pump ($\simeq$ 300 l/sec). With the lamp operating at 80 Torr He, and with a 500-μm entrance slit, a pressure of approximately 1×10^{-4} Torr is maintained in the monochromator. The intensity of the dispersed photon beam is

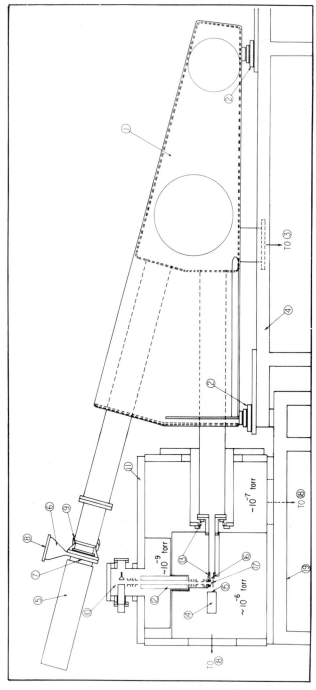

Fig. 5. Side view of the molecular-beam photoionization apparatus: ① McPherson Model 2253M VUV scanning monochromator; ② X–Y translational bench, ③ liquid-nitrogen-trapped 6-in. diffusion-pump system, ④ monochromator stand, ⑤ light source, ⑥ differential pumping arm (McPherson Model 820), ⑦ entrance slit, ⑧ to Roots blower pumping system (Leybold Heraeus Model WS500+DK100), ⑨ to ejector pump (CVC KS200), ⑩ Daly-type particle detector, ⑪ scattering chamber, ⑫ quadrupole mass spectrometer (Extranuclear Model 4-270-9), ⑬ flexible coupling bellows, ⑭ photon detector, ⑮ sodium-salicylate-coated quartz window, ⑯ exit slit, ⑰ photoionization center, ⑱ liquid-nitrogen-trapped 10-in. diffusion-pump system, ⑲ stand for scattering chamber.[24]

monitored with a sodium-salicylate-coated quartz window coupled to a photomultiplier tube in a water-cooled housing. The output of the photomultiplier is measured with a picoammeter which is subsequently connected to a
voltage-to-frequency converter. The digital signals are counted.

For high-resolution studies, it is necessary to isolate the system from
vibrations of the mechanical pumps. This was accomplished by connecting
the mechanical pumps to the diffusion pumps and the scattering chamber
with thin-wall flexible bellows.

Because of spatial considerations, the photon beam is modulated by a
150-Hz tuning-fork chopper instead of the more common practice of chop-
ping the molecular beam. In photoionization mass-spectrometric sampling of
high-temperature vapors, this arrangement allows correction for the background light from the oven by gating the photon detector. The chopper generates two gating signals corresponding to the photon beam being on and
off. Each gate activates one of two identical counters. The difference between the output from the two counters gives the net signal. Another gate is
used to activate a third counter, which measures the light intensity by counting the digital output from the voltage-to-frequency converter. A schematic
block diagram for the electronic automatic control system is shown in Fig.
6. The counters and the monochromator scan control and display unit
(McPherson Model 785) are interfaced to a microprocessor unit. After a
preset counting time, the teletype will print the current wavelength and the
readings of the three counters, as well as the calculated photoionization efficiency at that wavelength. This information is also recorded simultaneously with a cassette tape-deck system. At the end of a counting cycle, the
microprocessor sends a signal to the monochromator control and display
unit, which signals a stepping motor to advance the monochromator to the
next wavelength setting and starts a new counting cycle. The smallest wavelength increment which can be attained with the system is 0.0025 Å. After
an experiment, the data which were stored on the cassette tape are transferred to a computer where the PIE data and the corresponding light spectra
are plotted.

Recently Dehmer and Poliakoff[74] adopted a design which eliminates the
differential pumping in the molecular-beam production system so that a free
supersonic jet can be positioned as close as possible to the photon beam. The
method increases the density of the sample gas in the ionization region by
nearly two to three orders of magnitude over that obtainable from a skimmed
nozzle beam. The concurrent disadvantage of this design, due to the relatively high background pressure in the ionization chamber ($\sim 10^{-4}$ Torr), is
that secondary product ions formed by collision-induced ionization and dissociation often result in an ambiguous interpretation of the PIE spectra.

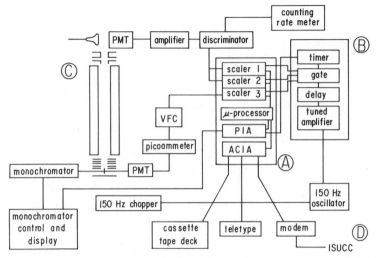

Fig. 6.   Block diagram for the automatic electronic control system. Unit A is an M6800 microprocessor equipped with three counters, peripheral interface adapters (PIA) for computer manipulation of the monochromator control and display, and asynchronous communication interface adapters (ACIA) for communication with the cassette tape deck and the teletype. It includes a modem (accoustic coupler) for phone-line communication with D, the Iowa State University Computation Center (ISUCC). Unit B is a molecular-beam timer and gate. Unit C is the detector assembly comprising a quadrupole mass spectrometer and scintillation detector. PMT and VFC are the photomultiplier tube and the voltage-to-frequency converter.[24]

III. EXPERIMENTAL RESULTS AND DISCUSSION

To date, the molecular-beam photoionization mass-spectrometric method has been used only in the studies of systems involving atoms and diatomic, triatomic, and simple polyatomic molecules. The following sections will be devoted to the discussion of some interesting results of these investigations.

A. Rare-Gas Systems

The rare-gas dimers and clusters represent the simplest type of van der Waals dimers and clusters. The photoionization studies of these systems have yielded significant insights into the excitation mechanism of these systems and more complicated molecular dimers and clusters.

1. Symmetric Rare-Gas Molecules Ar_2, Kr_2, and Xe_2

The feasibility of the direct examination of photoionization processes of van der Waals dimers was first demonstrated with the xenon dimer.[12] Since the ground states of Ar_2, Kr_2, and Xe_2 are[75-81] essentially repulsive states with shallow van der Waals wells and equilibrium internuclear distances approximately 1.2 Å greater than the corresponding equilibrium bond distances of the Ar_2^+, Kr_2^+, and Xe_2^+ ground states,[82-86] small Franck–Condon factors are expected. This is manifested by the very gradual increases in photoionization efficiency observed near the ionization thresholds of these systems (Fig. 7). The Franck–Condon envelopes obtained in the photoelectron spectra of Ar_2, Kr_2, and Xe_2[84-86] are also consistent with this expectation.

Using the measured ionization energies (IE) of the rare gas dimers $(R)_2$,[12,13,15,74] the known IEs of the rare-gas atoms R,[87] and the binding energies $D_0((R)_2)$ of $(R)_2$,[75-81] the dissociation energies $D_0((R)_2^+)$ of $(R)_2^+$ can be calculated from the relation

$$D_0((R)_2^+) = IE(R) + D_0((R)_2) - IE((R)_2) \qquad (3.1)$$

The calculated dissociation energies for Ar_2^+, Kr_2^+, and Xe_2^+ are listed in Table I and compared with results deduced from other experimental methods and theoretical calculations. The value for $D_0(Xe_2^+)$ derived from photoionization of Xe_2 is slightly greater than those obtained by other workers.[88-90] The value[13] for $D_0(Kr_2^+)$ is in excellent agreement with that obtained by Moseley and coworkers[91] in a photofragmentation study of Kr_2^+. A bond energy of 1.23 ± 0.02 eV determined by Ng et al.[13] for Ar_2^+ is in slight disagreement with a value of 1.31 ± 0.02 eV reported by Moseley et al.[92] However, the IE of Ar_2 remeasured recently by Dehmer et al.[74] is found to be consistent with the latter value. The favorable comparison of the photoionization results with values obtained using other techniques indicates that the IEs determined for Ar_2^+, Kr_2^+, and Xe_2^+ from direct photoionization studies of these van der Waals dimers are the true adiabatic IEs despite the very unfavorable Franck–Condon factors. The measured IEs[26,93] for Ar_3 and $Kr_{n=3,4}$ are only approximately 0.1–0.2 eV lower than those of the corresponding dimers.

In order to illustrate the gradual change in the excitation mechanism for Ar_2, Kr_2, and Xe_2, photoionization-efficiency data[12,13] for Ar_2^+, Kr_2^+, and Xe_2^+ were plotted in Fig. 7 on an energy scale (cm^{-1}) and were shifted so that the IEs of the atoms fall in a line. By examining the PIE curve for Ar_2^+, Kr_2^+, and Xe_2^+, Ng et al. concluded that autoionization is the predominant process for the formation of the rare-gas dimer ions at energies below the IEs of the rare-gas atoms. The formation of rare-gas dimer ions at the autoionization peaks consists of two discrete steps. The first one is the excitation of the dimer $(R)_2$ to a resonant molecular Rydberg state, and the second is the interaction between the excited Rydberg electron and the ion core, which results in the ejection of the electron and formation of $(R)_2^+$ in a discrete

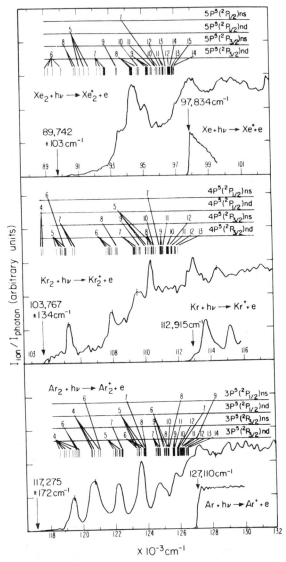

Fig. 7. Photoionization-efficiency curve for (a) Xe_2^+ in the energy range from 89 000 to 103 000 cm^{-1} (nozzle conditions: $P_0 \simeq 350$ Torr, $T_0 = 298$ K, $D = 0.0127$ cm[12]); (b) Kr_2^+ in the energy range from 103 000 to 118 000 cm^{-1} (nozzle conditions: $P_0 \simeq 500$ Torr, $T_0 = 298$ K, $D = 0.127$ cm[13]); (c) Ar_2^+ in the energy range from 117 000 to 132 000 cm^{-1} (nozzle conditions: $P_0 \simeq 500$ Torr, $T_0 = 298$ K, $D = 0.127$ cm[13]). The PIE curves for Xe_2^+, Kr_2^+, and Ar_2^+ were obtained with a photon bandwidth of 2.4 Å (FWHM).[12,13]

282

TABLE I

Dissociation Energies (eV) of the Symmetric Rare-Gas Dimers

Ions	Electron impact	Photoionization		Theoretical	Scattering[k]	Photofragmentation
		Gas cell	Supersonic beam			
Ar_2^+	1.1 ± 0.1[a]	1.049 ± 0.009[b]	1.23 ± 0.02[d] 1.30 ± 0.025[e]	1.25[g] 1.21[h] 1.24[i] 1.19[j]	1.25	1.31 ± 0.02[l]
Kr_2^+	1.0 ± 0.1[a]	0.995 ± 0.007[b] 1.13[c]	1.15 ± 0.016[d]	0.92[h] 1.23[i] 1.05[j]		1.14 ± 0.03[m]
Xe_2^+	0.9 ± 0.1[a]	0.968 ± 0.005[b] 0.99 ± 0.02[c]	1.03 ± 0.01[f]	0.65[h] 1.08[i] 0.79[j]	0.97	

[a]Reference 98.
[b]Reference 88.
[c]Reference 90.
[d]Reference 13.
[e]Reference 74.
[f]Reference 12.

[g]J. L. Gilbert and A. C. Wahl (unpublished SCF calculation).
[h]R. S. Mulliken, J. Chem. Phys. 52, 5170 (1970).
[i]Reference 82. Results without spin–orbit coupling.
[j]Reference 82. Results with spin–orbit coupling.
[k]Reference 83.
[l]Reference 92.
[m]Reference 91.

ionic level. The resonant molecular Rydberg state of $(R)_2$ in this energy region can be derived from the combination of one normal ground-state $(^1S_0)$ atom and one excited Rydberg atom. For the electronic states of the rare-gas dimer, the Hund–Mulliken case (c)[94,95] applied. The ground state of $(R)_2$ has a closed electronic shell, and thus is an 0_g^+ state. According to the selection rules of case (c) transitions from 0_g^+ to 0_u^+ or 1_u, excited molecular states are the only ones which are electric dipole allowed.

All the excited atomic states (except those with $J = 0$) can couple with the ground-state rare-gas atom ($np^6 \, ^1S_0$) (where $n = 5$ for Xe, $n = 4$ for Kr, and $n = 3$ for Ar) to give 0_u^+ and/or 1_u molecular state. In other words, all the excited atomic levels which have an energy higher than the potential-energy well of the ground electronic state of $(R)_2^+$ are available for autoionization. As shown in Fig. 7, good correlations were observed between the Rydberg atomic levels derived from $4p^5(^2P_{1/2,3/2})ns$ (or nd) $[3p^5(^2P_{1/2,3/2})ns$ (or nd)] for Kr [Ar] and the autoionization peaks resolved in the PIE curve for Kr_2^+ $[Ar_2^+]$. The correlation is not as obvious in the case of Xe_2. The difference was attributed to the lower dissociation energies of Ar_2 (12 meV)[77–79] and Kr_2 (17 meV)[75,80,71] than that of Xe_2 (24 meV).[75,76] Since the ground-state electronic configuration of a rare-gas atom is np^6 ($n = 4$ for Kr and $n = 3$ for Ar), this observation also suggests that autoionization of Kr_2 and Ar_2 follows the parity selection rule $\Delta l = \pm 1$. Nevertheless, this by no means excludes the autoionization of molecular Rydberg states which are derived from an excited Rydberg Kr [Ar] atom with the configuration $4p^5(^2P_{1/2,3/2})np$ (or nf) $[3p^5(^2P_{1/2,3/2})np$ (or nf)] and a normal ground-state Kr [Ar] atom. In fact, it is highly possible that the autoionization of these levels gives rise to a finite photoionization efficiency at the onset of these symmetric rare-gas dimer ions and allows the true adiabatic IEs to be determined by the photoionization method.

Recently, the high-resolution PIE curve for Ar_2^+ obtained by Dehmer et al.[74,96] has confirmed the previous conclusion that autoionization is the dominate mechanism for the formation of Ar_2^+ near the threshold. Detailed autoionization features resolved in this region are found to be lacking regular structure. This indicates that the autoionizing Rydberg-potential curves might be affected adversely by curve crossings and other perturbations near the threshold region.

2. Asymmetric Rare-Gas Molecules XeKr, XeAr, and KrAr

The IEs of XeKr, XeAr, and KrAr[15] are more distinct than those observed in the symmetric rare-gas systems, indicating that the Franck–Condon factors are more favorable for these systems. The measured IEs and the calculated dissociation energies for the ground states of $XeKr^+$, $XeAr^+$, and $KrAr^+$ using a relation similar to Eq. (3.1) are summarized and compared with results obtained by the electron-impact method[97,98] in Table II.

TABLE II
Ionization Energies and Dissociation Energies of the Asymmetric Rare-Gas Dimers[15]

| Ion | Ionization Energies (eV) | | Dissociation Energies (eV) |
	Electron impact	Photoionization[c]	Photoionization[c]
XeKr+	12.3 ± 0.1[a] 12.2[b]	11.757 ± 0.017	0.37 ± 0.02
XeAr+	13.5 ± 0.1[a] 13.5[b]	11.985 ± 0.017	0.14 ± 0.02
KrAr+	14.0 ± 0.1[a]	13.425 ± 0.02	0.59 ± 0.02

[a] Reference 98.
[b] Reference 97.
[c] Reference 15.

The dissociation energies of the asymmetric dimer ions are much smaller than those of the symmetric dimer ions.

Good correlations were found between the autoionization features in the PIE curves for XeKr+, XeAr+ (shown in Figs. 8 and 9) and KrAr+, and the s and d atomic lines of the corresponding rare-gas atoms. Thus, the auto-ionization of XeKr, XeAr, and KrAr obeys the pairty selection rule $\Delta l = \pm 1$, in agreement with the autoionization mechanism observed for Kr_2 and Ar_2.

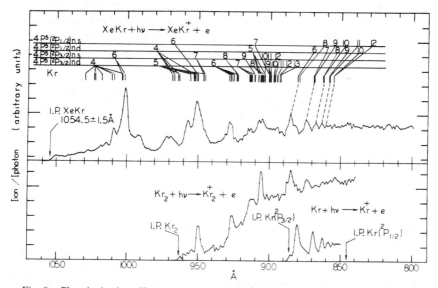

Fig. 8. Photoionization-efficiency curve for XeKr+ in the wavelength region from 800 to 1065 Å compared with the PIE curves for Kr+ and Kr_2^+. Wavelength resolution 2.4 Å (FWHM).[15]

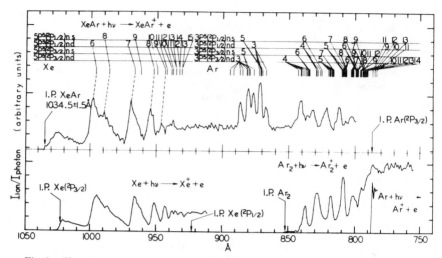

Fig. 9. Photoionization-efficiency curve for XeAr$^+$ in the wavelength region from 800 to 1045 Å compared with the PIE curves for Xe$^+$, and Ar$_2$$^+$. Wavelength resolution 2.4 Å (FWHM).[15]

In order to elucidate and rationalize this observation, schematic representations of the potential energy curves for $R_1 + R_2$, $R_1^* + R_2$, and $R_1^+ + R_2$ are depicted in Fig. 10. If the excited Rydberg state $(R_1 R_2)^*$ converges to the ionic state $(R_1 R_2)^+$, they are expected to be similar in shape. The equilibrium distances for XeKr$^+$, XeAr$^+$, and KrAr$^+$ are not known. Stemming from the stronger binding energies for these asymmetric dimer ions in com-

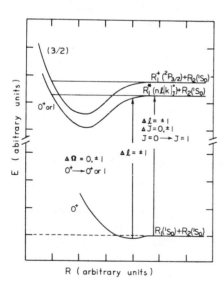

Fig. 10. Schematic representation of potential-energy curves of $R_1 + R_2$, $R_1^* + R_2$, and $R_1^+ + R_2$, where R_1 (or R_2), R_1^*, and R_1^+ are rare-gas atoms, an excited rare-gas atom, and an atomic rare-gas ion, respectively.[15]

parison with those of the neutral van der Waals molecules, the equilibrium bond distances for the neutral dimers should be longer than those for the corresponding dimer ions. Since the excitation is a vertical process and the excited molecular Rydberg dimer $(R_1R_2)^*$ formed is expected to be close to the asymptotic limit of the $R_1^* + R_2$ potential energy curve. In other words, the Rydberg electron in $(R_1R_2)^*$ is mainly in an excited atomic orbital associated with the excited atom, R_1^* (or R_2^*). Therefore, the two-step process for the formation of $(R_1R_2)^+$ in these experiments can be represented as

$$R_1 \cdot R_2 + h\nu \overset{(a)}{\rightarrow} R_1^* \cdot R_2 (\text{or } R_1 \cdot R_2^*) \overset{(b)}{\rightarrow} (R_1R_2)^+ + e^- \qquad (3.2)$$

In essence, the above discussion confirms the expectation that due to the weak interactions of these van der Waals molecules, to a first approximation, the excitation step (a) in Eq. (3.2) can be viewed as exciting only one moiety in the dimer. However, the results of the rare-gas dimer experiments show that as a consequence of the perturbation by its partner, the excitation selection rules of an atom in the dimer are actually relaxed from the atomic case ($\Delta l = \pm 1$, $\Delta J = +1$) to $\Delta l = \pm 1$. A comparison of the excitation selection rules in the atomic case, in the molecular case, and in the excitation region of these experiments is shown in Fig. 10. It is likely that step (b) in Eq. (3.2) proceeds by the ejection of the Rydberg electron from R_1^* (or R_2^*) by an electronic autoionization process. Strictly speaking, since the molecular Rydberg dimers, $R_1^* \cdot R_2$ (or $R_1 \cdot R_2^*$), formed by step (a) are highly vibrationally excited, it is also possible for step (b) to be a vibrational autoionization process. The formation of $(R_1R_2)^+$ follow the excitation selection rule of step (a), indicating that step (b) is much faster than step (a).

The other interesting result of this study is that the Rydberg series $5p^5(^2P_{1/2})ns(nd)$ for Xe, and $4p^5(^2P_{1/2})ns(nd)$ for Kr, which are manifested as autoionization peaks in the PIE curves for XeAr$^+$ and XeKr$^+$, respectively, were found to be red-shifted with respect to the positions in the PIE curves for Xe$^+$ and Kr$^+$. In the case of XeKr, this shift is approximately 0.08 eV. Using this value and the known binding energy (20 meV)[98] of the XeKr van der Waals molecule, the potential energy of the excited molecular Rydberg states, as derived by the combination of a normal Xe($^1S_0$) atom and an excited Kr $4p^5(^2P_{1/2})ns(nd)$ atom at the equilibrium interatomic distance (4.19 Å)[99] of XeKr, was calculated to be 0.10 eV. A value of 0.05 eV was obtained for the potential energies of the molecular Rydberg states formed by the interaction of Xe[$5p^5(^2P_{1/2})ns(nd)$] and Ar($^1S_0$) at the equilibrium distance (4.1 Å)[99] of XeAr. The charge-induced-dipole approximation yields the values of 0.09 and 0.04 eV for the cases of XeKr and XeAr, respectively. Taking into account the uncertainty of the experiment ($\simeq 0.017$ eV) and the fact that only a point-charge approximation was used, the agreement is quite good. The fact that only small shifts were identified for the Rydberg series is consistent with the above discussion concerning the excitation mechanism of this class of dimers.

B. Diatomic Molecules

1. NO, NO *dimer, and* NO *clusters*

Nitric oxide. Nitric oxide had for many years been considered the classic example which exhibits a step-function behavior for ionization. Recent results[11,25,100] have also revealed autoionization features superimposed on the steps. Ng et al.[11] found a more detailed autoionization structure in the PIE curve for NO^+ using the supersonic-beam method than had been measured in a gas cell[100] with the same optical resolution. More recently, a higher-resolution (0.14 Å FWHM) PIE curve for NO^+ in the region 1189–1340 Å was obtained by Ono et al.[25] using the molecular-beam photoionization method. The observed autoionization peaks can be satisfactorily correlated with the positions of Rydberg levels (ns, $np\sigma$, $np\pi$, $nd\sigma$, $nd\pi$, $nd\delta$, and nf) observed or calculated with quantum defects of lower members of the appropriate Rydberg series,[101] and series limits obtained from Miescher[102,103] and Alberti and Douglas.[104] Figure 11 shows the result in the region 1264–1340 Å, which spans the $v = 0$ and 1 vibrational steps of NO^+.

The comparison of this curve with medium-resolution[105] and high-resolution[102,105] absorption spectra has led to the conclusion that photoionization features corresponding to autoionization via $\Delta v < -1$ processes are present in NO and some of these features are as intense as peaks corresponding to $\Delta v = -1$ processes. According to the measured photoabsorption and photoionization cross sections of NO,[106] the ionization yield varies from approximately 20% in the $v = 0$ region to about 90% in the $v = 3$ region. This indicates that decay channels such as predissociation compete very effectively with autoionization in this region. The latter conclusion is supported by the fact that the relative intensities of neighboring peaks are often quite different in the PIE curve[25] and in the absorption spectrum.[105] Guisti and Jungen[107] pointed out that in the case of the peaks assigned as $np\pi$, $v \geqslant 3$ levels of $^2\Pi$ symmetry, which lie above the dissociation limit of the $B^2\Pi$ valence state, predissociation and autoionization are expected to be competing decay channels. Preliminary theoretical investigation on this system using the multichannel quantum-defect theory, extended to include dissociative channels, has been reported.[107] For the $5p\pi$, v superexcited levels, this calculation found, in agreement with the experiment,[25] that although predissociation is more efficient than autoionization for $v = 3$ and 4, the corresponding peaks still do appear clearly in the ionization spectrum. An interesting feature in the calculation is that autoionization is largely induced

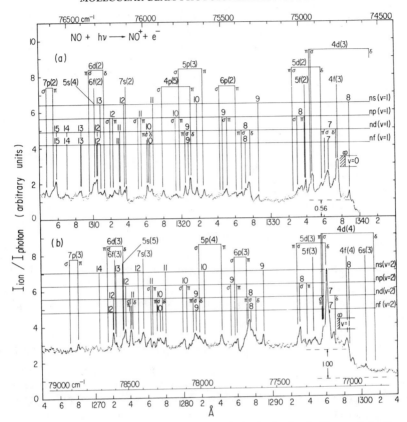

Fig. 11. Photoionization-efficiency curve for NO$^+$ obtained using the H$_2$ pseudocontinuum as the light source at a nozzle temperature of 150 K and a nozzle stagnation pressure of $\simeq$ 760 Torr in the region (a) 1304–1340 Å; (b) 1264–1304 Å. Wavelength resolution: 0.14 Å (FWHM).[25]

by the Rydberg–valence interaction as an indirect process whereby the Rydberg level is coupled to the ionization continuum via the dissociation channel. As a consequence $\Delta v = -1$ processes are not particularly favored.

One of the uncertainties introduced in deducing the relative Franck–Condon factors from the step heights measured in a photoionization experiment is the ambiguity in correcting for the contribution from autoionization. When autoionization features are not completely resolved in low-resolution studies, the Franck–Condon factors determined will suffer from appreciable uncertainties. In the photoionization study of NO by Ono et al., the autoionization band structure in the region 1242–1340 Å appears to be quite well resolved. Table III lists the relative Franck–Condon factors de-

TABLE III
Relative Franck–Condon Factors[a]

v	Photoionization[b] (± 0.03)	Photoelectron[c]	Calculated[d]
0	0.56	0.59	0.478
1	1.00	1.00	1.000
2	0.83	0.81	0.917
3	0.46^e	0.40	0.484

[a] Normalized for $v = 1$.
[b] Reference 25.
[c] Reference 108.
[d] References 109 and 110.
[e] The uncertainty of this value is bigger than ± 0.03. See Reference 25.

termined by Ono et al. and compares them with values derived from photoelectron spectroscopy[108] and theoretical calculations.[109,110] The photoionization results were found to be in good agreement with those obtained by photoelectron spectroscopy.

In favorable cases, it is possible to assess quantitatively the extent of relaxation of the thermally populated excited state by the analysis of the PIE curve. The $^2\Pi_{3/2}$ state of NO is 124 cm^{-1} above the ground $^2\Pi_{1/2}$ state.[111] If a Boltzmann distribution is assumed, the ratio of the population of NO($^2\Pi_{3/2}$) to that of NO($^2\Pi_{1/2}$) is $\simeq 54\%$ at 290 K and $\simeq 30\%$ at 150 K. If the expansion process does not relax NO($^2\Pi_{3/2}$), the relative intensity of the autoionization features originating from $^2\Pi_{1/2}$ and $^2\Pi_{3/2}$ should change correspondingly. Ono et al. found that the structure of the 290-K spectrum was very similar to that of the 150-K spectrum. A small step, observed about 124 cm^{-1} lower in energy than the first onset, was assigned to the NO$^+$ ($\tilde{X}^1\Sigma^+$, $v = 0$) $\leftarrow$ NO($^2\Pi_{3/2}$) transition. By comparing the height of the small step with the main step, Ono et al. concluded that the ratio of the population of NO($^2\Pi_{3/2}$) to that of NO($^2\Pi_{1/2}$) under their expansion conditions was 0.15 at the sampling region.

NO Dimer and Clusters. The PIE curve for (NO)$_2^+$ near the ionization threshold was first measured by Ng et al.[14] Two vibrational steps with the separation similar to the vibrational quantum of NO$^+$ were observed previously in their (NO)$_2^+$ spectrum. Figure 12 shows the PIE curve for (NO)$_2^+$ in the range 1235–1425 Å obtained by Linn et al.[30] with a better wavelength resolution (1.4 Å FWHM) than that used previously. As a consequence of substantial improvement in the signal-to-noise ratio, three vibrational steps are evident. These steps which are resolved in the dimer curve are quite

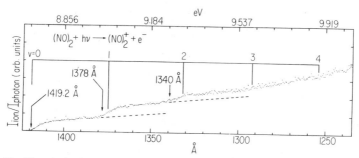

Fig. 12. Photoionization-efficiency curve for $(NO)_2^+$ in the region 1235–1425 Å. Experimental conditions: $P_0 \simeq 700$ Torr, $T_0 \simeq 150$ K, $D = 0.012$ cm, wavelength resolution 1.4 Å (FWHM).[30]

gradual and extend over $\simeq 6$ Å. The PIE data for $(NO)_2^+$ obtained using a resolution of 0.28 Å (FWHM) reveals finer steplike structures at the onset.[30] It was speculated that these finer structures might arise from the excitations of the low-frequency vibrational modes of $(NO)_2^+$.

To a first approximation, one expects the separations in energy of the IEs of $(NO)_2$ to $NO^+(\tilde{X}^1\Sigma^+, v=0)\cdot NO$, $NO^+(\tilde{X}^1\Sigma^+, v=1)\cdot NO$, and NO^+ $(\tilde{X}^1\Sigma^+, v=2)\cdot NO$ to be equal to the corresponding separations in the IEs of NO to $NO^+(\tilde{X}^1\Sigma^+, v=0)$, $NO^+(\tilde{X}^1\Sigma^+, v=1)$, and $NO^+(\tilde{X}^1\Sigma^+, v=2)$. In Fig. 12, the positions of the onsets for $NO^+(\tilde{X}^1\Sigma^+, v=0,1,2,3,4)$ were shifted by 0.529 eV, which is the difference of the IEs for $(NO)_2$ and NO, and compared with the structure observed in the PIE curve for $(NO)_2^+$. This comparison clearly suggests the second and third steps should be assigned as the onsets for $NO^+(\tilde{X}^1\Sigma^+, v=1)\cdot NO$ and $NO^+(\tilde{X}^1\Sigma^+, v=2)\cdot NO$, respectively. Interestingly, the vibrational spacing $\Delta v = 0–1$ and $1–2$ measured in the $(NO)_2^+$ spectrum were found to be greater than the corresponding spacings in $NO(\tilde{X}^2\Pi)$,[111] but less than those in $NO^+(\tilde{X}^1\Sigma^+)$.[25,104] Since the ionization of $NO(\tilde{X}^2\Pi)$ to form $NO^+(\tilde{X}^1\Sigma^+)$ involves the removal of an antibonding electron from the π^* antibonding molecular orbital, the greater vibrational spacing in $NO^+(\tilde{X}^1\Sigma^+)$ than in $NO(\tilde{X}^2\Pi)$ simply indicates an increase in bond strength in $NO^+(\tilde{X}^1\Sigma^+)$ as a result of the ionization process. Therefore, the comparison in Table IV supports the conclusion that the bonding of NO^+ in $NO^+\cdot NO$ is stronger than that of NO but weaker than that of NO^+.

The relatively sharp onsets observed for $(NO)_2^+$ can be considered as evidence that the geometry of $(NO)_2^+$ is similar to that of $(NO)_2$. Previous measurements[112–114] reveal a nearly cis rectangular configuration for $(NO)_2$. The relatively strong binding between the two NO molecules was attributed to the overlap of the highest occupied molecular orbital (π^*) of each NO.[113]

TABLE IV

Comparison of the vibrational spacings (eV) of
$NO^+(\tilde{X}^1\Sigma^+, v)\cdot NO$, $NO^+(\tilde{X}^1\Sigma^+, v)$, and $NO(\tilde{X}^2\Pi)$[30]

Δv	$NO^+(\tilde{X}^1\Sigma^+, v)\cdot NO)^a$	NO^{+b}	NO^c
0–1	0.261 ± 0.008	0.290	0.233
1–2	0.256 ± 0.020	0.287	0.229

[a] Reference 30.
[b] Reference 104.
[c] Reference 111.

Although the geometry for $NO^+\cdot NO$ is not known, it is reasonable to expect that a certain amount of overlap between the π^* orbitals of NO^+ and NO also exists in $NO^+\cdot NO$. This interaction would allow NO^+ to share the antibonding electron of NO. Based on this picture, one expects the bond strength of $NO^+(\tilde{X}^1\Sigma^+)$ in $(NO)_2^+$ to be between those of $NO^+(\tilde{X}^1\Sigma^+)$ and NO. This expectation is in accordance with the vibrational spacings measured for $NO^+(\tilde{X}^1\Sigma^+, v)\cdot NO$. As pointed out by Linn et al.,[30] part of the differences in the vibrational spacings observed might arise from the increase in reduced mass in $NO^+\cdot NO$. However, since the binding between NO^+ and NO is weak, this effect is expected to be quite small.

In view of the fact that the energy content ($\simeq 0.86$ eV) of $NO^+(\tilde{X}^1\Sigma^+, v = 3)\cdot NO$ is substantially higher than the binding energy (0.598 ± 0.006 eV)[30] of $NO^+\cdot NO$, intramolecular energy transfer should ultimately lead to the dissociation of the $NO^+(\tilde{X}^1\Sigma^+, v = 3)\cdot NO$ complex. The nonexistence of the fourth vibrational step in the PIE curve for $(NO)_2^+$ implies that intramolecular energy transfer in the dimer ion is faster than the flight time

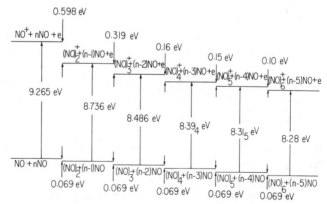

Fig. 13.   Energetics of the $(NO)_n^+$ ($n = 1-6$) system.[30]

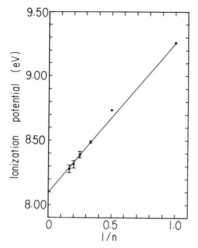

Fig. 14. Plot of the IE of $(NO)_n$ versus $1/n$.[30]

($\simeq 10^{-6}$ sec)[14] for the dimer ions to travel from the ionization region to the entrance of the mass spectrometer.

One of the goals of photoionization studies of clusters is to investigate the relationship between the energetics and size of the clusters. Using the measured IEs of $(NO)_{n=1-6}$,[30, 102] and by assuming the binding energy of $(NO)_n$, $n = 2-5$, with NO to be the same as that of $(NO)_2$,[113] the solvation energies for NO^+ by one, two, three, four, and five nitric oxides were calculated via the cycles shown in Fig. 13.

A plot of the measured IEs for $(NO)_n$, $n = 1-6$, as a function of $1/n$ is shown in Fig. 14. It was found that except for a small deviation of the IE of $(NO)_2$, the correlation of the IEs with cluster size is consistent with the prediction of the independent-systems model[115] for the cluster ion system. Similar agreements were observed in the cases for the acetone[18] and carbon disulfide[24] clusters. By assuming the IEs of $(NO)_n$, $n > 6$, to decrease linearly as a function of $1/n$, a value of approximately 8.1 eV was predicted for the bulk IE of nitric oxide.

2. O_2 Dimer and Clusters

Recently Anderson et al.[26] reported a photoionization mass-spectrometric study of oxygen clusters $(O_2)_{n=2-5}$ near the thresholds. From the IE for $(O_2)_2$ determined in their experiment, they concluded that $(O_2)_2^+$ is bound by 0.26 ± 0.02 eV. More recently Linn et al.[29] reexamined the threshold for $(O_2)_2^+$ and deduced a value of 0.42 ± 0.03 eV for the dissociation energy of $(O_2)_2^+$ which was found to be in agreement with the value, 0.457 ± 0.005 eV,

294   C. Y. NG

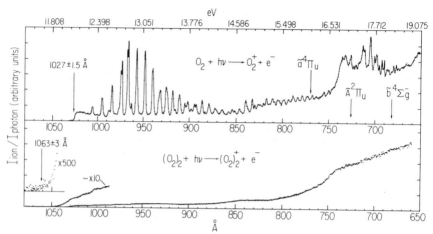

Fig. 15.   Photoionization-efficiency curve for O_2^+ and $(O_2)_2^+$ in the region 650–1080 Å. Experimental conditions: $P_0 \simeq 450$ Torr, $T_0 \simeq 110$ K, $D = 0.012$ cm, wavelength resolution 1.4 Å (FWHM).[29]

determined from equilibrium-constant data by Conway et al.[116] This suggests that the IEs measured by Anderson et al. are probably upper bounds of the true IEs for $(O_2)_{n=2-5}$.

Figure 15 shows the PIE curves for O_2^+ and $(O_2)_2^+$ obtained using an optical resolution of 1.4 Å (FWHM). The O_2^+ spectrum is dominated by intense autoionization features. Most of these structures have been identified as vibrational progressions or Rydberg series converging to electronic excited states of the oxygen molecular ion. However, in strong contrast to the O_2^+ spectrum, the PIE curve for $(O_2)_2^+$ is structureless. This observation indicates that the excited complex $O_2^*(n, v) \cdot O_2$ initially formed by photoionization in this region is strongly dissociative and that the $(O_2)_2^+$ ions observed are produced predominately by the direct ionization process, $O_2 \cdot O_2 + h\nu \rightarrow O_2^+ \cdot O_2 + e$.

Linn et al.[29] have also reported a study of the fragmentation of even-numbered oxygen-atom cluster ions yielding O_{2m+1}^+, $m = 1, 2, 3$. Similar to the observation in the PIE curves for $(O_2)_2^+$, the PIE spectra for O_3^+, O_5^+, $(O_2)_3^+$, O_7^+, and $(O_2)_4^+$ are also structureless, indicating that the contribution to these ions from autoionization process are negligible. The PIE curves for O^+ and O_3^+ are compared in Fig. 16. Because of the similarities between the PIE curves for O_5^+ and O_7^+, only the PIE curve for O_5^+ is shown here (Fig. 17b). When the PIE curves for $(O_2)_2^+$, $(O_2)_3^+$, and $(O_2)_4^+$ were normalized at a particular wavelength, these curves were found to be nearly superimpossible. This observation led Linn et al. to believe that the contamination of $(O_2)_2^+$ from fragmentations of higher clusters was not significant. In order to estimate the contribution to the O_3^+ spectrum from

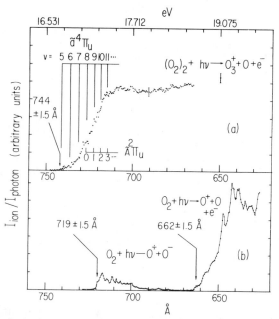

Fig. 16. Photoionization efficiency curves for (a) O_3^+ in the region 650–750 Å, (b) O^+ in the region 630–730 Å. Experimental conditions: $P_0 \simeq 450$ Torr, $T_0 \simeq 110$ K, $D = 0.012$ cm, wavelength resolution 1.4 Å (FWHM).[29]

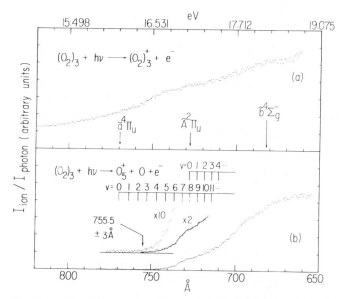

Fig. 17. Photoionization-efficiency curves for (a) $(O_2)_3^+$ in the region 660–820 Å, (b) O_5^+ in the region 650–780 Å. Experimental conditions: $P_0 \simeq 450$ Torr, $T_0 \simeq 110$ K, $D = 0.012$ cm, wavelength resolution 1.4 Å (FWHM).[29]

fragmentations of higher polymers during photoionization, a spectrum for O_3^+ was also obtained at a low pressure at which the concentrations of $(O_2)_3$ and $(O_2)_4$ are much less than that of $(O_2)_2$. The PIE curve for O_3^+ measured under such conditions was found to be in good agreement with that obtained at a higher pressure (Fig. 16a). The fact that only slight changes were observed in the values for $I(O_7^+)/I((O_2)_4^+)$, $I(O_5^+)/I((O_2)_3^+)$, and $I(O_3^+)/I((O_2)_2^+)$ as the nozzle stagnation pressure varied from $\simeq 200$ to 450 Torr supports the conclusion that O_3^+, O_5^+ and O_7^+ are fragment ions of $(O_2)_2^+$, $(O_2)_3^+$, and $(O_2)_4^+$, respectively. Here, $I(O_3^+)$, $I(O_5^+)$, $I(O_7^+)$, $I((O_2)_2^+)$, $I((O_2)_3^+)$, and $I((O_2)_4^+)$ represent the intensities of the respective ions.

From the measured appearance energy (AE), 16.66 ± 0.03 eV, for O_3^+ formation from $(O_2)_2$, the AE (18.73 eV) of the dissociative ionization process $O_2 + h\nu \rightarrow O^+ + O + e^-$, and the estimated binding energy (0.01 eV)[117] for $(O_2)_2$, the threshold energy for the ion–molecule reaction

$$O_2^+ + O_2 \rightarrow O_3^+ + O \qquad (3.3)$$

is deduced to be 4.58 ± 0.03 eV. Within experimental uncertainties, this value is in agreement with the thermochemical threshold for the reaction (3.3) with O_2^+ in the $\tilde{X}^2\Pi_g$ state. This measurement is consistent with the values reported in previous photoionization studies.[118,119] Using similar arguments and the measured AEs for O_5^+ and O_7^+, the stabilities of O_5^+ and O_7^+ were also determined. Figure 18 summarizes the energetics for the O_{2m+1}^+ ($m = 1, 2, 3$) system.

The AE of O_3^+ formation is approximately the energy for simultaneous ionization and excitation of O_2 to the $v = 5$ level of the $\tilde{a}^4\Pi_u$ state of O_2^+.[120] Since $(O_2)_2$ is a van der Waals dimer, it is likely that the actual absorption of a photon by an oxygen dimer involves only one of the O_2 molecules in $(O_2)_2$. As a result of the perturbation of the absorber by its partner, the excitation photon energy for the process

$$O_2 \cdot O_2 + h\nu \rightarrow O_2^+ \left(\tilde{a}^4\Pi_u, v \right) \cdot O_2 + e^- \qquad (3.4)$$

is expected to be slightly different from that for the excitation process

$$O_2 + h\nu \rightarrow O_2^+ \left(\tilde{a}^4\Pi_u, v \right) + e^- \qquad (3.5)$$

However, due to the weak binding energy ($\simeq 0.01$ and $\simeq 0.4$ eV for $(O_2)_2$ and $(O_2)_2^+$, respectively), the actual shift in energy for the above process in the dimer from that of monomer should be quite small. The positions corresponding to the excitation of O_2 to the $\tilde{a}^4\Pi_u$ ($v = 6, 7, 8, 9, 10, 11$) states of O_2 are shown in Fig. 16a.

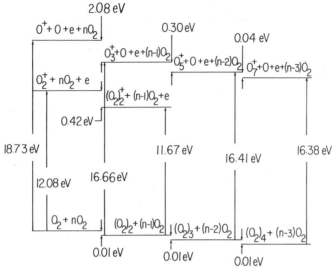

Fig. 18.   Energetics of the O_{2m+1}^+ ($m=1,2,3$) system.[29]

Assuming the validity of the conclusion that the O_3^+ ions are fragments of the $O_2^+(\tilde{a}^4\Pi_u, v)\cdot O_2$ complexes which are formed by the direct ionization process (3.4), the relative reaction probabilities (σ_v) for the half-reaction

$$O_2^+(\tilde{a}^4\Pi_u, v)\cdot O_2 \to O_3^+ + O \qquad (3.6)$$

can be obtained by normalizing the average increments in the PIE for O_3^+ of various vibrational states with the relative Franck–Condon factor for the reaction (3.5).[120] The onset of the $\tilde{A}^2\Pi_u$ ($v=0$) state of O_2^+ almost coincides with the $\tilde{a}^4\Pi_u$ ($v=8$) state. Therefore, at energies greater than the $O_2^+(\tilde{A}^2\Pi, v=0)$ converging limit, the reaction may also proceed via this state. However, since the Franck–Condon factors for transitions from the ground state of O_2 to the first few vibrational levels of the $\tilde{A}^2\Pi_u$ state are very small, the contribution from this state was neglected. The values for σ_v obtained by Linn et al. are listed in Table V and compared with the relative cross sections reported by Dehmer and Chupka.[119] The general trend for σ_v is similar to that observed for the relative cross sections, namely, it increases sharply as a function of vibrational energy and peaks at $v=10$. Dehmer et al. found that the total continuum for $v=11$ is approximately equal to that of $v=10$, which gives a cross section of zero for $v=11$. This was attributed to the dissociation of the energetically unstable O_3^+ ions into $O_2^+ + O$. The nearly constant PIE for O_3^+ beyond the threshold of the $v=11$ state observed by Linn et al. is consistent with this interpretation.

TABLE V

Relative Reaction Probabilities σ_v for the Ion–Molecule Half-reaction
$O_2^+(\tilde{a}^4\Pi_u, v) \cdot O_2 \rightarrow O_3^+ + O$ [29]

$O_2^+(\tilde{a}^4\Pi_u, v)$[a] (Å)	FCF[b]	σ_v[c]	Relative cross section[d]
$v = 5$ 742.08	0.13202	8	1.9
6 737.09	0.12152	28	11
7 732.28	0.10301	67	38.5
8 727.64	0.08192	88	66
9 723.17	0.06197	100	100
10 718.86	0.04451	92	89.5

[a]Reference 120.
[b]Franck–Condon factors for process (3.5). See Reference 120.
[c]Reference 29.
[d]Relative cross sections for the ion–molecule reaction O_2^+
$(\tilde{a}^4\Pi_u, v) + O_2 \rightarrow O_3^+ + O$ (Reference 119).

In a real collisional experiment, products can be formed by the direct processes as well as by processes which involve intermediate complexes. Since the configurations of the excited complexes prepared by photoionization are likely to be quite specific, and the division into direct and complex mechanisms is also quite arbitrary, it is difficult to relate the values for σ_v with those due to the complex mechanisms in real collisional experiments in which angular momenta are expected to play an important rule. Nevertheless, a more dramatic vibrational energy dependence for the relative reaction cross sections than for values of σ_v appears to reveal a stronger vibrational energy dependence for the reaction cross sections of the direct processes.

Using the same arguments as given above, the relative reaction probabilities σ_v for the reactions

$$O_2^+\left(\tilde{a}^4\Pi_u, v\right) \cdot (O_2)_m \rightarrow O_{2m+1}^+ + O, \qquad m = 2, 3 \qquad (3.7)$$

can also be derived as a function of vibrational energy. The values of σ_v obtained for the reactions (3.6) and (3.7) are compared in Fig. 19. As shown in the figure, σ_v for the reaction (3.6) has a stronger vibrational energy dependence than that of (3.7) with $m = 2$. The values of σ_v for the reaction (3.7) with $m = 3$ show only a weak dependence on vibrational energy. In view of the fact that a larger complex has more degrees of freedom in which to redistribute the energy after the excitation of one of the moieties in the cluster, it is reasonable to expect that vibrational energy is less effective in promoting the reaction as the size of the cluster increases.

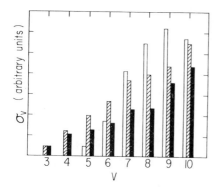

Fig. 19. Relative reaction probabilities σ_v, for the ion–molecule half-reactions $O_2^+(\tilde{a}^4\Pi_u, v)$ $\cdot(O_2)_m \rightarrow O_{2m+1}^+ + O$, $m = 1,2,3$, as a function of the vibrational quantum number v. Here $\sigma_{v=3}(O_5^+)$, $\sigma_{v=3}(O_7^+)$, and $\sigma_{v=5}(O_3^+)$ are arbitrarily normalized to the same value. Open bars: formation of O_3^+ from $(O_2)_2^+$; hatched bars: formation of O_5^+ from $(O_2)_3^+$; solid bars: formation of O_7^+ from $(O_2)_4^+$.[29]

The relation $I(O_7^+)/I((O_2)_4^+) \simeq I(O_5^+)/I((O_2)_3^+) \simeq 10[I(O_3^+)/I((O_2)_2^+))$ was found to be valid at $\simeq 730$ Å. Since the differences of $I(O_3^+)/I((O_2^+)_2)$ from $I(O_5^+)/I((O_2)_3^+)$ and $I(O_7^+)/I((O_2)_4^+)$ observed are too large to be caused by fragmentation, Linn et al. concluded that the reaction probabilities for (3.7) are substantially higher than for (3.6). The reaction probabilities for these reactions probably depend on solvation effects as well as the structures of these excited cluster ions. Stemming from the fact that the $O_2^+(\tilde{a}^4\Pi_u, v)$ molecular ions in $O_2^+(\tilde{a}^4\Pi_u, v)\cdot(O_2)_2$ and $O_2^+(\tilde{a}^4\Pi_u, v)\cdot(O_2)_3$ are in the proximity of two and three oxygen molecules, respectively, in comparison to only one oxygen molecule in O_2^+ $(\tilde{a}^4\Pi_u, v)\cdot O_2$, the result observed seems to be a reasonable one.

3. HF, HCl, HBr, and HI Dimers and Clusters

It has been shown in molecular-beam photoionization mass-spectrometry experiments[16,19] that hydrogen-containing dimers produced in a supersonic expansion are excellent parent molecules for the determination of absolute proton affinities. In the photoionization study of the hydrogen halide dimers, Tiedemann et al.[21] tried to assess the potentialities and limitations of this new method.

Thermochemical values for the HF, HCl, and HBr systems obtained by Tiedemann et al.[21] are summarized in Figs. 20, 21, and 22 respectively. The solvation energies of a proton by one HF, HCl, or HBr molecule were determined to be 4.09 ± 0.06 eV (94.3 ± 1.4 kcal/mol), 5.81 ± 0.04 eV (134 ± 1 kcal/mol), and 6.07 ± 0.04 eV (140 ± 1 kcal/mol), respectively.[121] Due to the high degree of rotational and low-frequency vibrational relaxation in the supersonic expansion, these values correspond to proton affinities for HF, HCl, and HBr at 0 K. In order to compare these values with those obtained by the ion–molecule equilibrium method, it is necessary to transfer the former values to 298 K. The proton affinities corrected to 298 K are 4.14 ± 0.06 eV

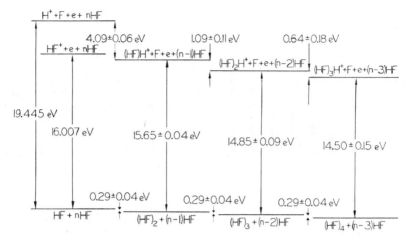

Fig. 20. Energy diagram for the HF system upon photoionization.[21]

(95.5 ± 1.4 kcal/mol) for HF, 5.85 ± 0.04 eV (135 ± 1 kcal/mol) for HCl, and 6.11 ± 0.04 eV (141 ± 1 kcal/mol) for HBr. The proton affinities of HCl and HBr agree very well with literature values: 135 ± 2 and 140 ± 1 kcal/mol, respectively. However, the proton affinity of HF is substantially lower than the value, 112 ± 2 kcal/mol, reported by Foster and Beauchamp.[122]

In the case of HF, the H_2F^+ ion intensity was found to be very high, but no $(HF)_2^+$ was observed. On the other extreme, in the case of HI, very little

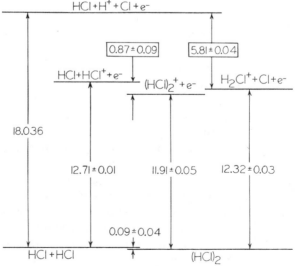

Fig. 21. Energy diagram for the HCl system upon photoionization (energies in eV).[21]

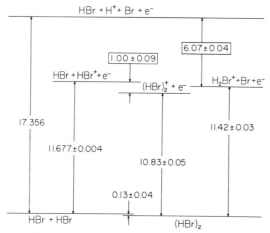

Fig. 22.   Energy diagram for the HBr system upon photoionization (energies in eV).

H_2I^+ was observed. Hydrogen chloride and hydrogen bromide were found to fall nicely in between, yielding both H_2X^+ and $(HX)_2^+$. Since the threshold for H_2X^+ formation can be considered as internal energy induced unimolecular decomposition of $(HX)_2^+$, Tiedemann et al. concluded that a reliable value for the proton affinity can be obtained from the measured AE for H_2X^+ formation if $(HX)_2^+$ is observed below the AE for H_2X^+. The fact that no $(HF)_2^+$ was observed can be accounted for by a zero potential-energy barrier for the reaction $(HF)_2^+ \rightarrow H_2F^+ + F$. This also suggests that the latter reaction is exothermic. If the production of H_2F^+ near the threshold indeed follows the two-step process

$$(HF)_2 + h\nu \rightarrow (HF)_2^+ + e^- \rightarrow H_2F^+ + F + e^-, \qquad (3.8)$$

the AE for H_2F^+ will be an upper bound. The direct formation of H_2F^+ from $(HF)_2$ is expected to be important for photon energies above the threshold for the formation of $H^+ + F + e^-$ from HF. Of course, the success of the molecular-beam photoionization method in the determination of proton affinities depends on the observation of the protonated molecules. In the case of HI, as a result of the very weak intensity of H_2I^+ produced by photoionization, the proton affinity for HI could not be determined.

4. CO and N_2 Dimers and Trimers

The pulsed high-pressure mass-spectrometric technique[123] is one of the most commonly used and accurate methods for determining the binding energy of a neutral molecule A and an ion B^+. This method generally involves the measurements of the temperature variation of the equilibrium constant K for the association reaction $A + B^+ \rightleftharpoons AB^+$. From the slope of a plot of $\log K$ versus $1/T$, the enthalpy change for the association process can be obtained. Since the accuracy for the measurements of K requires comparable concentrations of B^+ and AB^+, the experimental conditions are usually selected by raising the temperature of the ion source so that the forward and reverse reactions are very fast and equilibrium can be attained in a reasonably short time. However, due to the limited temperature range and delay time attainable with a particular instrument, this method is only applicable to dimer ions having low binding energies ($\leqslant 20$ kcal/mol). In a photoionization study of $(CO)_2$ and $(N_2)_2$, Linn et al.[30] deduced the bond energies for $CO^+ \cdot CO$ and $N_2^+ \cdot N_2$ from the measured IEs of $(CO)_2$ and $(N_2)_2$. Their experiment demonstrated that the molecular-beam photoionization method does not suffer from the above limitation and is an excellent method for the determination of ion–neutral interaction energies.

The bond energies of $CO^+ \cdot CO$ and $N_2^+ \cdot N_2$, together with those for $(O_2)_2^+$ and $(NO)_2^+$ determined by photoionization, are listed in Table VI and compared with values obtained by other techniques. The bond energies for $CO^+ \cdot CO$ and $N_2^+ \cdot N_2$ are consistent with the estimates by Meot-Ner et al.[124] and by Chong et al.[125] The value for the dissociation energy of $N_2^+ \cdot N_2$

TABLE VI
Bond Energies (kcal/mol) of $CO^+ \cdot CO$, $N_2^+ \cdot N_2$, $NO^+ \cdot NO$, and $O_2^+ \cdot O_2$.

Ion	Photoionization	Other technique[a]
$CO^+ \cdot CO$	22.4 ± 1.0[b]	26 ± 7[d]
		$\geqslant 21.2$[e]
$N_2^+ \cdot N_2$	20.8 ± 1.2[b]	27 ± 7[d]
		18.2[f]
$NO^+ \cdot NO$	13.79 ± 0.14[b]	13.6 ± 0.5[g]
$O_2^+ \cdot O_2$	9.7 ± 0.7[c]	10.38 ± 0.12[h]
		5.3 ± 0.5[i]

[a]Values from References 124, 125, and 126 have been converted to enthalpy change at 0 K using arguments similar to that used in Reference 24.
[b]Reference 30.
[c]Reference 29.
[d]Reference 125.
[e]Reference 124.
[f]Reference 126.
[g]Reference 14.
[h]Reference 116.
[i]Reference 26.

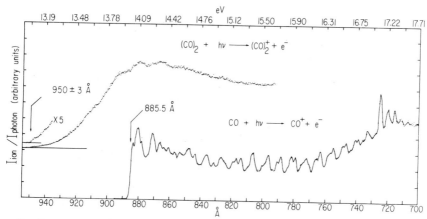

Fig. 23. Photoionization-efficiency curves for $(CO)_2^+$ and CO^+ in the region 700–955 Å. Experimental conditions: $P_0 \simeq 500$ Torr, $T_0 \simeq 110$ K, $D = 0.012$ cm, wavelength resolution 1.4 Å (FWHM).[30]

measured by Payzant et al.[126] was found to be slightly lower than that obtained by Linn et al. The comparison in Table VI suggests that antibonding electrons play an important role in the bonding of dimer ions. One can see a dramatic decrease in bond energy for the dimer ion as the number of antibonding electrons increases.

From the IEs of $(CO)_3$ and $(N_2)_3$,[30] the binding energies for $(CO)_2^+ \cdot CO$ and $(N_2)_2^+ \cdot N_2$ were deduced to be $\simeq 0.16$ and $\simeq 0.06$ eV, respectively. The charge–dipole interactions in $CO^+ \cdot CO$ and $(CO)_2^+ \cdot CO$ are probably responsible for the slightly higher bond dissociation energies observed for $CO^+ \cdot CO$ and $(CO)_2^+ \cdot CO$ than those for $N_2^+ \cdot N_2$ and $(N_2)_2^+ \cdot N_2$.

The PIE curves for $(CO)_2^+$ and $(N_2)_2^+$ obtained by Linn et al. are shown in Figs. 23 and 24, respectively. Similarly to the obversation in the Pie curves for $(O_2)_2^{+29}$ and $(NO)_2^{+127}$ in this wavelength region, the PIE curves for $(CO)_2^+$ and $(N_2)_2^+$, in sharp contrast with the monomer spectra, are structureless. This observation indicates that the rates for the dissociation of these excited molecular Rydberg dimers are much faster than those of the autoionization or association ionization processes such as $CO^*(n_1v) \cdot CO \rightarrow CO^+ \cdot CO + e^-$ which lead to formation of the dimer ions. Since the excited ionic states for CO^+, N_2^+, NO^+, and O_2^+ in this energy range have equilibrium bond distances quite different from those of the corresponding neutral ground states,§ Rydberg levels manifested as major autoionization peaks in the PIE of these molecular ions are likely to be vibrationally excited states.

§This conclusion is supported by the He I photoelectron spectra of CO, N_2, and O_2.[128]

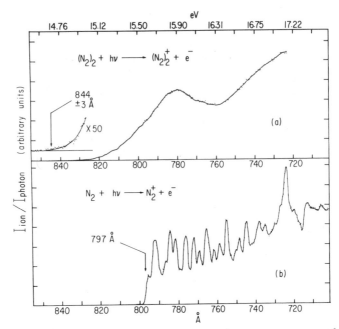

Fig. 24. Photoionization-efficiency curves for (a) $(N_2)_2{}^+$ in the region 720–855 Å; (b) $N_2{}^+$ in the region 700–800 Å. Experimental conditions: $P_0 \simeq 500$ Torr, $T_0 \simeq 110$ K, $D = 0.012$ cm, wavelength resolution 1.4 Å (FWHM).[30]

Stemming from this expectation, vibrational predissociation is probably the main dissociation mechanism for these excited Rydberg dimers.

The interaction of a Rydberg electron with the dimer ion core can be viewed as an electron-scattering process. The dissociation of the excited Rydberg dimer can also result from the transfer of energy from the Rydberg electron to the stretching vibrational mode between the two moieties of the dimer. A recent photoionization study[129] of N_2O and CO_2 shows evidence of this electronic predissociation process. Therefore, electronic predissociation cannot be excluded as an important dissociation mechanism for these excited Rydberg dimers.

5. H_2 Dimer and Clusters

In a photoionization mass-spectrometric study of H_2 dimer and clusters, Anderson et al.[26] only observed ions of odd-numbered hydrogen atoms. No evidence was found for the existence of stable $(H_2)_2{}^+$. Following similar arguments to those discussed above in the HF system, this observation can be rationalized by a zero potential-energy barrier for the ion–molecule reaction $H_2{}^+ + H_2 \rightarrow H_3{}^+ + H$. Many theoretical calculations[130–132] have shown that the ground-state surface has no barrier going down from $H_2^+ + H_2$ to $H_3^+ + H$, but has a shallow well at a geometry corresponding to a complex between $H_3^+ + H$. Previous experimental studies[134] of $H_2^+ + H_2 \rightarrow H_3^+ + H$ have also revealed no evidence of a barrier. Hence, the photoionization result is consistent with previous investigations.

According to previous high-resolution photoionization studies of H_2,[135,136] the PIE curve for H_2^+ is dominated by vibrational-induced autoionization structures. Typical lifetimes for $\Delta v = -1$ autoionization are 10^{-10}–10^{-11} sec, whereas for $\Delta v = -2$ autoionization they are on the order of 10^{-8} sec.[135] Even longer autoionization lifetimes are observed for $\Delta v = -3$ and -4. Despite the low rates for $\Delta v < -1$ autoionization, these states still appear as intense peaks in the PIE curve for H_2^+. In a low-resolution (~ 3.3 Å FWHM) experiment,[26] these autoionization structures are unresolved and appeared as strong peaks in the PIE curve (Fig. 25). The PIE curve for H_3^+ obtained with identical resolution is also shown in the figure. Many prominent peaks which appear in the PIE curve for H_2^+ are greatly suppressed or

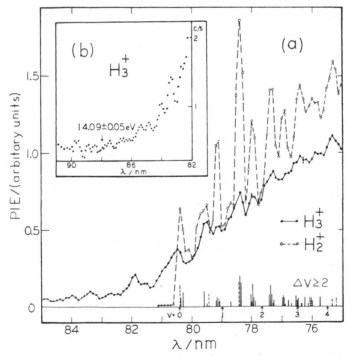

Fig. 25. Photoionization-efficiency curves for H_2^+ and H_3^+ in the region 720–900 Å. Experimental conditions (for the H_3^+ spectrum): $P_0 \simeq 18.3$ atm, $T_0 \simeq 78$ K, $D = 10$ μm, wavelength resolution $= 3.3$ Å (FWHM).[26]

simply missing in the H_3^+ curve. Anderson et al. speculated that vibrational predissociation of vibrationally excited dimer $H_2^*(n, v) \cdot H_2$ into $H_2^* + H_2$ would be the main mechanism for the depletion of H_3^+ intensity, particularly as the vibrational quantum number v increases. The rates, k_a and k_d, for the autoionization and predissociation processes, respectively, depend on n, v, and Δv:

$$H_2 \cdot H_2 \overset{hv}{\to} H_2 \cdot H_2^*(n, v) \overset{k_a}{\to} H_2 \cdot H_2^+(v') + e^- \to H_3^+ + H + e^- \qquad (3.9)$$

$$\searrow {\scriptstyle k_d}$$

$$H_2 + H_2^*(n, v'') \qquad (3.10)$$

Anderson et al. further argue that since the basic structure of the H_2^+ and H_3^+ spectra is the same, it seems that autoionization with $\Delta v = 1$ is faster than predissociation, that is, $k_d < k_a$ ($\Delta v = -1$). The states of H_2^* which must autoionize via $\Delta v < -2$ processes are marked at the bottom of Fig. 25. The good correlation of these states and the suppressed peaks in the H_3^+ curve led Anderson et al. to suggest that these peaks are missing because the process (3.10) is faster than (3.9) for $\Delta v < -1$, that is, $k_d > k_a$ ($\Delta v < -1$). Assuming the validity of the above argument, the predissociation lifetimes can be estimated if the rates of vibrational autoionization of H_2 are known and are similar to those of isolated molecules. From the autoionization rates obtained by Dehmer and Chupka, for some $v = 2$ states of H_2^* which must involve autoionization with $\Delta v = -2$, it was estimated $10^8 \text{ sec}^{-1} < k_d(\Delta v = -2) < 10^{10} \text{ sec}^{-1}$.

Recently, in a photoionization study of $(N_2O)_2$ and $(CO_2)_2$,[129] evidence was found that electronic predissociation might be an important predissociation mechanism for excited Rydberg dimers. In order to assess the importance of the electronic predissociation processes $H_2^*(n, v) \cdot H_2 \to H_2^*(n', v) + H_2$ as compared to vibrational predissociation and test the arguments given by Anderson et al., it is necesssary to perform a high-resolution photoionization experiment on this system.

C. Triatomic Molecules

1. CS_2, CS_2 Dimer, and CS_2 Clusters

At the present time, the carbon disulfide system is the most representative system studied which demonstrates the capabilities of the molecular-beam photoionization method.

Carbon Disulfide. The most accurate method for determining IEs has been the VUV absorption spectroscopy. In an absorption experiment, when high enough resolution can be employed to resolve the rotational structure of the Rydberg series whereby higher members of the series can be identified, rotational analysis of this structure can lead to very accurate values for the IEs. However, at sufficiently large n values, there are often strong interactions between Rydberg states of different n and l values. The coupling conditions and rotational structure of bands which result from this mixing are often complicated.[137] Although a substantial number of Rydberg series have been observed in the spectra of diatomic molecules, only a few of these have been analyzed. In the cases where the rotational structure has not been resolved, the band origins of members of Rydberg series will be difficult to establish, due to the finite rotational and vibrational temperatures of the sample gas, as well as the differences between the structure of higher Rydberg states and that of the ground state. Individual bands will be shaded either to the red or to the blue, depending on the sign and magnitude of the geometry change. Therefore, the IEs determined from fitting these Rydberg series will suffer from appreciable uncertainties. The photoionization experiments of CS_2[24] and OCS[28] have shown that using the molecular-beam method to relax the rotational envelope of sample gases results in high-resolution photoionization studies which provide a direct and accurate means of determining IEs.

Figure 26a shows the PIE curve for CS_2^+ near the threshold obtained by Ono et al.[24] using a wavelength resolution of 0.14 Å (FWHM). Autoionization peaks correspond to members of Rydberg series I[138,139] which converge to the spin–orbit state $^2\Pi_{1/2}$ of CS_2^+ were observed on the first vibrational step of the PIE curve for CS_2^+. One of the interesting results of this study is the identification of the Rydberg member $n = 16$ of series I as a strong autoionizing state. Since the position in energy of the level is 10.0768 eV, this measurement essentially supersedes any previous measurements[138-141] for the IE of CS_2 which are higher than this value. Under the nozzle-expansion conditions in the CS_2 experiment, the resolution was limited by the rotational temperature of CS_2 achieved in the expansion. The uncertainty of the initial threshold was found to be ± 0.25 Å (0.0020 eV), which corresponds to a rotational temperature of ~ 20 K. Thus, from the first and second steps resolved in the curve, the IEs for the $\tilde{X}^2\Pi_{3/2}$ and $^2\Pi_{1/2}$ states of CS_2^+ were determined to be 10.0685 ± 0.0020 and 10.1230 ± 0.0020 eV, respectively. Within experimental uncertainties, these values are in agreement with values derived from photoelectron spectroscopy[142] and photoionization studies.[22,143] The spin-orbit splitting (440 ± 23 cm^{-1}) is also consistent with previous measurements.[22,138,142,143]

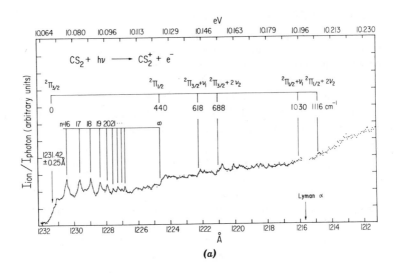

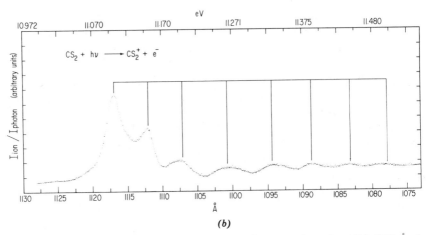

Fig. 26. Photoionization-efficiency curves for CS_2^+: (a) in the region 1210–1320 Å obtained with a photon bandwidth of 0.14 Å (FWHM) and using the Ar continuum as the light source; (b) in the region 1075–1130 Å obtained with a photon bandwidth of 0.14 Å (FWHM). The CS_2 molecular beam was produced by seeding CS_2 vapor ($\simeq$ 270 Torr) at approximately 290 K in 450 Torr of Ar and then expanding the mixture through a 0.07-mm-diameter stainless-steel nozzle.[24]

The slight steps at 618, 688, 1030, and 1116 cm^{-1} above the first onset were assigned to the onsets of $\tilde{X}\,^2\Pi_{3/2} + \nu_1$, $\tilde{X}\,^2\Pi_{3/2} + 2\nu_2$, $^2\Pi_{1/2} + \nu_1$, and $^2\Pi_{1/2} + 2\nu_2$, respectively. These assignments give values for $\nu_1(^2\Pi_{3/2})$, $\nu_2(^2\Pi_{3/2})$, $\nu_1(^2\Pi_{1/2})$, and $\nu_2(^2\Pi_{1/2})$ of 618 ± 32, 344 ± 32, ~ 590, and ~ 338 cm^{-1}, respectively, which are consistent with previous measurements of ν_1 and ν_2 of CS$_2$.[141, 144-146] According to optical selection rules, the excitation of a single quantum of bending vibration ν_2 is forbidden. However, a single-quantum excitation of ν_2 can be induced by vibronic coupling,[147, 148] and has been observed in the photoelectron spectrum of N$_2$O.[144, 149] The slight step observed at 1226.2 Å, which is 344 cm^{-1} above the first onset, can be assigned to the onset of $\tilde{X}\,^2\Pi_{3/2} + \nu_2$.

The high-resolution (0.14 Å FWHM) PIE curve [24] for CS$_2^+$ in the region 1075–1130 Å is depicted in Fig. 26b. Progressions of broad vibrational bands of CS$_2$ can be observed, and this structure was also evident in previous photoionization studies[22, 143, 150]. The first two resonances, at 1116.8 and 1112 Å, are coincident with the members of Rydberg series III, $n = 4$, observed in absorption.[138] The remaining features arise by autoionization from different vibrational states of the $\tilde{J}$ Rydberg state[151] of CS$_2$. The analysis of these bands gives an average vibrational spacing of 493 cm^{-1} for the $\tilde{J}$ state, which is in excellent agreement with a previous value.[151] However, autoionization features corresponding to absorption bands observed previously at 1126 and 1073 Å are absent in the PIE curve for CS$_2^+$.

CS$_2$ Dimer and Clusters. Photoionization-efficiency data for $(CS_2)_2^+$ at 1000–1295 Å were measured by Trott et al.[22] From the IE of CS$_2$ determined in their experiment, they deduced a value of 11.3 kcal/mol for the binding energy of CS$_2^+$ + CS$_2$, which is significantly below the enthalpy change at ~ 620 K ($-\Delta H_{620}^\circ = 21.9 \pm 2.5$ kcal/mol) for the association reaction

$$CS_2^+ + CS_2 \rightarrow CS_2^+ \cdot CS_2 \tag{3.11}$$

obtained in a pulsed high-pressure mass-spectrometric experiment by Meot-Ner et al.[152] In comparing the monomer and dimer PIE data, Trott et al. observed a close correspondence of autoionization features originating from Rydberg series IV, $n = 4$. In contrast, autoionization features arising from Rydberg series III, $n = 4$, were blue-shifted and broadly distributed. This was interpreted in terms of the geometry of the carbon disulfide dimer.

This system has been reexamined recently by Ono et al.[24] Figure 27a shows the PIE curve for $(CS_2)_2^+$ in the region 970–1350 Å. Due to substantial improvement in signal intensity in this experiment, the measured IE of $(CS_2)_2$ (9.36 ± 0.02 eV) was found to be ~ 0.3 eV lower than that reported previously.[22] This, together with the IE of CS$_2$ and the estimated binding energy

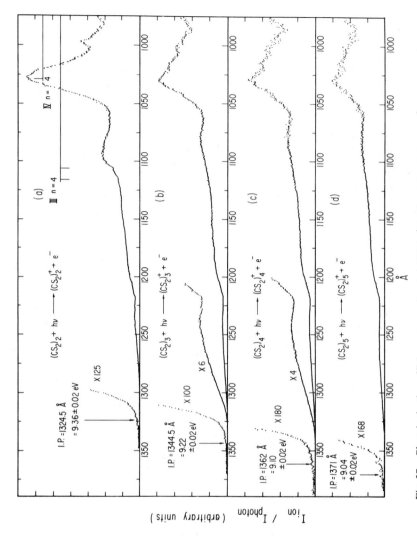

Fig. 27. Photoionization-efficiency curves for: (a) $(CS_2)_2{}^+$ in the region 970–1350 Å; (b) $(CS_2)_3{}^+$ in the region 970–1375 Å; (c) $(CS_2)_4{}^+$ in the region 970–1380 Å; (d) $(CS_2)_5{}^+$ in the region 970–1380 Å. Wavelength resolution: 1.4 Å (FWHM). The nozzle beam conditions are the same as described in Fig. 26.[24]

(0.05 eV) of $(CS_2)_2$, was used to calculate a binding energy of 0.76 ± 0.04 eV (17.5 ± 1 kcal/mol) between CS_2^+ and CS_2. Because of the high degree of rotational and low-frequency relaxation in the supersonic expansion, this value can be taken to be the enthalpy change for the reaction (3.11) at 0 K. In order to compare this value and ΔH_{620}° measured by Meot-Ner et al., Ono et al. converted the latter value to ΔH_0° by the relation

$$\Delta H_0^\circ = \Delta H_{620}^\circ - \int_0^{620} \left[C_p((CS_2)_2^+) - C_p(CS_2) - C_p(CS_2^+) \right] dT \quad (3.12)$$

where $C_p((CS_2)_2^+)$, $C_p(CS_2)$, and $C_p(CS_2^+)$ are the heat capacities at constant pressure. Assuming an ideal-gas model and excluding any vibrational and electronic contributions to C_p, a value of 18.2 ± 2.5 kcal/mol was obtained for ΔH_0°. Taking into account the uncertainties of these measurements, this value is within the limits of error of the binding energy deduced for $(CS_2)_2^+$ in the photoionization study. This comparison also indicates that the IE of $(CS_2)_2$ measured by Ono et al. is close to the true adiabatic IE.

The PIE curves for $(CS_2)_n^+$, $n = 3–5$ (Fig. 27b–d) have also been reported by Ono et al. The general profiles of these spectra are very similar. Thermochemical information for the carbon disulfide clusters deduced from the photoionization experiment is summarized in Fig. 28. Similarly to the observation in the photoionization studies of acetone[18] and nitric oxide[30] clusters, the measured IEs for $(CS_2)_n$, $n = 1–5$, except for a small deviation

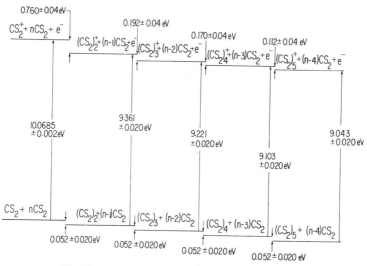

Fig. 28. Energetics of $(CS_2)_n^+$ ($n = 1–5$) system.[24]

of the IE of $(CS_2)_2$, fall squarely on a straight line when plotted as a function of $1/n$. By extrapolating the straight line in the IE-versus-$1/n$ plot, a value of ~ 8.77 eV was predicted for the bulk IE of CS_2.

Consistent with the observation by Trott et al., the PIE curve for $(CS_2)_2{}^+$ (Fig. 27a) clearly shows that series III, $n = 4$, at approximately 1116 Å is broadened and blue-shifted, whereas series IV, $n = 4$, at 1035 Å remains relatively unchanged. The broadening of the autoionization structure can be attributed partly to perturbation of the Rydberg orbital when a dimer is formed. It can be caused by fast dissociation channels of $CS_2^* \cdot CS_2$, which shorten the lifetime of the Rydberg complex. Since the binding energies for higher clusters (~ 0.1–0.2 eV) are substantially smaller than that of $(CS_2)_2{}^+$, it is expected that $(CS_2)_n{}^+$, $n > 2$, will be more susceptible to dissociation. The gradual smoothing of the autoionization structure as the size of the cluster increases is in accordance with this argument. The blue shift observed can be explained by the fact that the Rydberg electron in this low-lying Rydberg orbital is still relatively localized and the Pauli exclusion force possibly outweighs the attractive interactions in CS_2^*(III, $n = 4$)$\cdot CS_2$ at the equilibrium configuration of $(CS_2)_2$. Mullikan[153] points out that the value for the radius r_n of a Rydberg orbital at which the radial density is maximal is directly proportional to $(n^*)^2$, and can be approximated as $(n^*)^2 a_0$, where n^* and a_0 are the effective principal quantum number and the Bohr radius, respectively. Using the measured $n^* = 2.009$ for series III ($n = 4$), and $n^* = 2.312$ for series IV ($n = 4$),[154] the sizes of these Rydberg orbitals were estimated to be 2.14 and 2.83 Å, respectively. By comparison, the C—S bond distance in CS_2 is 1.554 Å.[155] Although the structure of the $(CS_2)_2$ dimer is unknown, the equilibrium separation between CS_2 and CS_2 can be estimated from the crystal structure[155] of solid CS_2 to be approximately 3.5–4.0 Å. Hence, while these two orbitals are large enough compared with the size of the CS_2 monomer to have hydrogenic characteristics, they are not large enough to remain unperturbed when the dimer is formed.

Higher members ($n > 4$) of Rydberg series III and IV were also observed in the PIE curve for $(CS_2)_2{}^+$ (Fig. 29b). Autoionization features arising from series III and IV ($n > 4$) are almost completely absent in the PIE curve for $(CS_2)_3{}^+$ (Fig. 29c). Similarly to the observation in the photoionization study of the rare gas dimers, the positions of these higher members of series III and IV are red-shifted from the corresponding series in the monomer spectrum (Fig. 29a). The most interesting observation is that the shifts $\Delta E(n)$ of Rydberg series IV are nearly a factor of 2 greater than those of Rydberg series III (Table VII). It is important to emphasize that the r_n's of all these Rydberg orbitals are greater than 4.8 Å, and are appreciably larger than or at least comparable to the size of $(CS_2)_2$. Therefore, an electron excited to one of these Rydberg orbitals will only play a very minor role in the bond-

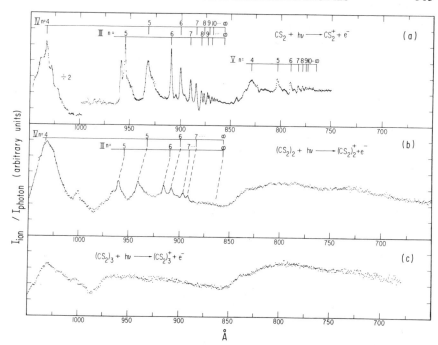

Fig. 29. Photoionization-efficiency curves for: (a) CS_2^+ in the region 750–1050 Å; (b) $(CS_2)_2^+$ in the region 600–1050 Å; (c) $(CS_2)_3^+$ in the region 650–1050 Å. Wavelength resolution: 1.4 Å (FWHM). The nozzle beam conditions are the same as described in Fig. 26.[24]

ing of an excited dimer, $CS_2^*(n > 4) \cdot CS_2$. These excited molecular Rydberg states which converge to the ionic state of $(CS_2)_2^+$ will have a potential-energy curve resembling that of $CS_2^+ + CS_2$. However, from a perturbation point of view, the existence of an electron in a particular Rydberg orbital mainly introduces the Pauli exclusion force and weakens the bonding of the ion core; hence the depth of the well in the Rydberg potential curve for $CS_2^*(n) + CS_2$ should be shallower than that of $CS_2^+ + CS_2$. Following similar arguments to those for the rare-gas dimer systems, a schematic representation showing the relationship of these potential-energy curves and that of $CS_2 + CS_2$ is depicted in Fig. 30, where $E_1^*(n)$ and $E_2^*(n)$ are the measured excitation energies from CS_2 to $CS_2^*(n)$ and from $(CS_2)_2$ to $CS_2^*(n) \cdot CS_2$, respectively. In view of the expectations that photoexcitation is a vertical process and the transition takes place primarily from the potential-energy well of $(CS_2)_2$, the potential energy $D^*(n)$ of $CS_2^*(n) \cdot CS_2$, *at the equilibrium nuclear configuration of the neutral dimer* can be calculated by the relationship

$$D^*(n) = E_1^*(n) - E_2^*(n) + 0.05 \text{ eV} \tag{3.13}$$

TABLE VII

Potential Energy of CS_2^*(III or IV, n)-CS_2 at the Equilibrium Nuclear Configuration of $(CS_2)_2$[24]

CS_2 Rydberg series[a] (eV)	n^{*b}	r_n (Å)	$E_1^*(n)^c$ (eV)	$E_2^*(n)^d$ (eV)	$\Delta E(n)^e$ (eV)	$D^*(n)^f$ (eV)
			III($ns\sigma$)			
$n=5$ 12.983	3.025	4.8	12.983	12.908	0.074	0.124
(955 Å)			(955 Å)	(960.5 Å)		
6 13.632	4.037	8.3	13.640	13.365	0.075	0.125
(909.5 Å)			(909 Å)	(914 Å)		
7 13.939	5.056	13.5	13.939	13.814	0.125	0.175
(889.5 Å)			(889.5 Å)	(897.5 Å)		
			IV($np\pi$)			
$n=5$ 13.303	3.386	6.1	13.303	13.155	0.148	0.198
(932 Å)			(932 Å)	(942.5 Å)		
6 13.776	4.405	10.3	13.776	13.640	0.136	0.186
(900 Å)			(900 Å)	(909 Å)		
7 14.010	5.402	15.5	14.018	13.876	0.142	0.192
(885 Å)			(884.5 Å)	(893.5 Å)		
			Series limit			
14.464			$E_1^*(n\to\infty)=$	$E_2^*(n\to\infty)=$	0.155	$D^*(n\to\infty)=$
(857.2 Å)	—	—	14.464	14.309		0.205
				(866.5 Å)^g		

[a] Reference 165.

[b] Reference 154.

[c] Autoionization peak observed in the PIE curve for CS_2^+.

[d] This value is interpreted as the energy difference between the potential energy curve of $(CS_2)_2$ and $CS_2^*(n)\cdot CS_2$ at the equilibrium geometry of $(CS_2)_2$.

[e] $\Delta E(n)=E_1^*(n)-E_2^*(n)$.

[f] The potential energy of CS_2^*(III or IV, n)$\cdot CS_2$ at the equilibrium nuclear configuration of $(CS_2)_2$.

[g] Reference 27.

Here, 0.05 eV is the estimated binding energy for $CS_2 + CS_2$. The calculated values for $D^*(n)$ are listed in Table VII. The series limit, $E_2^*(n\to\infty) = 14.309$ eV,[27] for series III and IV in the dimer was determined from the analysis of the CS_3^+ and S_2^+ PIE spectra which will be discussed later here. From this, the potential energy $D^*(n\to\infty)$ at the equilibrium nuclear geometry of $(CS_2)_2$ was calculated to be 0.205 eV.

The differences in value observed for $D^*(n)$ of series III and IV have provided insight into the geometry of the neutral carbon disulfide dimer. Based on the quantum defects of Rydberg series III and IV, Larzilliere and

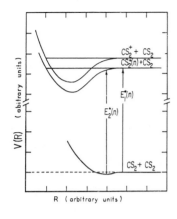

Fig. 30. Schematic diagram illustrating the relationship of the potential-energy curves of $(CS_2)_2$, $CS_2^*(n)\cdot CS_2$, and $(CS_2)_2^+$. Here R represents the separation between two CS_2 molecules and $V(R)$ the potential energy.[24]

Damany[156] suggest that series III is an $ns\sigma$ series while series IV is an $np\pi$ series. As indicated in Table VII, $D^*(n)$, $n = 5,6,7$, for series III are appreciably smaller than $D^*(n \to \infty)$, whereas $D^*(n)$, $n = 5,6,7$, deduced for series IV is very close to $D^*(n \to \infty)$. This difference implies that the perturbation introduced by an electron in an $ns\sigma$ orbital to the bonding of the ion core, $(CS_2)_2^+$, is more substantial than that induced by an electron in an $np\pi$ orbital. One expects an appreciable perturbation of an $ns\sigma$ orbital for any dimer geometry, and a linear or near-linear geometry for $(CS_2)_2$ is the most favorable configuration for minimizing the perturbation of an electron in an $np\pi$ orbital, from a symmetry point of view; thus the CS_2 dimer is expected to have a near-linear geometry.

To further explore the validity of the above argument, Ono et al.[24] also carried out a photoionization study of $Ar\cdot CS_2$ in the wavelength region 800–960 Å. Since Ar and CS_2 are held together mainly by long-range van der Waals interactions, and the size of Ar (covalent radius ~ 1.74 Å) is comparable to the equilibrium distance between the two sulfur atoms in CS_2, the most stable geometry for $ArCS_2$ should be for Ar to be close to the center of CS_2, so that both sulfur atoms are in the proximity of Ar. According to the symmetry picture, both the $np\pi$ and $ns\sigma$ orbitals will be perturbed. These interactions should give rise to similar shifts in energy for both series III and IV in the PIE curve for $ArCS_2^+$ with respect to series appeared in the CS_2^+ spectrum. Within experimental uncertainties, Ono et al. found the shifts in energy for members of series III and IV are indeed nearly the same. Therefore, this observation is supportive of the symmetry argument.

High Rydberg State and Ion–Molecule Reactions of CS_2. The interesting behavior of atoms and molecules in excited Rydberg states has been the subject of numerous experimental and theoretical investigations in recent years.[157] By far the greatest experimental effort has been in the area of inelastic collisions between Rydberg atoms and other atoms and molecules. The chemistry of this class of atoms and molecules, with the exception of metastable species, is still somewhat unexplored. One of the difficulties encountered in collisional experiments involving Rydberg atoms or molecules, besides the preparation of these species, is their short radiative lifetime. The lifetime consideration ($>10^{-6}$ sec) limits the quantitative studies of atoms and molecules by the scattering method to metastable and highly excited long-lived Rydberg states, usually with principal quantum numbers $n>10$. Recently, in a molecular-beam photoionization study of $(CS_2)_2$, Gress et al.[23] have demonstrated that molecular-beam photoionization of the van der Waals dimer is an excellent method for the quantitative study of the chemistry of excited molecules in short-lived radiative Rydberg states.

Following the discussion given previously, the excitation of a van der Waals molecule, such as $CS_2 \cdot CS_2$, can be viewed as the excitation of only one moiety making up the dimer. From the fact that there are two CS_2 molecules in a dimer, it is reasonable to assume that the cross sections for the absorption process $CS_2 \cdot CS_2 + h\nu \rightarrow CS_2^*(n) \cdot CS_2$ are twice that of the corresponding absorption process $CS_2 + h\nu \rightarrow CS_2^*$.[158] Knowing the absorption cross sections of the monomer will therefore make possible a quantitative estimate of the concentration of the excited dimer, $CS_2^*(n) \cdot CS_2$, initially prepared by the photoexcitation of the dimer. Since these excited molecular Rydberg dimers are formed in the asymptotic region of the interaction potential energy curve for $CS_2^*(n) + CS_2$, unimolecular reactions of these excited dimers can be regarded as high-Rydberg-state reactions of $CS_2^*(n) + CS_2$ proceeding at zero kinetic energy. The positions of the Rydberg series manifested as autoionization resonances in the PIE curves for the dimers and the fragment ions from the dimers should be identical. However, the slight shifts of these autoionizing levels in the PIE spectra for the fragment ions with respect to those in the PIE curve of the monomer will serve to distinguish the product ions formed from $CS_2^*(n) \cdot CS_2$ from ions due to secondary collisions of $CS_2^*(n)$ with CS_2 background molecules in the scattering chamber. The latter processes can be minimized by maintaining high vacuum in the ionization chamber and low jet density ($\leqslant 10^{-3}$ Torr) in the sampling region. By comparing the intensities of the product ions originating from different Rydberg levels and normalizing these to the corresponding known absorption cross sections of the monomer, not only can the relative chemiionization cross sections of a specific channel as a function of Rydberg level (n, l) be derived, but also the branching ratio of various product channels originating from a particular Rydberg level can be observed. Since the collisional lifetime of a van der Waals complex such as $CS_2^*(n) \cdot CS_2$ is expected to be $\lesssim 10^{-12}$ sec, which is much shorter than the radiative lifetime ($\sim 10^{-9}$ sec) of the Rydberg molecule $CS_2^*(n)$ in this energy range, the lifetime limitation should not be a problem for the experimental scheme described above.

A complete study of the following Rydberg-state reactions of carbon disulfide has been reported by Ono et al.[27]

$$CS_2^*(V, n) \cdot CS_2 \rightarrow CS_3^+ + CS + e \qquad (3.14)$$

$$\rightarrow S_2^+ + 2CS + e \qquad (3.15)$$

$$\rightarrow C_2S_3^+ + S + e \qquad (3.16)$$

$$\rightarrow (CS_2)_2^+ + e \qquad (3.17)$$

Photoionization efficiency data for CS_3^+, S_2^+, $C_2S_3^+$, and CS_4^+ are plotted in Figs. 31, 32, 33, and 34, respectively. Members of Rydberg series V which converge to the $\tilde{C}^2\Sigma_g^+$ state of CS_2^+ were observed in the PIE spectra for CS_3^+, S_2^+ and $C_2S_3^+$. As was expected, these autoionizing levels were found to be slightly shifted to the red, indicating that contributions to the product ions from background reactions of $CS_2^*(n) + CS_2$ are negligible. The fact that the positions of these autoionization peaks are the same in the PIE curves for CS_3^+, S_2^+, and $C_2S_3^+$ supports the conclusion that these product ions arise from a common intermediate complex, which is mainly the excited dimer, $CS_2^*(V, n) \cdot CS_2$, in this case. Although discrete structure corresponding to Rydberg series V (n) is not evident in the PIE curve for $(CS_2)_2^+$ (Fig. 29b), Ono et al.[27] were able to estimate the contributions of $(CS_2)_2^+$ forming the associative ionization process (3.17). The relative intensities for

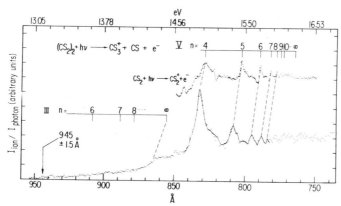

Fig. 31. Photoionization-efficiency curve for CS_3^+ from $(CS_2)_2$ in the region 735–950 Å. Wavelength resolution: 1.4 Å (FWHM). The nozzle beam conditions are the same as described in Fig. 26.[23,27]

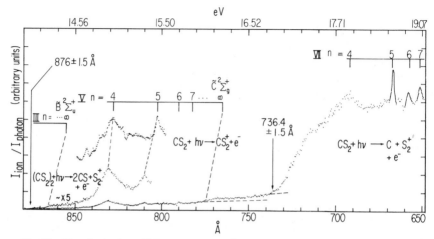

Fig. 32. Photoionization-efficiency curve for S_2^+ from $(CS_2)_2$ and CS_2 in the region 650–880 Å. Wavelength resolution: 1.4 Å (FWHM). The nozzle beam conditions are the same as described in Fig. 26.[27]

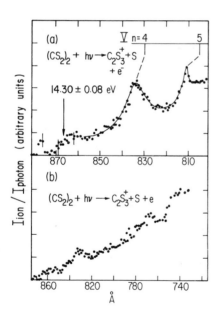

Fig. 33. Photoionization-efficiency curve for: (a) $C_2S_3^+$ from $(CS_2)_2$ in the region 800–885 Å (data plotted at intervals of 0.5 Å); (b) $C_2S_3^+$ in the region 715–880 Å (plotted at intervals of 1 Å). Wavelength resolution: 1.4 Å (FWHM). The nozzle beam conditions are the same as described in Fig. 26.[27]

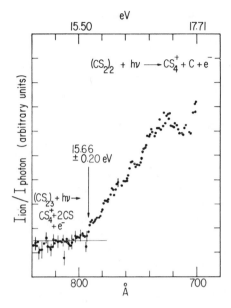

Fig. 34. Photoionization-efficiency curve for CS_4^+ in the region 700–820 Å. Wavelength resolution: 1.4 Å (FWHM). The nozzle beam conditions are the same as described in Fig. 26.[27]

$(CS_2)_2^+$, CS_3^+, S_2^+, $C_2S_3^+$, and CS_4^+ at 833 Å, which is the peak position of Rydberg series V ($n = 4$) in $(CS_2)_2$, were measured to be 1.0, 0.12, 0.02, 0.003, and 0.001, respectively. From this the PIE curves for $(CS_2)_2^+$, CS_3^+, S_2^+, and $C_2S_3^+$ were normalized. The relative reaction probabilities for the chemiionization processes (3.14)–(3.17) determined by Ono et al. are listed in Table VIII. The reaction probabilities, $\sigma CS_3^+(n)$, $\sigma C_2S_3^+(n)$, and

TABLE VIII
Relative Reaction Probabilities for the Chemiionization Processes
$CS_2^*(V, n) \cdot CS_2 \to CS_3^+ + CS + e$, $S_2^+ + 2CS + e$, $C_2S_3^+ + S + e$,
and $(CS_2)_2^+ + e^a$

n^b	$\sigma CS_3^+(n)$	$\sigma S_2^+(n)$	$\sigma C_2S_3^+(n)$	$\sigma C_2S_4^+(n)^c$
4	1	0.19	0.02	1.3
5	0.32	0.10	0.01	2.0
6	0.24	—	—	2.2
7	0.23	—	—	2.5
8	0.23	—	—	3.0

[a] Normalized for $\sigma CS_3^+(n = 4)$.[24]
[b] Rydberg series V $np\sigma$.
[c] Estimates.

$\sigma C_2 S_4^+ (n)$ for the formations of CS_3^+, S_2^+, $C_2 S_3^+$, and $C_2 S_4^+$, respectively, have been normalized for CS_3^+ (V, $n = 4$).

An interesting trend for the relative reaction probabilities of the chemi-ionization process derived from this experiment is that $\sigma CS_3^+ (n)$, $\sigma S_2^+ (n)$ and $\sigma C_2 S_3^+ (n)$ decrease as n increases, whereas $\sigma C_2 S_4^+ (n)$ increases. For the electronic energy of a Rydberg electron to be effective in promoting chemical reactions such as (3.14), (3.15), and (3.16), the Rydberg electron has to be able to couple with the ion core efficiently. Since the probability density of a Rydberg electron in the region of the ion core is approximately proportional to $(n^*)^{-3}$, the coupling between a Rydberg electron and the ion core will decrease as n increases. Therefore, $\sigma CS_3^+ (n)$, $\sigma S_2^+ (n)$, and $\sigma C_2 S_3^+ (n)$ should be larger when the electron is in the lower members of a Rydberg series. On the other hand, the energy which is needed to ionize the excited Rydberg dimers becomes smaller for an electron in a Rydberg level with higher n values; thus, the observed behavior of $\sigma C_2 S_4^+ (n)$ as a function of n is consistent with this interpretation.

Other than the product ions arising from excited Rydberg dimers, the CS_3^+, S_2^+, $C_2 S_3^+$, and CS_4^+ ions can also be produced by the unimolecular reactions of the excited $CS_2^+ (\tilde{A}\,^2\Pi_u, \tilde{B}\,^2\Sigma_u^+$, or $\tilde{C}\,^2\Sigma_g^+) \cdot CS_2$ ions which are formed by direct ionization of the neutral carbon disulfide dimer. The analysis of the PIE curves for these product ions has provided valuable information about the internal-energy effects and the energetics of the ion–molecule reactions of carbon disulfide,

$$CS_2^+ \left(\tilde{A}\,^2\Pi_u, \tilde{B}\,^2\Sigma_u^+ \text{ or } \tilde{C}\,^2\Sigma_g^+ \right) \cdot CS_2 \rightarrow CS_3^+ + CS \qquad (3.18)$$

$$\rightarrow S_2^+ + 2CS \qquad (3.19)$$

$$\rightarrow C_2 S_3^+ + S \qquad (3.20)$$

$$\rightarrow CS_4^+ + C \qquad (3.21)$$

The difficulty of studying these reactions in scattering experiments due to the short radiative lifetime of the $\tilde{A}\,^2\Pi_u$ and $\tilde{B}\,^2\Sigma_u^+$ states of CS_2^+ is expected to be alleviated by photoionization of the carbon disulfide dimers.

Both the CS_3^+ and S_2^+ have been observed previously in a gas-cell photoionization study of CS_2 by Eland and Berkowitz.[159] However, since these ions formed in the gas-cell experiment are promoted by internal energy as well as kinetic energy in secondary collisional processes involving CS_2^+, they were not able to locate the AEs for CS_3^+ and S_2^+ formations from the reaction of $CS_2^+ + CS_2$. The AEs for the reactions (3.18) and (3.19) as well as (3.20) and (3.21) have been measured by Ono et al.[27] (Figs. 31–34). Within the experimental uncertainties, the AE for (3.19) was found to be consistent with the thermochemical threshold for the reaction, $CS_2^+ (\tilde{X}\,^2\Pi_g) + CS_2 \rightarrow$

$S_2^+ + 2CS$. In other words, the activation energy for the latter reaction is zero. The energetics of the $CS_2^+ + CS_2$ system [23,24,27] is summarized in Fig. 35.

A steplike structure at 14.31 eV (866.5 Å) was observed in the PIE spectra for both CS_3^+ and S_2^+. The difference between the energy of this step and the converging limit (857.2 Å)[159] of series III and IV is 0.155 eV. Since this value is very close to the shifts in energy determined for the higher members of series III and IV in the PIE curve for $(CS_2)_2^+$ relative to the corresponding resonances resolved in the CS_2^+ spectrum, the step at 866.5 Å was assigned as the converging limit for Rydberg series III and IV in $(CS_2)_2$. This value has made possible the calculation of the potential energy, $D(\tilde{B}\,^2\Sigma_u^+)$ $= 0.205$ eV, for $CS_2^+(\tilde{B}\,^2\Sigma_u^+)\cdot CS_2$ at the equilibrium nuclear geometry of $(CS_2)_2$. The substantial increase in the yields of CS_3^+ and S_2^+ above the $\tilde{B}\,^2\Sigma_u^+$ state indicates that the reaction probabilities for the ion–molecule reactions (3.18) and (3.19) with CS_2^+ in the $\tilde{B}\,^2\Sigma_u^+$ state are much higher than those of the corresponding reactions with CS_2^+ in the $\tilde{A}\,^2\Pi_u$ state, an observation in agreement with the conclusions of previous experiments.[159]

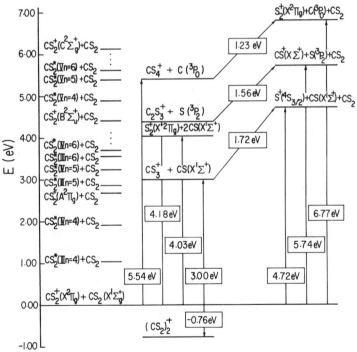

Fig. 35. Energetics of the CS_2^+ and $CS_2^+ + CS_2$ systems.[27]

Another interesting feature observed in the S_2^+ spectrum is the step at 16.03 eV (773.5 Å). The difference in energy of this step and the converging limit of series V of CS_2 (16.187 eV) is 0.158 eV. This value is comparable to the shifts in energy determined for the higher members of Rydberg series V resolved in the CS_3^+ spectrum, and Ono et al. have therefore identified the step at 16.03 eV as the converging limit of series V of $(CS_2)_2$. Using this value, they calculated a value of 0.210 eV for the potential energy $D(\tilde{C}\,^2\Sigma_g^+)$ for $CS_2^+(\tilde{C}\,^2\Sigma_g^+)\cdot CS_2$ at the equilibrium nuclear configuration of $(CS_2)_2$. Since one expects $D(\tilde{C}\,^2\Sigma_g^+)$ and $D(\tilde{B}\,^2\Sigma_u^+)$ to depend strongly upon charge multipole interactions between CS_2^+ and CS_2, and thus the charge distribution of CS_2^+, it is quite interesting to find that $D(\tilde{B}\,^2\Sigma_u^+)$ is essentially equal to $D(\tilde{C}\,^2\Sigma_g^+)$ within the experimental uncertainties. Furthermore, if the reaction probabilities for the reaction (3.19) are the same for CS_2^+ in the $\tilde{B}\,^2\Sigma_u^+$ and $\tilde{C}\,^2\Sigma_g^+$ states, one would expect to find a smooth transition in the region of the converging limit for series V, as is shown in the PIE curve for CS_2^+ in Fig. 29a. The observation of an increase in the PIE curve for S_2^+ at the converging limit of series V in CS_2 clearly reveals that the reaction probabilities for (3.19) with CS_2^+ in the $\tilde{C}\,^2\Sigma_g^+$ state is higher than when CS_2^+ is in the $\tilde{B}\,^2\Sigma_u^+$ state. Since no autoionization structure is evident in the PIE curve for CS_2^+ beyond the $\tilde{C}\,^2\Sigma_g^+$ state, it is a good approximation to assume that the relative populations of the $\tilde{X}\,^2\Pi_g$, $\tilde{A}\,^2\Pi_u$, $\tilde{B}\,^2\Sigma_u^+$, and $\tilde{C}\,^2\Sigma_g^+$ states resulting from photoionization with photon energy higher than the $\tilde{C}\,^2\Sigma_g^+$ state at ~ 765 Å are equal to the relative Franck–Condon factors derived from He I (584 Å) photoelectron spectroscopic data on CS_2. However, $CS_2^+(\tilde{X}\,^2\Pi_g, v)$ and $CS_2^+(\tilde{A}\,^2\Pi_u, v)$ produced by direct photoionization will not have sufficient energy for the reaction (3.19) to proceed, because the experimental AE for (3.19) is higher than the energy contents of $CS_2^+(\tilde{X}\,^2\Pi_g)$ or $CS_2^+(\tilde{A}\,^2\Pi_u)$ in vibrational states which are accessible to direct ionization processes. Therefore, it is reasonable to assume that the S_2^+ ions arise mainly from CS_2^+ populated in various vibrational levels of the $\tilde{B}\,^2\Sigma_u^+$ and $\tilde{C}\,^2\Sigma_g^+$ states. Since both the $\tilde{B}\,^2\Sigma_u^+$ and $\tilde{C}\,^2\Sigma_g^+$ states of CS_2 are nonbonding states, the $CS_2^+(\tilde{B}\,^2\Sigma_u^+$ or $\tilde{C}\,^2\Sigma_g^+)$ ions formed by direct ionization are predominately in the ground vibrational state $v = 0$. From the relative Franck–Condon factors for the $\tilde{B}\,^2\Sigma_u^+$ and $\tilde{C}\,^2\Sigma_g^+$ states obtained from photoelectron spectroscopy,[12] and attributing the increase in PIE for S_2^+ to higher reactivity of $CS_2^+(\tilde{C}\,^2\Sigma_g^+)$, the ratio of the reaction probability for the reaction (3.19) with CS_2^+ in the $\tilde{B}\,^2\Sigma_u^+$ state to that with CS_2^+ in the $\tilde{C}\,^2\Sigma_g^+$ state was deduced to be 0.34.

Starting at ~ 740 Å, an abrupt increase in the PIE for S_2^+ can be seen. The discrete structure resolved was found to match members of Rydberg series VIII ($n = 4$, 5, and 6) observed in previous absorption experiments,[138,154] which converge to a sixth IE at 19.38 eV. In addition, a slight step or a change in slope observed at 16.837 eV (736.4 ± 1.5 Å), was found to be equal to the thermochemical threshold for the formation of S_2^+ from CS_2. Thus, the S_2^+ ions produced in the region $\sim$650–740 Å can be attributed to the process $CS_2 + h\nu \rightarrow S_2^+ + C + e^-$.

The appearance of the $C_2S_3^+$ and CS_4^+ channels, and the monotonic increase in the PIE for these species towards high energy, are consistent with the decreasing trend of the yield for $(CS_2)_2^+$ in the same wavelength region. From the relative intensities of CS_3^+, S_2^+, and $C_2S_3^+$ at $\sim$855 Å, the relative reaction probabilities σCS_3^+, σS_2^+, and $\sigma C_2S_3^+$ for the ion–molecule half-reactions (3.18), (3.19), and (3.20) with CS_2^+ in the $\tilde{B}\,^2\Sigma_u^+$ state were deduced. Similarly, from the relative intensity of S_2^+ and CS_4^+ at $\sim$765 Å, σS_2^+ and σCS_4^+ for the reactions (3.19) and (3.21) can be derived. The relative values for σCS_3^+, σS_2^+, $\sigma C_2S_3^+$, and σCS_4^+ obtained by Ono et al. are listed in Table IX. Considering that the experimental AE for S_2^+ is greater than that for CS_3^+, it is likely that CS_3^+ is the intermediate for the formation of S_2^+. If $(CS_2)_2$ is indeed linear, as suggested by the symmetry argument, we might be sampling the reactions CS_2^+ [or $CS_2^*(n)$] + CS_2 colliding in a linear configuration. Higher values for $\sigma CS_3^+(n)$, $\sigma S_2^+(n)$, σCS_3^+, and σS_2^+ than for $\sigma CS_4^+(n)$, $\sigma C_2S_3^+(n)$, σCS_4^+, and $\sigma C_2S_3^+$ could probably result from a more favorable collisional complex prepared in this study. The geometry of the CS_3^+ ion is known. However, since the electronegativities for C and S are almost the same, and the size of S is larger than C, the S^+ ion is probably attached to an S atom of CS_2 in CS_3^+. A reasonable reaction pathway, which is in accordance with the above discussion is

$$(S-C-S)^+ \cdot S-C-S \rightarrow (S-S-C-S)^+ + C-S \rightarrow (S-S)^+ + 2C-S$$

$$(3.22)$$

A similar scheme can be written for the high-Rydberg-state reactions for $CS_2^*(n) \cdot CS_2$. Using this picture, one can also explain the observed relative reaction probabilities for the formation of CS_3^+ and S_2^+ with CS_2^+ in the

TABLE IX
Relative Reaction Probabilities for the Ion–Molecule Half-Reactions
$CS_2^+(\tilde{B}\,^2\Sigma_u^+$ or $C\,^2\Sigma_g^+) \cdot CS_2 \rightarrow CS_3^+ + CS$, $C_2S_3^+ + S$, and $CS_4^+ + C$ [a]

State	σCS_3^+	σS_2^+	$\sigma C_2S_3^+$	σCS_4^+
$\tilde{B}\,^2\Sigma_u^+$, $v = 0$	1	0.13	0.03	—
$\tilde{C}\,^2\Sigma_g^+$, $v = 0$	—	0.38	—	$\simeq 0.09$

[a] Normalized for $\sigma CS_3^+ (\tilde{B}\,^2\Sigma_u^+, v = 0)$.[24]

$\tilde{B}^2\Sigma_u^+$ and $\tilde{A}^2\Pi_u$ states. The $\tilde{B}^2\Sigma_u^+$ and $\tilde{A}^2\Pi_u$ states are formed by eject-
ing the electrons from σ_u and π_u molecular orbitals to the continuum, respec-
tively. The $CS_2^+(\tilde{B}^2\Sigma_u^+)$ ion, which has a vacant site in the σ-type orbital,
should be more favorable for the attack of the neutral CS_2 molecule than the
$CS_2^+(\tilde{A}^2\Pi_u)$ ion if CS_2 approaches along the molecular axis of CS_2^+.

Ono et al.[27] noted that the branching ratios and the relative reaction
probabilities derived in the molecular-beam photoionization experiment
might only apply to reactions that proceed from complexes in some specific
configurations. These values might be different from those obtained from real
collision experiments in which a large volume of phase space is sampled.

2. OCS, OCS·CS₂, and OCS Dimer and Trimer

Carbonyl Sulfide.   The high-resolution (0.14 Å FWHM) PIE curve for
OCS^+ near the threshold obtained by Ono et al.[28] using the molecular-beam
photoionization method is shown in Fig. 36. The first onset is very sharp,
with an uncertainty for the initial threshold of ± 0.15 Å (± 0.0015 eV), indicating that the rotational temperature of OCS achieved in the supersonic expansion is lower than 20 K. The next step is at 1104.95 Å. Thus, the IEs for
the $\tilde{X}^2\Pi_{3/2}$ and $^2\Pi_{1/2}$ states of OCS were determined to be 11.1736 ± 0.0015
and 11.2207 ± 0.0015 eV, respectively. These values are compared with those
obtained from photoelectron spectroscopy,[160] photoionization mass spectrometry,[143, 161] and absorption spectroscopy[161] in Table X. The spin–orbit
splitting (381 ± 17 cm^{-1}) determined by Ono et al. was found to be in agreement with previous measurements.[142–144, 160–162] As with the PIE curve for

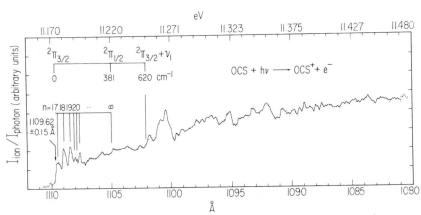

Fig. 36.   Photoionization-efficiency curve for OCS^+ in the region 1080–1111 Å obtained
using a wavelength resolution of 0.14 Å (FWHM) and the Ar continuum as the light source.
Nozzle conditions: $P_0 \simeq 800$ Torr, $T_0 \simeq 298$ K, $D = 0.05$ mm.[28]

TABLE X
Summary of Energetic Information for OCS, $(OCS)_2$, $(OCS)_3$, and $OCS \cdot CS_2$[28]

Ion	Molecular-beam photoionization[a]	Other techniques
	Ionization energy (eV)	
$OCS^+(\tilde{X}^2\Pi_{3/2})$	11.1736 ± 0.0015	$11.18(9) \pm 0.005^b$
		$11.18 \pm 0.01^{c,d}$
		11.190 ± 0.003^e
		11.174 ± 0.003^f
$OCS^+(^2\Pi_{1/2})$	11.2207 ± 0.0015	$11.23(3) \pm 0.005^b$
		$11.22 \pm 0.01^{c,d}$
		11.235 ± 0.003^e
		11.220 ± 0.003^f
$(OCS)_2^+$	10.456 ± 0.026	—
$(OCS)_3^+$	10.408 ± 0.026	—
$(OCS \cdot CS_2)^+$	9.858 ± 0.024	—
	Vibrational frequency ν_1 (cm^{-1})	
$OCS^+(\tilde{X}^2\Pi_{3/2})$	620 ± 24	650 ± 50^b
		610^c

[a]Reference 28.
[b]Reference 142.
[c]Reference 161.
[d]Reference 143.
[e]Reference 144.
[f]Reference 160.

CS_2^+,[27] members of the Rydberg series converging to the $^2\Pi_{1/2}$ state of OCS^+ were resolved on the first step of the OCS^+ spectrum. Assuming the converging limit of this series to be the $^2\Pi_{1/2}$ threshold at 90,502 cm^{-1}, the Rydberg equation

$$\nu_n = \left[90,502 - \frac{R}{(n+0.6)^2} \right] \text{cm}^{-1} \qquad (3.23)$$

was found to give a good fit to the observed autoionization peaks. The consistency of this analysis serves as strong support for the assignment of the second step in the PIE curve as the IE of the $^2\Pi_{1/2}$ state of OCS^+.

The step at 1102.05 Å was attributed to the threshold of $\tilde{X}^2\Pi_{3/2} + \nu_1$. This assignment gives a value of 620 ± 24 cm^{-1} for ν_1 which is in agreement with previous determinations[142, 161] (Table X). However, as a result of the superposition of strong autoionization structure in this region, the onsets of $\tilde{X}^2\Pi_{3/2} + 2\nu_1$ (1094.6 Å) and $\tilde{X}^2\Pi_{3/2} + \nu_3$ (1085.95 Å) suggested by Ono et al. are expected to have large uncertainties.

(OCS)₂ (OCS)₃ and OCS·CS₂. Previous photoionization studies[22,24] of $(CS_2)_2$ suggest that by measuring the shifts in energy of the Rydberg series of the monomer in the dimer, and from the symmetry of the Rydberg orbital, structural information of the neutral dimer can be obtained. Ono et al.[28] have extended this idea to $(OCS)_2$ and OCS·CS₂.

The PIE curves for OCS^+, $(OCS)_2^+$, and $(OCS)_3^+$ in the region 700–1200Å obtained using a wavelength resolution of 1.4 Å (FWHM) are shown in Fig. 37a, b, and c, respectively. These spectra were arbitrarily normalized at a structureless region ($\simeq 1050$ Å). From the observed IE (10.456 ± 0.026 eV) for $(OCS)_2$, and the estimated binding energy (0.029 eV),[117] a dissociation energy of 0.75 ± 0.04 eV (17.2 ± 1 kcal/mol) was deduced for $(OCS)_2^+$. Within the uncertainty of the experiment, this value is equal to the dissociation energy (0.76 ± 0.04 eV) for $(CS_2)_2^+$.[24] By assuming the binding energy of $(OCS)_2$ with OCS to be the same as in $(OCS)_2$ and using the measured IE (10.408 ± 0.026 eV) for $(OCS)_3$, the energy released by adding a carbonyl sulfide molecule to a carbonyl sulfide dimer ion was found to be only $\simeq 0.07$ eV.

As was expected, the positions of autoionization features resolved in the PIE spectrum for $(OCS)_2^+$ were found to be shifted with respect to those in the monomer spectrum. Rydberg series III, IV, V, and VI of OCS have been

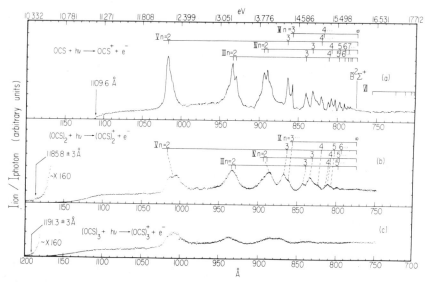

Fig. 37. Photoionization-efficiency curves for: (a) OCS^+ in the region 700–1115 Å; (b) $(OCS)_2^+$ in the region 750–1200 Å; (c) $(OCS)_3^+$ in the region 750–1200 Å. Experimental conditions: $P_0 \simeq 800$ Torr, $T_0 \simeq 298$ K, $D = 0.05$ mm, wavelength resolution 1.4 Å (FWHM).[28]

observed previously.[138,159,163] These Rydberg series have recently been assigned[159] as $np\sigma$(III), $nd\pi$(V), and $ns\sigma$(VI) with quantum defects equal to 0.65, 0.12, and 0.05, respectively. Since the lowest member of series V at 1015.8 Å was observed, Delwiche et al.[160] have interpreted this series as an $ns\sigma$ series with a quantum defect δ of 1.12 instead of $nd\pi$ with $\delta = 0.12$. In addition, they also suggested that the series labeled $ns\sigma$(VI) should be labeled $nd\pi$. The nature of series IV is not clear. According to its quantum defects of 0.48, this series is consistent with an $nd\sigma$ or an np series.[164] As shown in Fig. 36b, blue shifts were observed for series IV ($n = 2$) and series V ($n = 2$), whereas the peak position for series III ($n = 2$) in the PIE curve for $(OCS)_2^+$ is in close correspondence with that in the OCS^+ spectrum. This can be taken as evidence that the perturbation of series III ($n = 2$) (an $np\sigma$ orbital) in comparison with that of series V ($n = 2$) (an $ns\sigma$ or $nd\pi$ orbital) is small at the equilibrium configuration of $(OCS)_2$. From the symmetry point of view, the perturbation of the $np\sigma$ Rydberg orbital would be maximized if the perturber approached along the molecular axis of the absorber. Therefore, assuming the validity of the symmetry arguments, it is unlikely for $(OCS)_2$ to have a collinear configuration. Due to the qualitative nature of this method, this experiment cannot provide any specific structural parameters for $(OCS)_2$. However, the shifts observed for the Rydberg series of OCS in the $(OCS)_2^+$ spectrum are consistent with a nearly side-by-side or dihedral geometry for $(OCS)_2$. The symmetry properties of series IV are not known. If the carbonyl sulfide dimer indeed possesses a nearly side-by-side or dihedral bonding geometry, the blue shift observed for series IV ($n = 2$) is consistent with an $np\pi$ or an $nd\sigma$ assignment for series IV.

The PIE curve for $OCS \cdot CS_2^+$ in the region 730–1270 Å obtained by Ono et al. is shown in Fig. 38. The measured IE for $OCS \cdot CS_2$ is 9.858 ± 0.024 eV (1257.7 ± 3 Å). From this, the known IE for CS_2 (10.0685 ± 0.0020 eV),[24] and the estimated binding energy (0.039 eV)[§] between OCS and CS_2, the binding energy of $CS_2^+ + OCS$ was deduced to be 0.25 ± 0.04 eV (5.8 ± 1 kcal/mol). This value is much smaller than the dissociation energies of $(OCS)_2^+$ and $(CS_2)_2^+$.

From a comparison of the PIE curve for CS_2^+, $(CS_2)_2^+$,[24] $(OCS)_2^+$, and $(OCS \cdot CS_2)^+$, Ono et al. concluded that the prominent structure observed in the $(OCS \cdot CS_2)^+$ spectrum originates mainly from CS_2. Although the absorption and the ionization coefficients of CS_2[165] are similar in magnitude to those of OCS,[163] the structure related to OCS is almost unrecognizable in the PIE curve for $(OCS \cdot CS_2)^+$. However, the anomalous increases in

---

[§]The binding energies of $OCS \cdot CS_2$ are estimated to be $\{D[(CS_2)_2] \cdot D[(OCS)_2]\}^{1/2}$, where $D[(CS_2)_2]$ and $D[(OCS)_2]$ represent the dissociation energies of $(CS_2)_2$ and $(OCS)_2$, respectively.

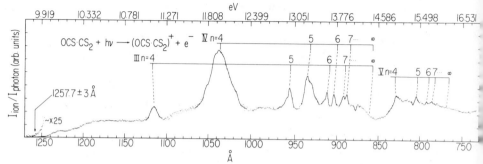

Fig. 38. Photoionization efficiency curve for $(OCS \cdot CS_2)^+$ in the region 730–1270 Å. Wavelength resolution: 1.4 Å. The $OCS \cdot CS_2$ molecule was prepared by seeding CS_2 vapor ($\simeq 130$ Torr) at 0°C in $\simeq 700$ Torr of OCS and then expanding the mixture through a 0.05-mm-diameter stainless-steel nozzle.[28]

strength of the peaks at approximately 1030, 930, and 940 Å, and so on, possibly have contributions from structures of both CS_2 and OCS. The $(OCS \cdot CS_2)^+$ ions observed at these peaks should consist of ions of the form $OCS^+ \cdot CS_2$ as well as $CS_2^+ \cdot OCS$. Interestingly, the trends in the shifts in energy of Rydberg series III, IV, and V of CS_2 in the $(OCS \cdot CS_2)^+$ spectrum were found to be almost identical to those found in $(CS_2)_2^+$, namely, the member $n = 4$ of series IV (an $np\pi$ series) appears to be unchanged in position, whereas the member $n = 4$ of series III (an $np\sigma$ series) is blue-shifted. In addition, larger shifts for series III ($n > 4$) than for series IV ($n > 4$) relative to the positions in the monomer spectrum were observed. Using symmetry arguments,[22,24] this observation supports a structure with OCS linked to CS_2 at an end sulfur. However, since the Rydberg series of OCS in the $(OCS \cdot CS_2)^+$ spectrum cannot be identified, this experiment does not provide information concerning the orientation of OCS with respect to the molecular axis of CS_2. In other words, the molecular axis of OCS could be parallel with or perpendicular to that of CS_2. If the interpretations of the shifts of the Rydberg series of OCS and CS_2 observed in $(OCS)_2$, $OCS \cdot CS_2$, and $(CS_2)_2$ are correct, it is interesting that $OCS \cdot SCS$ and $SCS \cdot SCS$ are end-on complexes, while $OCS \cdot OCS$ possesses a side-by-side geometry. The dipole moment of OCS is probably responsible for these differences.

The intermolecular distances for $(OCS)_2$ and $OCS \cdot CS_2$ are unknown. From the Lennard-Jones diameters of OCS and CS_2, the intermolecular distances for these dimers were estimated to be in the range $\simeq 4$–4.5 Å. Following the previous discussion, the size of a Rydberg orbital (r_n) can be estimated by $(n^*)^2 a_0$.[138] The analysis of Ono et al. shows that the r_n's of the Rydberg orbitals [series IV ($n = 2$) and series V ($n = 2$) of OCS, as well as series III ($n = 4$) and series V ($n = 4$) of CS_2], which were found to be blue-shifted in the $(OCS)_2^+$ and $(OCS \cdot CS_2)^+$ spectra, are either smaller than or comparable to the intermolecular distances of these dimers. Higher members of the Rydberg series having r_n's larger than the intermolecular distances of $(OCS)_2$ and $OCS \cdot CS_2$ were all found to be red-shifted. This observation supports the previous interpretation of the shifts of Rydberg series III and IV of CS_2 observed in the PIE curve for $(CS_2)_2^+$.[24] In order to elucidate this interpretation further, the radial probability densities $(r^2 R_{nl}^* R_{nl})$ of two Rydberg orbitals originating from the moiety A of a van der Waals dimer $A \cdot B$ are plotted in Fig. 39. The results of the $(OCS)_2$, $OCS \cdot CS_2$,[28] and $(CS_2)_2$[22,24] experiments are consistent with the following explanation. When an electron is excited to a Rydberg orbital having a radius[138] $r_{n_2} \simeq (n_2^*)^2 a_0$ larger than the intermolecular distance r_{AB}, the shielding of the ion core A^+ by the Rydberg electron becomes less effective, and charge–multipole attractive interactions and chemical forces will play more important roles in the bonding of the excited $A^*(n_2) \cdot B$ dimer. This makes the potential energy of $A^*(n_2) \cdot B$ at the equilibrium configuration of $A \cdot B$ greater than the dissociation energy of $A \cdot B$ and thus gives rise to a red shift in energy of the Rydberg transition in the dimer $A \cdot B$ from that in A. On the contrary, if an electron is in a Rydberg orbital with $r_{n_1} \simeq (n_1^*)^2 a_0$ smaller than r_{AB}, the exchange repulsive interaction will dominate and a blue shift will be observed in the dimer-ion spectrum. Although many other interactions, such as curve crossings, could cause the shifts of Rydberg transitions in the dimers, the above interpretation certainly is a reasonable one.

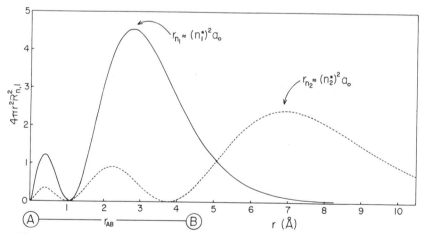

Fig. 39. Schematic representation of the radial probability density, $4\pi r^2 R_{nl} R_{nl}^*$, for Rydberg orbitals characterized by the effective principal quantum numbers n_1^* and n_2^*. Here r_{n_1} and r_{n_1} represent the radii of Rydberg orbitals characterized by n_1^* and n_2^*, respectively. r_{AB} is the intermolecular distance of a van der Waals molecule $A \cdot B$ (see text).

3. CO_2 and N_2O Dimers and Clusters

Photoionization efficiency data for CO_2^+, $(CO_2)_2^+$, and $(CO_2)_3^+$ obtained recently by Linn and Ng[129] using an optical resolution of 1.4 Å (FWHM) are plotted in Fig. 40a, b, and c, respectively. The stabilities of these cluster ions deduced from the measured IEs of $(CO_2)_n$, $n = 1, 2, 3, 4$, are listed and compared with values obtained by other techniques[152, 166–168] in Table XI. Values for the enthalpy change of the ion–molecule association reaction $CO_2^+ CO_2 \rightleftarrows CO_2^+ \cdot CO_2$, determined by Meot-Ner et al.[152] and J. V. Headley et al.,[166] after converting them to ΔH_0° using arguments similar to those used in Reference 24, were found to be in good agreement with the values of Linn et al.

The rich autoionization structure from 685 to 790 Å in the PIE curve for CO_2^+ is most readily attributable to Rydberg series converging to the $\tilde{B}^2\Sigma_u^+$ state of CO_2^+ which are labeled "sharp" and "diffuse" series by Tanaka and Ogawa.[138, 169, 170] Since the photoelectron spectrum[128, 140] of the $\tilde{B}^2\Sigma_u^+$ band

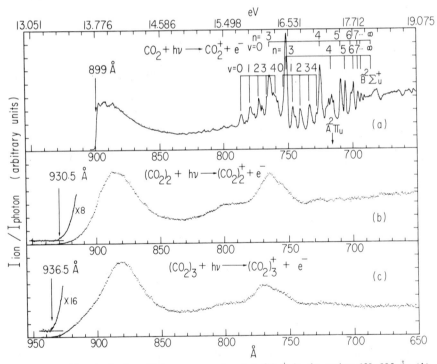

Fig. 40. Photoionization-efficiency curves for: (a) CO_2^+ in the region 650–905 Å; (b) $(CO_2)_2^+$ in the region 650–950 Å; (c) $(CO_2)_3^+$ in the region 650–945 Å. Experimental conditions: $P_0 \simeq 1000$ Torr, $T_0 \simeq 298$ K, $D = 0.05$ mm, wavelength resolution 1.4 Å (FWHM).[129]

TABLE XI
Bond Dissociation Energies of $(CO_2)_{1-3}^+ \cdot (CO_2)$, $N_2O^+ \cdot N_2O$ and $NO^+ \cdot N_2O$[139]

Ion	Molecular-beam photoionization (kcal/mol)	Other technique[a] (kcal/mol)
$CO_2^+ \cdot CO_2$	11.8 ± 1.0^b $\geqslant 9^c$	13.0 ± 1.5^d 12.7 ± 0.6^c 17.6 ± 2.8^e
$(CO_2)_2^+ \cdot CO_2$	3.3 ± 1.4^b	4.8 ± 1.5^d
$(CO_2)_3^+ \cdot CO_2$	2.8 ± 1.4^c	—
$N_2O^+ \cdot N_2O$	13.1 ± 0.9^b	—
$NO^+ \cdot N_2O$	$\geqslant 5^b$	—

[a] Values from References 152 and 166 have been converted to enthalpy change at 0 K using arguments similar to those used in Reference 24.
[b] Reference 129.
[c] Reference 166.
[d] Reference 152.
[e] Reference 168. Effective temperature unknown.

shows a much larger Franck–Condon factor for the $(0,0,0) \rightarrow (0,0,0)$ transition than for the other vibrational transitions, the "sharp" and "diffuse" series as shown in Fig. 38a have all been assigned to $v = 0$. Therefore, the autoionization mechanism for these series should be electronic rather than vibrational. Most of the remaining autoionization features in the CO_2^+ spectrum originate from the Tanaka–Ogawa series which converges to the $\tilde{A}^2\Pi_u$ state of CO_2^+.

Rydberg features observed in the PIE curve for $(CO_2)_2^+$ (Fig. 40b) are much weaker than those resolved in the PIE curve for CO_2^+, which indicates that excited Rydberg dimers formed by the photoexcitation process

$$CO_2 \cdot CO_2 + h\nu \rightarrow CO_2^*(n, v, J) \cdot CO_2 \qquad (3.24)$$

are strongly dissociative in this wavelength region. Here n, v, and J represent the principal, the vibrational, and the rotational quantum numbers, respectively. The redistribution of internal energies from the initially excited moiety of the complex to the van der Waals bond must eventually lead to dissociation of these excited dimers. The possible dissociation processes

$$CO_2^*(n, v, J) \cdot CO_2 \rightarrow CO_2^*(n, v, J') + CO_2 \qquad (3.25)$$
$$\rightarrow CO_2^*(n, v', J) + CO_2 \qquad (3.26)$$
$$\rightarrow CO_2^*(n', v, J) + CO_2 \qquad (3.27)$$

which involve the redistributions of rotational, vibrational, and electronic energies are known as rotational, vibrational and electronic predissociations, respectively. In view of the efficient rotational relaxation in a supersonic expansion and the rotational selection rules for photoexcitation processes, one expects the process (3.25) to be a minor dissociation mechanism. Due to the difference in geometry of $(CO_2)_2$ and $CO_2^*(n) \cdot CO_2$, the $CO_2^*(n) \cdot CO_2$ formed by the process (3.24) is vibrationally excited in the vibrational mode between $CO_2^*(n)$ and CO_2. The amount of energy ($\simeq 0.1$–0.2 eV) required to dissociate these excited Rydberg dimers is less than or of the order of a vibrational quantum of CO_2^+. Therefore, vibrational predissociation (3.26) is likely to be an important dissociation process for the excited Rydberg complex CO_2^* ($n, v \geqslant 1) \cdot CO_2$ formed by the process (3.24) involving vibrationally excited members of the Tanaka–Ogawa Rydberg series. However, for excited Rydberg complexes $CO_2^*(n, v = 0) \cdot CO_2$, with one moiety of the dimer excited to members of the "sharp" or "diffuse" series, vibrational predissociation should be excluded from consideration as an efficient dissociation mechanism. Following the above arguments, this experiment suggests electronic predissociation (3.27) to be a major dissociation mechanism of the $CO_2^*(n) \cdot CO_2$ excited Rydberg dimers.

For rare-gas dimers, both experiment[171] and theoretical calculations[172, 173] suggest the existence of potential barriers in their Rydberg states. If a potential-energy barrier exists in the interaction potential between $CO_2^*(n)$ and CO_2 at a shorter intermolecular distance than the equilibrium distance between CO_2 and CO_2, direct dissociation can also occur as a result of excitation to the side of the barrier outside the well region of the interaction potential for $CO_2^*(n) + CO_2$.

Furthermore, one needs to consider possible dissociation processes after ionization, that is,

$$CO_2^*(n, v) \cdot CO_2 \rightarrow CO_2^+(v') \cdot CO_2 + e^- \qquad (3.28)$$

If $CO_2^+(v')$ of the dimer ion formed by the autoionization process (3.28) is vibrationally excited ($v' \geqslant 1$), vibrational predissociation will again be a possible process. According to theoretical predictions,[174, 175] the vibrational energy distribution of $CO_2^+(v')$ resulting from autoionization is governed by Franck–Condon factors between the vibrational level of the Rydberg state and vibrational levels of the ionic state. For Rydberg levels converging to the $\tilde{A}^2\Pi_u$ and $\tilde{B}^2\Sigma_u^+$ states of CO_2^+, autoionization will most likely give rise to CO_2^+ in the $\tilde{X}^2\Pi_g$ ground state. Photoelectron data[128] for CO_2 show that the $\tilde{X}^2\Pi_g$ and $\tilde{B}^2\Sigma_u^+$ states are essentially nonbonding in nature; hence they are expected to have the same nuclear configuration. Because of the similarity between potential-energy curves of members of the "sharp" and "diffuse" series and that of the $\tilde{B}^2\Sigma_u^+$ state, the Franck–Condon factor for the $(0,0,0) \rightarrow (0,0,0)$ transition should be much larger than other transitions.

Since the initially populated Rydberg levels of the "sharp" and "diffuse" series are mainly in $v = 0$ states, autoionization from these states will produce mostly vibrationless $CO_2^+(\tilde{X}^2\Pi_g)$ ions. Assuming the autoionization process to be unperturbed in the excited dimer, the dimer ion formed by the process (3.28) will contain little vibrational energy. Hence, vibrational predissociation of $CO_2^+(\tilde{X}^2\Pi_g, v' = 0) \cdot CO_2$ is again an unlikely dissociation mechanism.

Nitrous oxide is isoelectronic with carbon dioxide, and the electronic structure of N_2O^+ is similar to that of CO_2^+. In accordance with the photoelectron spectrum of N_2O,[128,176] the $\tilde{X}^2\Pi$ and $\tilde{A}^2\Sigma^+$ states of N_2O^+ are also nonbonding in nature. As shown in Fig. 41a, Rydberg series III and IV,[138,177-180] which account for most of the prominent autoionization structures in the region 750–850 Å and converge to the $\tilde{A}^2\Sigma^+$ (0,0,0) state of N_2O^+, all correspond to the vibrational ground state. The general profiles of the PIE curves for N_2O^+ and $(N_2O)_2^+$ are similar except that structures originating from Rydberg series III and IV are almost completely unobservable in the PIE curve for $(N_2O)_2^+$ (Fig. 41b). Based on the same arguments used above to explain the dissociation mechanism of $CO_2^*(n) \cdot CO_2$, Linn and Ng[129] again concluded that electronic dissociation and/or predissociation are the principal processes responsible for the dissociation of the $N_2O^*(n, v = 0) \cdot N_2O$ excited Rydberg dimers.

Experimental observations for $(N_2O)_2^+$ and $(CO_2)_2^+$ are different from those of other linear triatomic systems such as $(CS_2)_2^+$,[24] $(OCS)_2^+$, and

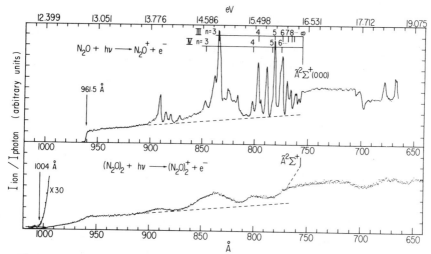

Fig. 41. Photoionization-efficiency curves for: (a) N_2O^+ in the region 660–965 Å; (b) $(N_2O)_2^+$ in the region 640–1020 Å. Experimental conditions: $P_0 \simeq 600$ Torr, $T_0 \simeq 220$ K, $D = 0.05$ mm, wavelength resolution 1.4 Å.[129]

$(OCS \cdot CS_2)^+$.[28] The PIE curves for $(CS_2)_2{}^+$, $(OCS)_2{}^+$, and $(OCS \cdot CS_2)^+$ show moderately resolved structures which correlate with Rydberg states of the monomers. Autoionization structures observed for $CS_2{}^+$ and OCS^+ are in general much stronger than those observed for N_2O^+ and $CO_2{}^+$. This is indicative of faster autoionization processes in CS_2 and OCS than those in N_2O and CO_2. If electronic predissociation is slow in comparison with auto-ionization, autoionization structures should be observed in the PIE curve for the dimer ion. This is probably the case for $(CS_2)_2{}^+$, $(OCS)_2{}^+$, and $(OCS \cdot CS_2)^+$. In the cases of $(CO_2)_2{}^+$ and $(N_2O)_2{}^+$, the strongly dissociative nature of $CO_2^*(n) \cdot CO_2$ and $N_2O^*(n) \cdot N_2O$ can be attributed to faster electronic predissociation than associative ionization processes, such as (3.28). Although vibrational predissociation is believed to be a major dissociation mechanism for $H_2^*(n, v) \cdot H_2$,[26] $O_2^*(n, v) \cdot O_2$,[29] $CO^*(n, v) \cdot CO$, $N_2^*(n, v) \cdot N_2$, and $NO^*(n, v) \cdot NO$,[30] electronic predissociation may still play a role in the dissociation of these excited Rydberg dimers.

From the measured IE (12.35 ± 0.02 eV) for $(N_2O)_2$, the known IE of N_2O (12.886 ± 0.002 eV),[177] and the estimated binding energy (0.02 eV),[181,182] Linn et al. calculated a value of 0.56 ± 0.04 eV for the bond-dissociation energy of $N_2O^+ \cdot N_2O$. The charge-induced dipole interaction is probably not the only interaction term which accounts for the stability of the dimer ions. Nevertheless, it is interesting to note the good correlation of the bond dissociation energies of $(CS_2)_2{}^+$ (0.76 eV),[24] $(OCS)_2{}^+$ (0.75 eV),[28] $(SO_2)_2{}^+$ (0.66 eV),[32] $(N_2O)_2{}^+$ (0.56 eV), and $(CO_2)_2{}^+$ (0.51 eV) with the polarizabilities of CS_2 (8.7×10^{-24} cm^3), OCS (5.7×10^{-24} cm^3), SO_2 (3.7×10^{-24} cm^3), N_2O (3×10^{-24} cm^3), and CO_2 (2.6×10^{-24} cm^3).[117]

The $N_3O_2{}^+$ ion is the only ion which originates from $(N_2O)_2$ and has a high enough intensity for the measurement of its spectrum. The PIE data for $N_3O_2{}^+$ is depicted in Fig. 42b and compared with the PIE curve for the NO^+ fragment from N_2O in Fig. 42a. Similarly to the observation for $S_2O_3{}^+$ from $(SO_2)_2$,[32] the PIE curve for $N_3O_2{}^+$ was found to have the same profile as that for NO^+ except that autoionization structures resolved in the NO^+ spectrum are unobservable in the $N_3O_2{}^+$ spectrum. This suggests that dissociations of N_2O^+ and $N_2O^+ \cdot N_2O$ to form $NO^+ + O$ and $NO^+ \cdot N_2O + O$, respectively, follow the same reaction pathway, and that the neutral moiety N_2O in $N_2O^+ \cdot N_2O$ acts like a spectator in the fragmentation process $N_2O^+ \cdot N_2O \rightarrow NO^+ \cdot N_2O + N$.

4. SO_2 and SO_2 Dimer

The ionization of sulfur dioxide has been studied previously by photoionization mass spectrometry[183–185] and many other methods.[128,140,186–194] Although consistent values for the IE of SO_2 have been obtained by various methods, considerable uncertainty still remains. There appears to be some

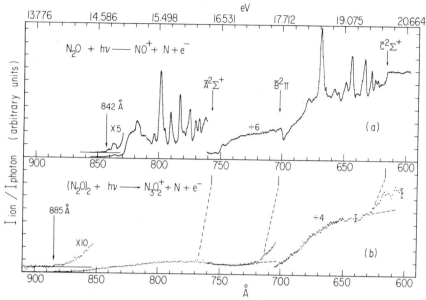

Fig. 42. Photoionization-efficiency curves for (a) NO^+ from N_2O in the region 600–865 Å; (b) $N_3O_2^+$ from $(N_2O)_2$ in the region 600–910 Å. Experimental conditions: $P_0 \simeq 600$ Torr, $T_0 \simeq 220$ K, $D = 0.05$ mm, wavelength resolution 1.4 Å.[129]

controversy about the energetics of the photodissociative ionization process

$$SO_2 + h\nu \rightarrow SO^+ + O + e^- \qquad (3.29)$$

From the value for the AE of the above process reported by Dibeler and Liston,[185] the value for the IE of SO was calculated to be 10.21 eV, which was found to be lower than the value (10.29 eV)[195,196] obtained by photoelectron spectroscopy. Recently, Erickson and Ng[32] have performed a high-resolution photoionization study of SO_2 using the molecular-beam method. They have been able to determine the IE of SO_2 and the AE of the reaction (3.29) with high accuracy and assign progressions of vibrational bands resolved in the PIE spectrum for SO_2^+.

Figure 43a shows the PIE curve for SO_2^+ obtained using an optical resolution of 1.4 Å (FWHM) at a stagnation pressure of $\simeq 125$ Torr. The first band in the photoelectron spectrum of SO_2 contains a progression with an average separation of 403 cm^{-1},[140] which was assigned as the vibrational quantum of the ν_2 bending mode of SO_2^+ in the $\tilde{X}^2A_1$ state. The excitation of ν_2 is manifested by a steplike structure at the ionization threshold of SO_2 with an average spacing of $\simeq 425$ cm^{-1}. The bending vibrational frequency

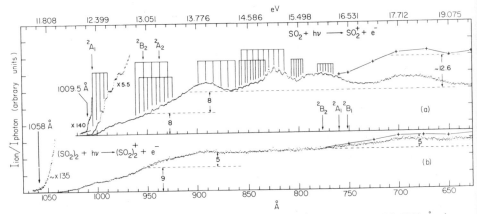

Fig. 43. Photoionization-efficiency curves for: (a) SO_2^+ in the region 625–1020 Å obtained using a wavelength resolution of 1.4 Å (FWHM) (nozzle conditions: $P_0 \simeq 125$ Torr, $T_0 \simeq$ 298 K, $D = 0.12$ mm); (b) $(SO_2)_2^+$ in the region 625–1070 Å obtained using a wavelength resolution of 1.4 Å (FWHM) (nozzle conditions: $P_0 \simeq 450$ Torr, $T_0 \simeq 298$ K, $D = 0.12$ mm).[32]

for the ground state of SO_2 is 518 cm^{-1}.[151] According to a Boltzmann distribution, the population of the $(0,1,0)$ vibrational excited state at room temperature is approximately 8% that of the $(0,0,0)$ state. Since a low nozzle stagnation pressure was used in this measurement, one might expect the relaxation of $(0,1,0)$ excitation to be inefficient. An extremely small step was observed at 1009.5 Å. The ratio of this step height to that of the next step at 1004.5 Å was found to be $\simeq 0.08$. Hence, the first small step was assigned to a hot band, and the IE of SO_2 determined by this low-resolution experiment was 12.34 eV (1004.5 Å).

The high-resolution (0.14 Å FWHM) PIE curve in the region 988–1006 Å obtained by Erickson et al. is shown in Fig. 44a. Again, the step-function behavior is evident. These steps are sharp, with an uncertainty of ± 0.15 Å. When the uncertainty of the wavelength calibration was included, the IE of SO_2 was determined to be 12.348 $\pm$ 0.002 eV (1004.08 $\pm$ 0.20 Å). This value is in good agreement with valued obtained by Golomb et al.[187] and Watanabe[184] (Table XII). The separations between adjacent steps resolved in Fig. 44a appear to be irregular. The third spacing, having a value of 497 ± 20 cm^{-1}, was found to be greater than the first and second spacings by $\simeq 60$ cm^{-1}. Similar observations have been reported for O_3^+ by Dyke et al.,[197] Frost et al.,[198] Brundle,[199] and Weiss et al.[200] As was the case in the analysis of the PIE curve for O_3^+, the observed irregular spacing ruled out any simple assignments of the first photoelectron band to the ν_2 bending mode of SO_2^+ accompanying ionization from the $8a_1$ orbital of SO_2.

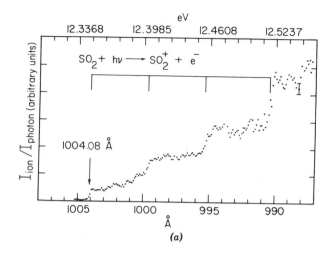

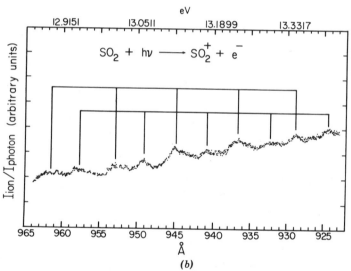

Fig. 44. Photoionization-efficiency curves for SO_2^+: (a) in the region 987–1006 Å obtained using a wavelength resolution of 0.14 Å (FWHM) and the hydrogen many-lined pseudocontinuum as the light source (data plotted at intervals of 0.1 Å); (b) in the region 922–965 Å (data plotted at intervals of 0.05 Å); (c) in the region 760–904 Å (data plotted at intervals of 0.05 Å). Nozzle conditions: $P_0 \simeq 450$ Torr, $T_0 \simeq 298$ K, $D = 0.12$ mm.[32]

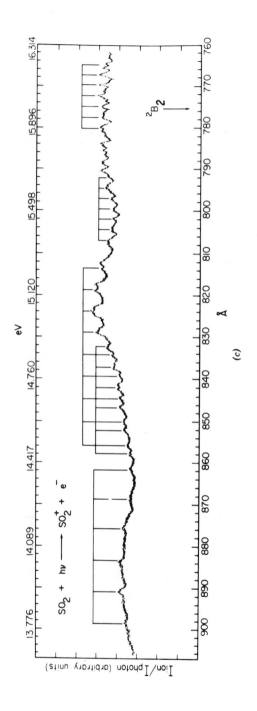

(c)

Weak features which arise by autoionization from different vibrational levels of Rydberg states are well resolved in the high-resolution (0.14 Å FWHM) PIE curves for SO_2^+ (Fig. 44b and c). Seven progressions of absorption bands, which were labeled as progressions I–VII, have been observed by Erickson et al. The average vibrational spacings for progressions I, II, III, IV, V, VI, and VII were found to be 911, 938, 956, 713, 386, and 428 cm^{-1}, respectively. Based on the average vibrational spacings determined for SO_2^+,[140, 192] $\Delta\nu$ for progressions I–V can probably be assigned to the symmetric stretching mode ν_1, and $\Delta\nu$ for progressions VI and VII can be assigned to the ν_2 bending mode of these Rydberg states.

Considerable structure was found in the PIE curve for the SO^+ fragment from SO_2 obtained by Dibeler and Liston[185] They also reported a weak but steplike structure at 784 Å (15.81 ± 0.02 eV), which was taken to be the AE of the reaction (3.29). Figure 45b shows the PIE curve for SO^+ obtained using a wavelength resolution of 1.4 Å (FWHM). Contrary to previous observations, the curve is smooth, with only a few weak autoionization features near the onset. A steplike structure was found at 775.5 instead of 784 Å. A high-resolution (0.14 Å FWHM) curve for SO^+ near the onset is shown in Fig.

TABLE XII
Summary of Photoionization Data for SO_2 and $(SO_2)_2$[32]

Ion	Threshold (eV)	Process	Thermodynamic properties[a] (kcal/mol)
SO_2^+	12.348 ± 0.002[a]	$SO_2 + h\nu \rightarrow SO_2^+ + e^-$	$\Delta H_{f0}^\circ(SO_2^+) = 214.43 \pm 0.05$
	12.30 ± 0.01[b]		
	12.34 ± 0.02[c]		
	12.42 ± 0.02[d]		
	12.32 ± 0.01[e]		
	12.34[f]		
SO^+	15.953 ± 0.010[a]	$SO_2 + h\nu \rightarrow SO^+ + O + e^-$	$\Delta H_{f0}^\circ(SO^+) = 238.6 \pm 0.2$
	15.93[g]		
S^+	16.228 ± 0.030[a]	$SO_2 + h\nu \rightarrow S^+ + O_2 + e^-$	$\Delta H_{f0}^\circ(S^+) = 303.9 \pm 0.7$
	16.334[g]		
$S_2O_3^+$	15.38 ± 0.06[a]	$(SO_2)_2 + h\nu \rightarrow S_2O_3^+ + O + e^-$	$\Delta H_{f0}^\circ(S_2O_3^+) = 154 \pm 1$
$(SO_2)_2^+$	11.72 ± 0.03[a]	$(SO_2)_2 + h\nu \rightarrow (SO_2)_2^+ + e^-$	$\Delta H_{f0}^\circ((SO_2)_2^+) = 129 \pm 1$

[a] Reference 32.
[b] Reference 140.
[c] Reference 184.
[d] Reference 183.
[e] Reference 185.
[f] Reference 187.
[g] Reference 194.

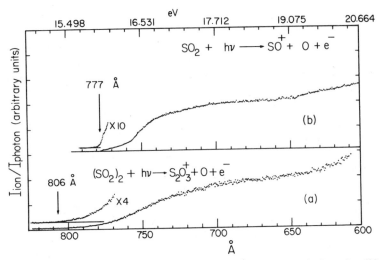

Fig. 45.   Photoionization efficiency curves for: (a) $S_2O_3^+$ from $(SO_2)_2$ in the region 600–806 Å; (b) SO^+ from SO_2 in the region 600–790 Å. Experimental conditions: $P_0 \simeq 450$ Torr, $T_0 \simeq$ 298 K, $D = 0.12$ mm, wavelength resolution 1.4 Å (FWHM).[32]

46. Weak autoionization structure observed in the SO^+ spectrum is similar to that resolved in the PIE curve for SO_2^+ in this region. The AE for the process (3.29) was determined to be 15.953 ± 0.010 eV (777.2 ± 0.5 Å). Using the known heat of formation of the oxygen atom (58.983 kcal/mol)[201] and $\Delta H_{f0}^{\circ}(SO_2^+)$ as well as the measured AE for (3.29), $\Delta H_{f0}^{\circ}(SO^+)$ was calculated to be 238.6 ± 0.2 kcal/mol (Table XII). Since $\Delta H_{f0}^{\circ}(SO^+)$ is the sum of $\Delta H_{f0}^{\circ}(SO)$ (1.6 ± 0.3 kcal/mol) and the IE of SO, the IE of SO was calculated to be 10.28 ± 0.02 eV, which is in excellent agreement with that (10.29 ± 0.01 eV) reported by Dyke et al.[196]

The onset of SO^+ formation is close to the thresholds of the $\tilde{C}^2B_2$, $\tilde{D}^2A_1$, and $\tilde{E}^2B_1$ states of SO_2^+. An earlier photoelectron–photon coincidence study[193] concluded that SO_2^+ formed in these states predissociates completely to $SO^+ + O$ in their ground states. However, later studies[194] showed that SO_2^+ prepared in this region only partially predissociates to SO^+ and S^+. The positions of some of the vibronic states[192] are shown in Fig. 46 in order to examine the correlation of these states with structures observed in the PIE curve for SO^+. Although the PIE curve is complicated by weak autoionization features near the onset, known thresholds of many vibronic states correlate well with steplike structures resolved in the curve. Relative values for the average increments in the PIE for SO^+ [$\Delta PIE(SO^+)$] of various vibronic states, after a correction for the contribution due to autoionization, are compared in Table XIII with the relative excitation probabilities

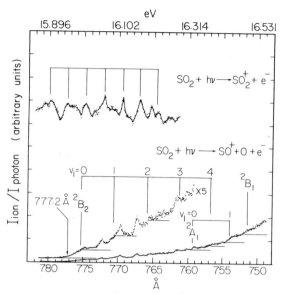

Fig. 46. Photoionization efficiency curve for SO^+ in the region 748–782 Å. Experimental conditions: $P_0 \simeq 450$ Torr, $T_0 \simeq 298$ K, $D = 0.12$ mm, wavelength resolution 0.14 Å (FWHM).[32]

TABLE XIII

Relative Fragmentation Cross Section for the

Vibrational Predissociation Process $SO_2{}^+ \xrightarrow{\sigma(SO^+)} SO^+ + O + e^-$[32]

Ionic state[a]	IE[a] (eV)	EP[b]	ΔPIE(SO^+)[c]	$\sigma(SO^+)$[d,e]
$\tilde{C}^2B_2(0,0,0)$	15.992	8	16	2.0
$\tilde{C}^2B_2(1,0,0)$	16.090	17	28	1.6
$\tilde{C}^2B_2(2,0,0)$	16.189	26	36	1.4
$\tilde{C}^2B_2(3,0,0)$	16.286	34	45	1.3
$\tilde{D}^2A_1(0,0,0)$	16.324	38	50	1.3
$\tilde{C}^2B_2(4,0,0)$	16.377	38	50	1.3
$\tilde{D}^2A_1(1,0,0)$	16.433	100	100	1.0
$\tilde{E}^2B_1(0,0,0)$	16.498	76	75	1.0

[a] Reference 192.

[b] Relative excitation probability (see text).

[c] Relative value for the average increment in the photoionization efficiency of SO^+.

[d] Relative fragmentation cross section for the formation of SO^+ from $SO_2{}^+$.

[e] $\sigma(SO^+)$ is estimated by the ratio ΔPIE(SO^+)/EP.

341

(EP) to vibronic states of SO_2^+ in this region. Values for the relative EP were estimated from the relative peak heights of corresponding vibrational bands in the photoelectron spectrum of SO_2.[192] Values for both the EP and $\Delta PIE(SO^+)$ have been normalized for the $\tilde{D}^2A_1(1,0,0)$ state. On this basis the general trend for values of $\Delta PIE(SO^+)$ was found to be similar to those of the EP. This observation can be taken as evidence that the SO^+ ions are formed predominantly by vibrational predissociation processes, $SO_2 + h\nu \rightarrow SO_2^+(\tilde{C}^2B_2, \tilde{D}^2A_1, \text{ or } \tilde{E}^2B_1(\nu_1, \nu_2, 0)) + e^- \rightarrow SO^+ + O + e^-$, in this wavelength region.

The observation of resolved vibrational structure in the third band in the photoelectron spectrum of SO_2 indicates that SO_2^+ formed in these excited vibronic states must have a lifetime longer than one vibrational period. If the formation of SO^+ is indeed dominated by vibrational predissociation from SO_2^+, relative reaction probabilities $\sigma(SO^+)$ for the process (3.29) can be estimated from the ratio $\Delta PIE(SO^+)/EP$. The relative values calculated for $\sigma(SO^+)$ of the $\tilde{C}^2B_2(0,0,0)$ and $\tilde{C}^2B_2(1,0,0)$ states were found to be greater than for other states. Above the $\tilde{C}^2B_2(2,0,0)$ state, $\sigma(SO^+)$ depends only slightly on internal energy.

The cross section for the photodissociative ionization process,

$$SO_2 + h\nu \rightarrow S^+ + O_2 + e^- \qquad (3.30)$$

is much smaller than that of (3.29). The ratio of the intensity of S^+ to that of SO^+ was found to be less than 0.02 in the region 584–760 Å. This observation can be considered as evidence that the coupling between the symmetric and the antisymmetric stretching vibrations is much stronger than that with the bending vibrational mode of SO_2^+. Because of the low S^+ signal, the PIE curve for S^+ was found to be contaminated by O_2^+ produced by background O_2. After the correction for the contribution from O_2, the PIE curve for S^+ is smooth (Fig. 47). Using the observed AE for the process (3.30), $\Delta H_{f0}^\circ(S^+)$ was calculated to be 303.9 ± 0.7 kcal/mol (Table XII).

Interestingly, broad autoionization structures observed in the PIE spectrum for SO_2^+ are hardly discernible in the PIE curve for $(SO_2)_2^+$ (Fig. 43b). Since Rydberg states converging to excited ionic states of SO_2^+ in this region are mostly vibrationally excited states, it is reasonable to believe that the lack of autoionization structure in the PIE curve for $(SO_2)_2^+$ is due to more efficient vibrational predissociation than associative ionization of vibrationally excited Rydberg dimers. The PIE for $(SO_2)_2^+$ was found to increase gradually from the onset and then level off in the Franck–Condon gap region between the first and the second bands. A further increase in the $(SO)_2^+$ spectrum occurs at the onset of the third band. This observation implies that the $(SO_2)_2^+$ ions are formed predominantly by direct photoioniza-

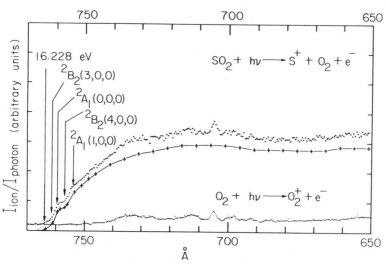

Fig. 47.   Photoionization-efficiency curve for S^+ from SO_2 in the region 650–770 Å. Experimental conditions: $P_0 \simeq 450$ Torr, $T_0 \simeq 298$ K, $D = 0.12$ mm, wavelength resolution 1.4 Å (FWHM). ●: PIE data for S^+ before correcting for the contribution of background O_2^+; ∙: PIE data for O_2^+; +: PIE data for S^+ from SO_2 after correcting for background O_2^+.[32]

tion processes. From the measured IE for $(SO_2)_2^+$, the binding energy for $SO_2^+ + SO_2$ was calculated to be 0.66 ± 0.04 eV (Table XII).

The relative intensities of the three bands in the photoelectron spectrum of SO_2 starting from the first ionization onset are $8:9:12$.[193] Since a PIE curve represents integrated data for a photoelectron spectrum, the relative intensities of these three bands can be estimated by the relative increments (ΔPIE) in PIE provided contributions to ΔPIE from autoionization can be corrected. As shown in Fig. 43a, the relative values for ΔPIE of the three bands, after correcting for contributions from autoionization and taking into account the fragmentation channels $SO^+ + O + e^-$ and $S^+ + O_2 + e^-$, were found to be $8:8:12.6$, in good agreement with previous measurements.[193] Similar measurements for the relative values for ΔPIE of the three bands in the PIE curve for $(SO_2)_2^+$, after including the contribution of the $S_2O_3^+ + O + e^-$ channel, gives $9:5:5$. There are many dissociative channels accessible to the $(SO_2)_2^+$ ion, such as

$$SO_2 \cdot SO_2 + h\nu \rightarrow SO_2^+(v) \cdot SO_2 + e^- \rightarrow SO_2^+(v') + SO_2 + e^- \qquad (3.31)$$

$$\rightarrow SO^+(v) \cdot SO_2 + O + e^- \rightarrow SO^+(v') + SO_2 + O + e^-$$

$$(3.32)$$

which cannot be measured in the photoionization experiment. Since the cross sections of these reactions are likely to increase with internal excitation, the relative values for ΔPIE of the second and the third bands obtained from the PIE curves, without taking into account contributions of these processes, are expected to be lower than that of the first band.

The similarity in the PIE curves observed for SO^+ and $SO^+ \cdot SO_2$, as shown in Fig. 45a and b, indicates that fragmentations of SO_2^+ and $SO_2^+ \cdot SO_2$ to form $SO^+ + O$ and $SO^+ \cdot SO_2 + O$, respectively, follow the same reaction pathway. Similar observations were found in the PIE curves for NO^+ and $NO^+ \cdot N_2O$ from N_2O^+ and $(N_2O)_2^+$, respectively.[129] From the measured AE for the formation of $S_2O_3^+$ from $(SO_2)_2$, the bond dissociation energy for $SO^+ \cdot SO_2$ was deduced to be 0.60 ± 0.07 eV.

5. H₂O Dimer

The PIE curves for $(H_2O)_2^+$, H_2O^+, and H_3O^+ from $(H_2O)_2$ in the region 950–1100 Å obtained by Ng et al.[16] with a wavelength resolution of 2.5 Å (FWHM) are shown in Fig. 48. The IE of $(H_2O)_2$ was determined to be 11.21 ± 0.09 eV, which is shifted 1.4 eV with respect to the IE of H_2O.[203] Using the IEs of $(H_2O)_2$ and H_2O as well as the binding energy between H_2O

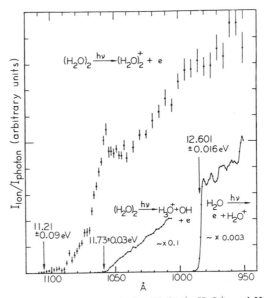

Fig. 48. Photoionization-efficiency curves for $(H_2O)_2^+$, H_3O^+, and H_2O^+ in the region 950–1120 Å, scaled for relative signals at 950 Å. Experimental conditions: wavelength resolution 2.5° (FWHM). The H_2O molecular beam was produced by seeding water vapor at 89°C in 150 Torr of Ar and then expanding through a 0.15-mm-diameter Pyrex nozzle.[16]

TABLE XIV
Proton Affinity of H_2O[16]

Value (kcal/mol)	Temperature (K)	Reference
165.8 ± 1.8	0	16
169 ± (1 or 2)	600	204
166 ± 2	—	205
165 ± 3	323–373	206
168.2 ± 3.4	340	207
164 ± 4	—	208
167.5	—	209

and H_2O, the bond dissociation energy for $H_2O^+ \cdot H_2O$ was deduced to be 1.58 ± 0.13 eV.

The PIE for $(H_2O)_2$ increases gradually above the ionization onset. At 1055 Å a very sharp break in its general trend occurs which coincides with the AE (11.73 ± 0.03 eV) of H_3O^+ formation. This suggests that H_3O^+ is formed by the following processes:

$$H_2O \cdot H_2O + h\nu \rightarrow (H_2O^+)^{\ddagger} \cdot H_2O + e^- \rightarrow H_3O^+ + OH + e^-, \quad (3.33)$$

and 0.52 ± 0.12 eV of internal excitation of the dimer ion is necessary for the formation of H_3O^+. The second product was assumed to be OH instead of OH^-. This assumption stemmed from the fact that H_2O does not dissociate via an ion-pair channel in this energy range. The measured AE for the process (3.33) makes possible the calculation of the proton affinity of water at 0 K. The value (165.8 ± 1.8 kcal/mol) for the proton affinity of water determined by Ng et al. is compared with values obtained by other methods[204-209] in Table XIV.

D. Polyatomic Molecules

1. Acetone, Acetone Dimer, and Acetone Clusters

For polyatomic molecules, there are many low-frequency vibrational modes which are populated significantly at room temperature. It is interesting to explore the effect of partially relaxed vibrational hot bands on high-resolution studies and the nozzle expansion conditions with which considerable relaxation of these vibrational excitations can be achieved.

Acetone and acetone-d_6 are the most complex molecules which have been studied today by the molecular-beam photoionization method.[18] The PIE curves for acetone and acetone-d_6 obtained using a total nozzle stagnation pressure in the range 350–750 Torr at room temperature show sharp and distinct steplike structures near the ionization onsets. The sharpness of these steps was found to be limited by the optical bandwidth (1.5 Å FWHM) used in the photoionization study, indicating that the rotational temperature of the sample is $\lesssim 100$ K. Similarly, vibrational temperatures are apparently lowered to such an extent that significant population of states occurs only within a narrow energy band above the ground state. However, since the methyl-group torsional frequencies, $\nu_{12} = 105.3$ cm^{-1} and $\nu_{24} = 108.4$ cm^{-1},[210,211] are comparable to the resolution used in the acetone experiment, the contributions to the initial thresholds by these low-frequency modes cannot be assessed. The vibrational spacings resolved in the PIE curves for acetone and acetone-d_6 are listed in Table XV. Trott et al. have not given specific assignments to these transitions. Based on the vibrational selection rules and the comparison of these spacings with known vibrational frequencies in acetone and acetone-d_6, it is likely that the onsets of 320 and 260 cm^{-1} correspond to the A_1 symmetric C—C—C deformation mode (ν_8) in acetone and acetone-d_6, respectively. The onset of 695 cm^{-1} might correspond to the symmetric $C—C$ stretch (ν_7) or $2\nu_8$. A more definite assignment of these structures will require a higher-resolution photoionization study in the future.

   Table XV summarizes the energetic information for the acetone and acetone-d_6 dimer and cluster ions. Trott et al. found that the measured IEs for $(CH_3COCH_3)n$ fall squarely on a straight line when plotted as a function of

TABLE XV
Energetic Information for Acetone and Acetone-d_6
Dimer and Cluster Ions

Ion	Ionization or appearance potential[a] (eV)
$(CH_3COCH_3)_2{}^+$	9.26 ± 0.03
$(CD_3COCD_3)_2{}^+$	9.25 ± 0.03
$(CH_3COCH_3)_3{}^+$	9.10 ± 0.03
$(CH_3COCH_3)_4{}^+$	9.02 ± 0.03
$(CH_3COCH_3) \cdot CH_3CO^+$	10.08 ± 0.05

Ion	Vibrational frequency (cm^{-1})
$CH_3COCH_3{}^+$	320 ± 50
	695 ± 50
$CD_3COCH_3{}^+$	260 ± 50

[a]Reference 18.

$1/n$. This observation is consistent with a simple independent-systems model[115] for the cluster-ion system. Assuming the IEs for $(CH_3COCH_3)_n$, $n > 4$, continue to decrease linearly as a function of $1/n$, a value of 8.8 eV was deduced for the bulk IE of acetone. In view of the uncertainty as to whether the measured cluster-ion thresholds correspond to adiabatic transition, the latter value is probably only an upper bound.

2. Ammonia Dimer and Clusters

Due to the rapid exothermic reaction $NH_3^+ + NH_3 \rightarrow NH_4^+ + NH_2$, the $(NH_3^+) \cdot NH_3$ intermediate complex has not been observed previously. In a molecular-beam photoionization study of $(NH_3)_2$, Ceyer et al.[19] have been able to produce this complex by the direct photoionization of the ammonia dimer. The IE of $(NH_3)_2$ was determined to be 9.54 ± 0.05 eV. Using this value, the binding energy (3.5 kcal/mol)[212] between NH_3 and NH_3, and the IE (10.162 ± 0.008 eV)[213] of NH_3, a value of 0.79 ± 0.05 eV (18.1 ± 1 kcal/mol) was deduced for the dissociation energy of $(NH_3^+) \cdot NH_3$.

When the internal excitation of the ammonia dimer ion is increased to 0.052 eV, the NH_4^+ fragment is observed. The AE (9.59 ± 0.02 eV)[19] of NH_4^+ formation from $(NH_3)_2$, the IE of H, and the NH_2-H bond energy,[214] and the well depth of $(NH_3)_2$ were employed to calculate the proton affinity of NH_3 as 8.76 ± 0.06 eV (202.1 ± 1.3 kcal/mol) at 0 K. Assuming that only rotational degrees of freedom are excited at 298 K, a value of 203.6 ± 1.3 kcal/mol was obtained for the proton affinity of NH_3 at 298 K. This value was found to be in good agreement with a recent measurement of 202.3 kcal/mol obtained in ion-cyclotron-resonance thermal-equilibrium experiments at 298 K.[215] The previously accepted experimental value for the proton affinity of NH_3 was 207 ± 3 kcal/mol.[216,217]

The AEs of the $(NH_4^+) \cdot NH_3$ and $(NH_4^+) \cdot (NH_3)_2$ fragment ions from $(NH_3)_3^+$ and $(NH_3)_4^+$, respectively, have also been measured by Ceyer et al. These measurements allow the calculation of the solvation energy of an ammonium ion by one or two ammonia molecules. These values obtained by Ceyer et al. are listed in Table XVI and compared with literature values[218–221] for the solvation energies. As shown in the table, the agreement between the photoionization results and the literature value is poor. This comparison suggests that the AEs for $(NH_4^+) \cdot NH_3$ and $(NH_4^+) \cdot (NH_3)_2$ by photoionization might be upper bounds of the true AEs.

3. Acetylene

There has been some controversy concerning the thermochemistry of the decomposition process $C_2H_2^+ \rightarrow C_2H^+ + H$. This not only introduces an ambiguity into the interpretation of the kinetic behavior, but also gives rise

TABLE XVI
Energetics of Proton Solvation[a] by NH_3[19]

$n, n-1$	Molecular-beam photoionization[b] (ΔH_0°, kcal/mol)	Other techniques (ΔH_{298}°, kcal/mol)
1,0	13.8 ± 1.4	18.4[c]
		21.5[d]
		24.8[e]
		27.0[f]
2,1	6.4 ± 1.6	16.2[d]
		17.5[e]
		17.0[f]

[a]$(NH_4^+)\cdot(NH_3)_n \rightleftarrows (NH_4^+)\cdot(NH_3)_{n-1} + NH_3$.
[b]Reference 19.
[c]Reference 218.
[d]Reference 219.
[e]Reference 220.
[f]Reference 221.

to inconsistent values of $\Delta H_f^\circ(C_2H)$, $\Delta H_f^\circ(C_2H^+)$, and the IE of C_2H obtained by different experimental methods.[§] In an early photoionization study, Botter et al.[222] reported an AE of 17.22 eV for the photodissociative ionization process

$$C_2H_2 + h\nu \rightarrow C_2H^+ + H + e^- \qquad (3.34)$$

In a later study, Dibeler et al.[223] reported a value of 17.36 ± 0.01 eV for the AE of the process (3.34) at 0 K. Using this value, and the well-established heats of formation of C_2H_2 (54.33 kcal/mol) and H (51.63 kcal/mol), a value of 402.8 ± 0.2 kcal/mol was deduced for $\Delta H_f^\circ(C_2H^+)$. A direct measurement from a study of the high-temperature reaction of graphite with various hydrocarbons yields $\Delta H_{f298}^\circ(C_2H) = 130 \pm 3$ kcal/mol.[224] The IE for C_2H has been determined by electron impact to be 11.6 ± 0.5 eV.[224] From the measured photodissociative ionization threshold for $C_2H^+ + CN + e^-$ and the photodissociation threshold for $C_2H + CN$ from C_2HCN, Okabe and Dibeler[225] reported an estimate of 11.96 ± 0.05 eV for $IE(C_2H)$. However, in another indirect measurement by Miller and Berkowitz[226] based on the difference in photodissociative ionization thresholds for the production of C_2H^+ and Br^+, a value of 11.51 eV was obtained for $IE(C_2H)$. It is argued by Berkowitz[§] that since the latter value depends only on processes involved

[§]See Reference 33, pp. 285–290.

in simple bond ruptures and the values obtained are in good agreement with the direct electronic-impact result (but on the low-energy side), 11.51 eV is a better estimate for $IE(C_2H)$. Using $IE(C_2H) = 11.51$ eV and the AE for (3.34) as reported by Dibeler et al., the best previous estimate for $\Delta H_{f0}(C_2H)$ derived by the photoionization method is 137.6 kcal/mol, which is $\simeq 8$ kcal higher than the value obtained by Wyatt et al. after correcting to 0 K. A plausible explanation for this descrepancy is that the process (3.34) required an excess energy of approximately 0.5 eV.

The PIE curve for C_2H^+ formation[222,223] from C_2H_2 obtained previously was found to decrease dramatically at $\simeq 714$ Å. Dibeler et al.[223] attributed the finite ion yield and the tailing structure below 714 Å to hot-band effects. They obtained the AE for (3.34) at 0 K by extrapolating the rapidly rising edge of the PIE curve for C_2H^+. Recently, Ono et al. have reexamined the threshold region of the process (3.34) using the molecular-beam photoionization method. The PIE curve for C_2H^+ in the wavelength region 690–750 Å obtained by Ono et al. is shown in Fig. 49. A slight step at 738.5 ± 1.4 Å (16.79 ± 0.3 eV), which is 0.57 eV lower in energy than the AE for C_2H^+ reported by Dibeler et al., can be seen in the figure. Since one would expect appreciable rotational and low-frequency vibrational relaxations in a supersonic expansion, this slight structure is unlikely to be due to hot bands. The absence of this structure in the previous study is partly due to the hot-band effects. Since this steplike structure is very weak, the C_2H^+ ions formed by collisional dissociation of $C_2H_2^+$ in the gas cell can also smear out this feature. If the position of this step at 738.5 Å is taken to be the AE for the process (3.34), it is found to be in excellent agreement with the value (16.72 ± 0.12

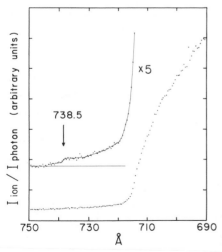

Fig. 49. Photoionization-efficiency curve for C_2H^+ from C_2H_2 in the region 690–750 Å. Experimental conditions: $P_0 \simeq 1000$ Torr, $T_0 \simeq 298$ K, $D = 0.125$ mm, wavelength resolution 1.4 Å.[31]

eV) obtained by Maier[227] using the rare-gas ion-impact method. Using this new value (16.79 ± 0.03 eV) as the AE for the process (3.34), $\Delta H^\circ_{f0}(C_2H^+)$ was calculated to be 389.9 ± 0.8 kcal/mol. From the value $\Delta H^\circ_{f0}(C_2H) = 129 \pm 3$ kcal/mol,[224] a value of 11.31 ± 0.13 eV was deduced for IE(C_2H), which is more consistent with the estimate obtained by Miller and Berkowitz. If IE(C_2H) is taken as 11.51 eV, then $\Delta H^\circ_{f0}(C_2H)$ becomes 124.4 kcal/mol, which is consistent with the previous literature value for $\Delta H^\circ_{f0}(C_2H)$ within the combined error estimate of both experiments.

4. Ethylene Dimer and Trimer

The ion–molecule reaction of $C_2H_4^+ + C_2H_4$ has been studied extensively in the past.[228–242] The major product ions observed for this reaction are $C_3H_5^+$ and $C_4H_7^+$. The results of the previous investigations support the conclusion that the reaction proceeds through some long-lived complexes, $[C_4H_8^+]^\ddagger$,[229,234,243] that is,

$$C_2H_4^+ + C_2H_4 \rightleftharpoons [C_4H_8^+]^\ddagger \rightarrow C_3H_5^+ + CH_3 \qquad (3.35)$$

$$\rightarrow C_4H_7^+ + H \qquad (3.36)$$

$$\xrightarrow{M} C_4H_8^+ \qquad (3.37)$$

The $[C_4H_8^+]^\ddagger$ complex can be stabilized by collisions in the gas cell [reaction (3.37)]. Since both reactions (3.35) and (3.36) are exothermic, previous studies have yielded no information about the threshold behavior of these processes. It is highly possible that the loosely bound dimer ion $C_2H_4^+ \cdot C_2H_4$ is the precursor to the complex $[C_4H_8^+]^\ddagger$ provided there is a potential-energy barrier between the two complexes. The thermochemical thresholds for both reactions (3.35) and (3.36) are known. If the dissociation energy for the $C_2H_4^+ \cdot C_2H_4$ loose complex is greater than the exothermicities of the latter reactions, the AEs for these processes can be measured in a study of the unimolecular decomposition of $C_2H_4^+ \cdot C_2H_4$ to form the $C_3H_5^+$ and $C_4H_7^+$ ions as a function of internal-energy content of the $C_2H_4^+ \cdot C_2H_4$ complex. From the measured AE, the highest potential-energy barrier along the reaction coordinate can be determined.

Ceyer et al.[20] have performed such a study of this system. In their study, the stabilized collision complex was prepared by the direct photoionization of ethylene dimer, and the internal energy of the complex was controlled by dispersed VUV photons. The AE for the reaction (3.36) measured by Ceyer et al. was found to be consistent with the accepted thermochemical threshold. However, the AE for (3.35) was found to be $\simeq 7.7$ kcal/mol greater than the accepted thermochemical threshold.[244,245] This was interpreted to imply a potential-energy barrier of this magnitude in the exit channel for the formation of $C_3H_5^+ + CH_3$. This interpretation is in conflict with the translational-energy distribution for $C_3H_5^+$ measured by Kemper et al.[246,247] Ceyer et al.[20] have also observed $C_5H_9^+$ and $C_6H_{11}^+$ and supposed these ions to be the fragment ions from $(C_2H_4)_3^+$.

Recently, this system has been reinvestigated by Ono and Ng.[248] They found that the AEs of these fragments are dependent upon the nozzle stagnation pressure. At high nozzle stagnation pressure, where the concentrations of $(C_2H_4)_3$ and heavier clusters are high, the thresholds of all these fragment ions were found to shift to lower energy, indicating that ethylene trimers and heavier clusters can give rise to the same product ions as the ethylene dimers. The AEs for $(C_2H_4)_2^+$, $C_3H_5^+$, and $C_4H_7^+$ measured at a low nozzle stagnation pressure, where the concentrations of $(C_2H_4)_3$ and heavier clusters are negligible, are 9.84 ± 0.04 eV (1260 ± 5 Å), 10.21 ± 0.04 eV (1214 ± 5 Å), and 10.05 ± 0.04 eV (1234 ± 5 Å), respectively. The values for the AEs of $(C_2H_4)_2^+$ and $C_4H_7^+$ from $(C_2H_4)_2$ obtained by Ono and Ng are higher than that of Ceyer et al., where the concentration of $(C_2H_4)_3$, as measured by the intensity of $(C_2H_4)_3^+$, is approximately 30% that of $(C_2H_4)_2$.

The new AEs for $C_4H_7^+$ and $C_3H_5^+$ of Ono et al. are still higher than the thermodynamic thresholds for the reactions (3.35) and (3.36) by approximately 3 kcal/mol. These differences are nevertheless within the combined error estimate of the thermochemical thresholds of the reactions (3.35) and (3.36) and the AEs of Ono et al. From the new IE of $(C_2H_4)_2$, a value of 15.5 ± 1 kcal/mol was deduced for the binding energy between $C_2H_4^+$ and C_2H_4. At high nozzle stagnation pressures where the concentrations of heavier ethylene clusters are appreciable, the AEs for $C_4H_7^+$ and $C_3H_5^+$ were found to be lower than the thermochemical thresholds of (3.35) and (3.36).

It is surprising that the AEs of the fragment ions from a heavier cluster such as $(C_2H_4)_3$ are lower than those of the fragment ions from $(C_2H_4)_2$. For example, if the mass-56 ion produced by the photodissociative ionization process has the structure of a loose complex, that is,

$$(C_2H_4)_3 + h\nu \rightarrow C_2H_4^+ \cdot C_2H_4 + C_2H_4 + e \qquad (3.38)$$

then the AE of the process (3.38) is expected to be higher than the threshold of the process

$$(C_2H_4)_2 + h\nu \rightarrow C_2H_4^+ \cdot C_2H_4 + e \qquad (3.39)$$

This expectation is in accordance with previous findings concerning the threshold measurements of the clusters of simple inorganic molecules such as Ar_2, Kr_2, $(H_2O)_2$, and $(CS_2)_{n=2-5}$. A logical explanation for the fact that the AEs observed for the mass-56 ions originating from $(C_2H_4)_3$ or heavier ethylene clusters are lower than the IE of $(C_2H_4)_2$ is that these mass-56 ions are stable $C_4H_8^+$ molecular ions having tight structures. Since a stable $C_4H_8^+$ ion is much more stable than a loose $C_2H_4^+ \cdot CH_4$ ion formed by the process (3.39), the IE of $(C_2H_4)_2$ is thus higher than the AE of processes such as $(C_2H_4)_3 + h\nu \rightarrow C_4H_8^+ + C_2H_4 + e^-$. This interpretation is consistent with the picture that the loose complexes such as $(C_2H_4)_3^+$, which are formed by photoionization of the trimer, rearranges to some tight complexes $[C_6H_{12}^+]^\ddagger$ before fragmenting. Thus

$$(C_2H_4)_3 + h\nu \rightarrow (C_2H_4)_3^+ + e^- \rightarrow [C_6H_{12}^+]^\ddagger \rightarrow C_4H_8^+ + C_2H_4$$

$$(3.40)$$

$$\rightarrow C_4H_7^+ + C_2H_5 \qquad (3.41)$$

$$\rightarrow C_3H_5^+ + C_3H_7 \qquad (3.42)$$

and so on. Based on this scheme, the AEs for $C_4H_7^+$ and $C_3H_5^+$ from $(C_2H_4)_3$ should be lower than those from $(C_2H_4)_2$.

IV. CONCLUSION AND FUTURE DEVELOPMENTS

Most of the experimental works described in this chapter are beyond the scope of conventional gas-cell or effusive-beam photoionization mass-spectrometric experiments. The new experimental areas which have been made possible by the combination of the supersonic molecular beam method and photoionization mass spectrometry are summarized as follows:

A. High-Resolution Photoionization Studies of Molecules and Dimers

In the high-resolution photoionization experiments on NO, CS_2, OCS, and SO_2, the resolution attained is approximately 1.5 meV. Judging from the sharpness of the first onset in the PIE curve for CS_2, the resolution was actually limited by the rotational temperature ($\simeq 25$ K) instead of the spectral bandwidth ($\simeq 10$ K). The rotational population of these and other sample gases can be further reduced by the seeded-helium supersonic-beam technique. Since monomers are the major constituent in the molecular beam, the PIE spectrum for a monomer ion is expected to be less susceptible to the influence of secondary collisional processes. Therefore, the sensitivity of ion detection can be improved by relaxing the vacuum requirements and increasing the number density of the sample gas in the ionization region. A simple way of accomplishing this is to shorten the distance between the nozzle and the sampling region. With proper optimization of the molecular-beam intensity and a given pumping capacity, it is possible to resolve rotational autoionization structures in the PIE spectra for molecular ions such as NO. The higher photon intensity promised by synchrotron-radiation sources and VUV lasers assures very high-resolution photoionization studies of rotationally cooled molecules in the near future.

Due to rapid predissociation processes in excited molecular Rydberg dimers, autoionization structures resolved in most of the previous PIE curves for molecular dimer and cluster ions are much broader than those observed in the PIE spectra for the monomer ions. Higher-resolution studies of these species probably will not provide any more information than those obtained from low-resolution studies. Nevertheless, the high-resolution PIE curves for atomic rare-gas dimer ions do reveal rich autoionizing features. This may be partly due to a more uniform dimer geometry for atomic than for molecular dimers. The important class of dimers which are likely to reveal rich structure in their PIE curves are metal dimers. At the present time, there is little information available on the electronic structure of this class of dimer ions. Since the binding energies of metal dimers are usually much larger than those of the rare-gas dimers, structures resolved in the high-resolution PIE curves for the metal dimer ions will provide information not only on electronic structures of the dimer ions but also on the bonding of neutral dimers.

One other fruitful area which has not been explored is high-resolution photoionization studies of radicals. Very often, radical species are produced by thermal dissociation or in electrical discharge with substantial populations in the rotational and vibrational degrees of freedom. The supersonic-beam technique should help to reduce these excitations and enhance the possibility of high-resolution studies.

B. Thermochemistry

The accuracy of thermochemical data obtained in a photoionization experiment is directly tied to the resolution achieved in the study. As demonstrated by many previous high-resolution photoionization experiments, the sharp onset observed in the PIE curves for a molecular ion provided more accurate thermochemical data, such as the heat of formation of the molecular ion, than those determined previously in conventional photoionization studies.

The photoionization of dimers and clusters synthesized by supersonic expansion has made possible the determination of the stabilities of many dimer, cluster, and radical ions which cannot be measured by other methods.

With further improvement in sensitivity, the molecular-beam photoionization mass-spectrometric method is likely to become the most important and accurate source of thermochemical data for gaseous ions.

C. Molecular Dynamics

The ability to determine the highest potential-energy barriers along the reaction coordinates of bimolecular ion–molecule reactions, the stabilities of collision complexes, and the reactivities of vibrational and electronic excited states of molecules and molecular ions has made the molecular-beam photoionization mass-spectrometric method a valuable technique in molecular dynamics.

In order to study the high-Rydberg-state reactions of an atom or a molecule, it is necessary to resolve the Rydberg series of an atom or molecule in the PIE curves for the dimer ion or the fragment ions from the dimer. Furthermore, the relative absorption cross sections of the Rydberg levels must be known. Since many atoms, such as the rare-gas ones, satisfy the latter requirement, it is most likely that the Rydberg-state chemistry of these atoms can be examined by this method.

The charge-transfer method[249] has been one of the most commonly used techniques for the study of the reactivity of electronic-excited molecular ions. This method relies on the measurements of the rates for subsequent reactions in the gas cell after the charge transfer. Therefore, if the radiative lifetime of the ionic state is short, experimental results will be affected by the partial decay by radiation of the reactant ions. As discussed previously, the short radiative lifetimes of excited atomic and molecular ions are not expected to be a problem in the molecular-beam photoionization method. In addition, the internal excitation of the molecular ions can be varied almost continuously in photoionization, whereas in charge transfer, the internal excitation of the molecular ions depends on the available incident ions and thus only discrete internal excitation of the molecular ions can be changed. According to the above discussion, the molecular-beam photoionization of van der Waals dimer is a better method than the charge-exchange technique for the investigation of the reactivity of internally excited molecular ions.

As the excitation energy of a dimer or cluster ion decreases towards the threshold for a fragmentation process, the lifetime for the unimolecular decomposition of the dimer or cluster ion will increase. For large polyatomic molecular dimers or clusters when the lifetimes of these complexes at the thresholds exceed the flight time from the ionization region to the entrance of the mass spectrometer ($\sim 10^{-5}$ sec), the fragment ions from these ion complexes will not be detected. The AEs for the fragment ions form a dimer or cluster, and thus the potential-energy barrier for bimolecular ion–molecule reactions measured by the molecular-beam photoionization method should be an upper bound. That the AEs for $(NH_4^+) \cdot NH_3$ and $(NH_4^+) \cdot (NH_3)_2$ obtained by photoionization are lower than those from equilibrium data might be a result of this effect.

One of the ambiguities in probing the ionization and fragmentation processes of dimers or clusters by this method is the difficulty in eliminating the contributions from heavier clusters. Very often, the clusters are produced in a broad distribution of sizes, making it impractical to select certain cluster sizes by just changing the nozzle expansion conditions. Part of the difficulty can be overcome by the photoelectron–photoion coincidence technique. A study of the photoelectron spectrum of Xe_3 by this method has been reported recently by Poliakoff et al.[250] With substantial improvement in the sensitivity of this method in the future, it will become an important method for the study of the state-selected unimolecular decompositions of dimer or cluster ions.

References

1. W. Frost, *Theory of Unimolecular Reactions*, Academic, New York, 1973.

2. P. J. Robinson and K. A. Holbrook, *Unimolecular Reactions*, Wiley-Interscience, New York, 1972.

3. P. W. Dehmer and W. A. Chupka, *J. Chem. Phys.* **65**, 2243 (1976).

4. M. G. Liverman, S. M. Beck, and R. E. Smalley, *J. Chem. Phys.* **70**, 192 (1979).

5. R. E. Smalley, L. Wharton, and D. H. Levy, *Acc. Chem. Res.* **10**, 139 (1977).

6. D. H. Levy, L. Wharton, and R. E. Smalley, in C. B. Moore, Ed., *Chemical and Biochemical Applications of Lasers*, Vol. 2, Academic, New York, 1977, p. 1.

7. L. Wharton, D. Auerbach, D. H. Levy, and R. E. Smalley, in A. H. Zewail, Ed., *Advances in Laser Chemistry*, Springer, New York, 1978, p. 408.

8. D. H. Levy, in B. S. Robinovitch, J. M. Schurr, and H. L. Strauss, Eds., *Annual Review of Physical Chemistry*, Vol. 31, Annual Review Inc., Palo Alto, Calif., 1980, p. 197.

9. G. R. Parr and J. W. Taylor, *Rev. Sci. Instrum* **44**, 1578 (1973).

10. G. R. Parr and J. W. Taylor, *Int. J. Mass Spectrom. Ion Phys.* **14**, 467 (1974).

11. C. Y. Ng, B. H. Mahan, and Y. T. Lee, *J. Chem. Phys.* **65**, 1965 (1976).

12. C. Y. Ng, D. J. Trevor, B. H. Mahan, and Y. T. Lee, *J. Chem. Phys.* **65**, 4327 (1976).

13. C. Y. Ng, D. J. Trevor, B. H. Mahan, and Y. T. Lee, *J. Chem. Phys.* **66**, 446 (1977).

14. C. Y. Ng, P. W. Tiedemann, B. H. Mahan, and Y. T. Lee, *J. Chem. Phys.* **66**, 3985 (1977).

15. C. Y. Ng, P. W. Tiedemann, B. H. Mahan, and Y. T. Lee, *J. Chem. Phys.* **66**, 5737 (1977).

16. C. Y. Ng, D. J. Trevor, P. W. Tiedemann, S. T. Ceyer, P. L. Kronebush, B. H. Mahan, and Y. T. Lee, *J. Chem. Phys.* **67**, 4235 (1977).

17. G. G. Jones and J. W. Taylor, *J. Chem. Phys.* **68**, 1768 (1978).

18. W. M. Trott, N. C. Blais, and E. A. Walter, *J. Chem. Phys.* **69**, 3150 (1978).

19. S. T. Ceyer, P. W. Tiedemann, B. H. Mahan, and Y. T. Lee, *J. Chem. Phys.* **70**, 14 (1979).

20. S. T. Ceyer, P. W. Tiedemann, C. Y. Ng, B. H. Mahan, and Y. T. Lee, *J. Chem. Phys.* **70**, 2138 (1979).

21. P. W. Tiedemann, S. L. Anderson, S. T. Ceyer, T. Hirooka, C. Y. Ng, B. H. Mahan, and Y. T. Lee, *J. Chem. Phys.* **71**, 605 (1979).

22. W. M. Trott, N. C. Blais, and E. A. Walters, *J. Chem. Phys.* **71**, 1692 (1979).

23. M. E. Gress, S. H. Linn, Y. Ono, H. F. Prest, and C. Y. Ng, *J. Chem. Phys.* **72**, 4242 (1980).

24. Y. Ono, S. H. Linn, H. F. Prest, M. E. Gress, and C. Y. Ng, *J. Chem. Phys.* **73**, 2523 (1980).

25. Y. Ono, S. H. Linn, H. F. Prest, C. Y. Ng, and E. Miescher, *J. Chem. Phys.* **73**, 4855 (1980).

26. S. L. Anderson, T. Hirooka, P. W. Tiedemann, B. H. Mahan, and Y. T. Lee, *J. Chem. Phys.* **73**, 4779 (1980).

27. Y. Ono, S. H. Linn, H. F. Prest, M. E. Gress, and C. Y. Ng, *J. Chem. Phys.* **74**, 1125 (1981).

28. Y. Ono, E. A. Osuch, and C. Y. Ng, *J. Chem. Phys.* **74**, 1645 (1981).

29. S. H. Linn, Y. Ono, and C. Y. Ng, *J. Chem. Phys.* **74**, 3348 (1981).

30. S. H. Linn, Y. Ono, and C. Y. Ng, *J. Chem. Phys.* **74**, 3342 (1981).

31. Y. Ono and C. Y. Ng, *J. Chem. Phys.* **74**, 6985 (1981).

32. J. Erickson and C. Y. Ng, *J. Chem. Phys.* **75**, 1650 (1981).

33. J. Berkowitz, *Photoabsorption, Photoionization, and Photoelectron Spectroscopy*, Academic, New York, 1979.

34. W. A. Chupka, in C. Sandorf, P. J. Ausloos, and M. B. Robin, Eds., *Chemical Spectroscopy and Photochemistry in the Vacuum-Ultraviolet*, NATO–Advanced Study Institutes Series C, Vol. 8, Reidel, Boston, 1973, p. 433; J. Berkowitz, *ibid.*, pp. 75, 93.

35. W. A. Chupka, in J. L. Franklin, Ed., *Ion – Molecule Reaction*, Vol. 1, Plenum, New York, 1972, p. 33.

36. G. V. Marr, *Photoionization Process in Gases*, Academic, New York, 1967.

37. J. A. R. Samson, *Techniques of Vacuum Ultraviolet Spectroscopy*, Wiley, New York, 1967.

38. J. J. Hopfield, *Phys. Rev.* **35**, 1133 (1930).

39. J. J. Hopfield, *Phys. Rev.* **36**, 784 (1930).

40. J. J. Hopfield, *Astrophysics* **72**, 133 (1930).

41. Y. Tanaka, *J. Opt. Soc. Am.* **45**, 710 (1955).

42. R. E. Huffman, J. C. Larrabee, L. Tanaka, *J. Opt. Soc. Am.* **52**, 101 (1965).

43. R. E. Huffman, J. C. Larrabee, and D. Chambers, *Appl. Opt.* **4**, 1145 (1965).

44. Circuit developed by J. R. Haumann, Argonne National Laboratory.

45. W. A. Chupka, P. W. Dehmer, and W. J. Jivery, *J. Chem. Phys.* **63**, 3929 (1975).

46. C. Y. Ng, Ph.D. Thesis, 1976, University of California, Berkeley, Calif.

47. Velonix Model V-2403.

48. C. Kunz, Ed., *Synchrotron Radiation: Techniques and Application*, Springer, New York, 1979.

49. K. Radler and J. Berkowitz, *J. Opt. Soc. Am.* **68**, 1181 (1978).

50. M. Seya, *Sci. Light* **2**, 8 (1952).

51. T. Namioka, *Sci. Light* **3**, 15 (1954).

52. T. Namioka, *J. Opt. Soc. Am.* **49**, 951 (1959).

53. B. Vodar, *Rev. Opt.* **21**, 97 (1942).

54. S. Robin, *J. Phys. Radium* **14**, 551 (1953).

55. R. Tousey, Thesis, 1933, Harvard University.

56. McPherson 1 M (Model 225) and 3 M (Model 2253) normal-incidence monochromator.

57. J. A. R. Samson, *J. Opt. Soc. Am.* **54**, 6 (1964).

58. J. A. R. Samson, unpublished data (1965); see p. 215 of Reference 37.

59. E. C. Bruner, Jr., Thesis, 1964 University of Colorado.

60. C. F. Giese, *Rev. Sci. Instrum.* **30**, 260 (1959).

61. A. Kantrowitz and J. Grey, *Rev. Sci. Instrum.* **22**, 328 (1951).

62. G. B. Kistiakowsky and W. P. Slichter, *Rev. Sci. Instrum.* **22**, 333 (1951).

63. E. W. Becker and K. Bier, *Z. Naturforsch.* **9A**, 975 (1954).

64. H. Pauly and J. P. Toennies, in B. Bederson and W. L. Fite, Eds., *Methods of Experimental Physics*, Vol. 7A, Academic, New York, 1968, p. 227.

65. J. B. Anderson, R. P. Andres, and J. B. Fenn, in J. Ross, Ed., *Advances in Chemical Physics*, Vol. 10, Wiley, New York, 1966, p. 275.

66. J. B. Anderson, in P. P. Wegener, Ed., *Molecular Beams and Low Density Gasdynamics*, Marcel Dekker, New York, 1974, p. 1; O. F. Hagena, *ibid.* p. 93.

67. H. W. Liepmann and A. Roshko, *Elements of Gas Dynamics*, Wiley, New York, 1957.

68. H. Ashkenhas and F. S. Sherman, in J. H. de Leeuw, Ed., *Rarefied Gas Dynamics, 4th Symposium*, Vol. 2, Academic, New York, 1966, p. 84.

69. J. B. Anderson and J. B. Fenn, *Phys. Fluids* **8**, 780 (1965).

70. E. W. Becker, K. Bier, and W. Henkes, *Z. Phys.* **146**, 333 (1956).

71. R. E. Leckenby, E. J. Robbins, and P. Trevalion, *Proc. R. Soc. London Ser. A* **280**, 409 (1964).

72. Y. Ono and C. Y. Ng, unpublished data, 1979.

73. A. van Deursen and J. Reuss, *Int. J. Mass Spectrom. Ion Phys.* **23**, 109 (1977).

74. P. M. Dehmer and E. D. Poliakoff, *Chem. Phys. Lett.* **77**, 326 (1981).

75. J. A. Baker, R. O. Watts, J. K. Lee, T. P. Schafer, and Y. T. Lee, *J. Chem. Phys.* **61**, 308 (1974).

76. G. C. Maitland and E. B. Smith, *Chem. Phys. Lett.* **22**, 443 (1973).

77. J. M. Parson, P. E. Siska, and Y. T. Lee, *J. Chem. Phys.* **56**, 1511 (1972).

78. J. A. Barker, R. A. Fisher, and R. O. Watts, *Mol. Phys.* **21**, 657 (1971).

79. G. C. Maitland and E. B. Smith, *Mol. Phys.* **22**, 861 (1971).

80. D. W. Gough, E. B. Smith, and G. C. Maitland, *Mol. Phys.* **27**, 867 (1974).

81. U. Buck, M. G. Dondi, U. Valbusa, M. L. Klein, and G. Scoles, *Phys. Rev. A* **8**, 2409 (1973).

82. W. R. Watts, *J. Chem. Phys.* **68**, 402 (1978).

83. D. C. Lorents, R. E. Olsen, and G. M. Conklin, *Chem. Phys. Lett.* **20**, 589 (1973).

84. P. M. Dehmer and J. L. Dehmer, *J. Chem. Phys.* **67**, 1774 (1977).

85. P. M. Dehmer and J. L. Dehmer, *J. Chem. Phys.* **68**, 3462 (1978).

86. P. M. Dehmer and J. L. Dehmer, *J. Chem. Phys.* **69**, 125 (1978).

87. C. E. Moore, *Atomic Energy Levels*, Natl. Bur. Stand. (U.S.) 467, Vols. I–III, 1949.

88. R. E. Huffman and D. H. Katayama, *J. Chem. Phys.* **45**, 138 (1966).

89. M. S. Munson, J. L. Franklin, and H. F. Field, *J. Phys. Chem.* **67**, 1542 (1963).

90. J. A. R. Samson and R. B. Cairns, *J. Opt. Soc. Am.* **56**, 1140 (1966).

91. B. A. Huber, R. Abouaf, P. C. Cosby, R. P. Saxon, and J. Moseley, *J. Chem. Phys.* **68**, 2406 (1978).

92. J. T. Moseley, R. P. Saxon, B. A. Huber, P. C. Cosby, R. Abouaf, and M. Tadjeddine, *J. Chem. Phys.* **67**, 1659 (1977).

93. T. Hirooka, S. A. Anderson, P. W. Tiedemann, and Y. T. Lee, Lawrence Berkeley Laboratory, Materials and Molecular Research Division, Annual Report, 1978, p. 328.

94. R. S. Mulliken, *Rev. Mod. Phys.* **3**, 89 (1931).

95. R. S. Mulliken, *Rev. Mod. Phys.* **4**, 1 (1932).

96. P. M. Dehmer, in S. Datz, Ed., *Abstracts of Contributed Papers*, XII International Conference on the Physics of Electronic and Atomic Collisions, Gatlinburg, Tenn. Vol. 1, 1981, p. 93.

97. W. Kaul and R. Taubert, *Z. Naturforsch.* **179**, 88 (1962).

98. R. J. Buss, C. H. Becker, and Y. T. Lee, unpublished results.

99. J. M. Parson, T. P. Schaefer, P. E. Siska, F. P. Tully, Y. C. Wong, and Y. T. Lee, *J. Chem. Phys.* **53**, 3755 (1970).

100. P. C. Killogoar, Jr., G. E. Leroi, J. Berkowitz, and W. A. Chupka, *J. Chem. Phys.* **58**, 803 (1973).

101. E. Miescher and K. P. Huber, in *International Review of Science, Physical Chemistry*, Vol. 3, Butterworths, London, 1976, p. 37.

102. E. Miescher, *Can. J. Phys.* **54**, 2074 (1976).

103. E. Miescher, *Helv. Phys. Acta* **29**, 135 (1956).

104. F. Alberti and A. E. Douglas, *Can. J. Phys.* **53**, 1179 (1975).

105. E. Miescher and F. Alberti, *J. Phys. Chem. Ref. Data* **5**, 309 (1976).

106. K. Watanabe, *J. Chem. Phys.* **22**, 1564 (1954).

107. A. Guisti and Ch. Jungen, in S. Datz, Ed., *Abstracts of Contributed Papers*, XII International Conference on the Physics of Electronic and Atomic Collisions, Gatlinburg, Tenn., Vol. 1, 1981, p. 71.

108. D. A. S. Vroom, Ph.D. Thesis, 1966, University of British Columbia.

109. M. E. Wacks, *J. Chem. Phys.* **41**, 930 (1964).

110. M. Halmann and L. Lanlicht, *J. Chem. Phys.* **43**, 1503 (1965).

111. R. H. Gillete and E. H. Eyster, *Phys. Rev.* **56**, 1113 (1939).

112. C. E. Dinerman and G. E. Ewing, *J. Chem. Phys.* **53**, 626 (1970).

113. B. L. Blaney and G. E. Ewing, in B. S. Rabinovitch, J. M. Schurr, and H. L. Strauss, Eds., *Annual Review of Physical Chemistry*, Vol. 27, Annual Review, Inc., 1976, p. 553.

114. W. N. Lipscomb, F. E. Wang, W. R. May, and E. Lippert, Jr., *Acta Crystallogr.* **14**, 1100 (1961).

115. W. T. Simpson, *Theories of Electrons in Molecules*, Prentice-Hall, Englewood Cliffs, N.J., 1962, Chapter 4.

116. D. C. Conway and G. S. Janik, *J. Chem. Phys.* **53**, 1859 (1970).

117. J. O. Hirschfelder, C. F. Curtiss, and R. B. Bird, *Molecular Theory of Gases and Liquids*, Wiley, New York, 1964, p. 1111.

118. J. M. Ajello, K. D. Pang, and K. M. Monahan, *J. Chem. Phys.* **61**, 3152 (1974).

119. P. M. Dehmer and W. A. Chupka, *J. Chem. Phys.* **62**, 2228 (1975).

120. P. Krupenie, *J. Phys. Chem. Ref. Data* **1**, 423 (1972).

121. C. W. Polley and B. Munson, *Int. J. Mass Spectrom. Ion Phys.* **26**, 49 (1978).

122. M. S. Foster and J. L. Beauchamp, *Inorg. Chem.* **14**, 1229 (1975).

123. J. J. Solomon, M. Meot-Ner, and F. H. Field, *J. Am. Chem. Soc.* **96**, 3727 (1974).

124. M. Meot-Ner and F. H. Field, *J. Chem. Phys.* **61**, 3742 (1974).

125. S. L. Chong and J. L. Franklin, *J. Chem. Phys.* **54**, 1487 (1971).

126. J. D. Payzant and P. Kebarle, *J. Chem. Phys.* **53**, 4723 (1970).

127. S. H. Linn, Y. Ono, and C. Y. Ng, unpublished data, 1980.

128. D. W. Turner, C. Baker, A. D. Baker, and C. R. Brundle, *Molecular Photoelectron Spectroscopy*, Wiley, New York, 1970.

129. S. H. Linn and C. Y. Ng, *J. Chem. Phys.* **75**, 4921 (1981).

130. R. D. Poshusta and F. A. Matsen, *J. Chem. Phys.* **47**, 4795 (1967).

131. R. D. Pshusta and D. F. Zetik, *J. Chem. Phys.* **58**, 118 (1973).

132. M. E. Schwartz and L. J. Srhaad, *J. Chem. Phys.* **48**, 4709 (1968).

133. K. Morokuma et al., *Annual Review*, Institute for Molecular Science, Okazaki, Japan, 1979.

134. C. H. Douglass, D. J. McClure, and W. R. Gentry, *J. Chem. Phys.* **67**, 4931 (1977).

135. P. M. Dehmer and W. A. Chupka, *J. Chem. Phys.* **65**, 2243 (1976).

136. W. A. Chupka and J. Berkowitz, *J. Chem. Phys.* **51**, 4244 (1969).

137. A. E. Douglas, in C. Sandorfy, P. J. Ausloos, and M. B. Robin, Eds., *Chemical Spectroscopy and Photochemistry in the Vacuum-Ultraviolet*, Vol. 8, NATO–Advanced Study Institutes Series C, 1973, p. 113.

138. Y. Tanaka, A. S. Jursa, and F. J. LeBlanc, *J. Chem. Phys.* **32**, 1205 (1960).

139. W. C. Price and D. M. Simpson, *Proc. R. Soc. London Ser. A* **165**, 272 (1938).

140. J. H. D. Eland and C. J. Danby, *Int. J. Mass Spectrom. Ion Phys.* **1**, 111 (1968).

141. M. J. Schirmer, W. Domeke, L. S. Cedarbaum, W. von Niessen, and L. Asbrink, *Chem. Phys. Lett.* **61**, 30 (1979).

142. C. R. Brundle and D. W. Turner, *Int. J. Mass Spectrom. Ion Phys.* **2**, 195 (1969).

143. V. H. Dibeler and J. A. Walker, *J. Opt. Soc. Am.* **57**, 1007 (1967).

144. R. Frey, B. Gotchov, W. B. Peatman, H. Pollak, and E. W. Schlag, *Int. J. Mass Spectrom. Ion Phys.* **26**, 137 (1978).

145. V. E. Bondybey, J. H. English, and T. A. Miller, *J. Chem. Phys.* **70**, 1621 (1979).

146. J. Wayne Rabalais, *Principles of Ultraviolet Photoelectron Spectroscopy*, Wiley, New York, 1977.

147. H. Köppel, L. S. Cederbaum, W. Domcke, and W. von Niessen, *Chem. Phys.* **37**, 303 (1979).

148. Ch. Jungen and A. J. Merer, in R. N. Rao, Ed., *Molecular Spectroscopy: Modern Research*, Vol. 2, Academic, New York, 1976, p. 127.

149. P. M. Dehmer, J. L. Dehmer, and W. A. Chupka, *J. Chem. Phys.* **73**, 126 (1980).

150. P. Coppens, J. C. Reynaert, and J. Drowart, *J. Chem. Soc. Faraday Trans.*, 2, **75**, 292 (1979).

151. G. Herzberg, *Molecular Spectra and Molecular Structure. III. Electronic Spectra and Electronic Structure of Polyatomic Molecules*, Van Nostrand, Princeton, N.J., 1966, p. 600.

152. M. Meot-Ner (Mautner) and F. H. Field, *J. Chem. Phys.* **66**, 4527 (1977).

153. R. S. Mulliken, *J. Am. Chem. Soc.* **86**, 3183 (1964).

154. M. Ogawa and H. C. Chang, *Can. J. Phys.* **48**, 2455 (1970).

155. N. C. Baenziger and W. L. Daux, *J. Chem. Phys.* **48**, 2974 (1968).

156. M. Larzilliere and N. Damany, *Can. J. Phys.* **56**, 1150 (1978).

157. R. F. Stebbings, in D. R. Bates and B. Bederson, Eds., *Advances in Atomic and Molecular Physics*, Vol. 15, Academic, New York, 1979, p. 77.

158. J. W. Otvos and D. P. Stevenson, *J. Am. Chem. Soc.* **78**, 546 (1956).

159. J. H. D. Eland and J. Berkowitz, *J. Chem. Phys.* **70**, 5151 (1979).

160. J. Delwiche, M. J. Hubin-Franskin, G. Caprace, and P. Natalis, *J. Electr. Spectrosc. Relat. Phenom.* **21**, 205 (1980).

161. F. M. Matsunaga and K. Watanabe, *J. Chem. Phys.* **46**, 4457 (1967).

162. D. C. Frost, S. T. Lee, and C. A. McDowell, *J. Chem. Phys.* **59**, 5484 (1973).

163. G. R. Cook and M. Ogawa, *J. Chem. Phys.* **51**, 647 (1969).

164. E. Lindholm, *Ark. Fys.* **40**, 97 (1969).

165. G. R. Cook and M. Ogawa, *J. Chem. Phys.* **51**, 2419 (1969).

166. J. V. Headley, R. S. Mason, and K. R. Jennings, presented at the 28th Annual Conference on Mass Spectrometry and Applied Topics, New York, N.Y., May 25–30, 1980.

167. G. G. Jones and J. W. Taylor, *J. Chem. Phys.* **68**, 1768 (1978).

168. A. B. Rakshit and P. Warneck, *Int. J. Mass Spectrom. Ion Phys.* **35**, 23 (1980).

169. Y. Tanaka and M. Ogawa, *Can. J. Phys.* **40**, 879 (1962).

170. H. J. Henning, *Ann. Phys.* (*Leipz.*) **13**, 599 (1932).

171. Y. Tanaka and W. C. Walker, *J. Chem. Phys.* **74**, 2760 (1981).

172. J. S. Cohen and B. Schneider, *J. Chem. Phys.* **61**, 3230 (1974).

173. R. S. Mulliken, *J. Chem. Phys.* **52**, 5170 (1970).

174. J. N. Bradsley, *Chem. Phys. Lett.* **2**, 329 (1968).

175. A. L. Smith, *Phil. Trans. R. Soc. London A* **268**, 169 (1970).

176. P. Natalis and J. E. Collin, *Int. J. Mass Spectrom. Ion Phys.* **2**, 222 (1969).

177. J. Berkowitz and J. H. D. Eland, *J. Chem. Phys.* **67**, 2740 (1977).

178. P. Coppens, J. Smets, M. G. Fishel, and J. Drowart, *Int. J. Mass Spectrom. Ion Phys.* **14**, 57 (1974).

179. V. H. Dibeler and J. A. Walker, *Adv. Mass Spectrom.* **4**, 767 (1967).

180. V. H. Dibeler, J. A. Walker, and S. K. Liston, *J. Res. Natl. Bur. Stand.* **71A**, 371 (1967).

181. H. L. Johnston and K. E. McCloskey, *J. Phys. Chem.* **44**, 1038 (1940).

182. M. Trautz and F. Kurz, *Ann. Phys.* (5) **9**, 981 (1931).

183. E. C. Y. Inn, *Phys. Rev.* **91**, 1194 (1953).

184. K. Watanabe, *J. Chem. Phys.* **26**, 542 (1957).

185. V. H. Dibeler and S. K. Liston, *J. Chem. Phys.* **49**, 482 (1968).

186. W. C. Price and D. M. Sampson, *Proc. R. Soc. London A* **165**, 272 (1938).

187. D. Golomb, K. Watanabe, and F. F. Marmo, *J. Chem. Phys.* **36**, 958 (1962).

188. H. D. Smyth and D. W. Mueller, *Phys. Rev.* **43**, 121 (1933).

189. R. M. Reese, V. H. Dibeler, and J. L. Franklin, *J. Chem. Phys.* **29**, 880 (1958).

190. R. Hagemann, *C. R. Acad. Sci.* **255**, 1102 (1962).

191. R. Botten, R. Hagemann, G. Nief, and E. Roth, in *Advances in Mass Spectrometry*, Vol. 3, The Institute of Petroleum, London, 1966, p. 951.

192. D. R. Lloyd and P. J. Roberts, *Mol. Phys.* **26**, 225 (1973).

193. B. Brehm, J. H. D. Eland, R. Frey, and A. Küstler, *Int. J. Mass Spectrom. Ion Phys.* **12**, 197 (1973).

194. M. J. Weiss, T-C. Hsieh, and G. G. Meisels, *J. Chem. Phys.* **71**, 567 (1979).

195. N. Jonathan, D. J. Smith, and K. J. Ross, *Chem. Phys. Lett.* **9**, 217 (1971).

196. J. M. Dyke, L. Golob, N. Jonathan, A. Morris, M. Okuda, and D. J. Smith, *J. Chem. Soc. Faraday Trans. 2* **70**, 1818 (1974).

197. J. M. Dyke, L. Golob, N. Jonathan, A. Morris, and M. Okuda, *J. Chem. Soc. Faraday Trans. 2* **70**, 1828 (1974).

198. D. C. Frost, S. T. Lee, and C. A. McDowell, *Chem. Phys. Lett.* **24**, 4525 (1974).

199. C. R. Brundle, *Chem. Phys. Lett.* **26**, 25 (1974).

200. M. J. Weiss, J. Berkowitz, and E. H. Appelman, *J. Chem. Phys.* **66**, 2049 (1977).

201. D. D. Wagman, W. H. Evans, V. B. Parker, I. Halow, S. M. Bailey, and R. H. Schumm, Eds., *Selected Values of Chemical Thermodynamic Properties. Tables for the first Thirty-Four Elements in the Standard Order of Arrangements*, Natl. Bur. Stand. (U.S.) Tech. Note 270-3, U.S. GPO, Washington, D.C., 1968.

202. "JANAF Thermodynamical Tables," *Natl. Stand. Ref. Data Ser.* Natl. Bur. Stand. (U.S.), **37** (1971).

203. L. Karlsson, L. Mattsson, R. Jadrny, R. G. Albridge, S. Pinchas, T. Bergmark, and K. Siegbahn, *J. Chem. Phys.* **62**, 4745 (1975).

204. R. Yamdagni and P. Kebarle, *J. Am. Chem. Soc.* **98**, 1320 (1976).

205. R. J. Cotter and W. S. Koski, *J. Chem. Phys.* **59**, 784 (1973).

206. J. Long and B. Munson, *J. Am. Chem. Soc.* **95**, 2427 (1973).

207. S. Chong, R. A. Myers, Jr., and J. L. Franklin, *J. Chem. Phys.* **56**, 2417 (1972).

208. J. L. Beauchamp and S. E. Buttrill, Jr., *J. Chem. Phys.* **48**, 1783 (1968).

209. P. A. Kollman and C. F. Bender, *Chem. Phys. Lett.* **21**, 271 (1973).

210. J. Chao and B. J. Zwolinski, *J. Phys. Chem. Ref. Data* **5**, 319 (1976).

211. G. Dellepiane and J. Overend, *Spectrochim. Acta* **22**, 593 (1966).

212. J. S. Rowlinson, *Discuss. Faraday Soc.* **45**, 974 (1949).

213. V. H. Dibeler, J. A. Walker, and H. M. Rosenstock, *J. Res. NBS* **70A**, 459 (1966).

214. D. K. Bohme, R. S. Hemsworth, and H. W. Rundle, *J. Chem. Phys.* **59**, 77 (1973).

215. J. F. Wolf, R. H. Staley, I. Koppel, M. Taagepera, R. T. McIver, Jr., J. L. Beauchamp, and R. W. Traft, *J. Am. Chem. Soc.* **99**, 5417 (1977).

216. M. A. Haney and J. L. Franklin, *J. Chem. Phys.* **50**, 2028 (1969).

217. M. A. Haney and J. L. Franklin, *J. Phys. Chem.* **73**, 4328 (1969).

218. H. Wincel, *Int. J. Mass Spectrom. Ion Phys.* **9**, 267 (1972).

219. M. R. Arshadi and J. H. Futrell, *J. Phys. Chem.* **78**, 1482 (1974).

220. J. D. Payzant, A. J. Cunningham, and P. Kebarle, *Can. J. Chem.* **51**, 3242 (1973).

221. S. K. Searles and P. Kebarle, *J. Phys. Chem.* **72**, 742 (1968).

222. R. Botter, V. H. Dibeler, J. A. Walker, and H. M. Rosenstock, *J. Chem. Phys.* **44**, 1271 (1966).

223. V. H. Dibeler, J. A. Walker, and K. E. McCulloh, *J. Chem. Phys.* **59**, 2264 (1973).

224. J. R. Wyatt and F. E. Stafford, *J. Phys. Chem.* **76**, 1913 (1972).

225. H. Okabe and V. H. Dibeler, *J. Chem. Phys.* **59**, 2430 (1973).

226. S. I. Miller and J. Berkowitz, unpublished observations, 1971, described in Reference 33.

227. W. B. Maier, II, *J. Chem. Phys.* **42**, 1790 (1965).

228. G. G. Meisels, *J. Chem. Phys.* **42**, 2328 (1965).

229. G. G. Meisels, *J. Chem. Phys.* **42**, 3237 (1965).

230. P. S. Gill, Y. Inel, and G. G. Meisels, *J. Chem. Phys.* **54**, 2811 (1971).

231. P. Kebarle and R. M. Haynes, *J. Chem. Phys.* **47**, 1676 (1967).

232. A. A. Herod and A. G. Harrison, *Int. J. Mass Spectrom. Ion Phys.* **4**, 415 (1970).

233. P. Warneck, *Ber. Bunsenges. Phys. Chem.* **76**, 421 (1972).

234. L. W. Sieck and P. Ausloos, *J. Res. Natl. Bur. Stand. Sect. A* **76**, 253 (1972).

235. I. H. Suzuki and K. Maeda, *Int. J. Mass Spectrom. Ion Phys.* **17**, 249 (1975).

236. P. G. Miasek and A. G. Harrison, *J. Am. Chem. Soc.* **97**, 714 (1975).

237. M. T. Bowers, D. D. Elleman, and J. L. Beauchamp, *J. Phys. Chem.* **72**, 3599 (1968).

238. S. E. Buttrill, Jr., *J. Chem. Phys.* **52**, 6174 (1970).

239. M. L. Gross and J. Norbeck, *J. Chem. Phys.* **54**, 3651 (1971).

240. W. T. Huntress, *J. Chem. Phys.* **56**, 5111 (1972).

241. A. J. Ferrer-Correia and K. R. Jennings, *Int. J. Mass Spectrom. Ion Phys.* **11**, 111 (1973).

242. P. R. LeBreton, A. D. Williamson, J. L. Beauchamp, and W. T. Huntress, *J. Chem. Phys.* **62**, 1623 (1975).

243. Z. Herman, A. Lee, and R. Wolfgang, *J. Chem. Phys.* **51**, 452 (1969).

244. F. P. Lossing, *Can. J. Chem.* **50**, 3973 (1972).

245. F. P. Lossing, *Can. J. Chem.* **49**, 356 (1971).

246. P. A. M. Kemper and M. T. Bowers (unpublished data).

247. W. J. Chesnavich, L. Bass, T. Su, and M. T. Bowers, *J. Chem. Phys.* **74**, 2228 (1981).

248. Y. Ono and C. Y. Ng (to be published).

249. E. Lindholm, *Advan. Chem. Ser.* **58**, 1 (1966).

250. E. D. Poliakoff, P. M. Dehmer, J. L. Dehmer, and R. Stockbauer, *J. Chem. Phys.* **75**, 1568 (1981).

RANDOM WALKS:
THEORY AND SELECTED APPLICATIONS

GEORGE H. WEISS

*National Institutes of Health
Bethesda, Maryland 20205*

ROBERT J. RUBIN

*National Bureau of Standards
Washington, D.C. 20234*

CONTENTS

I. INTRODUCTION

A random walk in discrete time is a sum of random variables:

$$\mathbf{R}_n = \mathbf{r}_1 + \mathbf{r}_2 + \cdots + \mathbf{r}_n \qquad (1.1)$$

The study of such sums arises naturally in any analysis of games of chance. Thus, one might say that random-walk theory had its origins in 17th-century analyses of gambling.[1] The identification and formulation of random-walk problems per se is a product of the early 20th century. In spite of the simple definition of random walks in Eq. (1.1) (see the book by Spitzer[2] for an alternative definition in terms of transition probabilities), few mathematical models have attained as high a degree of utility in both physical and social sciences. The original statement of a random-walk problem by Pearson[3,4] arose in the context of a biological problem. The formulation of random-walk problems can involve a discrete lattice, as in random walks in solids, or a continuum, as in models of diffusion. One of the more interesting early applications of random-walk theory was by Bachelier[5] in a theory of stock-market behavior. The theory of random walks has enriched and been enriched by contact with applications: in the physical sciences new areas of mathematical investigation have been suggested; new applications of mathematical results are continually being made. In this chapter we review portions of the theory of random walks, particularly developments of the past thirty years, together with several applications to problems in chemical physics and one area of biology. It would be presumptuous to claim completeness for this chapter; we have concentrated on topics close to our own interests, which have been in lattice random walks, and applications of the theory to polymer physics. The very important subject of excluded-volume random walks will not be discussed; indeed, it would require a series of monographs to begin to do it justice, and Domb and Green[6] have edited just such a series. Similarly we do not go into many of the mathematical problems discussed at great length in the monograph by Spitzer.[2]

The first, rather informal description of a random walk was posed as a problem by Karl Pearson[3] in a letter to *Nature* in 1905. He requested the solution to a problem phrased as follows: "A man starts from a point O and walks a distance a in a straight line, he then turns through any angle whatever and walks a distance a in a second straight line. He repeats this process n times. I require the probability that after these n stretches he is at a distance between r and $r + dr$ from his starting point at O." A complete solution to this problem was furnished by Kluyver[7] in 1906 for any arbitrary set of step lengths $a_1, a_2, \ldots, a_n$. Rayleigh[8] subsequently provided an analysis of the three-dimensional problem. His argument is more accessibly reproduced by Chandrasekhar[9] for the case of equal displacements, and for the more general case the result is to be found in the book by Watson.[7] The first mention of random walks on a lattice was apparently made by Polya[10] in 1921, in a paper considering the problem of finding the probability that a random walker initially at the origin of coordinates would, in the course of his or her excursions, eventually return to that point. On the more applied side G. I. Taylor[11] discussed the dynamics of turbulent diffusion in 1921 by passing to the limit of infinitely small steps in a random walk. A real impetus to study mathematical properties of random walks was provided by the attempt to characterize the configuration of polymer chains in terms of random flights by Kuhn[12] in 1934 and by Kuhn and Grün[13] in 1942. A similar analysis was given by Treloar[14] in 1946. A good account of the early history of this subject is contained in the book by Volkenshtein.[15] These early models of polymer configurations all make the assumption that the overall chain dimensions are so much larger than the dimensions of a monomer that the random walk can be treated as a random flight, that is, one can pass to a continuum limit. A considerable body of research exists on polymer configuration models which retain some structure for the polymer chain. The so-called rotational-isomer approximation replaces the continuous potential function describing the interaction between successive monomers of a chain by a discrete one which restricts the possible angles between adjacent monomer-monomer bonds to a small number.[16] Another such model is that of the wormlike chain which allows continuous curvature. There is a considerable body of literature in polymer physics which makes use of such models, and interesting accounts of some of the mathematical problems suggested by them have been given by Kac[17] and Daniels.[18] On the purely mathematical side, from the 1920s to the present, great progress has been made in elucidating the central limit theorem, which predicts the asymptotic form of the sum $\mathbf{R}_n$ in Eq. (1.1) from very general properties of the distribution of the $\mathbf{r}_i$. A good summary of much of this research is to be found in the monograph by Gnedenko and Kolmogoroff.[19] Much of the exposition in later sections in which we derive asymptotic distributions of end-to-end distances of the random walk can be considered as heuristic derivations of results usually obtained rigorously by much more complicated methods. A substantial portion of the theory to be described has been developed since the middle 1950s, when Montroll's work[20] reawakened interest in properties of lattice random walks.

This chapter is divided into two major parts, the first being the general theory, and the second giving selected applications. Section II.A presents some of the introductory identities for lattice random walks, both for regular discrete intervals of time between successive steps and for random intervals of time occurring between successive steps. The latter type are now referred to as continuous-time random walks (or CTRWs) in the chemical physics literature. We also develop formalism for so-called persistent random walks in which the direction in which a given step is taken depends on the direction of a fixed number of the preceding steps. In Section II.B we pass to the continuum limit in both time and space, deriving the diffusion limit for random walks, in which the variance of a single step is finite, and for which the average time between steps, for a CTRW, is also finite. The corresponding continuum limit for persistent random walks is shown to be described by a telegraph (rather than a diffusion) equation. Modifications to these results arising from either infinite single-step variances or infinite mean times between successive steps are also discussed. Section II.C contains material on asymptotic properties of lattice random walks. These involve a number of properties that can be calculated using Tauberian theorems for power series (these relate the asymptotic behavior of the coefficients of a power series to the analytic properties of the underlying function near its singularities). The properties that can be calculated using this technique include the expected number of points visited by an n-step walk, all moments of the number of times a given point (or set of points) has been visited by an n-step walk, and statistical properties of the number of points visited a specified number of times by an n-step walk. In Section II.D we discuss the effects of different types of boundaries on properties of random walks. This section includes a derivation of common types of boundary conditions, and an analysis of the spans, which are the lengths of the edges of the smallest parallelepiped with sides parallel to the coordinate planes that completely encloses the random walk.

Four classes of applications are described in the concluding part of the chapter. Section III.A is a review of several selected problems in polymer physics. The plethora of random-walk applications in this area (there are a number of monographs on the subject) prevents us from giving an exhaustive discussion of most of them, but we describe recent calculations of the radius of gyration, measures of asymmetry of random-walk models, the wormlike chain, some techniques for calculating the probability density of the end-to-end distance of short polymer chains, and models of polymer adsorption on surfaces. In Section III.B we describe the formalism for the recently developed theory of random walks in which the walker can be in any one of a discrete number of states. These models have been applied to the analysis of chromatographic processes and to diffusion in solids. Section III.C contains a description of several applications of random-walk theory to transport in solids. Particular attention is paid to work on exciton migration and trapping phenomena. We briefly describe some of the analyses of hopping transport in amorphous solids, a subject that has engaged several investigators recently. The final Section III.D is devoted to random-walk models for the motion of microorganisms on surfaces. Advances in technology have improved the ability to collect information on the parameters of this motion, and the language of random walks provides a natural description for these results.

It is nearly impossible to cover the complete spectrum of applications of random-walk ideas in the physical sciences in a single article. Nevertheless we hope that the sampling of topics contained here will indicate some of the strengths and limitations of random-walk theory as applied to physical problems, as well as some areas for future research.

II. GENERAL THEORY

A. Elementary Properties

1. Discrete Steps and Discrete Time

We consider first the simplest possible situation, a random walk in one dimension (1-D) on a lattice infinite in both directions. To begin with we shall assume that steps are taken periodically. The probability that a single step is of displacement j will be denoted by $p(j)$, where $j = \ldots -2, -1, 0, 1, 2, \ldots$. The probability that the random walker is at lattice point r given that it was initially at r_0 will be written as $P_n(r|r_0)$. These probabilities satisfy the recurrence relation

$$P_{n+1}(r|r_0) = \sum_{l=-\infty}^{\infty} p(r-l)P_n(l|r_0) \tag{2.1}$$

subject to $P_0(r|r_0) = \delta_{r, r_0}$. Since the sum on the right-hand side of this equation is in convolution form, Fourier methods can be used to solve for the

$P_n(r|r_0)$. If we introduce the generating functions

$$\lambda(\theta) = \sum_{r=-\infty}^{\infty} p(r)\exp(ir\theta)$$

$$\Gamma_n(\theta, r_0) = \sum_{r=-\infty}^{\infty} P_n(r|r_0)\exp(ir\theta) \qquad (2.2)$$

then Eq. (2.1) is equivalent to $\Gamma_{n+1}(\theta, r_0) = \lambda(\theta)\Gamma_n(\theta, r_0)$, or

$$\Gamma_n(\theta, r_0) = \lambda(\theta)\Gamma_{n-1}(\theta, r_0) = \lambda^n(\theta)\exp(ir_0\theta) \qquad (2.3)$$

The inversion of this formula yields the explicit expression

$$P_n(r|r_0) = \frac{1}{2\pi}\int_{-\pi}^{\pi} \lambda^n(\theta)e^{i(r_0-r)\theta}\,d\theta \qquad (2.4)$$

Since $P_n(r|r_0)$ depends only on the difference $r - r_0$, we shall henceforth assume that $r_0 = 0$ and use the notation $P_n(r)$ for the state probability or propagator. The function $\lambda(\theta)$ will be known as the structure function. Its analytic properties will be seen to be intimately related to many interesting properties of the random walk. The extension of the above results to random walks in D dimensions is straightforward and leads to the result

$$P_n(\mathbf{r}) = \frac{1}{(2\pi)^D}\int_{-\pi}^{\pi}\cdots\int \lambda^n(\boldsymbol{\theta})\exp(-i\mathbf{r}\cdot\boldsymbol{\theta})\,d^D\theta \qquad (2.5)$$

where

$$\lambda(\boldsymbol{\theta}) = \sum_{\mathbf{j}} p(\mathbf{j})\exp(i\mathbf{j}\cdot\boldsymbol{\theta})$$

Various elementary properties of the random walk follow immediately from Eqs. (2.1)–(2.5). One can immediately calculate various moments of the displacement after n steps from Eq. (2.4) or its D-dimensional analogue, if it is noted that the moments of the single-step transition probabilities are

$$\langle n^l(1)\rangle = \sum_{r=-\infty}^{\infty} r^l p(r) = (-i)^l \frac{d^l\lambda}{d\theta^l}\bigg|_{\theta=0}, \qquad l=1,2,\ldots \qquad (2.6)$$

when these moments are finite. Hence, using the representation in Eq. (2.4), we find

$$\langle n(k) \rangle = k \langle n(1) \rangle$$
$$\sigma^2(k) = k\sigma^2 \qquad (2.7)$$

in $1-D$. We shall use the convention that σ^2 without any argument always denotes the variance of a single step of the random walk. Later we shall also be interested in the variance of other quantities. These we shall denote by $\sigma^2(\ )$, where the parentheses include the quantity of interest. In higher dimensions one can calculate correlation properties from the structure function. For example, if we define

$$\langle n_1^{l_1}(k) n_2^{l_2}(k) \ldots n_D^{l_D}(k) \rangle = \sum_{\{n\}} n_1^{l_1} n_2^{l_2} \ldots n_D^{l_D} P_k(\mathbf{n}) \qquad (2.8)$$

then

$$\langle n_i(k) \rangle = k \langle n_i(1) \rangle, \qquad i = 1, 2, \ldots, D$$
$$\langle n_i(k) n_j(k) \rangle - \langle n_i(k) \rangle \langle n_j(k) \rangle = k \{ \langle n_i(1) n_j(1) \rangle - \langle n_i(1) \rangle \langle n_j(1) \rangle \}$$
$$\qquad (2.9)$$

In general the cumulants of the k-step random walk are k times those of the single-step walk.

2. Discrete Steps and Continuous Time

So far we have assumed that the times between successive steps of the random walk are equal. Montroll and Weiss[21] introduced the notion of the continuous-time random walk (henceforth to be abbreviated CTRW, as is common in the literature) in which times between successive steps are random variables. We develop the theory in a slightly generalized form suggested by Scher and Lax.[22] Let us define $p(\mathbf{r}, t) dt$ to be the probability that the time between two successive steps is between t and $t + dt$, and the transition that follows is $\mathbf{r}$ lattice points. This function obviously satisfies the normalization condition

$$\sum_{\mathbf{r}} \int_0^\infty p(\mathbf{r}, t) dt = 1 \qquad (2.10)$$

Montroll and Weiss[21] made the specific assumption that the waiting time and size of step were independent, in which case $p(\mathbf{r}, t)$ can be factored as

$$p(\mathbf{r}, t) = p(\mathbf{r}) \psi(t) \qquad (2.11)$$

where $\psi(t)$ is the probability density of the waiting time, and $p(\mathbf{r})$ is the transition probability. We shall call a CTRW with the factorizing property of this last equation a *separable* random walk. Most applications use only separable random walks, but in at least one application of the theory by Scher and Wu[23] the full formalism is required. In either case one can always define a probability density for the waiting time by

$$\psi(t) = \sum_{\mathbf{r}} p(\mathbf{r}, t) \qquad (2.12)$$

which is properly normalized according to Eq. (2.10). In addition to $p(\mathbf{r}, t)$, we also define a joint probability density $p_1(\mathbf{r}, t)$ corresponding to the time between $t = 0$ and the first jump. This distinction is required because $t = 0$ need not coincide with the start of a waiting period. This idea occurs naturally in the theory of renewal processes[24] and was first discussed in the context of the CTRW by Tunaley.[25] A waiting-time density $\psi_1(t)$ can be found from $p_1(\mathbf{r}, t)$ through the relation in Eq. (2.12). We also define $\Psi(t)$ to be the probability that a complete waiting time (i.e., the time between successive jumps) is $\geqslant t$, so that

$$\Psi(t) = \int_t^\infty \psi(\tau)\, d\tau \qquad (2.13)$$

with a similar relation holding between $\Psi_1(t)$ and $\psi_1(t)$.

We wish to calculate $P(\mathbf{r}, t)$, the probability of being at $\mathbf{r}$ at time t. It is convenient to define $\eta(\mathbf{r}, t)\,dt$ to be the probability that a jump was made to $\mathbf{r}$ between times t and $t + dt$. Then $P(\mathbf{r}, t)$ can be expressed as

$$P(\mathbf{r}, t) = \delta_{\mathbf{r}, 0}\Psi_1(t) + \int_0^t \eta(\mathbf{r}, \tau)\Psi(t - \tau)\, d\tau \qquad (2.14)$$

corresponding to the two possibilities that (1) the random walker has not moved from the origin, or (2) a jump occurred to $\mathbf{r}$ at time $\tau \leqslant t$ and no further transitions took place. Thus we see that $P(\mathbf{r}, t)$ depends on $\eta(\mathbf{r}, t)$. This function, in turn, satisfies

$$\eta(\mathbf{r}, t) = p_1(\mathbf{r}, t) + \sum_{\mathbf{r}'} \int_0^t \eta(\mathbf{r}', \tau)p(\mathbf{r} - \mathbf{r}', t - \tau)\, d\tau. \qquad (2.15)$$

Equations (2.14) and (2.15) can be formally resolved by using a joint generating function and Laplace transform. Define the joint transform of $\eta(\mathbf{r}, t)$ by $\eta^*(\boldsymbol{\theta}, s)$, where

$$\eta^*(\boldsymbol{\theta}, s) = \int_0^\infty e^{-st}\left(\sum_{\mathbf{r}} \eta(\mathbf{r}, t)\exp(i\mathbf{r}\cdot\boldsymbol{\theta})\right) dt \qquad (2.16)$$

and similarly denote the joint transform of any other function by the same symbol with an asterisk and the appropriate arguments. It is easy to verify from Eq. (2.15) that

$$\eta^*(\boldsymbol{\theta}, s) = \frac{p_1^*(\boldsymbol{\theta}, s)}{1 - p^*(\boldsymbol{\theta}, s)} \tag{2.17}$$

and from Eq. (2.14) that

$$P^*(\boldsymbol{\theta}, s) = \Psi_1^*(s) + \frac{\Psi^*(s)p_1^*(\boldsymbol{\theta}, s)}{1 - p^*(\boldsymbol{\theta}, s)} \tag{2.18}$$

When $p_1(\mathbf{r}, t) = p(\mathbf{r}, t)$ this formula simplifies somewhat to

$$P^*(\boldsymbol{\theta}, s) = \frac{\Psi^*(s)}{1 - p^*(\boldsymbol{\theta}, s)} \tag{2.19}$$

If we suppose that the random walk is initially observed at a random time, by which we mean that there is no correlation between the time of observation and the time of the transition just prior to it, then a relation can be established between $\psi_1(t)$ and $\psi(t)$. To demonstrate this relation we need to assume that

$$\mu = \int_0^\infty t\psi(t)\,dt < \infty \tag{2.20}$$

or that the mean time between jumps is finite. Suppose that a jump occurs at $t = 0$, and the initial observation is made at some $t > 0$, it being assumed that there are no intervening jumps. Then, since complete ignorance is assumed, the probability that the observation is made in a specific interval $(t, t + dt)$ is $A\,dt$, where A is a constant with the dimensions of inverse time that is to be determined by normalization. Since no jump may occur in $(0, t)$ we have

$$\psi_1(t) = A\Psi(t) = \frac{\Psi(t)}{\mu} \tag{2.21}$$

where A has been found by requiring that $\psi_1(t)$ integrate to 1. The Laplace transform of $\psi_1(t)$ is

$$\psi_1^*(s) = \frac{1 - \psi^*(s)}{\mu s} \tag{2.22}$$

The two functions $\psi(t)$ and $\psi_1(t)$ are identical only for the particular form

$\psi(t) = (1/T)\exp(-t/T)$. This identity may be expressed by saying that the negative-exponential waiting-time density has no memory. A further generalization of random-walk theory can be made, to consider random walks in discrete time in which the times between successive transitions are integer random variables. The development of this theory is quite similar to the development of CTRWs, and in the formal development one replaces the Laplace transforms by generating functions.

3. Relation Between the CTRW and the Generalized Master Equation

Kenkre, Montroll, and Shlesinger[26] have established the relation between CTRWs and the master-equation description of the system. This relation can be established using the joint generating-function–Laplace-transform of the state probabilities as in Eq. (2.16). The master equation for a translationally invariant random walk can be written

$$\frac{dP(\mathbf{r}, t)}{dt} = \sum_{\rho} \int_0^t P(\rho, \tau) K(\mathbf{r} - \rho, t - \tau)\, d\tau \qquad (2.23)$$

and our object is to provide an expression for the kernel $K(\mathbf{r}, t)$. This is not easy to do directly, but if we introduce the function $K^*(\boldsymbol{\theta}, s)$ as

$$K^*(\boldsymbol{\theta}, s) = \sum_{\rho} \int_0^{\infty} K(\rho, t) \exp(i\rho \cdot \boldsymbol{\theta} - st)\, dt \qquad (2.24)$$

we find

$$P^*(\boldsymbol{\theta}, s) = \left[s - K^*(\boldsymbol{\theta}, s) \right]^{-1} \qquad (2.25)$$

Equating this and Eq. (2.18), we find

$$K^*(\boldsymbol{\theta}, s) = s - \frac{1 - p^*(\boldsymbol{\theta}, s)}{\Psi_1^*(s)\left[1 - p^*(\boldsymbol{\theta}, s)\right] + \Psi^*(s) p_1^*(\boldsymbol{\theta}, s)} \qquad (2.26)$$

This expression differs from that given by Kenkre, Montroll, and Shlesinger[26] because they do not allow for the possibility that the first sojourn differs from succeeding ones, that is, they assume that $t = 0$ always coincides with the beginning of a sojourn. The simplest example of the use of the relation just proved is a separable random walk in which $\psi(t) = (1/T)\exp(-t/T)$. Under these conditions we find from Eq. (2.26) that

$$K^*(\boldsymbol{\theta}, s) = -\frac{1}{T}\left[1 - \lambda^*(\boldsymbol{\theta})\right] \qquad (2.27)$$

With this simple form for the transformed kernel one can invert to find an explicit expression for $K(\mathbf{r}, t)$:

$$K(\mathbf{r}, t) = -\frac{1}{T}[\delta_{\mathbf{r},0} - p(\mathbf{r})]\delta(t) \qquad (2.28)$$

so that

$$\frac{dP(\mathbf{r}, t)}{dt} + \frac{1}{T}P(\mathbf{r}, t) = \frac{1}{T}\sum_{\mathbf{l}}P(\mathbf{l}, t)p(\mathbf{r}-\mathbf{l}) \qquad (2.29)$$

Several other cases in which explicit expressions for $K(\mathbf{r}, t)$ can be found are given in Reference 26. If one restricts oneself to separable random walks and to the case in which $t = 0$ coincides with the beginning of a sojourn, then

$$K(\mathbf{r}, t) = -[\delta_{\mathbf{r},0} - p(\mathbf{r})]k(t), \qquad (2.30)$$

where $k(t)$ is given in terms of an inverse Laplace transform

$$k(t) = \mathcal{L}^{-1}\left\{\frac{s\psi^*(s)}{1 - \psi^*(s)}\right\} \qquad (2.31)$$

If $p^*(\mathbf{0}, s)$ cannot be factored, then the kernel $K(\mathbf{r}, t)$ does not factor as it does in Eq. (2.30).

Recently Klafter and Silbey[27] have derived the CTRW for the transport of excitations on a lattice randomly occupied by guests, under the assumption that the transport in any configuration is described by a master equation. They use a projection-operator formalism to derive a generalized master equation for transport, averaged over all configurations, thereby showing that the CTRW provides a correct description. The crucial assumption in the calculation is that of the master-equation description for particular configurations. Bedeaux, Lindenberg, and Shuler[28] have elaborated on the relationship between the master equation and random walks in discrete time in the context of the factorized case. They showed that when $\psi(t)$ has a finite first moment μ, the solution for $P(\mathbf{r}, t)$ can be approximated by using the asymptotic form of $k(t)$. Since for small $|s|$, $\psi^*(s)$ has the expansion $\psi^*(s) \sim 1 - \mu s + o(s)$ when $\mu < \infty$, the Laplace transform $k^*(s)$ of $k(t)$ goes to $1/\mu$ as $s \to 0$, and we might expect that

$$k(t) \sim \mathcal{L}^{-1}\left\{\frac{1}{\mu}\right\} = \frac{1}{\mu}\delta(t) \qquad (2.32)$$

The validity of this approximation and the times for which it may be used are the topics treated by the cited authors. They find that the approximation can be expected to be useful for all times greater than the maximum of $(\mu_n/n!)^{1/n}$, where μ_n is the nth moment of $\psi(t)$.

Several authors[29-31] have also discussed the equivalence between propagator and master-equation formalisms in the context of models for transport in solids. Their contribution will be described in greater detail in Section III.C on that topic. Shugard and Reiss[32] have extended the notion of the equivalence between random walks and master equations to random walks which do not necessarily have translational invariance. Alley and Alder[33,34] have used the theory in a study of transport properties of a Lorentz gas. In those papers the authors calculate a waiting-time density from the velocity autocorrelation function determined from computer simulation of the dynamics of the Lorentz gas. We follow their analysis. Let the displacement of the random walk at time t be $\mathbf{r}(t)$, and let the Laplace transform of mean squared displacement be $\mathcal{L}\{\langle r^2(t)\rangle\}$. Although one cannot define a velocity for brownian motion,‡ one can define a formal velocity autocorrelation function $\rho(t)$ through its Laplace transform

$$\rho^*(s) = \frac{s^2 \mathcal{L}\{\langle r^2(t)\rangle\}}{\langle v^2\rangle} \tag{2.33}$$

where $\langle v^2\rangle$ is a constant obtained from experimental data. This function is the one that would result for processes for which the velocity autocorrelation function is definable. Notice that although neither $\rho(t)$ nor $\langle v^2\rangle$ exists for random walks, these functions can be estimated from the simulation results. If we consider the simplest situation of a separable random walk in which $t = 0$ coincides with the beginning of a waiting time, the joint generating-function–Laplace-transform of $P(\mathbf{r}, t)$ is

$$P^*(\boldsymbol{\theta}, s) = \frac{1 - \psi^*(s)}{s} \frac{1}{1 - \lambda(\boldsymbol{\theta})\psi^*(s)} \tag{2.34}$$

The Laplace transform of the moments can be obtained by differentiation of $P^*(\boldsymbol{\theta}, s)$:

$$\mathcal{L}\{\langle r^2(t)\rangle\} = -\sum_i \left.\frac{\partial^2 P}{\partial \theta_i^2}\right|_{\boldsymbol{\theta}=0} = \frac{\sigma^2 \psi^*(s)}{s[1 - \psi^*(s)]} \tag{2.35}$$

‡Since one can show that $\langle [r(t+\Delta t) - r(t)]^2\rangle$ is $O(\Delta t)$ for $\Delta t \to 0$. This problem does not arise in correlated random walks as they are described in later paragraphs, nor for Ornstein–Uhlenbeck processes.

The combination of Eqs. (2.33) and (2.35) allows us to establish the relation between the properties of the random walk and autocorrelation function. We find

$$\rho^*(s)[1 - \psi^*(s)] = \frac{\sigma^2 s \psi^*(s)}{\langle v^2 \rangle} \tag{2.36}$$

in which Alder[33] uses the relation

$$\psi(t) = \int_0^t K(\tau)[1 - \psi(t - \tau)] \, d\tau \tag{2.37}$$

where

$$K(t) = \frac{\langle v^2 \rangle}{\sigma^2} \int_0^t \rho(\tau) \, d\tau \tag{2.38}$$

Thus, given the experimentally determined $\rho(t)$ and $\langle v^2 \rangle$, one can calculate a sojourn-time density $\psi(t)$. Higher moments of the displacement can be determined for the random-walk model, in terms of $K(t)$ and lower moments. It is the asymptotic behavior of these higher moments that formed the object of Alder's investigation.

A theory parallel to that described in the preceding paragraphs can be developed for random walks on lattices with periodic boundary conditions. If the period is assumed to be N in each lattice direction, then the result in Eq. (2.5) is replaced by

$$\mathbf{P}(\mathbf{r}) = \frac{1}{N^D} \sum_{\{s\}} \lambda^n \left(\frac{2\pi \mathbf{s}}{N} \right) \exp \left(-\frac{2\pi i \mathbf{r} \cdot \mathbf{s}}{N} \right) \tag{2.39}$$

where each component of $\mathbf{s}$ is to be summed from 0 to $N - 1$.

4. Persistent Random Walks

Before proceeding to a discussion of the properties of random walks that can be derived from the formalism given so far, we shall develop the relevant equations for so-called persistent random walks. These were first introduced by Taylor[11] in an analysis of diffusion by continuous motion. The most complete analysis of such random walks on a lattice was given by Goldstein[35]; related results are to be found in papers by many other authors.[36]

Persistent random walks, in which the transition probabilities depend on some property of the last transition, can be classified as multistate walks in which the state of the walk is defined by the last transition. A more detailed account of generalizations of two-state random walks have been given by Weiss,[37] and some asymptotic results will be mentioned in Section III.B. More recently Landman, Shlesinger, and Montroll[38] have analyzed multi-state random walks as models for the motion of clusters on crystalline surfaces.

In our description of persistent random walks we treat the 1-D case only. The simplest, and original, version of this random walk is one in which steps can be made to nearest neighbors only. Suppose that step $n-1$ has been made. Then step n will be made in the same direction with probability α, and a reversal occurs with probability $1-\alpha$. One can put this type of random walk into a more general context by using the idea of a random walk that can exist in one of two states. If it is in state 1, the transition probabilities are $p(j)$; if in state 2, they are $p(-j)$. Thus the structure functions for these two states are $\lambda(\theta)$ and $\lambda(-\theta)$ respectively. The random walk is symmetric in the sense that at any site it can be in one of two states that have mirror-image symmetry. At each step the probability of remaining in the same state is α, and that of switching states is $\beta = 1 - \alpha$. If we introduce state probabilities $P_{1,n}(r)$ and $P_{2,n}(r)$ for the two states, the generalization of Eq. (2.1) can be written

$$P_{1,n+1}(r) = \alpha \sum_{l=-\infty}^{\infty} P_{1,n}(l)p(r-l) + \beta \sum_{l=-\infty}^{\infty} P_{2,n}(l)p(l-r)$$
$$P_{2,n+1}(r) = \beta \sum_{l=-\infty}^{\infty} P_{1,n}(l)p(r-l) + \alpha \sum_{l=-\infty}^{\infty} P_{2,n}(l)p(l-r)$$
(2.40)

If we define the generating functions

$$\Gamma_{j,n}(\theta) = \sum_{r} P_{j,n}(r)e^{ir\theta}, \qquad j = 1,2$$
(2.41)

Eq. (2.40) can be transformed into the following recursion relation:

$$\begin{pmatrix} \Gamma_{1,n+1} \\ \Gamma_{2,n+1} \end{pmatrix} = \begin{pmatrix} \alpha\lambda(\theta) & \beta\lambda(-\theta) \\ \beta\lambda(\theta) & \alpha\lambda(-\theta) \end{pmatrix} \begin{pmatrix} \Gamma_{1,n} \\ \Gamma_{2,n} \end{pmatrix}$$
(2.42)

which can be solved by standard techniques for the elements of

$$\begin{pmatrix} \Gamma_{1,n} \\ \Gamma_{2,n} \end{pmatrix} = \begin{pmatrix} \alpha\lambda(\theta) & \beta\lambda(-\theta) \\ \beta\lambda(\theta) & \alpha\lambda(-\theta) \end{pmatrix}^{n} \begin{pmatrix} \Gamma_{1,0} \\ \Gamma_{2,0} \end{pmatrix}$$
(2.43)

The exact values of the Γ's are not as much of interest as their asymptotic properties, which will be discussed in the next section. Perhaps the most interesting observation possible at this point is that there are two initial data: the state and the position. We shall see that this implies a different continuum limit for the equations describing the evolution of the state probabilities than for the diffusion equation. Various generalizations of the two-state problem are easily developed. These include the continuous time version of the model just described, as well as an $n(>2)$-state model with equations similar to Eq. (2.40) for the state probability vector. An account of the continuum-limit properties of persistent random walks will be given in Section II.B, since they are intimately related to asymptotic properties which will be discussed in the next section.

B. Asymptotic Properties: Continuum Results

1. Central-Limit Results

If we start from the representation of the state probability in Eq. (2.5) or (2.18), then there are few random walks for which exact computationally useful evaluation of the integrals is possible. The simplest case is that of the symmetric random walk to nearest neighbors only, for which $\lambda(\theta) = \cos\theta$. A solution is available either by direct evaluation or by a combinatorial argument. Another model for which exact results can be found and for which longer-range transitions can occur is that specified by the transition probabilities

$$p(j) = \tfrac{1}{2}(e^a - 1)e^{-a|j|}, \qquad j = \pm 1, \pm 2, \pm 3, \ldots \qquad (2.44)$$

This model was first analyzed by Lindenberg and Shuler[39] for random walks on unbounded lattices and by Lindenberg[40] for finite lattices. The results obtained by these authors are of interest if one wishes to have a model which allows the examination of transient phenomena, but most applications of random-walk theory make use of asymptotic results. In the asymptotic regime, random walks with negative-exponential transition probabilities have no special features of interest, and the statistical properties of these random walks can be found from the general asymptotic theory to be outlined below.

If we consider the CTRW with separable $p(\mathbf{r}, t)$, there are two functions that specify the properties of these random walks: the $p(\mathbf{j})$ and $\psi(t)$. Different asymptotic results are obtained, depending on the assumptions made about the moments of these functions. The four possibilities that might be considered are:

(1) $\Sigma_j j_i^2 p(\mathbf{j}) < \infty$ for all i; $\mu = \int_0^\infty t\psi(t)\,dt < \infty$.
(2) $\Sigma_i j_i^2 p(\mathbf{j}) = \infty$ for at least one i; $\mu < \infty$.
(3) $\Sigma_j j_i^2 p(\mathbf{j}) < \infty$ for all i; $\mu = \infty$.
(4) $\Sigma_j j_i^2 p(\mathbf{j}) = \infty$ for at least one i; $\mu = \infty$.

Variants of all four cases have appeared or will appear[41] in the random walk literature; but the fourth has been of lesser interest for applications. Let us first consider the simplest possible version of case 1, a 1-D random walk in discrete time [i.e., $\psi(t) = \delta(t-1)$]. The state probability for this case is given in Eq. (2.5). We shall show that the representation in that equation, together with the assumption of a finite variance, implies that $P_n(r)$ approaches a gaussian density as $n \to \infty$, and that with further restriction on the finiteness of moments of $p(j)$ one can calculate corrections to the gaussian approximation. In the mathematical literature these types of results are related to developments of the central limit theorem.[19] Since there is such a voluminous literature on this subject, we shall suggest proofs rather than try to give a rigorous account of the theory. To this end we seek a transformation of variables in Eq. (2.5), $\theta = A_n \rho$, such that $\lim_{n \to \infty} |\lambda^n(A_n \rho)| = f(\rho) \neq \infty$ and $\neq 0$ except at a denumerable set of points. It is convenient for this purpose to define the moments

$$\mu_l = \sum_{j=-\infty}^{\infty} j^l p(j) \qquad (2.45)$$

as well as σ^2, the variance of a single step. The function $\lambda(\theta)$ can be expanded around $\theta = 0$ as

$$\lambda(\theta) \sim 1 + i\mu_1\theta - \frac{\mu_2\theta^2}{2} + o(\theta^2) \qquad (2.46)$$

so that if we set $\theta = \rho/\sqrt{n}$, we find that for n large

$$\lambda^n(\theta) \sim \exp\left[\frac{ni\mu_1\rho}{\sqrt{n}} - \frac{\sigma^2\rho^2}{2}\right] \qquad (2.47)$$

which satisfies the requirement set above. In this approximation we find

$$\begin{aligned} P_n(r) &\sim \frac{1}{2\pi\sqrt{n}} \int_{-\pi\sqrt{n}}^{\pi\sqrt{n}} \exp\left\{-\frac{\sigma^2\rho^2}{2} - i(r - n\mu_1)\frac{\rho}{\sqrt{n}}\right\} d\rho \\ &\sim \frac{1}{2\pi\sqrt{n}} \int_{-\infty}^{\infty} \exp\left\{-\frac{\sigma^2\rho^2}{2} - i(r - n\mu_1)\frac{\rho}{\sqrt{n}}\right\} d\rho \\ &= \frac{1}{\sigma\sqrt{2\pi n}} \exp\left\{-\frac{(r - n\mu_1)^2}{2n\sigma^2}\right\} \end{aligned} \qquad (2.48)$$

which is a gaussian density with mean equal to μ_1 and variance equal to $n\sigma^2$. Corrections to this approximation are easily calculated when it is known that higher-order moments are finite. For example, if μ_3 is assumed to be finite,

$$\lambda^n(\theta) \sim \exp\left[\frac{ni\mu_1\rho}{\sqrt{n}} - \frac{\sigma^2\rho^2}{2} - \frac{i\kappa_3\rho^3}{6\sqrt{n}}\right]$$

$$\sim \exp\left[\frac{ni\mu_1\rho}{\sqrt{n}} - \frac{\sigma^2\rho^2}{2}\right]\left\{1 - \frac{i\kappa_3\rho^3}{6\sqrt{n}}\right\} \qquad (2.49)$$

where κ_3 is the third cumulant,

$$\kappa_3 = \mu_3 - 3\mu_1\mu_2 + 2\mu_1^2 \qquad (2.50)$$

When this approximation to $\lambda^n(\theta)$ is substituted into the integral representation of $P_n(r)$, we find

$$\mathbf{P}(r) \sim \frac{1}{\sigma\sqrt{2\pi n}}\exp\left[-\frac{(r-n\mu_1)^2}{2n\sigma^2}\right]$$

$$\times\left\{1 - \frac{\sqrt{2\pi}\,\kappa_3}{6n\sigma^4}(r-n\mu_1)\left[3 - \frac{(r-n\mu_1)^2}{n\sigma^2}\right] + \cdots\right\} \qquad (2.51)$$

The first correction term goes like $n^{-1/2}$, since the gaussian limit is valid when $|r - n\mu_1| = O(\sqrt{n})$. One can see this on the heuristic consideration that if $|r - n\mu_1|$ were to be of higher order, the expression for $P_n(r)$ in this last equation could become negative. Higher-order correction terms require the existence of higher moments and go in powers of $n^{-1/2}$.

The asymptotic result just derived in Eq. (2.48) is easily extended to higher dimensions by starting from Eq. (2.5). The small-θ expansion of $\lambda(\theta)$ to quadratic terms is

$$\lambda(\theta) \sim 1 + i\mu_1\cdot\theta - \frac{\theta'\cdot\mu_2\cdot\theta}{2} \qquad (2.52)$$

where μ_1 is a row vector whose jth element is the average displacement in a single step along the jth axis, and μ_2 is a matrix with elements

$$(\mu_2)_{il} = \sum_{\mathbf{j}} j_i j_l p(\mathbf{j}) \qquad (2.53)$$

Following the steps that lead from Eq. (2.46) to Eq. (2.48), we find

$$P_n(\mathbf{r}) \sim \frac{1}{(2\pi n)^{D/2}} \frac{1}{|\mathbf{M}|} \exp\left\{ -\frac{1}{2n} (\mathbf{r} - n\boldsymbol{\mu}_1)' \mathbf{M}^{-1} (\mathbf{r} - n\boldsymbol{\mu}_1) \right\} \quad (2.54)$$

where $\mathbf{M}$ is the covariance matrix with elements

$$\mathbf{M}_{kl} = (\boldsymbol{\mu}_2)_{kl} - (\boldsymbol{\mu}_1)_k (\boldsymbol{\mu}_1)_l \quad (2.55)$$

and $|\mathbf{M}|$ is the determinant of $\mathbf{M}$.

One can also derive a gaussian approximation for the CTRW. For simplicity we consider this result for separable CTRWs in 1-D only. The principal idea is easily seen if we write the displacement at time t as

$$r(t) = r_1 + r_2 + \cdots + r_{n(t)} \quad (2.56)$$

where $n(t)$ is the (random) number of steps at time t. When t/μ is large, where μ is the mean time between steps, $r(t)$ is the sum of a large number of random variables, and according to the results just derived for random walks in discrete time, the distribution of $r(t)$ tends towards a gaussian. To completely characterize this gaussian we need only find the mean and variance of $r(t)$ in the large-t/μ limit. For this purpose we use the value of $P^*(\theta, s)$ given in Eq. (2.19) to generate moments following Alder's[33] procedure. In this way we find

$$\mathcal{L}\{\langle r(t)\rangle\} = -i\frac{\partial P^*}{\partial \theta}\bigg|_{\theta=0} = \frac{\mu_1\psi^*(s)}{s[1-\psi^*(s)]}$$

$$\mathcal{L}\{\langle r^2(t)\rangle\} = \frac{\mu_2\psi^*(s)}{s[1-\psi^*(s)]} + \frac{2}{s}\left[\frac{\mu_1\psi^*(s)}{1-\psi^*(s)}\right]^2 \quad (2.57)$$

where the μ_i are the ith moments of the step length. The second line differs slightly from Alder's result in Eq. (2.35) because μ_1 can differ from 0. To find the asymptotic forms of the moments we expand these transforms in powers of s and retain the lower-order terms. In this way we find[‡] that

$$\langle r(t)\rangle \sim \frac{\mu_1 t}{\mu}$$

$$\sigma^2(r(t)) \sim \frac{(\sigma^2\mu^2 + \sigma_t^2\mu_1^2)t}{\mu^3} \quad (2.58)$$

[‡]One must also assume that $\langle r(t)\rangle$ and $\langle r^2(t)\rangle$ are monotonic functions[42] to assure the validity of Eq. (2.58).

in which σ_t^2 is the variance calculated from $\psi(t)$ and assumed finite. Notice that there are two contributions to the variance of displacement: one due to the variance of step size, and the second due to that of the waiting time. If both of these are equal to zero, then $r(t)$ is determinate and $\sigma^2(r(t)) = 0$. However, if one process is determinate and the other not, $r(t)$ still has an asymptotically gaussian distribution with a nonzero variance; e.g., if the random walker always takes one step to the right, but does so at random times, the variance will be totally due to the variance in waiting time. The factor t/μ is the asymptotic mean number of steps taken in time t, as one easily finds by the methods of renewal theory.[24] In the case of a nonseparable CTRW, i.e., when $p(r, t)$ does not factor, there is an additional term in $\sigma^2(r(t))$ due to the correlation between r and t.

2. The Continuum Limit

Another very useful type of limiting process is the diffusion limit, in which a partial differential equation is obtained. Let us set $t = n\,\Delta T$ and $x = r\Delta L$, where we shall eventually let ΔL and $\Delta T \to 0$ while r and $n \to \infty$ in such a way that x and t are $O(1)$. In order to obtain the diffusion limit we shall see that ΔL and ΔT cannot go to 0 independently but must satisfy the condition

$$\lim_{\Delta L, \Delta T \to 0} \frac{(\Delta L)^2}{\Delta T} = \text{constant} \qquad (2.59)$$

To see how this comes about, let us return to the integral representation of $P_n(r)$ in Eq. (2.5), and let us set $P_n(r) = P(x, t)\Delta L$, where $x = r\Delta L$, so that $P(x, t)$ is a probability density. If we expand $\lambda^n(\theta) = \exp[n \ln \lambda(\theta)]$, using Eq. (2.46), and retain up to quadratic terms in θ, we find

$$P(x, t)\Delta L = \frac{1}{2\pi} \int_{-\pi}^{\pi} \exp\left\{ i\left(\frac{\mu_1 t}{\Delta T} - \frac{x}{\Delta L} \right)\theta - \frac{\sigma^2 t \theta^2}{2\,\Delta T} \right\} d\theta$$

$$= \frac{\Delta L}{2\pi} \int_{-\pi/\Delta L}^{\pi/\Delta L} \exp\left\{ i\left(\frac{\mu_1 \Delta L}{\Delta T} - x \right)v - \frac{\sigma^2 t (\Delta L)^2}{2\,\Delta T} v^2 \right\} dv$$

$$(2.60)$$

In the limit $\Delta L = 0$ the limits of integration go to $\pm\infty$. Furthermore, if the exponent in the integrand is not to go either to 0 or ∞ for $\sigma^2 t = O(1)$, we require that the condition in Eq. (2.59) be satisfied. If we also require the coefficient of t in the cosine term to remain finite, we must have $\mu_1 = O(\Delta L)$, which is also reasonable from the physical point of view. When all of these limits are taken one finds that $P(x, t)$ satisfies the partial differential equation

$$\frac{\partial P}{\partial t} = D\frac{\partial^2 P}{\partial x^2} - v\frac{\partial P}{\partial x} \qquad (2.61)$$

where

$$D = \lim_{\Delta L, \Delta T \to 0} \frac{\sigma^2 (\Delta L)^2}{\Delta T}, \qquad v = \lim_{\Delta L, \Delta T \to 0} \frac{\mu_1 \Delta L}{\Delta T} \qquad (2.62)$$

Since the transition probabilities were assumed to be translationally invariant, both v and D are constant. Equation (2.61) represents diffusion in a homogeneous field with a constant force. It has been used extensively in the study of chromatographic processes in the absence of imposed gradients.[43] In dimensions greater than 1, one can derive, by similar means,

$$\frac{\partial P}{\partial t} = \nabla \cdot (\mathbf{D} \cdot \nabla P - \mathbf{v}P) \qquad (2.63)$$

where $\mathbf{D}$ is a constant matrix that allows for anisotropic effects, and $\mathbf{v}$ is a constant vector. Notice that Eq. (2.63) does not depend on the type of lattice used as a starting point and is a continuum approximation.

It is instructive to pass to the continuum limit in the case of persistent random walks that satisfy Eq. (2.42), because the resulting partial differential equation is a telegraph equation rather than a diffusion equation. It follows from Eq. (2.42) that both $\Gamma_{1,n}(\theta)$ and $\Gamma_{2,n}(\theta)$ satisfy

$$\Gamma_{n+2}(\theta) = \alpha[\lambda(\theta) + \lambda(-\theta)]\Gamma_{n+1}(\theta) + (1-2\alpha)\lambda(\theta)\lambda(-\theta)\Gamma_n(\theta) \qquad (2.64)$$

which can also be written in terms of the difference operator, $\Delta\Gamma_n = \Gamma_{n+1} - \Gamma_n$, as

$$\Delta^2\Gamma_n(\theta) + \{2 - \alpha[\lambda(\theta) + \lambda(-\theta)]\}\Delta\Gamma_n(\theta)$$
$$+ \{1 - \alpha[\lambda(\theta) + \lambda(-\theta)] + (2\alpha - 1)\lambda(\theta)\lambda(-\theta)\}\Gamma_n(\theta) = 0. \qquad (2.65)$$

In order to obtain a continuum limit we first let $t = n\,\Delta T$ and note that

$$\Delta\Gamma_n \sim (\Delta T)\frac{\partial\Gamma}{\partial t} \qquad (2.66)$$

The spatial derivatives are introduced by expanding the λ's to second order in θ and making use of the correspondence between the θ and the x domains,

$$\theta^2\Gamma_n(\theta) \to -(\Delta L)^2\frac{\partial^2 P}{\partial x^2} \qquad (2.67)$$

If we make these substitutions and keep lowest-order terms, we find

$$(\Delta T)^2\frac{\partial^2 P}{\partial t^2}+2(1-\alpha)\Delta T\frac{\partial P}{\partial t}=\left[\alpha\mu_2-(2\alpha-1)\sigma^2\right](\Delta L)^2\frac{\partial^2 P}{\partial x^2} \quad (2.68)$$

We next consider the matter of scaling. The two terms on the left side will be of the same order in ΔT if $1-\alpha=\Delta T/(2T)$, where T is a constant. The limits ΔL and $\Delta T \to 0$ must be taken so that

$$\lim_{\Delta L, \Delta T \to 0} \frac{\Delta L}{\Delta T}=\text{constant} \qquad (2.69)$$

When these substitutions are made, we find that P satisfies

$$\frac{\partial^2 P}{\partial t^2} + \frac{1}{T}\frac{\partial P}{\partial t} = c^2\frac{\partial^2 P}{\partial x^2} \qquad (2.70)$$

which is the telegraph rather than the diffusion equation. In general two initial conditions are required for its solution, corresponding to the need to specify both an initial state and an initial position.

Morse and Feshbach[44] and Goldstein[35] give somewhat detailed analyses of the solution to Eq. (2.70). The solution in free space that corresponds to the gaussian in Eq. (2.48) is

$$P(x,t)=\frac{1}{cT}e^{-t/2T}I_0\left(\frac{t}{2T}\sqrt{1-\frac{x^2}{c^2t^2}}\right)H(ct-|x|) \qquad (2.71)$$

where $H(x)$ is the step function $[H(x)=1$ for $x>0$ and $H(x)=0$ for $x<0]$ and $I_0(x)$ is a Bessel function. Several curves of $P(x,t)$ are shown in Fig. 1. These show the general qualitative behavior of solutions to the telegraph equation. Because the equation is hyperbolic, the speed of propagation is finite, that is, $P(x,t)=0$ for $x>ct$. This stands in contrast to the diffusion equation, which has an infinite speed of propagation. However, if one fixes x and lets $ct \to \infty$ in Eq. (2.71), then the gaussian limit of Eq. (2.48) is obtained. Goldstein has also developed the theory of persistent random walks with traps.[35]

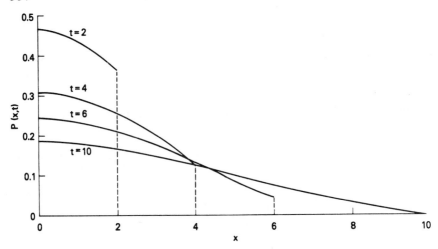

Fig. 1.   Curves of $P(x, t)$ as generated from Eq. (2.71) with $c = T = 1$ and $x = 0$. The curves are symmetric around $x = 0$ and are shown as a function of x for fixed values of t. As t increases the curves tend to a gaussian shape.

3. Random Walks with Steps of Infinite Variance

When both the variance of step length and the average time between steps are finite, we have shown that the distribution of end-to-end distance of a random walk can be approximated by the solution of a partial differential equation. When either of the two parameters just mentioned is infinite, one is perforce required to analyze the exact solution given in Eq. (2.5) or (2.39). It is remarkable that there is a single distribution—the gaussian—which gives a good approximation to the distribution of $P_n(\mathbf{r})$ for large n whatever the form of $p(\mathbf{j})$, provided only that moments of sufficiently high order exist. When this condition is not satisfied, the results depend more intimately on the form of $p(\mathbf{j}, t)$. The family of distributions which has been most used for studying the effects of infinite moments has been that of the stable law.[45] A detailed account of some properties of stable laws has been given by Gnedenko and Kolmogoroff,[19] Feller,[46] and Montroll and West.[47]

Let us first consider the case of lattice random walks in discrete time with infinite-variance steps. The stable law in 1-D can be defined by requiring that

$$p(j) \sim \frac{A}{|j|^{\alpha+1}} \tag{2.72}$$

asymptotically. When $\alpha \leqslant 2$ the second moment for this distribution is infinite. In order to investigate the properties of $P_n(r)$ for this stable law one is led to the problem of investigating $\lambda(\theta)$ for small values of $|\theta|$. To do this let us write

$$\lambda(\theta) = \lambda_0(\theta) + 2A \sum_{j=1}^{\infty} \frac{\cos j\theta}{j^{\alpha+1}} \tag{2.73}$$

where $\lambda_0(\theta)$ is assumed to have a second derivative at $\theta = 0$. It is evident that for $\alpha \leqslant 2$ the second derivative of $\lambda(\theta)$ does not exist at $\theta = 0$, and for $\alpha < 1$ neither does the first. We must therefore find the behavior of the function

$$G(\theta) = \sum_{j=1}^{\infty} \frac{1 - \cos j\theta}{j^{1+\alpha}} \tag{2.74}$$

near $\theta = 0$. Gillis and Weiss[48] have carried out this analysis, and we shall briefly sketch their calculation. If we use the integral representation

$$\frac{1}{j^{1+\alpha}} = \frac{1}{\Gamma(1+\alpha)} \int_0^{\infty} t^{\alpha} e^{-jt} dt \tag{2.75}$$

in Eq. (2.74) and interchange orders of summation and integration, we find that

$$G(\theta) = \frac{1 - \cos\theta}{\Gamma(1+\alpha)} \int_0^{\infty} \frac{t^{\alpha} e^{-t}(1 + e^{-t})}{(1 - e^{-t})\left[(1 - e^{-t})^2 + 2(1 - \cos\theta)e^{-t}\right]} dt$$

$$\sim \frac{\theta^2}{2\Gamma(1+\alpha)} \int_0^{\infty} \frac{t^{\alpha} e^{-t}(1 + e^{-t}) dt}{(1 - e^{-t})\left[(1 - e^{-t})^2 + \theta^2 e^{-t}\right]} \tag{2.76}$$

where the second line is valid for $\theta \to 0$. If we look only at the integral we see that it diverges when $\theta \to 0$ because of the singularity at $t = 0$. Hence the strategy is to approximate the integrand accurately near $t = 0$. This amounts to our setting $1 - e^{-t} = t$ in the denominator and $1 + e^{-t} = 2$ in the numerator, thus arriving at

$$G(\theta) \sim \frac{\theta^{\alpha}}{\Gamma(1+\alpha)} \int_0^{\infty} \frac{u^{\alpha-1} e^{-\theta u}}{u^2 + 1} du \tag{2.77}$$

for $\theta > 0$. When $\alpha < 2$ we find that

$$G(\theta) \sim \frac{\pi \theta^{\alpha}}{2\Gamma(1+\alpha)\sin(\pi\alpha/2)} \tag{2.78}$$

When $\alpha = 2$ we use an Abelian theorem[‡] for Laplace transforms,[42] to find from Eq. (2.77) that

$$G(\theta) \sim -\frac{\theta^2}{2}\ln\theta \qquad (2.79)$$

Therefore the asymptotic expression for $P_n(r)$ is

$$P_n(r) \sim \frac{1}{2\pi}\int_{-\infty}^{\infty} e^{-Bn|\theta|^\alpha}\cos r\theta\, d\theta \qquad (2.80)$$

in which B is a constant that is easily calculated. Analysis of this integral, shows that,

$$P_n(r) \sim \frac{\alpha Bn}{\pi r^{\alpha+1}}\Gamma(\alpha)\sin\left(\frac{\pi\alpha}{2}\right) \qquad (2.81)$$

for $\alpha < 2$, and for $r \to \infty$ with n large but fixed. When $\alpha = 1$ the integral in Eq. (2.80) can be evaluated exactly, leading to the result

$$P_n(r) \sim \frac{1}{\pi}\frac{Bn}{r^2 + B^2 n^2} \qquad (2.82)$$

which satisfies Eq. (2.81) when $|r| \gg Bn$. Thus we see that stable laws lead to long-tailed distributions for the end-to-end distances. Some curves of the densities for stable laws are shown in Fig. 2. Holt and Crow, and Tunaley,[49]

[‡]The Abelian theorem states: Let $h(t)$ be absolutely integrable in some neighborhood of $t = 0$ and let it be expressible, for sufficiently large t, as

$$h(t) = At + h_1(t)$$

where the Laplace transform $\mathcal{L}\{h_1(t)\}$ has an abscissa of convergence $\beta_1 < 0$. Then $\mathcal{L}\{h(t)\}$ exists in the right half plane and can be expressed as

$$h^*(s) = \begin{cases} \dfrac{A\Gamma(\alpha+1)}{s^{\alpha+1}} + g(s), & \alpha \neq -1, -2, -3,\ldots \\[2ex] \dfrac{(-1)^n As^{n-1}\ln s}{(n-1)!} + g(s), & \alpha = -n, \quad n = 1, 2, 3,\ldots \end{cases}$$

where $g(s)$ has no singularity in $\mathrm{Re}\, s > \beta_1$.

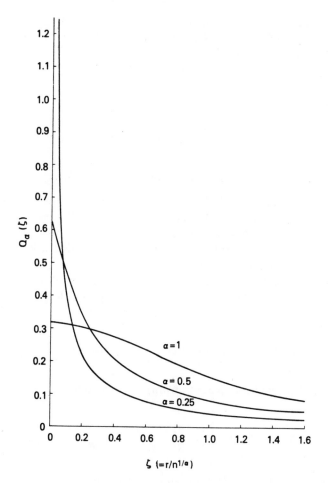

Fig. 2. Curves of $n^{1/\alpha} P_n(r) = Q_\alpha(\zeta)$, where ζ is the variable $r/n^{1/\alpha}$, for $P_n(r)$ the probability density of a stable law. The function $Q_\alpha(\zeta)$ is

$$Q_\alpha(\zeta) = \frac{1}{\pi} \int_0^\infty e^{-v^\alpha} \cos(\zeta v)\, dv$$

and is symmetric around $\zeta = 0$.

have given tables and a more extensive set of curves. The results just cited are valid for $1-D$ random walks. Gillis and Weiss[48] have given some results for the 2-D transition probabilities

$$p(j,k) \sim \frac{B}{\left(\varepsilon^2 + j^2 + k^2\right)^{1+\beta}} \qquad (2.83)$$

where $0 < \beta \leqslant 1$, but no 3-D long-tailed distributions have been studied.

4. Pausing-Time Densities with Infinite Mean

Considerable attention has been given to continuous-time random walks in which the waiting-time density $\psi(t)$ has a long tail, that is, one for which $\int_0^\infty t\psi(t)\,dt = \infty$. One physical situation in which this arises is in charge transport in amorphous insulators.[50] Some of these models will be described in Section III.C. Without going into detail about the solid-state models, we derive some results for a particular class of $\psi(t)$ with an infinite mean. These results were first given by Shlesinger.[50] Let us suppose that for $t \to \infty$, $\Psi(t)$ satisfies

$$\Psi(t) = \int_t^\infty \psi(\tau)\,d\tau \sim \frac{A}{t^\beta}, \qquad 0 < \beta \leqslant 1 \qquad (2.84)$$

An Abelian theorem for Laplace transforms[42] then allows us to conclude that

$$\psi^*(s) \sim \begin{cases} 1 - A\Gamma(1-\beta)s^\beta, & \beta \neq 1 \\ 1 + As\ln s, & \beta = 1 \end{cases} \qquad (2.85)$$

in a neighborhood of $s = 0$. We shall consider some of the consequences of having $\psi(t)$ belonging to the class of functions satisfying Eq. (2.84) and calculate the asymptotic properties of the first two moments using the formulas of Eq. (2.57). The small-$|s|$ behavior of those transforms yields

$$\mathcal{L}\{\langle r(t)\rangle\} \sim \frac{\mu_1}{A\Gamma(1-\beta)} \frac{1}{s^{\beta+1}}$$

$$\mathcal{L}\{\langle r^2(t)\rangle\} \sim \frac{2\mu_1^2}{A^2\Gamma^2(1-\beta)} \frac{1}{s^{2\beta+1}} + \frac{\mu_1^2 + \sigma^2}{A\Gamma(1-\beta)s^{\beta+1}} \qquad (2.86)$$

A Tauberian theorem for Laplace transforms[42] together with the observation that $\langle r(t) \rangle$ and $\langle r^2(t) \rangle$ are monotonic in time implies that

$$\langle r(t) \rangle \sim \frac{\mu_1 t^\beta}{A\Gamma(1-\beta)\Gamma(1+\beta)}, \qquad \beta \neq 1$$

$$\sigma^2(r(t)) \sim \begin{cases} \dfrac{\sigma^2 t^\beta}{A\Gamma(1-\beta)\Gamma(1+\beta)}, & \mu_1 = 0 \\[3mm] \dfrac{\mu_1^2}{A^2\Gamma^2(1-\beta)}\left[\dfrac{2}{\Gamma(1+2\beta)} - \dfrac{1}{\Gamma^2(1+\beta)}\right]t^{2\beta}, & \mu_1 \neq 0, \quad \beta < 1 \end{cases}$$

$$(2.87)$$

at sufficiently large times. When $\beta = 1$,

$$\langle r(t) \rangle \sim \frac{\mu_1 t}{A \ln t}$$

$$\sigma^2(r(t)) \sim \begin{cases} \dfrac{\sigma^2 t}{A \ln t}, & \mu_1 = 0 \\[3mm] \dfrac{\mu_1^2 t^2}{A^2 \ln^3 t}, & \mu_1 \neq 0 \end{cases} \qquad (2.88)$$

and when $\beta > 1$,

$$\langle r(t) \rangle \sim \frac{\mu_1 t}{\mu}$$

$$\sigma^2(r(t)) \sim \left\{ \langle n \rangle^2 \left(\frac{\mu_2}{\mu_1^2} - 1\right) + \sigma^2 \right\} \frac{t}{\mu} \qquad (2.89)$$

but there may be noninteger powers in the higher-order terms in the expansion of $\langle r(t) \rangle$ and $\sigma^2(t)$, depending on the assumptions made about the asymptotic behavior[50] of $\psi(t)$.

In order to give results for the asymptotic behavior of $P(r, t)$ derived from the small-$|s|$ behavior of $P^*(\theta, s)$ in Eq. (2.19), we shall need to introduce the standard stable distribution of order β. We define this to be the function that is the inverse Laplace transform of

$$\psi_\beta^*(s) = \exp(-s^\beta) \qquad (2.90)$$

The inverse transform will be denoted by $\psi_\beta(t)$, and $\Psi_\beta(t)$ will be the cumulative distribution corresponding to $\psi_\beta(t)$. The integral representation in Eq. (2.19) can yield asymptotic information on $P(r, t)$ if we pass to the limit of small $|s|$ and θ. When $\mu_1 = 0$ we find

$$\mathcal{L}\{P(r, t)\} \sim \frac{as^{\beta-1}}{\pi} \int_{-\infty}^{\infty} \frac{e^{-i|r|\theta}}{\sigma^2\theta^2 + 2as^\beta} d\theta$$

$$= \frac{1}{\sigma}\sqrt{\frac{a}{2}} s^{(\beta/2)-1}\exp\left(-\frac{(2as^\beta)^{1/2}|r|}{\sigma}\right) \qquad (2.91)$$

If we calculate the Laplace transform of the cumulative distribution, we have

$$\int_r^\infty \mathcal{L}\{P(\rho, t)\} d\rho \sim \frac{1}{2s}\exp\left(-\frac{(2as^\beta)^{1/2}|r|}{\sigma}\right) \qquad (2.92)$$

or

$$\int_r^\infty P(\rho, t) d\rho \sim \frac{1}{2}\left[1 - \Psi_{\beta/2}\left(\left(\frac{\sigma}{r}\right)^{2/\beta}\frac{t}{(2a)^{1/\beta}}\right)\right], \qquad r > 0$$

By symmetry one also has

$$\int_r^\infty P(\rho, t) d\rho = \int_{-\infty}^{-r} P(\rho, t) d\rho \qquad (2.93)$$

When $\mu_1 \neq 0$ we need only expand $\lambda(\theta)$ to first order in θ in the integral representation for $P^*(\theta, s)$ in Eq. (2.19). This procedure yields

$$\mathcal{L}\{P(r, t)\} \sim \frac{as^{\beta-1}}{2\pi} \int_{-\infty}^{\infty} \frac{e^{-ir\theta}}{as^\beta - i\mu\theta} d\theta$$

$$= \frac{as^{\beta-1}}{\mu_1}\exp\left(-\frac{as^\beta r}{\mu_1}\right) \qquad (2.94)$$

which allows us to conclude that when $\mu_1 > 0$,

$$\int_r^\infty P(\rho, t) d\rho \sim 1 - \Psi_\beta\left(\left(\frac{\mu_1}{ar}\right)^{1/\beta}t\right), \qquad r > 0$$

$$\int_{-\infty}^{-r} P(\rho, t) d\rho \sim 0 \qquad (2.95)$$

When $\beta = 1$ the integrals in Eqs. (2.91) and (2.94) can be evaluated exactly, and we can find expressions in closed form for the asymptotic value of $P(r, t)$. These results are

$$\mu_1 = 0: \quad P(r, t) \sim \frac{1}{\sigma} \sqrt{\frac{a}{2\pi t}} \exp\left(-\frac{ar^2}{2\sigma^2 t} \right)$$

$$\mu_1 \neq 0: \quad P(r, t) \sim \delta\left(r - \frac{\mu_1 t}{a} \right)$$

(2.96)

There are different expansions available for the calculation of the function $\psi_\beta(t)$. Several of these have been reviewed by Feller,[46] Montroll and West,[47] and Brockwell and Brown.[51] The technique used in deriving the results just discussed is essentially due to Tunaley.[49] A more detailed account is given in that reference, as well as numerically calculated values of $\psi_\beta(t)$ for several values of β. Similar material for specific choices of $\psi_\beta(t)$ was discussed by Montroll and Scher.[52] There do not appear to be any theoretical results available for long-tailed distributions not asymptotic to the stable laws.

C. Asymptotic Properties: Discrete Results

1. The Propagator and Return to the Origin

There are a number of properties of random walks on lattices that do not necessarily have analogues in the continuum limit. The earliest of these to have been analyzed is the problem of return to the origin suggested and solved by Polya[10] in 1921. We shall pose the problem in slightly more general terms than Polya did. Suppose that a random walker is initially at the origin at $n = 0$. What is the probability that the random walker will eventually reach a given point, $\mathbf{r}$? We shall answer this question for discrete-time random walks, and show that the answer depends critically on the dimensionality of the lattice. It is convenient, for purposes of analysis, to introduce a new function $F_n(\mathbf{r})$, the probability of reaching $\mathbf{r}$ for the first time at step n. This function differs from $P_n(\mathbf{r})$ because the latter allows the possibility of having been at $\mathbf{r}$ a number of times in the past. The probability of reaching $\mathbf{r}$ at some time during the random walk is

$$F(\mathbf{r}) = \sum_{n=0}^{\infty} F_n(\mathbf{r})$$

(2.97)

But this function can be related to the structure function of the random walk by expressing $F_n(\mathbf{r})$ in terms of $P_n(\mathbf{r})$, which has the integral representation given in Eq. (2.5). This can be done most conveniently in terms of generating functions, since a simple enumeration of the ways in which a random walker can reach **r** at step n shows that

$$P_n(\mathbf{r}) = \delta_{n,0}\delta_{\mathbf{r},0} + \sum_{j=0}^{n} F_j(\mathbf{r})P_{n-j}(\mathbf{0}) \qquad (2.98)$$

Hence if we introduce the generating functions

$$P(\mathbf{r}; z) = \sum_{n=0}^{\infty} P_n(\mathbf{r})z^n$$
$$F(\mathbf{r}; z) = \sum_{n=0}^{\infty} F_n(\mathbf{r})z^n \qquad (2.99)$$

we find from Eq. (2.98) that

$$F(\mathbf{r}; z) = \frac{P(\mathbf{r}; z) - \delta_{\mathbf{r},0}}{P(\mathbf{0}; z)} \qquad (2.100)$$

Since an expression is available for $P_n(\mathbf{r})$ [Eq. (2.5)], we can find an integral representation for $P(\mathbf{r}; z)$ as

$$P(\mathbf{r}; z) = \frac{1}{(2\pi)^D} \int_{-\pi}^{\pi} \cdots \int \frac{\exp(i\mathbf{r}\cdot\boldsymbol{\theta})}{1 - z\lambda(\boldsymbol{\theta})} d^D\boldsymbol{\theta} \qquad (2.101)$$

in D dimensions. Notice that the probability of reaching **r** at some time can be written in terms of the generating function $F(\mathbf{r}; z)$ as

$$F(\mathbf{r}) = F(\mathbf{r}; 1) \qquad (2.102)$$

With this formula we can immediately give an answer to Polya's original question of whether return to the origin is certain or occurs with some probability <1. From Eq. (2.100) one sees that $F(\mathbf{0}) = 1 - P^{-1}(\mathbf{0}; 1)$, so that return to the origin occurs with probability 1 when $P(\mathbf{0}; 1) = \infty$ and is less than 1 when $P(\mathbf{0}; 1)$ is finite. We must therefore examine the properties of

$$P(\mathbf{0}; 1) = \frac{1}{(2\pi)^D} \int_{-\pi}^{\pi} \cdots \int \frac{d^D\boldsymbol{\theta}}{1 - \lambda(\boldsymbol{\theta})} \qquad (2.103)$$

in different numbers of dimensions. The question of return to the origin is mainly of interest for symmetric random walks, since otherwise systematic drift tends to move the random walker away from the origin. Since the integrals in Eq. (2.103) have finite limits, the only way in which $P(0; 1)$ can be infinite is to have the denominator of the integrand vanish. One point at which it vanishes is at $\theta = 0$. We assume that this is the only point at which $\lambda(\theta) = 1$. The significance of the assumption is discussed by Gnedenko and Kolmogoroff[19] for 1-D random walks. In the neighborhood of $\theta = 0$ we can expand $\lambda(\theta)$ as

$$\lambda(\theta) = 1 - \frac{1}{2} \sum_{l=1}^{D} \mu_2(l)\theta_l^2 \qquad (2.104)$$

where

$$\mu_2(l) = \sum_j j_l^2 p(\mathbf{j}) \qquad (2.105)$$

is assumed to be finite. To examine the convergence of the integral in Eq. (2.103) we introduce polar coordinates. The Jacobian of the transformation leads to a weighting factor $\theta^{m-1} d\theta$ in the numerator, where $\theta = (\boldsymbol{\theta} \cdot \boldsymbol{\theta})^{1/2}$, the remaining factors being unimportant. We see that convergence or divergence of the multiple integral for $P(0; 1)$ depends on the convergence of

$$\int_0 \theta^{D-3} d\theta$$

in any bounded region including the origin. Thus, the integral diverges in 1 or 2 dimensions and converges in 3 or more dimensions. It is interesting to observe that in 1-D, a random walk whose transition probabilities are asymptotically equal to those of a stable law [i.e., $p(j) \sim A|j|^{-(\beta+1)}$], returns to the origin with probability 1 when β, the index of the stable law, is > 1, but the return probability is less than 1 otherwise. When $P(0, 1)$ is infinite, so that return to the origin occurs with probability 1, the probability of reaching an arbitrary point is also equal to 1. This is a consequence of Eq. (2.101), since the singularity at $\theta = 0$ is the same in the integral representation of $P(\mathbf{r}; z)$ as it is in that for $P(0; z)$. Foldes and Gabor[53] have considered the problem of finding the probability of k random walkers simultaneously returning to the origin. They show that two random walkers simultaneously return to the origin in 1-D with probability 1, but that the probability is less than 1 for three or more random walkers. The calculation is made by observing that the random walk of n random walkers in 1-D is equivalent to that of a single random walker in n-D, from which the conclusion follows immediately.

Although the probability of returning to the origin is equal to 1 in one and two dimensions, the average return time is infinite. This can be verified starting from the formula for the average return time,

$$\langle n_R \rangle = \sum_{n=0}^{\infty} n F_n(\mathbf{0}) = \frac{dF(\mathbf{0}; z)}{dz}\bigg|_{z=1}$$

$$= \frac{1}{P^2(\mathbf{0}; z)}\bigg|_{z=1} \frac{dP(\mathbf{0}; z)}{dz}\bigg|_{z=1} \tag{2.106}$$

In 1-D, when the variance of the transition probabilities is finite, we can evaluate $P(0; z)$ in a neighborhood of $z = 1$ as

$$P(0; z) = \frac{1}{2\pi} \int_{-\pi}^{\pi} \frac{d\theta}{1 - z\lambda(\theta)} \sim \frac{1}{2\pi} \int_{-\infty}^{\infty} \frac{d\theta}{1 - z + \frac{1}{2}\sigma^2\theta^2}$$

$$= \frac{1}{\sigma\sqrt{2(1 - z)}} \tag{2.107}$$

from which it follows that $\langle n_R \rangle = \infty$. We expect the same to be the case in 2-D, since there is more space per step. Formally, we can write

$$P(0; z) \sim \frac{1}{(2\pi)^2} \iint_{-\pi}^{\pi} \frac{d\theta_1 \, d\theta_2}{1 - z + \frac{1}{2}\sigma_1^2\theta_1^2 + \frac{1}{2}\sigma_2^2\theta_2^2} \tag{2.108}$$

for $z \to 1$. We can get an alternative representation of the integral by making use of the identity

$$u^{-1} = \int_0^{\infty} \exp(-ut) \, du \tag{2.109}$$

to find

$$P(0; z) \sim \frac{1}{(2\pi)^2} \int_0^{\infty} e^{-(1-z)t} G_1(t) G_2(t) \, dt \tag{2.110}$$

where

$$G_i(t) = \int_{-\pi}^{\pi} \exp\left(-\frac{\sigma_i^2\theta^2 t}{2}\right) d\theta \tag{2.111}$$

To find the asymptotic behavior of $P(0; z)$ we now appeal to an Abelian theorem for Laplace transforms[42] that relates the behavior of the integral in Eq. (2.110) as $z \to 1$ to the asymptotic behavior of $G_1(t)G_2(t)$ as $t \to \infty$. This is easily determined to be

$$G_1(t)G_2(t) \sim \frac{2\pi}{\sigma_1\sigma_2 t} \qquad (2.112)$$

so that

$$P(0; z) \sim -\frac{1}{2\pi\sigma_1\sigma_2}\ln(1-z) \qquad (2.113)$$

From this result it is evident that $\langle n_R \rangle = \infty$ in 2-D. In three or more dimensions return to the origin is not certain, the probability of return being given by Eq. (2.100) with $z = 1$, the expression for $P(0; 1)$ being that in Eq. (2.103). Values of the probability of returning to the origin have been determined in closed form for random walks with transitions to nearest neighbors for the three cubic lattices.[20,54] The results are:

$$F = 0.3405 + \quad \text{(simple cubic)}$$
$$F = 0.2822 + \quad \text{(body-centered cubic)}$$
$$F = 0.2563 + \quad \text{(face-centered cubic)}$$

All of the preceding results are for the case of a single random walker starting at the origin and ending either at the origin or at some other designated point. It we suppose that $k > 1$ random walkers start from the origin, then we can inquire about the average time for at least one of them to reach the origin.[55] The expression for this expected time to reach a designated point is

$$\langle n(\mathbf{R}) \rangle = \sum_{n=1}^{\infty} \left\{ \sum_{j=n+1}^{\infty} F_j(\mathbf{R}) \right\}^k \qquad (2.114)$$

When the variance of single-step transitions is finite, it is not too difficult to show, starting from Eq. (2.107), that in 1-D

$$\sum_{j=n}^{\infty} F_j(R) \sim \sqrt{\frac{2}{\pi n}} \frac{|R|}{\sigma} \qquad (2.115)$$

so that the series in Eq. (2.114) converges if $k \geq 3$. This might seem paradoxical at first, since the average time for any one of the random walkers to reach R is infinite. However, it is well known that the distribution of the minimum of a set of identically distributed random variables is not the same as that for an arbitrary member of the set. In 2-D the average time to reach a designated point is infinite, independent of the number of random walkers, since

$$\sum_{j=n}^{\infty} F_j(\mathbf{R}) \sim \frac{A}{\ln n} \tag{2.116}$$

where A is a constant and $\sum_{n=2}^{\infty} \ln^{-k} n = \infty$ for all real k. In 3-D, return to the origin is not certain, so that the average time to reach a designated point will necessarily be infinite. However, if one calculates this average time for random walkers constrained to reach the target point at some time, then the average time is infinite for $k = 1$ or 2 walkers and finite otherwise. It is interesting, also, to consider this equation for $D > 3$. Lindenberg et al.[55] show that in 4-D, conditional on reaching a target point, two random walkers ensure reaching the point in a finite average time, while in five or more dimensions any random walker who reaches a target point will do so in a finite mean time. While this might seem paradoxical at first, it should be borne in mind one is conditioning on much smaller sets of random walkers as the dimensionality increases. That is to say, if one increases the number of dimensions, any random walker who reaches a target point will do so very quickly or not at all.

2. The Number of Distinct Sites Visited by an n-Step Walk

There are a number of other parameters describing lattice random walks that can be obtained by a systematic application of Tauberian theorems for power series. The first that we describe is the expected number of distinct lattice points visited during the course of an n-step walk. This quantity was first obtained by Dvoretzky and Erdös,[56] and a later more heuristic derivation was given by Vineyard.[57] Our derivation follows one originally presented by Montroll.[20,21] Let S_n be the number of distinct points visited during the course of an n-step walk, and let Δ_n be the expected number of new points visited on the nth step. With these definitions we have

$$\langle S_n \rangle = 1 + \sum_{j=1}^{n} \Delta_j$$
$$\Delta_j = \sum_{\mathbf{r}} F_j(\mathbf{r}) \tag{2.117}$$

If we let $\Delta(z)$ be the generating function of the Δ_n,

$$\Delta(z) = \sum_{n=0}^{\infty} \Delta_n z^n \qquad (2.118)$$

then we can combine Eqs. (2.99), (2.117), and (2.118) to find that

$$\Delta(z) = \frac{z}{(1-z)P(\mathbf{0}; z)} \qquad (2.119)$$

This result, together with Eq. (2.117), implies that the generating function of the $\langle S_n \rangle$ is

$$\sum_{n=0}^{\infty} \langle S_n \rangle z^n = \frac{z}{(1-z)^2 P(\mathbf{0}; z)} \qquad (2.120)$$

The derivation of the asymptotic values of $\langle S_n \rangle$ requires the use of a Tauberian theorem for power series.[58] The theorem is as follows: Let $U(y)$ be defined by

$$U(y) = \sum_{n=0}^{\infty} a_n \exp(-ny) \qquad (2.121)$$

where $a_n > 0$. Let $U(y)$ have the asymptotic form, as $y \to 0$,

$$U(y) \sim \varphi(y^{-1}) = y^{-\gamma} L(y^{-1}) \qquad (2.122)$$

where $L(y)$ is a slowly varying function,‡ and $x^\gamma L(x)$ is a positive increasing function of x for sufficiently large x. Then as $n \to \infty$

$$a_1 + a_2 + \cdots + a_n \sim \frac{\varphi(n)}{\Gamma(\gamma + 1)} \qquad (2.123)$$

---

‡A slowly varying function at $x = \infty$ is defined to be one for which

$$\lim_{x \to \infty} \frac{L(cx)}{L(x)} = 1$$

for every $c > 0$. An example of such a function is $L(x) = \ln x$. A lucid account of some of the properties of slowly varying functions has been given by Feller.[46]

If the a's are monotonic and $\varphi(x)$ is differentiable, we can infer that

$$a_n \sim \frac{\varphi'(n)}{\Gamma(\gamma+1)} \tag{2.124}$$

for the individual a's.

The theorem just cited can be applied to the problem of finding the large-n behavior of $\langle S_n \rangle$. When $\sigma^2 < \infty$ we have

1-D: $\qquad\qquad\qquad P(0,z) \sim \dfrac{1}{\sigma\{2(1-z)\}^{1/2}}$

2-D: $\qquad\qquad\qquad P(\mathbf{0},z) \sim -(2\pi\sigma_1\sigma_2)^{-1}\ln(1-z) \tag{2.125}$

3-D: $\qquad\qquad\qquad P(\mathbf{0},z) \sim P(\mathbf{0},1)$

where $P(\mathbf{0},1)$ is convergent in three or more dimensions. These results imply that for this case the $\langle S_n \rangle$ are asymptotically given by

1-D: $\qquad\qquad\qquad \langle S_n \rangle \sim \sigma \left(\dfrac{8n}{\pi} \right)^{1/2}$

2-D: $\qquad\qquad\qquad \langle S_n \rangle \sim \dfrac{\pi\sigma_1\sigma_2 n}{\ln n} \tag{2.126}$

3-D: $\qquad\qquad\qquad \langle S_n \rangle \sim \dfrac{n}{P(\mathbf{0},1)}$

These results are valid when one is interested in the expected number of distinct points visited on an entire lattice. The same formalism suffices to find the expected number of distinct points visited on some subset of points on the lattice.[59] If one assumes the special subset, which we denote by T, to be periodic, then a unit cell can be defined. If we allow the dimensions of this unit cell to increase without limit, and the number of points of T to increase so that a density can be defined (i.e., $\lim_{N \to \infty} N_T/N = \rho =$ constant, where N is the number of lattice points in the unit cell and N_T is the number of points of T in the unit cell), then Weiss and Shlesinger[59] have shown that $\langle S_n \rangle_T \sim \rho \langle S_n \rangle$, where $\langle S_n \rangle$ is given in Eq. (2.126). They also derived results for specific subsets T. For example, in 3-D, they showed that for symmetric random walks the expected number of distinct points visited on a plane (in a simple cubic lattice) by an n-step walk is asymptotically proportional to $n^{1/2}$, and the number of visits to a line is asymptotic to $\ln n$.

The technique for obtaining the results described above cannot be extended to find expressions for higher moments of S_n or the distribution function for S_n. Using much more difficult techniques, Jain and his collaborators[60] have found the asymptotic variance in two or more dimensions and the asymptotic distribution function in three or more dimensions. In 2-D they find that $\mathrm{var}\, S_n \sim cn^2/\ln^4 n$, where c is a specified constant, while in 3-D, when the mean step length is equal to 0, they find that

$$\mathrm{var}\, S_n \sim c_1 n \ln n \qquad (2.127)$$

where c_1 is a constant. The asymptotic distribution for the random variable $(S_n - \langle S_n \rangle)/\mathrm{var}\, S_n$ is found to be a gaussian with mean 0 and variance 1. A different distribution function is found to hold in 1-D, and the distribution function in 2-D appears not be known at this time. These results are consistent with Monte Carlo calculations by Beeler and Delaney[61] and by Beeler[62] for different 3-D lattices. They found indications of the asymptotic gaussian distribution for $n > 5000$ steps in the 3-D case, but were not able to make a definitive statement for the 2-D square lattice. These results are of interest in studies of energy transport in the presence of traps. Some of this work will be described in Section III.C of this paper.

Gillis and Weiss[48] extended the calculations for $\langle S_n \rangle$ to cover the case where $p(j) \sim A/|j|^{1+\beta}$ for large values of j. They showed that in 1-D

$$\langle S_n \rangle \sim \begin{cases} \dfrac{n}{P(0,1)}, & 0 < \beta < 1 \\[2ex] \dfrac{3An}{\ln n}, & \beta = 1 \\[2ex] An^{1/\beta}, & 1 < \beta \leq 2 \\[2ex] \left[\dfrac{An\ln n}{\zeta(3)} \right]^{1/2}, & \beta = 2 \end{cases} \qquad (2.128)$$

In 2-D they analyzed the set of transition probabilities in Eq. (2.83), showing that

$$\langle S_n \rangle \sim \begin{cases} \dfrac{n}{P(\mathbf{0};1)}, & 0 < \beta < 1 \\[2ex] \dfrac{\pi^2 Bn}{8\ln\ln n}, & \beta = 1 \end{cases} \qquad (2.129)$$

3. Occupancy

Another way of characterizing properties of random walks is by means of the so-called occupancy. This property was first analyzed by Erdös and Taylor,[63] and later by Montroll and Weiss.[21] We follow the latter analysis.

The occupancy of a lattice site $\mathbf{r}$ in the course of an n-step walk is defined to be the number of visits that the random walker has made to that site. We have defined $F_j(\mathbf{r})$ to be the probability that the random walker reached $\mathbf{r}$ for the first time at step j starting from 0. We likewise define $F_j^{(l)}(\mathbf{r})$ to be the probability that a random walker reaches $\mathbf{r}$ for the lth time at step j. The $F_j^{(l)}(r)$ satisfy the recurrence relation

$$F_j^{(l+1)}(\mathbf{r}) = \sum_{s=0}^{j} F_s^{(l)}(\mathbf{r}) F_{j-s}(\mathbf{0}) \tag{2.130}$$

The probability that a random walker has visited $\mathbf{r}$ at least s times during an n-step walk is

$$\sum_{j=1}^{n} F_j^{(s)}(\mathbf{r}), \qquad \mathbf{r} \neq \mathbf{0}$$

$$\sum_{j=1}^{n-1} F_j^{(s-1)}(\mathbf{r}), \qquad \mathbf{r} = \mathbf{0}$$

where a distinction needs to be made between $\mathbf{r} = \mathbf{0}$ and $\mathbf{r} \neq \mathbf{0}$ because the random walker started at the origin. If we let $\beta_n^{(s)}(\mathbf{r})$ be the probability that the random walker visited $\mathbf{r}$ exactly s times in an n-step walk, then

$$\beta_n^{(s)}(\mathbf{r}) = \begin{cases} \sum_{j=1}^{n} \left[F_j^{(s)}(\mathbf{r}) - F_j^{(s+1)}(\mathbf{r}) \right], & \mathbf{r} \neq \mathbf{0} \\ \sum_{j=1}^{n} \left[F_j^{(s-1)}(\mathbf{0}) - F_j^{(s)}(\mathbf{0}) \right], & \mathbf{r} = \mathbf{0} \end{cases} \tag{2.131}$$

If one fixes $\mathbf{r}$ and wishes to find statistical properties of the number of visits to $\mathbf{r}$, it is convenient to introduce the generating functions

$$F^{(s)}(\mathbf{r}; z) = \sum_{n=0}^{\infty} F_n^{(s)}(\mathbf{r}) z^n$$

$$\beta^{(s)}(\mathbf{r}; z) = \sum_{n=0}^{\infty} \beta_n^{(s)}(\mathbf{r}) z^n \tag{2.132}$$

Equation (2.130) has as its consequence

$$F^{(s)}(\mathbf{r}; z) = \left[F(\mathbf{0}; z) \right]^{s-1} F(\mathbf{r}; z) \tag{2.133}$$

where $F(\mathbf{r}; z)$ is the generating function defined in Eq. (2.99). The combination of Eqs. (2.131)–(2.133) allows us to write

$$\beta^{(s)}(\mathbf{r}; z) = \begin{cases} (1-z)^{-1}[1-F(\mathbf{0}; z)][F(\mathbf{0}; z)]^{s-1}F(\mathbf{r}; z), & \mathbf{r} \neq \mathbf{0} \\ (1-z)^{-1}[1-F(\mathbf{0}; z)][F(\mathbf{0}; z)]^{s-1}, & \mathbf{r} = \mathbf{0} \end{cases}$$

(2.134)

The average number of visits to $\mathbf{r}$ after n steps is

$$M_n(\mathbf{r}) = \sum_{s=1}^{\infty} s\beta_n^{(s)}(\mathbf{r})$$

(2.135)

which has the generating function

$$M(\mathbf{r}; z) = \sum_{n=0}^{\infty} M_n(\mathbf{r})z^n = \sum_{s=1}^{\infty} s\beta^{(s)}(\mathbf{r}; z)$$

$$= (1-z)^{-1}P(\mathbf{r}; z)$$

(2.136)

as is obvious in any case.

We see that the asymptotic properties of the $M_n(\mathbf{r})$ can be determined from analytic properties of $P(\mathbf{r}; z)$ in a neighborhood of the singularity at $z = 1$. In 1-D when σ^2 is finite we can determine the analytic behavior from

$$P(r; z) \sim \frac{1}{2\pi} \int_{-\infty}^{\infty} \frac{e^{-ir\theta}d\theta}{1 - z - i\mu\theta + \frac{1}{2}\sigma^2\theta^2}$$

(2.137)

where μ is the average step length. This integral has no singularity if $\mu \neq 0$; hence we can infer that the average number of visits to r is finite in that case. When $\mu = 0$ we find that

$$P(r; z) \sim \frac{1}{2\pi} \int_0^{\infty} e^{-(1-z)t}dt \int_{-\infty}^{\infty} e^{-ir\theta - \sigma^2\theta^2 t/2}\, d\theta$$

$$= \frac{1}{\sigma\sqrt{2\pi}} \int_0^{\infty} e^{-(1-z)t - r^2/2\sigma^2 t}\frac{dt}{\sqrt{t}} = \frac{\exp\left\{-(r/\sigma)\sqrt{2(1-z)}\right\}}{\sigma\sqrt{2(1-z)}}$$

(2.138)

The combination of this relation, Eq. (2.136), and the Tauberian theorem quoted earlier leads to the result

$$M_n(r) \sim \frac{1}{\sigma}\sqrt{\frac{2n}{\pi}} \qquad (2.139)$$

To a first approximation the average number of visits to r goes up as $n^{1/2}$, and is independent of r. A calculation of higher correction terms becomes somewhat more complicated but leads to terms that depend on r. In 3-D, $P(\mathbf{r}; 1)$ is finite provided that the second moments of the step length are finite so that to a first approximation

$$M_n(\mathbf{r}) \sim P(\mathbf{r}; 1) \qquad (2.140)$$

The corrections to this result go in powers of $n^{-1/2}$, since it can be shown that

$$P(\mathbf{r}; z) \sim P(\mathbf{r}; 1) + a_1(\mathbf{r})(1-z)^{1/2} + a_2(\mathbf{r})(1-z) + \cdots \qquad (2.141)$$

Several authors have calculated the coefficients $a_1(0), a_2(0), \ldots$ for specific models describing symmetric random walks on different cubic lattices[64] with transitions to nearest neighbors only. The most complete studies along this line are those of Joyce,[65] which treat nearest-neighbor random walks. A parallel analysis for random walks with more general transition probabilities seems to be much more difficult to carry out.

The analysis of the last few paragraphs is directed towards the statistical properties of the occupancy of a single point. Recently Rubin and Weiss[66] developed a formalism for calculating properties of the occupancy of a set of points, $T = \{\mathbf{R}_i\}$, $i = 1, 2, \ldots, m$. They start from a set of recursion relationships for the probabilities $P(\mathbf{r}|l_1, l_2, \ldots, l_m)$ defined as the probability of being at r at step n, conditional on having been at $\mathbf{R}_j$ l_j times, $j = 1, 2, \ldots, m$. If one defines the multiple generating functions

$$\Gamma(\boldsymbol{\theta}, \mathbf{x}, z) = \sum_{\mathbf{r}} e^{i\mathbf{r}\cdot\boldsymbol{\theta}} \sum_{n=0}^{\infty} z^n \sum_{l_1=0}^{\infty} \cdots$$

$$\sum_{l_m=0}^{\infty} P_n(\mathbf{r}|\mathbf{l}) x_1^{l_1} x_2^{l_2} \cdots x_m^{l_m} \qquad (2.142)$$

it can be shown that

$$\Gamma(\boldsymbol{\theta}, \mathbf{x}, z) = \frac{1 - \dfrac{1}{Q}\displaystyle\sum_{j=1}^{m}(1-x_j)Q_j e^{i\boldsymbol{\theta}\cdot\mathbf{R}_j}}{1 - z\lambda(\boldsymbol{\theta})} \qquad (2.143)$$

where Q is the $m \times m$ determinant with elements

$$
\left.\begin{array}{l}
Q_{jj} = x_j + (1 - x_j) P(0; z) \\
Q_{jk} = (1 - x_k) P(\mathbf{R}_j - \mathbf{R}_k; z)
\end{array}\right\} \quad j, k = 1, 2, \ldots, m \quad (2.144)
$$

and Q_j in Eq. (2.143) is obtained from Q by replacing the jth column with a vector whose kth component is $P(\mathbf{R}_k; z)$. A number of known results can be obtained immediately from Eq. (2.143), including those for the occupancy of a single point. One can also find expressions for such quantities as the generating function for the probability of a random walker going, in n steps, from $\mathbf{0}$ to $\mathbf{R}$ with no intermediate visits to either of those points. For a symmetric random walk this generating function is

$$
\Pi(z) = \frac{P(\mathbf{R}; z)}{P^2(0; z) - P^2(\mathbf{R}; z)} \qquad (2.145)
$$

Similar techniques have been applied by Rubin[67] in a study of the configuration of polymers at an interface.

Another parameter whose asymptotic properties are calculable in much the same way as $M_n(\mathbf{r})$ is the expected number of points visited exactly s times in an n-step random walk. This quantity is given by

$$
V_n^{(s)} = \sum_{\mathbf{r}} \beta_n^{(s)}(\mathbf{r}) \qquad (2.146)
$$

Using Eqs. (2.131) and (2.132), we find that

$$
V_s(z) \equiv \sum_{n=0}^{\infty} V_n^{(s)} z^n = \frac{1}{(1-z)^2 P^2(0; z)} \left\{ 1 - \frac{1}{P(0; z)} \right\}^{s-1} \qquad (2.147)
$$

In 1-D with $\mu = 0$, $\sigma^2 < \infty$, we have from Eq. (2.138)

$$
V_s(z) \sim \frac{2\sigma^2}{1-z} \left[1 - \sigma\sqrt{2(1-z)} \right]^{s-1} \qquad (2.148)
$$

for $z \to 1$. This representation, together with the Tauberian theorem cited earlier, implies that

$$
V_1^{(s)} + V_2^{(s)} + \cdots + V_n^{(s)} \sim 2\sigma^2 n \left(1 - \sigma\sqrt{\frac{2}{n}} \right)^{s-1} \qquad (2.149)
$$

for large n. One can also show that this implies the result

$$V_n^{(s)} \sim 2\sigma^2 \tag{2.150}$$

for sufficiently large n. Thus, for sufficiently large n the number of points visited exactly s times tends to a constant for all s until $s = O(n^{1/2})$, at which point it begins to fall off with increasing s. For symmetric random walks in 2-D with finite variances one likewise finds that

$$V_n^{(s)} \sim \frac{4\pi^2\sigma_1^2\sigma_2^2 n}{\ln^2 n} \left(1 - \frac{2\pi\sigma_1\sigma_2}{\ln n}\right)^{s-1} \tag{2.151}$$

This result was obtained for $s = 1$ by Erdös and Taylor.[63] In the 3-D case under similar assumptions one finds

$$V_n^{(s)} \sim \frac{n}{P^2(\mathbf{0};1)} \left(1 - \frac{1}{P(\mathbf{0};1)}\right)^{s-1} \tag{2.152}$$

which can also be written in terms of F, the probability of returning to the origin [Eq. (2.102)], as $V_n^{(s)} \sim n(1-F)^2 F^{s-1}$.

The results just presented are relevant for the occupancy of single points. There has been some attention devoted to occupancy problems for finite sets. A good introduction to this material is found in the book by Spitzer.[2] A considerable portion of this research area is devoted to occupancy problems for a half line, which arise naturally in the context of queueing theory.[2,46,68] Quantitative results analogous to those that we have discussed in the last few sections are more difficult to obtain for sets rather than single points.[66] There is one calculation applicable to occupancy problems in more than one dimension by Darling and Kac[69] that will be summarized here for its methodological interest. We follow the original derivation in passing to the continuum limit, that is, replacing the recurrence relation describing the random walk by a partial differential equation. Let $\mathbf{r}(t)$ be the position of a random walker at time t, and let $V(x)$ be a nonnegative function. Darling and Kac study the limiting distribution of the functional

$$\mathcal{V}(t) = \frac{1}{u(t)} \int_0^t V(\mathbf{r}(\tau)) \, d\tau \tag{2.153}$$

where $u(t)$ is a normalizing function that is found as part of the analysis. The interest in finding the distribution of $\mathcal{V}(t)$ is that if $V(\mathbf{r}) = 1$ for some set A and $V(\mathbf{r}) = 0$ otherwise, then

$$\int_0^t V(\mathbf{r}(\tau))\, d\tau \qquad (2.154)$$

represents the occupancy of A in time t.

The proof proceeds by finding asymptotic expressions, as $s \to 0$, of the Laplace transforms of the moments $\mu_n(t)$, which are defined to be

$$\mu_n(t) = \left\langle \left\{ \int_0^t V(\mathbf{r}(\tau))\, d\tau \right\}^n \right\rangle$$

$$= \int_0^t \cdots \int \langle V(\mathbf{r}(\tau_1)) V(\mathbf{r}(\tau_2)) \ldots V(\mathbf{r}(\tau_n)) \rangle\, d\tau, d\tau_2 \ldots d\tau_n \qquad (2.155)$$

We will suppose that the probability density for $\mathbf{r}(t)$ satisfies a diffusion equation with no drift, i.e., the random walk has an average displacement equal to zero, and a finite variance associated with the displacement. The propagator for free diffusion in m dimensions can be written

$$P(\mathbf{r}, t | \mathbf{r}_0, 0) = \frac{1}{(4\pi Dt)^{m/2}} \exp\left\{ -\frac{(\mathbf{r} - \mathbf{r}_0)^2}{4Dt} \right\} \qquad (2.156)$$

In order to put Eq. (2.155) into a more useful form it is convenient to time-order the t's. This leads to the expression

$$\mu_n(t) = n! \int_0^t d\tau_n \int_0^{\tau_n} d\tau_{n-1} \cdots \int_0^{\tau_2} d\tau_1 \langle V(\mathbf{r}(\tau_1)) V(\mathbf{r}(\tau_2)) \ldots V(\mathbf{r}(\tau_n)) \rangle$$

$$(2.157)$$

The averages can now be written explicitly in terms of the $P(\mathbf{r}, t | \mathbf{r}_0, 0)$ as

$$\langle V(\mathbf{r}(\tau_1)) V(\mathbf{r}(\tau_2)) \ldots V(\mathbf{r}(\tau_n)) \rangle = \int_{-\infty}^{\infty} \cdots \int V(\mathbf{r}_1) V(\mathbf{r}_2) \ldots V(\mathbf{r}_n)$$

$$\times P(\mathbf{r}_1, \tau_1 | 0, 0) P(\mathbf{r}_2, \tau_2 | \mathbf{r}_1, \tau_1) \ldots$$

$$\times P(\mathbf{r}_n, \tau_n | \mathbf{r}_{n-1}, \tau_{n-1})\, d\mathbf{r}_1\, d\mathbf{r}_2 \ldots d\mathbf{r}_n$$

$$(2.158)$$

But because of the Markov character of the random walk, $P(\mathbf{r}, t | \mathbf{r}', t') = P(\mathbf{r}, t - t' | \mathbf{r}', 0)$, which means that the time integrals in Eq. (2.157) are in the form of a convolution. If we denote the Laplace transform of $P(\mathbf{r}, t|\mathbf{r}_0, 0)$ with respect to t by $P^*(\mathbf{r}, s|\mathbf{r}_0, 0)$, then it follows that

$$\int_0^\infty \mu_n(t) e^{-st}\, dt = \frac{n!}{s} \int \cdots \int_{-\infty}^\infty V(\mathbf{r}_1) V(\mathbf{r}_2) \dots V(\mathbf{r}_n)$$

$$\times P^*(\mathbf{r}_1, s|0, 0) P^*(\mathbf{r}_2, s|\mathbf{r}_1, 0) \dots P^*(\mathbf{r}_n, s|\mathbf{r}_{n-1}, 0)\, d\mathbf{r}_1\, d\mathbf{r}_2 \dots d\mathbf{r}_n$$

$$(2.159)$$

To derive this last result we need make no assumptions other than that of a Markov property for the random walk, and the possibility of interchanging the order of averaging and the integrations. At this point we shall follow Darling and Kac in evaluating Eq. (2.159) in the limit $s = 0$ for the 2-D case. The Laplace transform of Eq. (2.156) in two dimensions is

$$P^*(\mathbf{r}, s|\mathbf{r}_0, 0) = \frac{1}{2\pi D} K_0 \left(|\mathbf{r} - \mathbf{r}_0| \sqrt{\frac{s}{D}} \right)$$

$$\sim -\frac{1}{2\pi D} \ln \left(|\mathbf{r} - \mathbf{r}_0| \sqrt{\frac{s}{D}} \right) \qquad (2.160)$$

where the last line is the small-$|s|$ approximation. In order to keep the results dimensionally correct we introduce the unit of time T. Then Eq. (2.160) becomes

$$P^*(\mathbf{r}, s|\mathbf{r}_0, 0) \sim \frac{1}{4\pi D} \ln \frac{1}{sT} \qquad (2.161)$$

and the asymptotic value of the Laplace transform of $\mu_n(t)$ is

$$\int_0^\infty e^{-st} \mu_n(t)\, dt \sim \frac{n!}{(4\pi D)^n} \frac{C^n}{sT} \left(\ln \frac{1}{sT} \right)^n \qquad (2.162)$$

in which

$$C = \int_{-\infty}^\infty V(\mathbf{r})\, d\mathbf{r} \qquad (2.163)$$

is the area of the set A when $V(\mathbf{r}) = 1$ for $\mathbf{r} \in A$ and $V(\mathbf{r}) = 0$ otherwise. Since, on physical grounds, $\mu_n(t)$ must be a monotonically increasing function of t, the asymptotic value of $\mu_n(t)$ for large t/T is

$$\mu_n(t) \sim \frac{n!}{(4\pi DT)^n} C^n \left(\ln \frac{t}{T} \right)^n \qquad (2.164)$$

Finally, in order to transform this estimate into an expression for the distribution function of $\mathcal{V}(t)$, we calculate an asymptotic form for the characteristic function $\langle \exp(-\zeta \int_0^t V(\mathbf{r}(\tau))\,d\tau)\rangle$ which can be inverted in closed form. We find for this function

$$\left\langle \exp\left(-\zeta \int_0^t V(\mathbf{r}(t))\,d\tau\right)\right\rangle = \sum_{n=0}^{\infty} \frac{(-1)^n}{n!}\mu_n(t)\zeta^n$$

$$\sim \left[1 + \frac{\zeta C}{4\pi DT}\ln\frac{t}{T}\right]^{-1} \tag{2.165}$$

Since the inversion with respect to ζ is simple, we find

$$\lim_{t/T \to \infty} \Pr\left\{\frac{4\pi DT}{C\ln(t/T)}\int_0^{t/T} V(\mathbf{r}(\tau))\,d\tau < \alpha\right\}$$

$$= \lim_{t/T \to \infty} \Pr\left\{\mathcal{V}\left(\frac{t}{T}\right) < \alpha\right\} = 1 - e^{-\alpha} \tag{2.166}$$

Darling and Kac have also discussed further generalizations of this analysis, in particular extending the present analysis to the case of Markov chains in discrete time. Related results are given in a companion paper by Kac.[70]

D. Boundary Conditions

1. One Trapping Point

So far we have concentrated on unrestricted random walks. In the following paragraphs we shall discuss the effects of various kinds of restrictions or boundaries on random walks. These may be of several varieties, but the most important may be broadly categorized as being absorbing or reflecting. In the former a random walk impinging on an absorbing surface ends there, and in the latter the impinging random walker is redirected to some other point. In addition to these, one can have partially absorbing surfaces in which trapping of the random walk occurs with some probability less than 1.

The simplest example of an absorbing boundary is that of a single trapping point. When a trapping point exists, it is natural to inquire about the statistical properties of the time till trapping occurs. There are several ways of modeling traps, and some of these will be discussed in Section III.C. One way of modeling traps, originally discussed by Montroll,[71] starts by considering a random walk on a lattice with a periodic sublattice of traps. In the latter model one can find the statistics of the trapping time from the first passage time to reach the trapping point. The relevant probabilities are the

$F_n(\mathbf{r})$, whose generating function is contained in Eq. (2.100). The simplest version of the Montroll trapping model is that in which there is a single trapping point, $\mathbf{r}$, and one wishes to find the moments of the first passage time to reach that point. Since Montroll assumes a periodic sublattice of trapping sites, one can replace the random walk on an infinite lattice by a random walk on a torus, i.e., by a random walk with periodic bounday conditions. An expression for the probability of being at $\mathbf{r}$ at step n (on a uniform lattice) is given in Eq. (2.39) for a random walk starting at the origin. The probability of being trapped at step n is $F_n(\mathbf{r})$. We have shown that the generating function of the $F_n(\mathbf{r})$ are related to those of the state probabilities. We shall find it expedient to separate the singular and nonsingular parts of the generating function $P(\mathbf{r}; z)$ by writing

$$
\begin{aligned}
P(\mathbf{r}; z) &= \frac{1}{N^D(1-z)} + \frac{1}{N^D} \sum_{\mathbf{s}}{}' \frac{\exp(2\pi i \mathbf{r} \cdot \mathbf{s}/N)}{1 - z\lambda(2\pi \mathbf{s}/N)} \\
&= \frac{1}{N^D(1-z)} + \varphi(\mathbf{r}; z),
\end{aligned}
\tag{2.167}
$$

where the prime on the sum excludes the $\mathbf{s} = \mathbf{0}$ term. Under the assumption that $\lambda(2\pi \mathbf{r}/N) = 1$ only when $\mathbf{r} = \mathbf{0}$, the terms in $\varphi(\mathbf{r}; z)$ remain bounded when $z = 1$. Expressions for $F(\mathbf{r}; z)$ can be found from Eq. (2.100), and the moments of the first passage time to any given point are given by

$$
\langle n(\mathbf{r}) \rangle = z \frac{\partial}{\partial z} F(\mathbf{r}; z) \Big|_{z=1}, \qquad \langle n^2(\mathbf{r}) \rangle = \left(z \frac{\partial}{\partial z} \right)^2 F(\mathbf{r}; z) \Big|_{z=1} \tag{2.168}
$$

In particular, we find for the first two moments

$$
\langle n(\mathbf{r}) \rangle = \begin{cases} N^D[\varphi(\mathbf{0}; 1) - \varphi(\mathbf{r}; 1)], & \mathbf{r} \neq \mathbf{0} \\ N^D, & \mathbf{r} = \mathbf{0} \end{cases}
$$

$$
\langle n^2(\mathbf{r}) \rangle = \begin{cases} [2N^D\varphi(\mathbf{0}; 1) + 1]\langle n(\mathbf{r}) \rangle \\ \quad + 2N^D \left[\dfrac{\partial \varphi(\mathbf{0}; z)}{\partial z} - \dfrac{\partial \varphi(\mathbf{r}; z)}{\partial z} \right] \Big|_{z=1}, & \mathbf{r} \neq \mathbf{0} \\ 2N^{2D}\varphi(\mathbf{0}; 1) + N^D, & \mathbf{r} = \mathbf{0} \end{cases} \tag{2.169}
$$

It is interesting to note that the expected number of steps to return to the origin is equal to the number of lattice points in the unit cell, independent of the transition probabilities and the lattice structure. Detailed properties of the random walk are contained in the function $\varphi(\mathbf{r}; 1)$ and show up in the remaining moment expressions. It is considerably more difficult to analyze the case of more than one trapping site per unit cell. On this point it is instructive to see the difficulties of calculating properties of random walks on an infinite lattice in 2-D with one forbidden site, as analyzed by Rubin.[72] Related analysis has been given by Weiss[73] for this model. More recently, Scher and Wu[23] gave a detailed analysis of random walks on lattices with a periodic array of traps. Shuler, Silver, and Lindenberg[74] showed that the mean time to trapping on a toroidal lattice bears a relation to the expected number of sites visited, of the form $\langle S_{\langle n \rangle} \rangle \sim \gamma N$, where N is the total number of lattice points and γ is a parameter that is $O(1)$.

It is easy to calculate the first passage time to a single trapping point on a 1-D continuum.[75] If the mean drift is equal to 0, the trapping point is $r_0 > 0$, and the initial point is at $r = 0$, then the probability density for the random walker to be at $r < r_0$ at time t is obtained by solving the diffusion equation subject to $p(r,0) = \delta(r)$ and $p(r_0, t) = 0$, as will be justified later. The probability of absorption at some time $\geq t$ is easily shown to be

$$G(t) = \int_{-\infty}^{r_0} p(r,t)\, dr = 2\Phi\left(\frac{r_0}{\sqrt{2Dt}}\right) - 1 \qquad (2.170)$$

where $\Phi(x)$ is the error function $\Phi(x) = (2\pi)^{-1/2} \int_{-\infty}^{x} \exp(-u^2/2)\, du$. Therefore the probability density for the trapping time $g(t)$ is given by

$$g(t) = -\frac{dG}{dt} = \frac{r_0}{\sqrt{4\pi Dt^3}} \exp\left(-\frac{r_0^2}{4Dt}\right) \qquad (2.171)$$

This function is seen to be asymptotic to a stable law having no finite moments. The result in Eq. (2.171) is consistent with that of Eq. (2.169) in that the mean first passage time is infinite in the limit $N = \infty$ in the latter equation.

One can extend the calculation whose results are given in Eq. (2.169) to cover the situation in which there is a single partially absorbing trap. If we suppose that a random walk landing on point $\mathbf{r}$ is absorbed with probability α or continues with probability $1 - \alpha$, then the probability of being absorbed at step n is

$$A_n(\mathbf{r}) = \sum_{j=1}^{\infty} \alpha(1-\alpha)^{j-1} F_n^{(j)}(\mathbf{r}) \qquad (2.172)$$

The generating function of the $A_n(\mathbf{r})$ (with respect to n) is easily found using

Eq. (2.133). This relation implies

$$A(\mathbf{r}; z) \equiv \sum_{n=0}^{\infty} A_n(\mathbf{r}) z^n = \frac{\alpha F(\mathbf{r}; z)}{1 - (1 - \alpha) F(\mathbf{0}; z)} \qquad (2.173)$$

The first two moments of the time to absorption are

$$\langle n_\alpha(\mathbf{r}) \rangle = \langle n_1(\mathbf{r}) \rangle + \frac{1 - \alpha}{\alpha} \langle n_1(\mathbf{0}) \rangle$$

$$\sigma_\alpha^2(\mathbf{r}) = \sigma_1^2(\mathbf{r}) + \frac{1 - \alpha}{\alpha} \langle n_1^2(\mathbf{0}) \rangle + \frac{(1 - \alpha)^2}{\alpha^2} \langle n_1(\mathbf{0}) \rangle^2 \qquad (2.174)$$

where $\langle n_\alpha(\mathbf{r}) \rangle$ is the first-passage time to $\mathbf{r}$ when the trapping probability is equal to α, and $\sigma_\alpha^2(r) = \langle n_\alpha^2(r) \rangle - \langle n_\alpha(r) \rangle^2$.

The usual way of describing an absorbing boundary is to say that a random walk ends whenever it impinges on the boundary. This is not a unique description for lattice random walks in which jumps can be made to other than nearest neighbors. For example, in 1-D one may have $r = 0$ as an absorbing point but specify that the random walk does not stop unless the random walker is exactly at $r = 0$, continuing if it is at any other lattice site. A different possibility is for any point $r \leq 0$ to be absorbing, so that the random walk is terminated at any point in the left half line. The solutions to the two problems just described are, of course, different. The method of images is available for the solution of the first problem, but not for that of the second. The simplest problems on a lattice involving barriers are those in which steps may be taken to nearest neighbors only. When the barriers are absorbing, one can use the first-passage-time formalism relating the $F_n(\mathbf{r})$ to the $P_n(\mathbf{r})$. The great advantage of this possibility is that the $P_n(\mathbf{r})$ are calculated for unrestricted random walks. An alternative procedure is one in which a boundary condition is used. The appropriate one for an absorbing barrier is

$$P_n(\mathbf{r}) = 0 \qquad (2.175)$$

for all $\mathbf{r}$ on the barrier surface and for all n. To see how this boundary condition arises, let us consider the simplest case of a lattice random walk in 1-D, with an absorbing point at $r = 0$. The equation governing the random walk at an arbitrary point $r \neq 1$ is

$$P_{n+1}(r) = p P_n(r - 1) + q P_n(r + 1) \qquad (2.176)$$

where $p + q = 1$ are the transition probabilities. When $r = 1$ one has

$$P_{n+1}(1) = qP_n(2) \tag{2.176a}$$

since no random walker can come from $r = 0$. But this last relation is the same as Eq. (2.176) if we add $pP_n(0)$ ($= 0$) to the right side of this relation. When there can be longer-range transitions, say transitions to both nearest- and next-nearest-neighbor points, two equations must be modified to have an absorbing boundary at $r = 0$. In the diffusion limit the condition for an absorbing boundary is $P(\mathbf{r}, t) = 0$ for all $\mathbf{r}$ on the boundary. Since the diffusion limit is valid for all random walks for which $\sigma^2 < \infty$, it is difficult to see how the single boundary condition can be obtained from the lattice model because of the long-range transitions. Van Kampen and Oppenheim[76] have examined the passage to the limit critically for 1-D random walks with transitions to nearest and next nearest neighbors. They consider a systematic expansion of the dynamic equations in powers of a small parameter ε, by using the scaling $r = \varepsilon n$, $\tau = \varepsilon^2 t$, $p_n(t) = \varepsilon P(r, \tau)$. It is assumed that in the limit $\varepsilon = 0$, r, τ, and $P(r, \tau)$ are all $O(1)$. Van Kampen and Oppenheim found that to zeroth order in ε the appropriate boundary condition for absorption is $P(r, \tau) = 0$ as one would expect, but that there are corrections in higher orders of ε. Cox and Miller[77] prove the validity of the boundary condition starting in the diffusion limit.

2. One-Dimensional Walks Between Absorbing Points

For random walks with nearest-neighbor transitions only, one can use the method of images when the absorbing boundaries are planes parallel to the coordinate axes. In 1-D, if $r = -a$ and $r = b$ are absorbing points, the solution obtained by this means is

$$U_n(r; -a, b|r_0) = \sum_{l=-\infty}^{\infty} \{ P_n(r - r_0 + 2l(a+b)) \\ - P_n(r + r_0 + 2a + 2l(a+b)) \} \tag{2.177}$$

where $U_n(r; -a, b|r_0)$ is the solution of the restricted random walk with a starting point at r_0, and $P_n(r)$ is the solution to the unrestricted random walk. Similar results can be obtained in higher dimensions. The sum in this last equation is convenient to evaluate for small values of n, but it is slowly convergent when n is large. A more convenient form for U_n is obtained for large n by substituting the integral representation of Eq. (2.4) into Eq. (2.177) and performing the sum over l using the identity[78]

$$\sum_{l=-\infty}^{\infty} \exp(-2ilm\theta) = \sum_{l=-\infty}^{\infty} \delta\left(\frac{m\theta}{\pi} - l\right) \tag{2.178}$$

In this way one finds that U_n can be represented as

$$U_n(r; -a, b|r_0) = \frac{1}{a+b} \sum_{|l|<a+b} \lambda^n\left(\frac{\pi l}{a+b}\right) \sin\frac{\pi l(r_0+a)}{b+a} \sin\frac{\pi l(r+a)}{b+a}$$

(2.179)

which is quickly convergent for large n but poorly convergent for small n. Although the expression in Eqs. (2.177) and (2.179) can be proved easily for nearest-neighbor random walks, they are in fact asymptotically correct, in the limit of large n, for random walks with longer-range transitions. One can prove this, for example, by using the asymptotic theory of Toeplitz matrices.[79] Spitzer and Stone[80] and Kesten[81] (cf. also Reference 2) made interesting use of techniques based on Toeplitz forms to derive results on the occupancy of points for random walks in 1-D in an interval bounded by two absorbing barriers. Spitzer and Stone show, for example, that if one has a symmetric random walk in an interval $(-n, n)$ with a finite variance, the expected number of distinct points visited in the interval before leaving it satisfies $\lim_{n \to \infty}\langle M_n\rangle/(2n) = \ln 2$. This result is remarkable because it is independent of σ^2. Many other results for random walks in 1-D are obtainable rather elegantly using the theory of Toeplitz forms.

The probability of not being absorbed by step n can be written in terms of $U_n(r; -a, b|r_0)$ as

$$Q_n(r_0) = \sum_{r=-a+1}^{b-1} U_n(r; -a, b|r_0)$$

$$= \frac{1}{a+b} \sum_{|2j+1| \leq a+b} \lambda^n\left(\frac{\pi(2j+1)}{a+b}\right)$$

$$\times \sin\left(\frac{\pi(2j+1)(r_0+a)}{b+a}\right)\cot\left(\frac{\pi(j+\frac{1}{2})}{b+a}\right) \qquad (2.180)$$

If one assumes that the initial position is uniformly distributed between $r = -a+1$ and $r = b-1$, then the mean time to trapping at one of the absorbing points is

$$\langle n_A\rangle = \frac{1}{(b+a)^2} \sum_{|2j+1| \leq b+a} \frac{\cot^2\left(\dfrac{\pi(j+\frac{1}{2})}{b+a}\right)}{1-\lambda\left(\dfrac{\pi(2j+1)}{b+a}\right)} \qquad (2.181)$$

Let us consider the case $a = 0$, $b = N$. If we let μ_1 be the average displacement in a single step and we assume that $\sigma^2 < \infty$, then under the further assumption that either (1) $N \gg \mu_1 \neq 0$ or (2) $N \gg \sigma$, $\mu_1 = 0$, the following asymptotic limits can be established[82] for $\langle n_A \rangle$:

$$\langle n_A \rangle \sim \begin{cases} \dfrac{N}{2\mu_1}, & \mu_1 \neq 0 \\[2ex] \dfrac{N^2}{6\sigma^2}, & \mu_1 = 0 \end{cases} \tag{2.182}$$

In the first case the process behaves as if there was no diffusion present. For stable laws in which $\lambda(\theta) \sim 1 - |\theta/\theta_0|^\alpha$, with $\alpha \leqslant 2$, Eq. (2.181) leads to

$$\langle n_A \rangle \sim \frac{8N^\alpha \theta_0^\alpha}{\pi^{\alpha+2}} \sum_{j=0}^{\infty} \frac{1}{(2j+1)^{\alpha+2}} \tag{2.183}$$

One other type of problem that can be analyzed by techniques similar to those just described is the determination of the probability of being absorbed by one of several possible absorbing surfaces. Suppose that a random walk takes place in a region R with a boundary B that can be decomposed into two absorbing boundaries B_1 and B_2. If we let $H(\mathbf{r})$ be the probability that absorption occurs at B_1 rather than B_2, then $H(\mathbf{r}) = 1$ for all $\mathbf{r} \in B_1$ and $H(\mathbf{r}) = 0$ for all $\mathbf{r} \in B_2$. Further, the $H(\mathbf{r})$ satisfy

$$H(\mathbf{r}) = \sum_{\mathbf{s}} H(\mathbf{s}) p(\mathbf{r} - \mathbf{s}) \tag{2.184}$$

subject to the boundary conditions given above. When the second moments are finite and the average step size is small compared to typical dimensions of R, one can derive an approximating partial differential equation analogous to Eq. (2.63). For example, in $1 - D$, if there are two absorbing points, one at $x = 0$ and one at $x = r$, then the probability of absorption at 0 satisfies

$$D\frac{d^2 H}{dx^2} - v\frac{dH}{dx} = 0 \tag{2.185}$$

where D and v are defined in Eq. (2.62). This equation is to be solved subject to $H(0) = 1$ and $H(r) = 0$. The solution is easily found to be

$$H(x) = \begin{cases} \dfrac{1 - e^{-(v/D)(r-x)}}{1 - e^{-vr/D}}, & v \neq 0 \\[2ex] 1 - \dfrac{x}{r}, & v = 0 \end{cases} \tag{2.186}$$

3. Wald's Identity for First-Passage-Time Problems in One Dimension

Another approach to this class of problems for 1-D random walks was introduced by Wald.[83] His approach allows both exact and approximate results to be obtained for many such problems. Although Wald's method is well known to statisticians, it has not appeared in any of the chemical-physics literature. The endpoint of the analysis is an identity that relates the time to absorption and the place of absorption. Different approximate results can then be derived from the identity. We follow the derivation by Cox and Miller[77] for continuous random walks in discrete time, noting that the appropriate generalizations to processes in continuous time can be made. Let us suppose that the random walk takes place on the interval $(-a, b)$, both ends being absorbing, and that the probability density for the displacement in a single step is $p(x)$. Let N be the step at which absorption first occurs (i.e., the first-passage time), and let X_N be the position at which absorption occurs, so that $X_N \geqslant b$ or $X_N \leqslant -a$. We define $f_n(x)\,dx$ to be the joint probability that absorption occurs at step n and $x < X \leqslant x + dx$ for a random walker initially at the origin. The $f_n(x)$ clearly satisfy

$$f_n(x) = \int_{-a}^{b} f_{n-1}(y)p(x-y)\,dy \qquad (2.187)$$

We define $f^*(s)$ to be the two-sided Laplace transform of $f_n(x)$,

$$f_n^*(s) = \int_{-\infty}^{\infty} f_n(x)e^{-sx}\,dx \qquad (2.188)$$

and let $K(s, z)$ be the joint generating function and Laplace transform,

$$K(s, z) = \sum_{n=0}^{\infty} z^n \int_{-a}^{b} f_n(u)e^{-su}\,du \qquad (2.189)$$

We shall calculate the quantity $\langle z^N e^{-sX_N} \rangle$ and show that it can be expressed in terms of the quantities defined above. Then one can calculate averages of first-passage-time probabilities by differentiating and choosing z and s appropriately.

By definition we have

$$
\begin{aligned}
\langle z^N e^{-sX_N} \rangle &= \sum_{n=1}^{\infty} z^n \left\{ \int_{-\infty}^{-a} f_n(u)e^{-su}\,du + \int_{b}^{\infty} f_n(u)e^{-su}\,du \right\} \\
&= \sum_{n=1}^{\infty} z^n \left\{ \int_{-\infty}^{\infty} f_n(u)e^{-su}\,du - \int_{-a}^{b} f_n(u)e^{-su}\,du \right\} \\
&= \sum_{n=1}^{\infty} z^n f_n^*(s) - [K(s, z) - 1] \qquad (2.190)
\end{aligned}
$$

If we take the two-sided Laplace transform of Eq. (2.187), we find

$$f_n^*(s) = \int_{-a}^{b} e^{-sy} f_{n-1}(y)\, dy \int_{-\infty}^{\infty} e^{-s(x-y)} p(x-y)\, dx$$
$$= p^*(s) \int_{-a}^{b} e^{-sy} f_{n-1}(y)\, dy \tag{2.191}$$

Substituting this expression into Eq. (2.190), we find that

$$\langle z^N e^{-sX_N} \rangle = 1 - \left[1 - z p^*(s) \right] K(s, z) \tag{2.192}$$

Finally, setting $z = [p^*(s)]^{-1}$, we find

$$\langle [p^*(s)]^{-N} e^{-sX_N} \rangle = 1 \tag{2.193}$$

which is known in the statistical literature as Wald's identity. The usefulness of this formula is due, in part, to the fact that $p^*(s)$ is the Laplace transform of the unrestricted probabilities and the boundaries do not appear explicitly. Let H_1 be the probability of absorption at $-a$, and H_2 be the probability of absorption at b. An equivalent formulation of Eq. (2.193) is

$$H_1 \langle [p^*(s)]^{-N} e^{-sX_N} | X_N \leqslant -a \rangle + H_2 \langle [p^*(s)]^{-N} e^{-sX_N} | X_N \geqslant b \rangle = 1 \tag{2.194}$$

where $\langle \alpha | \beta \rangle$ means the expectation of α conditional on β.

One can use this last equation to find approximate values for H_1 and H_2. The usual approximation that is made is to assume that X_N is equal to either $-a$ or b, ignoring jumps beyond the boundaries. If one does this and chooses S_0 to be the root of $p^*(s) = 1$ (uniqueness of the root can be proven), then one finds

$$H_1 e^{aS_0} + H_2 e^{-bS_0} = 1 \tag{2.195}$$

which, together with $H_1 + H_2 = 1$, yields the approximation

$$H_1 = \frac{1 - e^{-bS_0}}{e^{aS_0} - e^{-bS_0}} \tag{2.196}$$

When the average displacement in a single step is 0, then $S_0 = 0$ and

$$H_1 = \frac{b}{a+b} \tag{2.197}$$

which can be found from Eq. (2.196) by passing to the limit. Wald's identity can also be used to provide an approximation to the moments of the absorption time. To see this we expand $[p^*(s)]^{-N}\exp(-sX_N)$ in powers of s, noting that the average that appears in Wald's identity is equal to 1. This implies that the coefficient of each power of s vanishes identically. Equation (2.193) is, to second order in s,

$$\left\langle 1 - s(X_N - N\mu_1) + \frac{s^2}{2}\left[(X_N - N\mu_1)^2 - \sigma^2 N\right] - \cdots \right\rangle = 1 \quad (2.198)$$

from which we infer

$$\langle X_N \rangle = \langle N \rangle \mu_1$$
$$\langle (X_N - N\mu_1)^2 \rangle = \sigma^2 \langle N \rangle \qquad (2.199)$$

These relations allow us to write

$$\langle N \rangle = \begin{cases} \langle X_N \rangle / \mu_1, & \mu_1 \neq 0 \\ \langle X_N^2 \rangle / \sigma^2, & \mu_1 = 0 \end{cases} \qquad (2.200)$$

The final step in this derivation is to incorporate the approximations

$$\langle X_N \rangle \sim H_2 b - H_1 a$$
$$\langle X_N^2 \rangle \sim H_2 b^2 + H_1 a^2 \qquad (2.201)$$

where H_1 is found from Eq. (2.196). Higher moments of the time to absorption can also be found by this technique.

4. Maximum Displacements and Spans

The formalism for random walks with absorbing barriers can be used to specify several interesting parameters that characterize the dimensions and shape of the random walk. Two of these are the maximum displacement of the random walk from its starting point and the spans of the random walk. The spans are defined to be the dimensions of the smallest box with sides parallel to the coordinate axes that entirely contain the random walk. The notion of the span is illustrated in Fig. 3 for a random walk in 2-D. A third parameter that would be of considerable interest but is surprisingly difficult to calculate is the distribution of the diameter of the smallest sphere entirely containing the random walk. The first analysis of spans was that given by Daniels.[84] Feller[85] and Darling and Siegert[86] considered statistical properties of the span in 1-D, Weiss and Rubin[87] considered the spans in somewhat greater generality, and Rubin, Mazur, and Weiss[88] gave the solution to

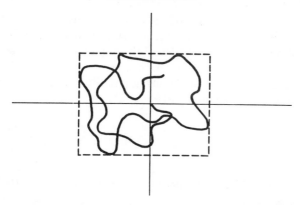

Fig. 3. Schematic diagram of the spans (dashed lines) of a random walk in 2-D.

several problems suggested by the configuration of polymers. In what follows we present some theoretical results related to the maximum displacement and the span in 1-D. The extension of these ideas presents no difficulties.

The problem of finding the distribution of the maximum displacement of the random walk and that of finding the span distribution can be combined using the probability of remaining entirely within an interval $(-a, b)$ in an n-step walk. In this section we shall denote this function by $Q_n(r_0 | -a, b)$ to emphasize the dependence on the boundaries. The probability that the maximum displacement of a random walk starting from the origin is less than $a > 0$ after n steps will be denoted by $U_n(a)$. It is clear that

$$U_n(a) = Q_n(0 | -a, a) \qquad (2.202)$$

The probability that the maximum displacement is exactly equal to a is given by

$$V_n(a) = U_n(a+1) - U_n(a) \qquad (2.203)$$

This may be seen by rewriting it as $U_n(a+1) = U_n(a) + V_n(a)$, i.e., if the maximum displacement is less than $a + 1$, it is strictly less than a or else is exactly equal to a. The span of a 1-D random walk is defined in terms of the maximum and minimum displacement, $r_{\max}(n)$ and $r_{\min}(n)$, by

$$\text{span} = r_{\max}(n) - r_{\min}(n) \qquad (2.204)$$

In order to calculate the distribution of the span we must be able to vary the endpoints of the interval independently, since the maximum and minimum displacements are not necessarily equal. The probability that the maximum positive displacement is less than b and the minimum negative displacement is greater than $-a$ is $Q_n(0|-a, b)$. Hence by an extension of the argument used to find the distribution of the maximum displacement, the probability that the span is exactly equal to S is

$$W_n(S) = -\sum_{a=0}^{S} \Delta_a \Delta_b Q_n(0|-a, b)\big|_{b=S-a} \qquad (2.205)$$

in which Δ_a is the difference operator: $\Delta_a f(a) = f(a+1) - f(a)$. The upshot of these calculations is that the properties of both $V_n(a)$ and $W_n(S)$ depend on $Q_n(0|-a, b)$.

Before analyzing this function we remark that one of the first papers which study functionals of the maximum displacement in 1-D is that of Erdös and Kac.[89] They not only found the limiting density of the maximum displacement, but also that of the maximum positive displacement as well as

$$\lim_{n \to \infty} \Pr\left\{ \frac{r^2(1) + r^2(2) + \cdots + r^2(n)}{n^2} < \alpha \right\}$$

and

$$\lim_{n \to \infty} \Pr\left\{ \frac{|r(1)| + |r(2)| + \cdots + |r(n)|}{n^{3/2}} < \alpha \right\}$$

where $r(j)$ is the displacement at step j. Erdös and Kac treated only the case where the average step size is zero and where the variance of the step size is finite. The extension of their results to an arbitrary mean step size is not difficult but is of lesser interest. Our analysis of the distribution $V_n(a)$ differs from that of Erdös and Kac, but their paper is well worth reading for its introduction of the invariance principle, which shows that the limiting distribution of the functions that they calculate is independent of the underlying distribution of the step size. That assertion allowed them to use the simplest random walk satisfying the required assumptions to carry out their calculations. The invariance principle was later generalized by Donsker[90] and others, and is a powerful tool for tackling many difficult problems. Our analysis, more on the heuristic side, does essentially the same thing by using the small-$|\theta|$ properties of the structure function.

We first consider the case $\sigma^2 < \infty$ and $\mu_1 = 0$. Since the theory is only valid in an asymptotic sense [Eq. (2.177) is exact for nearest-neighbor random walks but only valid asymptotically for the case $\sigma^2 < \infty$] we shall assume that n and the parameters a and b both get large in such a way that $n\sigma^2/(a+b)^2$ approaches a constant. If we set $n\sigma^2 = \tau$, then we may pass to the appropriate limit in Eq. (2.180) to find that

$$Q_\tau(0 \mid -a, b) \sim \frac{2}{\pi} \sum_{l=-\infty}^{\infty} \frac{\exp\left\{-\dfrac{\pi^2 \tau (2l+1)^2}{2(a+b)^2}\right\}}{2l+1} \sin \frac{\pi(2l+1)a}{a+b}$$

(2.206)

An equivalent expression to this is obtained by using a Poisson transformation

$$Q_\tau(0 \mid -a, b) \sim \sum_{l=-\infty}^{\infty} (-1)^l \left\{ \Phi\left(\frac{a+l(a+b)}{\sqrt{2\tau}}\right) + \Phi\left(\frac{a-l(a+b)}{\sqrt{2\tau}}\right) - 1 \right\}$$

(2.207)

where $\Phi(x)$ is the error function defined after Eq. (2.170). The probability density for the maximum displacement will now be written

$$V(a, \tau) \sim \frac{\partial Q_\tau(0 \mid -a, a)}{\partial a}$$

(2.208)

where $Q_\tau(0 \mid -a, a)$, the asymptotic distribution function, is found from Eq. (2.207). The expression for the probability that the maximum displacement is $\leq a$ follows directly from Eq. (2.206) and is

$$Q_\tau(0 \mid -a, a) \sim \frac{4}{\pi} \sum_{l=0}^{\infty} \frac{(-1)^l}{2l+1} \exp\left\{-\frac{\pi^2 \tau (2l+1)^2}{8a^2}\right\}$$

(2.209)

as found originally by Erdös and Kac. The average value of the maximum displacement is most easily found by starting from Eq. (2.207). We find

$$\langle a(\tau) \rangle = \sqrt{\pi \tau}$$

(2.210)

and for the variance one has

$$\sigma^2(a) = (4G - \pi)\tau = (0.52227+)\tau$$

(2.211)

where G is Catalan's constant, $G = 0.916966+$. These results show that the ratio of standard deviation to average displacement (the coefficient of varia-tion) is asymptotically constant and equal to $0.4077+$. If we let $a/\sqrt{\tau}=\zeta$, then the probability density function for ζ is given by

$$v(\zeta)=\frac{2}{\sqrt{\pi}}\sum_{l=0}^{\infty}(-1)^{l}(2l+1)\exp\left\{-\frac{\zeta^{2}}{4}(2l+1)^{2}\right\}\qquad(2.212)$$

independent of time. Figure 4 shows a plot of $v(\zeta)$ as a function of ζ. It has a maximum at $\zeta=1.28$, or $a_{m}=1.28\sqrt{\tau}$, and is skewed to the right.

The same formalism can be used to examine the case in which the transi-tion probabilities are asymptotic to those of a stable law, and for which $\lambda(\theta)\sim1-|\theta/\theta_{0}|^{\alpha}$ for θ small. In that case the asymptotic analysis leads to

$$Q_{\tau}(0|-a,a)\sim\frac{4}{\pi}\sum_{l=0}^{\infty}\frac{(-1)^{l}}{2l+1}\exp\left\{-\frac{\pi^{\alpha}\tau(2l+1)^{\alpha}}{(2a)^{\alpha}}\right\}\qquad(2.213)$$

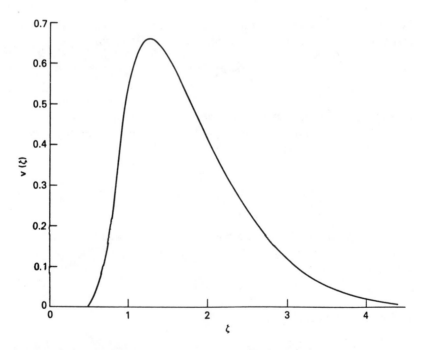

Fig. 4. The probability density $v(\zeta)$ for the normalized maximum displacement of a 1-D random walk with a finite variance. If a is the maximum displacement, then ζ is defined by $\zeta=a/\sqrt{\tau}$.

where τ is now defined by $\tau = n/|\theta_0|^\alpha$. It is difficult to work with this expression for arbitrary values of $\alpha \leqslant 2$, but for $\alpha = 1$ one can perform a Poisson transformation of Eq. (2.180) to find that

$$Q_\tau(0|-a,a) \sim \frac{4}{\pi} \sum_{l=0}^{\infty} (-1)^l \tan^{-1}\left[(2l+1)\frac{a}{\tau}\right] \qquad (2.214)$$

which implies that the probability density of the maximum displacement is

$$V(a,\tau) \sim \frac{\tau/a^2}{\cosh(\pi\tau/2a)} \qquad (2.215)$$

If we let $a = \Gamma\tau$, where $0 \leqslant \Gamma < \infty$, then the probability density of the scaling parameter Γ is $v(\Gamma) = [\Gamma^2 \cosh(\pi/2\Gamma)]^{-1}$, which is plotted in Fig. 5. The resulting curve has a long tail that goes to zero as Γ^{-2}. It is evident, in this

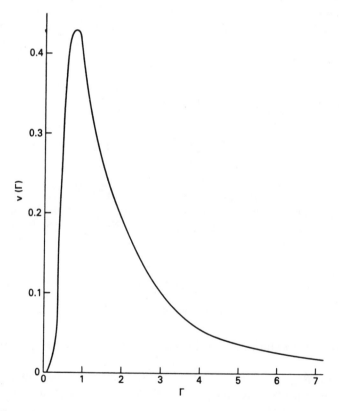

Fig. 5.   The probability density $v(\Gamma)$ for the normalized maximum displacement of a 1-D random walk following a stable law with $\alpha = 1$. The scaling is defined by $\Gamma = a/\tau$.

example, that the integer moments of a are infinite. If we wish to compare some characteristic parameter for this random walk with the same parameter for a finite-variance random walk, one possible choice is that of a fractional moment. When $\alpha = 1$, any moment of the form $\langle a^\beta(\tau) \rangle$ will exist for all β less than 1. For convenience we choose $\beta = \frac{1}{2}$, in which case we find

$$\langle a^{1/2}(\tau) \rangle \sim \sqrt{\frac{8\tau}{\pi}} \int_0^\infty \frac{dv}{\cosh v^2} \sim 1.889\tau^{1/2} \qquad (2.216)$$

while in the finite-variance case

$$\langle a^{1/2}(\tau) \rangle \sim 1.306\tau^{1/4} \qquad (2.217)$$

Thus the time dependence of this fractional moment clearly distinguishes between the two cases. Similar results for other values of α appear to be much more difficult to obtain.

The analysis of the span distribution can be calculated starting from the representation in Eq. (2.205). In the asymptotic regime (large number of steps), when $\sigma^2 < \infty$ and $\mu_1 = 0$, the probability density of the span can be written in either of two forms:

$$p(S, \tau) \sim \frac{8}{\sqrt{2\pi\tau}} \sum_{j=1}^\infty (-1)^{j+1} j^2 \exp\left(-\frac{j^2 S^2}{2\tau} \right)$$

$$= \frac{8\tau}{S^3} \sum_{j=0}^\infty \left[\frac{\pi^2 (2j+1)^2 \tau}{S^2} - 1 \right] \exp\left(-\frac{\pi^2 \tau (2j+1)^2}{2S^2} \right) \qquad (2.218)$$

in which $\tau = \sigma^2 n$. If we let $S = \rho\sqrt{\tau}$, then we can write $v_s(\rho) = p(\rho\sqrt{\tau}, \tau)\sqrt{\tau}$ for the probability density of ρ. This function is plotted in Fig. 6. The moments are easily calculated from Eq. (2.218). One finds for the mean and variance of the span

$$\langle S(\tau) \rangle = \left(\frac{8\tau}{\pi} \right)^{1/2}$$

$$\sigma^2(S(\tau)) = 4\left(\ln 2 - \frac{2}{\pi} \right)\tau \qquad (2.219)$$

Explicit representations for the higher moments were given by Darling and Siegert.[86]

When $\lambda(\theta) \sim 1 - |\theta/\theta_0|$ for small $|\theta/\theta_0|$, one can derive an asymptotically valid expression for the probability density of the span, which is

$$p(S, \tau) \sim \frac{4\tau^2}{S^3 \sinh(\pi\tau/S)} \qquad (2.220)$$

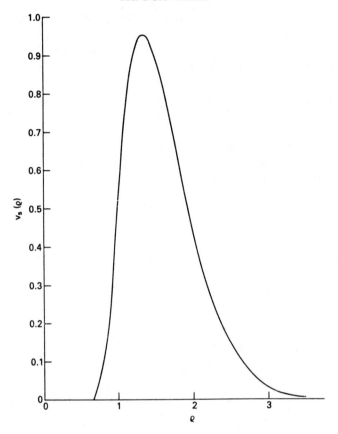

Fig. 6. Probability density for the normalized span of a symmetric 1-D random walk with finite variance. The normalized variable is defined by $\rho = S/\sqrt{\tau}$.

where we have set $\tau = n/\theta_0$. One can show that, analogously to Eqs. (2.216) and (2.217), the fractional moments for the span satisfy

$$
\begin{aligned}
\langle S^{1/2}(\tau)\rangle &\sim 0.7031\tau^{1/4} \qquad (\sigma^2 < \infty) \\
\langle S^{1/2}(\tau)\rangle &\sim 2.101\tau^{1/2} \qquad (\lambda \sim 1 - |\theta/\theta_0|)
\end{aligned}
\tag{2.221}
$$

In general, when $\lambda(\theta) \sim 1 - |\theta/\theta_0|^\alpha$ for small $|\theta/\theta_0|$, Weiss and Rubin[87] have shown that

$$
p(S,\tau) \sim \frac{8}{S^3} \sum_{l=0}^{\infty} \frac{d^2}{d\theta^2}\left(e^{-\tau\theta^\alpha}\right)\Bigg|_{\theta = \pi(2l+1)/S}
\tag{2.222}
$$

in which $\tau = n/\theta_0$. If we let $S = \rho\tau^{1/\alpha}$, then the probability density function for ρ is $h(\rho) = p(\rho\tau^{1/\alpha}, \tau)\tau^{1/\alpha}$, which depends only on ρ and α. Curves of $h(\rho)$ are shown in Fig. 7 for $\alpha = 0.5$, 1, and 1.5. These curves are seen to be quite asymmetric. The function $h(\rho)$ goes like $\rho^{-(1+\alpha)}$ for large ρ, so that the smaller the value of α, the longer is the tail.

Our results so far describe the properties of the span averaged over all extremities of the random walk. Some simulation results by Rubin and Mazur[91] suggested that it would be interesting to study the internal configurations of a span-constrained random walk. That is to say, one can ask for the probability distribution of the end-to-end distance of a random walk conditional on its having a span S after n steps in discrete time. This function was dis-

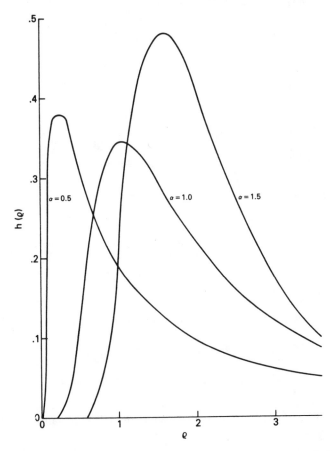

Fig. 7.   Probability densities for the normalized span of symmetric random walks in 1-D that follow a stable law, $\alpha = 0.5, 1, 1.5$. The pdf is denoted by $h(\rho)$, where $\rho = S/\tau^{1/\alpha}$.

cussed by Weiss and Rubin.[92] We derive an expression for this distribution by starting from the probability distribution of the position, at step n, of a random walker moving between absorbing barriers at $-a$ and b. This probability will be denoted by $U_n(r; -a, b|r_0)$ and can be represented by the series in Eq. (2.179). The probability that the random walker is at r at step n, having been at $r = b$ at least once but never at $r = b + 1$, is $U_n(r; -a, b + 1|r_0) - U_n(r; -a, b|r_0) = \Delta_b U_n(r; -a, b|r_0)$. Therefore, it follows by a simple argument that the analogous probability conditional on having been both at $r = a$ and $r = b$ before or at step n is $\Delta_a \Delta_b U_n(r; -a, b|r_0)$. Since one is interested in fixing the span S, it proves convenient to work with the expected number of random walkers at r at step n, summed over all starting points. This function will be denoted by $Q_n(r|-a, b)$; we have

$$Q_n(r|-a, b) = \sum_{r_0 = -a}^{b} U_n(r; -a, b|r_0) \qquad (2.223)$$

Finally, the expression for the conditional probability for being at r at step n given that the span is S at that time is

$$p_n(r|S) \frac{\Delta_a \Delta_b Q_n(r|-a, b)}{\sum_{r = -a}^{b} \Delta_a \Delta_b Q_n(r|-a, b)}\Bigg|_{a=0, b=S} \qquad (2.224)$$

When the step variance is finite, $\sigma^2 < \infty$, we can write $r = \theta S$ where $0 \leqslant \theta \leqslant 1$, and let $\tau = n\sigma^2$. The resulting conditional density for θ will be written as $p_\tau(\theta|S)$, where

$$p_\tau(\theta|S) = \frac{S^2}{4\tau} \frac{\sum\limits_{l=-\infty}^{\infty} (-1)^l l(l+1)(l+\theta) \exp\left(-\frac{S^2}{4\tau}(l+\theta)^2\right)}{\sum\limits_{l=-\infty}^{\infty} (-1)^l l^2 \exp\left(-\frac{S^2 l^2}{4\tau}\right)} \qquad (2.225)$$

In the limit $S^2/\tau \to 0$ one finds the limiting form $p_\infty(\theta) = (\pi/2)\sin \pi\theta$. Some curves of $p_\tau(\theta|S)$ as a function of θ are shown in Fig. 8 for different values of the parameter $S^2/(4\tau)$. The conclusion of interest from these curves is that two different types of curves appear, depending on the magnitude of $S^2/(4\tau)$. When $S^2/(4\tau)$ is greater than 1, the random walker tends to be found at points away from the extremes, and for $S^2/(4\tau)$ less than 1, the end of the random walk is most likely to be found at the extremes.

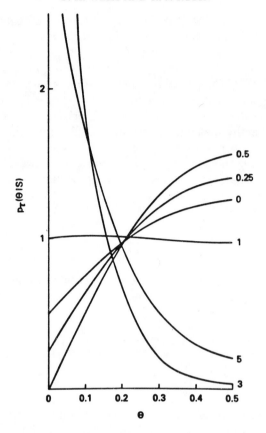

Fig. 8. Curves of the conditional probability density $p_\tau(\theta|S)$, where $\theta = r/S$ is the displacement normalized by the span. The parameters labeling each curve in this figure represent constant values of $S^2/(4\tau)$.

Another approach to this property of random walks starts from the joint distribution of the span and end-to-end distance r, at time τ. When $\sigma^2 < \infty$ the continuum limit of this function is

$$p_\tau(r, S) = \frac{1}{\sqrt{4\pi\tau^3}} \sum_{l=-\infty}^{\infty} (-1)^{l+1} l(l+1)(r+lS) \exp\left\{ -\frac{1}{4\tau}(r+lS)^2 \right\}$$

(2.226)

If we now set $r = \theta S$ where $0 \leq \theta \leq 1$ and integrate over all S, we find that

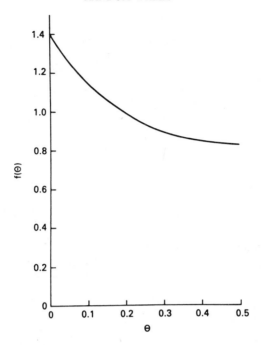

Fig. 9. The unconditional probability density of $\theta = r/S$ plotted as a function of θ. Notice that $f(\theta)$ is the same both for the case of finite variance random walks and for the stable-law case in which the structure function has the form $\lambda(\theta) \sim 1 - |\theta/\theta_0|$ for $|\theta/\theta_0|$ small.

the probability density function of θ is

$$f(\theta) = \int_0^\infty p_\tau(\theta S, S) S \, dS$$
$$= \sum_{l=1}^\infty (-1)^{l+1} \left\{ \frac{l(l+1)}{(l+\theta)^2} - \frac{l(l-1)}{(l-\theta)^2} \right\} \qquad (2.227)$$

A graph of this function is shown in Fig. 9, where it is seen that when one averages over the possible variation of the span, the end of the random walk tends to be found at the extreme points. Results similar to those just outlined are valid when the transition probabilities are asymptotically equal to a stable law. That is, the curves of $P_\tau(r|S)$ are U-shaped or bell-shaped, depending on the relation of S to τ. Weiss and Rubin[92] also showed that when $p(j) \sim A/j^2$ the function $f(\theta)$ is exactly that given in Eq. (2.227). It is not known whether Eq. (2.227) is valid more generally than for the particular cases examined.

5. Reflecting Boundary Conditions

A derivation of the boundary condition usually taken to characterize a reflecting barrier can be given in terms of the simple random walk whose equations are given in (2.176). To give a complete specification of what is meant by "reflecting" we must say exactly what happens to a random walker that lands on the reflecting barrier. In the 1-D example in Eq. (2.176) we shall say that a random walker that reaches $r = 0$ is instantly brought out to $r = 1$. This assumption allows us to write

$$
\begin{aligned}
P_{n+1}(r) &= pP_n(r-1) + qP_n(r+1), \qquad r > 2 \\
P_{n+1}(1) &= qP_n(1) + qP_n(2) \\
&= pP_n(0) + qP_n(2) + \big[qP_n(1) - pP_n(0) \big]
\end{aligned}
\tag{2.228}
$$

The resulting equations can be handled in one of two ways. The general set of equations can be expressed as in Eq. (2.176) but subject to the boundary condition $qP_n(1) = pP_n(0)$. A second way of dealing with the $r = 1$ terms in Eq. (2.176) is to incorporate the term in brackets into the set of equations. This leads to

$$
P_{n+1}(r) = pP_n(r-1) + qP_n(r+1) + \big[qP_n(1) - pP_n(0) \big] \delta_{r,1} \tag{2.229}
$$

Northrup and Hynes[93] have elaborated this approach in a study of the coupling of translational and reactive dynamics. The continuum limit of the boundary conditions is obtained from the scaling procedures leading to Eq. (2.61). One finds that the boundary condition for a reflecting point is

$$
D \frac{\partial P}{\partial x} = vP \tag{2.230}
$$

at that point. Alternatively one can write, following Northrup and Hynes,

$$
\frac{\partial P}{\partial t} = D \frac{\partial^2 P}{\partial x^2} - v \frac{\partial P}{\partial x} + \delta(x - \epsilon) \left\{ D \frac{\partial P}{\partial x} - vP \right\}, \qquad x > 0
$$
$$
P(0, t) = 0 \tag{2.231}
$$

where the reflecting point is at $x = 0$. In this formulation the effect of the reflecting point is contained in a source term that instantaneously injects the random walker (in the continuum limit) at $x = \epsilon$ after it is absorbed at $x = 0$. The formulation of boundary conditions in Eq. (2.231) is especially useful for considering problems in which a reaction occurs when two random walkers (molecules) collide.

Equation (2.61), the continuum limit, can be written as a conservation equation

$$\frac{\partial P}{\partial t} = -\frac{\partial J}{\partial x} \tag{2.232}$$

where $J = vP - D(\partial P/\partial x)$ is a flux. The boundary condition at a reflecting point, Eq. (2.230), is equivalent to the requirement that the flux vanish at that point. In higher numbers of dimensions the diffusion equation is

$$\frac{\partial P}{\partial t} = -\nabla \cdot \mathbf{J} \tag{2.233}$$

where

$$\mathbf{J} = v P - \mathbf{D} \cdot \nabla P \tag{2.234}$$

If R is a reflecting boundary, the boundary condition is $\mathbf{J} \cdot \mathbf{n} = 0$, where $\mathbf{n}$ is the unit normal to R. It is also possible to put Eq. (2.228) into the form of a conservation equation. It is easy to verify that Eq. (2.228) is equivalent to

$$\Delta_n P_n(r) = -\Delta_r J_n(r-1) \tag{2.235}$$

where $J_n(r) = (p-q)P_n(r) - q\Delta_r P_n(r)$. The condition for a reflecting point at $r = 0$ becomes $J_n(0) = 0$. By starting from the model embodied in Eq. (2.235) we can also derive the boundary conditions at a partially absorbing point. Let us suppose that if a random walker reaches the point $r = 0$ it is either absorbed with probability α or else it is set at $r = 1$ instantly. Then similar reasoning to that already used serves to establish that the boundary condition is

$$\alpha P_n(0) + (1-\alpha)J_n(0) = 0 \tag{2.236}$$

which contains the known conditions at the limits $\alpha = 0$ and $\alpha = 1$.

In the 1-D case when the average displacement is equal to zero and the random walk is to nearest neighbors only, a solution can be obtained by using the method of images when there are reflecting points at $-a$ and b. If the solution to this case is denoted by $W_n(r; -a, b|r_0)$, then

$$W_n(r; -a, b|r_0) = \sum_{l=-\infty}^{\infty} \{P_n(r - r_0 + 2l(a+b)) + P_n(r + r_0 + 2l(a+b))\}$$

$$\tag{2.237}$$

where $P_n(r)$ is the random-walk propagator in the absence of boundaries. This expression is analogous to Eq. (2.177) for absorbing boundary conditions, except that the propagators are added rather than subtracted. A Poisson transformation of this equation also allows us to write

$$W_n(r; -a, b|r_0) = \frac{1}{a+b} \sum_{|l| \leq a+b} \lambda^n \left(\frac{\pi l}{a+b} \right) \cos \frac{\pi l(r_0+a)}{b+a} \cos \frac{\pi l(r+a)}{b+a}$$

$$(2.238)$$

The diffusion limit is easily obtained from these equations when $\sigma^2 < \infty$.

The reader will notice that the theory that we have derived starting from Eq. (2.234) corresponds to a particular specification of what is meant by a reflecting point. In general, a change in the definition of reflection for lattice random walks will lead to different sets of boundary conditions. It is intuitively clear that a large class of these boundary conditions must pass to the known continuum limit. To our knowledge there has been no demonstration that this is so starting from a variety of lattice models, and indeed not even the start on this question that van Kampen and Oppenheim[76] have made on the analogous problem for absorbing boundary conditions. An interesting account of the relation between discrete and continuum boundary conditions from an engineering point of view is contained in an article by Zvirin and Shinnar.[94]

III. APPLICATIONS

In the sections to follow we illustrate how random-walk models have been applied to a variety of problems in the physical and biological sciences, as well as discuss some extensions of random-walk theory suggested by these applications. It is not possible to be exhaustive; new applications appear at a great rate. As in other parts of this chapter, we discuss applications mainly related to our own research interests.

A. Polymer Physics

There are few research areas in which random-walk methodology plays as central a role as that of polymer physics. The earliest investigations of polymer configurations were phrased in terms of random walks,[12,13,95] and over the years random-walk theory and the theory of polymer configurations have been mutually enriched by their interactions. However, the symbiotic relation between these two areas has given rise to a number of excellent monographs on the subject[15,96-98] that renders superfluous an extensive exposition in this chapter. We therefore confine ourselves to describing some of the major models for describing polymer configurations and some recent problems in the area. As noted in the Introduction, we omit completely any discussion of the excluded-volume random walk, which enjoys an enormous literature of its own.

1. Freely Jointed Chains

Kuhn[12] and Guth and Mark[95] appear to have been the first to discuss polymer configuration in terms of the random-walk model. Their analysis is based on a much earlier paper by Rayleigh[8] on random walks in three dimensions. Rayleigh's model is the direct generalization of Pearson's original formulation of the random-walk problem quoted in our Introduction. The Rayleigh model is known as the freely jointed chain in polymer physics and consists of straight-line segments, in which the length of any step or segment is a random variable and for which one end of any step is uniformly distributed over a sphere whose center lies at the opposite end of the segment. If $p(\mathbf{r})$ denotes the probability density for the length of a single step, the probability density for the end-to-end distance of an n step walk can be written

$$p_n(\mathbf{r}) = \frac{1}{8\pi^3} \iiint_{-\infty}^{\infty} C^n(\boldsymbol{\rho}) \exp(-i\boldsymbol{\rho} \cdot \mathbf{r}) \, d^3\boldsymbol{\rho} \qquad (3.1)$$

where

$$C(\boldsymbol{\rho}) = \iiint_{-\infty}^{\infty} p(\mathbf{r}) \exp(i\boldsymbol{\rho} \cdot \mathbf{r}) \, d^3\mathbf{r} \qquad (3.2)$$

In the Rayleigh model $p(\mathbf{r}) = p(r)$, where $r = |\mathbf{r}|$, so that

$$C(\boldsymbol{\rho}) = C(\rho) = \frac{4\pi}{\rho} \int_0^{\infty} r p(r) \sin \rho r \, dr \qquad (3.3)$$

The freely jointed chain is defined by

$$p(r) = \frac{1}{4\pi l^2} \delta(r - l) \qquad (3.4)$$

which implies that

$$C(\rho) = \frac{\sin \rho l}{\rho l} \qquad (3.5)$$

and

$$p_n(r) = \frac{1}{2\pi^2 r} \int_0^{\infty} \rho \left(\frac{\sin \rho l}{\rho l} \right)^n \sin \rho r \, d\rho \qquad (3.6)$$

This integral can be evaluated exactly,[9,14,99] but the resulting expression is useless for computation except at small values of n. When n is large one can use a central-limit-theorem argument to derive the asymptotic gaussian approximation

$$p_n(r) \sim \left(\frac{2\pi n l^2}{3} \right)^{-3/2} \exp\left(-\frac{3r^2}{2nl^2} \right) \qquad (3.7)$$

This expression is valid for $nl \gg r$. When the end-to-end distance of the random walk is of the order of magnitude of the maximum possible displacement, another approach must be used to find an approximation to $p_n(r)$. Dvorak[100] and Yamakawa[98] have derived accurate approximations to this function using the method of steepest descents. A slightly flawed approximation was first given by Kuhn and Grün,[13] and the error in their derivation, important for small chains, was pointed out by Jernigan and Flory.[101] Yamakawa's result[98] is expressed in terms of the dimensionless variable β which is the solution to the equation

$$\coth \beta - \frac{1}{\beta} = \frac{r}{nl} \qquad (3.8)$$

as

$$p_n(r) \sim \frac{n\beta^2}{(2\pi n)^{3/2} l^2 r \{1 - (\beta \operatorname{csch} \beta)^2\}^{1/2}} \left(\frac{\sinh \beta}{\beta} \right)^n \exp\left(-\frac{r\beta}{l} \right) \qquad (3.9)$$

The importance of using this more accurate version of the probability density for studying configurational statistics of short chains has been stressed by Jernigan and Flory.[101] When real polymer chains are modeled in terms of freely jointed chains, an equivalence must be established that implies that more than a single bond of the real molecule is modeled by one bond of the freely jointed chain. The equivalence is somewhat arbitrary, since different criteria can be used to establish the relation. For example, Jernigan and Flory[102] establish a relation between a real structured chain with n bonds each of length l and a freely jointed chain with n' bonds of length l'. The two parameters n' and l' are chosen so that $n'l' = nl\cos(\theta/2)$ (see Fig. 10 for the definition of the angle θ) and $n'(l')^2 = nl^2$. These conditions ensure that the lengths at maximum extension of the two chains coincide, and that the mean square end-to-end distances coincide. For a polymethylene chain Jernigan and Flory find that $n' = n/10$ and $l' = 8.3l$. The gaussian form for $p_n(r)$ appears to be quite accurate for freely jointed random walks of 20 or more steps.

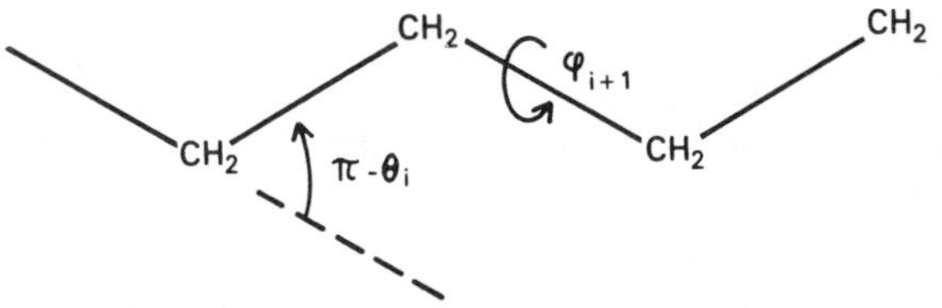

Fig. 10. Typical portion of a polymer chain (in this case a polymethylene chain). The bond angle is denoted by θ_i for the ith bond, and φ_i denotes the bond rotation angle. This notation will be used in the remainder of this section.

2. Size and Shape Parameters of Random-Flight Chains

We have described the freely jointed chain because it has played a considerable role in the theory of polymer configurations. Real polymer configurations differ from those described by the freely jointed chain because they do not have totally random bond angles. Historically three types of models have been used to describe polymer configurations (in the absence of excluded-volume effects). The first is the so-called random-flight model, which is equivalent to a diffusion limit. The use of this limit can be justified for very long chains independent of microscopic details because of central limit theorems for Markov processes. The second is that of a random walk with some parameters (e.g., bond angles) fixed. For this model it is feasible to calculate moments of the end-to-end distance directly from a knowledge of intermolecular potentials, bond legths, and bond angles. This approach is admirably summarized in the monograph by Flory.[97] Although this microscopic technique furnishes moments of the distribution of end-to-end distances, it does not directly allow the calculation of the associated probability density. Several attempts have been made to expand the probability density in a series of Hermite polynomials multiplied by an appropriate gaussian.[101, 102] This technique, long in use in classical statistics, suggests itself because the coefficients of the Hermite polynomials are expressible in terms of moments. We shall have more to say about this later. An approach intermediate between the two mentioned above is that of trying to retain some microscopic features in a continuum model. These models are generally referred to as wormlike chains, and their study was initiated by Kratky and Porod,[103] Daniels,[104] and Hermans and Ullman.[105] The work of Kratky and Porod has been used extensively for the interpretation of light-scattering experiments.

The random-flight model is most often used to describe gross configurational properties such as size and shape. When the random flight takes place in a force-free field, the end-to-end distance has a gaussian distribution given in Eq. (3.7). Other functions that can be used to describe size and shape and lend themselves to a computation are the maximum extension of the random walk, the radius of gyration, and the spans. If we concentrate for the moment on the diffusion limit, then the probability density of the maximum extension is rather simply obtained. If, after n steps, the random walk has remained within a sphere of radius R centered at the starting point, the maximum extension is $\leqslant R$. Let $F_n(R)$ be the probability of remaining within a sphere of radius R, and let the probability density for the maximum extension be $f_n(R)$. The relation between these two functions is $f_n(R) = dF_n(R)/dR$. In the random-flight model one passes to the continuum limit, in which case the probability density for being in the volume $(r, r + dr)$ after n steps satisfies the equation $\partial p / \partial n = (l^2/6)\nabla^2 p$ when the average step length is equal to l. If the random walk remains within a sphere of radius R, this equation must be solved subject to $p = 0$ on the boundary. Hollingsworth[106] has treated a number of related problems in some detail, as have Rubin, Mazur, and Weiss.[107] If one sets the step length equal to 1, the probability that the random flight remains within a sphere of radius R is

$$F_n(R) = 2 \sum_{j=1}^{\infty} (-1)^{j+1} \exp\left(-\frac{\pi^2 j^2 n}{6R^2}\right) \qquad (3.10)$$

from which it follows that the average value of the maximum excursion is $\langle R_n \rangle = (\pi^3 n / 24)^{1/2}$. More recently Rubin and Weiss[108] have generalized these calculations, considering the maximum extension of a chain whose end-to-end distance is constrained to be equal to r. They show that the probability that the maximum extension is between R and $R + dR$ at time $\tau = n/6$ is $s_\tau(R)dR$ where

$$s_\tau(R) = \frac{1}{r} \sum_{j=-\infty}^{\infty} j\left[2 - \frac{2jR+r}{\tau}\right]\exp\left\{-\frac{jR}{\tau}(r+jR)\right\} \qquad (3.11)$$

and the average end-to-end distance is

$$\frac{\langle R \rangle}{r} = 1 + \frac{\tau}{r^2}\left\{1 - \sum_{j=1}^{\infty} \frac{1}{j(j+1)}\exp\left[-\frac{j(j+1)r^2}{\tau}\right]\right\} \qquad (3.12)$$

From this expression it is possible to determine the limiting behavior at $r = 0$ to be $\langle R_n \rangle = (\pi n/6)^{1/2} = 0.724 n^{1/2}$, in contrast to Hollingsworth's[106] result $\langle R_n \rangle = 1.137 n^{1/2}$ for the unconstrained case. When $r^2 \gg n/6$ one finds $\langle R_n \rangle \sim r$. Rubin and Weiss also calculate the probability density for the maximum extension of a random walk in three dimensions in which the initial position lies on an impenetrable plane. The expression for the probability density is rather complicated, but the first moment is found to be $\langle R_n \rangle = 1.308 n^{1/2}$. Many calculations of similar quantities are to be found in the earlier polymer literature. For example Katchalsky, Kuenzle, and Kuhn[109] have determined the probability density of the distance between the first and an arbitrary atom on a gaussian chain with fixed ends. If atom 1 is located at the origin, atom N is at $(x, y, z) = (0, 0, \rho)$, and the mean end-to-end distance of a single step is equal to 1, then the probability density for the location of atom k is shown to be

$$p_k(\mathbf{r}) = \left(\frac{3N}{2\pi k(N-k)} \right)^{3/2} \exp\left\{ -\frac{3N}{2k(N-k)} \left[x^2 + y^2 + \left(z - \frac{k\rho}{N} \right)^2 \right] \right\}$$

(3.13)

The mean squared distance from the origin to this atom is

$$\langle r_k^2 \rangle = \left(\frac{k}{N} \right)^2 \rho^2 + k\left(1 - \frac{k}{N} \right)$$

(3.14)

There is a value of k other than $k = N$ that maximizes this expression when $\rho^2 < N$, i.e., when $\rho^2 < \langle r_N^2 \rangle$. Otherwise $\langle r_k^2 \rangle$ increases monotonically with k. It should be noted that all of the calculations refer to gaussian random walks, so that the results are only qualitatively correct for more realistic polymer models. Recently Domb[110] has discussed the general problem of calculating averages for lattice random walks subject to constraints—for example, having fixed end-to-end distances.

Another parameter that has received considerable attention, as a description both of the size and of the asymmetry of random walks, is the radius of gyration, which we denote by R_G. This parameter arises naturally in the study of light scattering by polymers.[111] It is defined in terms of the locations of the vectors to all the atomic positions by

$$R_G^2 = \frac{1}{n} \sum_{i=1}^{n} (\mathbf{r}_i - \boldsymbol{\mu})^2$$

(3.15)

for an n-step walk in which $\boldsymbol{\mu}$ is the location of the center of mass,

$$\boldsymbol{\mu} = \frac{1}{n} \sum_{i=1}^{n} \mathbf{r}_i$$

(3.16)

One can similarly define the radius of gyration with respect to different axes.

For example

$$R^2_{G,x} = \frac{1}{n} \sum_{i=1}^{n} (x_i - \mu_x)^2 \tag{3.17}$$

is the x component of the radius of gyration. According to its definition, R_G is a random variable. By taking averages one can readily verify that $\langle R^2_G \rangle = \langle r^2_n \rangle / 6$, where $\langle r^2_n \rangle$ is the mean squared end-to-end distance of the n-step walk. Higher-order moments of R_G are readily obtained from the definition in Eq. (3.15). The calculation of the probability density of R_G or its components is a problem that has been studied by several investigators, starting with the analysis by Fixman.[112] The starting point of all of the investigations has been the gaussian random walk, and formally similar analytical techniques have been used in all of the investigations. To give the flavor of these calculations we first observe that R^2_G can be written as a quadratic form in the r_i:

$$R^2_G = \sum_j \sum_k g_{jk} \mathbf{r}_j \cdot \mathbf{r}_k \tag{3.18}$$

If $P_n(R^2_G)$ denotes the probability density of R^2_G and $p_n(\mathbf{r})$ denotes the joint distribution of the $\{\mathbf{r}_i\}$, then a formal expression for $P_n(R^2_G)$ can be written:

$$P_n(R^2_G) = \int_{-\infty}^{\infty} \cdots \int \delta\left(R^2_G - \sum_j \sum_k g_{jk} \mathbf{r}_j \cdot \mathbf{r}_k \right) p_n(\mathbf{r}) \, d^3\mathbf{r} \tag{3.19}$$

The substitution of the Fourier representation for the delta function into this expression leads to the much simpler looking

$$P_n(R^2_G) = \frac{1}{2\pi} \int_{-\infty}^{\infty} K(s) \exp(-isR^2_G) \, ds \tag{3.20}$$

where

$$K(s) = \int_{-\infty}^{\infty} \cdots \int \exp\left(is \sum_j \sum_k g_{jk} \mathbf{r}_j \cdot \mathbf{r}_k \right) p_n(\mathbf{r}) \, d^3\mathbf{r} \tag{3.21}$$

When $p_n(\mathbf{r})$ is gaussian, $K(s)$ is just an integral of the exponential of a quadratic form. The resulting integrals have been evaluated in different ways. The most comprehensive calculations appear to be those of Fujita and Norisuye.[113] They gave an exact result for $P_n(R^2_G)$ in terms of an infinite series of Bessel functions. Letting $t = 6R^2_G/n = R^2_G/\langle R^2_G \rangle$, Fujita and

Norisuye find that for small t,

$$\frac{n}{6}P_n(R_G^2) \sim 9\sqrt{\frac{6}{\pi}} \frac{1}{t^3} e^{-2.25/t}\left(1 - \tfrac{19}{36}t + \cdots\right) \tag{3.22}$$

while in the limit of large t

$$\frac{n}{6}P_n(R_G^2) \sim \left(\frac{\pi^5 t}{2}\right)^{1/2} e^{-\pi^2 t/4}\left(1 + \frac{9}{4\pi^2 t} + \cdots\right) \tag{3.23}$$

Coriell and Jackson[114] obtained analogous results for the probability density of $R_{G,x}$. Flory and Fisk[115] have given an empirical approximation to $P_n(R_G)$ that is

$$P_n(R_G) = \frac{343}{15}\left(\frac{14}{\pi}\right)^{1/2} t^3 e^{-3.5t} \tag{3.24}$$

which is in good qualitative agreement with the more exactly calculated values. Hoffman and Forsman[112] have dealt with some computational aspects of finding the probability density.

Solč[116] and Solč and Stockmayer[117] have taken these calculations further in a study of the asymmetry to be expected in 3-D random-flight models. Although the parameters that specify a random walk may indicate that on average the random walk is symmetric (e.g., the three components of the mean squared end-to-end distance are equal), the instantaneous configuration may be far from symmetric. That is to say, the joint probability density for the three components of end-to-end distance

$$X_n^2 = \frac{1}{n}\sum_{i=1}^{n} x_i^2, \qquad Y_n^2 = \frac{1}{n}\sum_{i=1}^{n} y_i^2, \qquad Z_n^2 = \frac{1}{n}\sum_{i=1}^{n} z_i^2 \tag{3.25}$$

may give significant weight to regions in which the three parameters differ considerably from one another. This asymmetry was first pointed out for some special parameters by Kuhn[12] in 1934. In Solč's first paper on the subject[116] he considers the general problem of finding the probability density of the linear combination

$$Q = C_1 R_{G,x}^2 + C_2 R_{G,y}^2 + C_3 R_{G,z}^2 \tag{3.26}$$

where the C_i are arbitrary constants. In particular, if one wants to study deviations from isotropy, one can choose $C_1 + C_2 + C_3 = 0$, so that $\langle Q \rangle = 0$.

Using the methods developed by Coriell and Jackson,[114] Solč derived an integral representation for the probability density of Q:

$$P(Q) = \frac{1}{2\pi N^{3/2}} \int_{-\infty}^{\infty} e^{i\lambda Q} \prod_{k=1}^{3} U_{N-1}^{-1/2}\left(1 + \frac{i\lambda\sigma^2}{3N}C_k\right) d\lambda \qquad (3.27)$$

where $U_N(x)$ is the Chebyshev polynomial of the second kind. The case considered by Coriell and Jackson corresponds to $C_1 = 1$, $C_2 = C_3 = 0$. One can evaluate the integral in terms of a sum of residues when $C_1 = C_2$ and C_3 is of the opposite sign or equal to zero. In that case Solč shows that

$$P(Q) = \frac{6N^{5/2}}{|C_1|} \sum_{m=1}^{N-1} (-1)^{m+1} v_m^{-N} \sin^2\left(\frac{\pi m}{N}\right)$$

$$\times \left[\frac{y_m(1+y_m^2)^{1/2}}{1-v_m^{-4N}}\right]^{1/2} \exp\left(\frac{6N^2 y_m^2 Q}{C_3}\right) \qquad (3.28)$$

where

$$y_m = \left(-\frac{C_3}{C_1}\right)^{1/2} \sin\left(\frac{m\pi}{2N}\right)$$

$$v_m = y_m + \left(1 + y_m^2\right)^{1/2} \qquad (3.29)$$

Since $P(Q)$ in Eq. (3.27) is expressed as a Fourier integral, the moments of $P(Q)$ can be found in closed form by differentiating the product in the integrand and setting $\lambda = 0$. In particular Solč considers the case

$$Q^* = 2R_{G,x}^2 - R_{G,y}^2 - R_{G,z}^2 \qquad (3.30)$$

for which one can show quite straightforwardly that

$$\langle Q^* \rangle = 0, \qquad \langle Q^{*2} \rangle = \tfrac{8}{15}\langle R_G^2 \rangle^2, \qquad \langle Q^{*3} \rangle = \tfrac{128}{315}\langle R_G^2 \rangle^3 \qquad (3.31)$$

The magnitude of these numbers suggests that the instantaneous nonsphericity of the random walk can be appreciable. Solč then goes on to separate the effects of shape distribution and random orientation with respect to the fixed (x, y, z) coordinate system. He does this by calculating statistical properties of randomly oriented chains with ordered components of the radius of gyration $L_1 \le L_2 \le L_3$, as well as by simulation. Typical results for a chain with $N = 101$ links are

$$\langle L_3^2 \rangle : \langle L_2^2 \rangle : \langle L_1^2 \rangle = 11.7 : 2.7 : 1$$

which illustrates the extent of departure from symmetry. Solč and Gobush[116] have considerably enlarged on these calculations by treating ring-shaped, cyclic, starlike and comblike chains. Kranbuehl, Verdier, and Spencer[118] dealt, by simulation, with the same type of problem, considered as a function of time. They used a technique developed by Verdier[119] to study brownian motion in polymer chains with a fixed number of links. When the components of the radius of gyration for this system are averaged over a sufficiently large time, they tend towards equality. The authors show that the relaxation of the individual components to a common value takes place in a time comparable with the longest relaxation times that characterize the motion of the polymer chain.

Weiss and Rubin[120] have also peripherally dealt with the question of asymmetry within the framework of the theory of spans, which we have introduced in Section II.D. The spans were defined to be the sides of the smallest rectangular box entirely containing the random walk, with sides parallel to the coordinate axes. If we let $P_n(\mathbf{r})$ be the probability (for a lattice random walk) that the random walker is at $\mathbf{r}$ at step n, then the probability that the spans are $S_1, S_2, \ldots, S_D$, where D is the number of dimensions, is asymptotically given by

$$\rho_n(\mathbf{S}) = 8^D \sum_{j_1=1}^{\infty} \cdots \sum_{j_D=1}^{\infty} (-1)^{j_1 + j_2 + \cdots + j_D + D} j_1^2 j_2^2 \cdots j_D^2$$
$$\times P_n\big(j_1(S_1+1), j_2(S_2+1), \ldots, j_D(S_D+1)\big) \qquad (3.32)$$

When steps along each of the coordinate axes are uncorrelated and each of the variances

$$\sigma_j^2 = \sum_{\mathbf{r}} r_i^2 p(\mathbf{r}) \qquad (3.33)$$

is finite, where $p(\mathbf{r})$ is the single-step transition probability, then $P_n(\mathbf{r})$ is asymptotically factorizable in the limit of large n and takes the form

$$P_n(\mathbf{r}) \sim (2\pi n)^{-D/2} \exp\left\{ -\frac{1}{2n} \sum_{i=1}^{D} \frac{r_i^2}{\sigma_i^2} \right\} \qquad (3.34)$$

When this is substituted into Eq. (3.32), the function $\rho_n(\mathbf{S})$ likewise factors into $\rho_n(\mathbf{S}) = \rho_n(S_1)\rho_n(S_2)\ldots\rho_n(S_D)$ where $\rho_n(S)$ is the one-dimensional span distribution. This factorization property allows us, for example, to write the probability that the smallest span is equal to S,

$$\rho_n^{(1)}(S) \sim D\rho_n(S)\left\{ \sum_{j=S}^{\infty} \rho_n(j) \right\}^{D-1} \qquad (3.35)$$

Similar expressions can be given for the distribution of each ordered span. In three dimensions the ratio of the mean of the largest to the mean of the smallest span in this approximation is 1.64, which does not indicate as large a degree of asymmetry as is given by Solč's analysis. However, the span is a different parameter than those considered by Solč. Mazur and Rubin[121] made more detailed investigations of the spans of polymer chains with and without excluded-volume restrictions using a combination of simulation and exact computations. They showed that the ratio of the largest mean span to the smallest goes from 2.41 for a 10-step walk to a figure close to 1.64 for a 100-step walk. The contrasting figures for a self-avoiding random walk on a simple cubic lattice are 2.38 for a 10-step walk and 1.81 for a 150-step walk. The near-equality of the two figures for the 10-step walks is to be expected, since very few configurations are excluded at that small number of steps.

Rubin and Weiss[122] have analyzed the spans of star-branched polymer chains using the continuous analogue of Eq. (2.205). Let us start with a random walk whose state probability density is $P_n(\mathbf{r})$ conditional on having remained in a box B described by $0 \le x_1 \le S_1$, $0 \le x_2 \le S_2,\ldots$, $0 \le x_D \le S_D$, and averaged over all starting points in B. Further, let us define the quantity

$$\psi_n(\mathbf{S}) = \int_0^{S_1} \cdots \int_0^{S_D} P_n(\mathbf{r})\, d^D\mathbf{r} \qquad (3.36)$$

The probability density function of the spans, $\rho_n(\mathbf{S})$, can be expressed in terms of ψ_n as

$$\rho_n(\mathbf{S}) = \frac{\partial^{2D}\psi_n}{\partial S_1^2 \partial S_2^2 \ldots \partial S_D^2} \qquad (3.37)$$

This relation was first stated by Daniels,[123] who derived it in a slightly different manner than in our presentation. Equation (3.37) simplifies considerably when ψ can be written as a product of factors, $\psi^i(S_i)$, that depend on the S_i separately. This proves to be the case for star-branched molecules, in which the $P_n(\mathbf{r})$ can be found as solutions to $\partial_n P_n(\mathbf{r}) = \frac{1}{6}\nabla^2 P_n(\mathbf{r})$. The results so derived relate to a random-walk model for gel permeation chromatography discussed by Casassa and Tagami,[124] Guttman and DiMarzio,[125] and Casassa.[126]

3. Determination of Distribution of End-to-End Distance from Moments

All of the problem dealt with so far in this section relate to idealized models of polymers. The principal distinction between these models and real polymers is the appearance of structure and stiffness in polymer chains.

These introduce correlations in the random-walk models. In the absence of excluded-volume effects, the central limit theorem guarantees a gaussian limit for the probability density of end-to-end distance for a large enough number of bonds. Two general approaches have been adopted to incorporate details of structural information into random-walk models of polymers. The first is to calculate moments of end-to-end distance and use these directly to compute physically interesting parameters, and the second is to pass to a continuum limit while retaining some structural features in the formulation of relevant equations that characterize the random walk. For light scattering, for example, one needs to calculate the quantity[127]

$$J = \left\langle \sum_{i \neq j} \sum \frac{\sin \mu r_{ij}}{\mu r_{ij}} \right\rangle \tag{3.38}$$

where r_{ij} is the vector connecting atom i to atom j on the chain and $\mu = (4\pi/\lambda)\sin(\theta/2)$, where λ is the wavelength of the light and θ is the difference in angle between incident and scattered beams. This expression can be converted to a sum over the moments of r_{ij} by expanding $\sin \mu r_{ij}$ in a power series. There are many studies that relate moments of different features of polymer configurations to molecular parameters, since this was one of the earliest topics to engage polymer physicists. Many of these are admirably summarized in the monographs by Volkenshtein,[15] Birshtein and Ptitsyn,[96] Flory,[97] and Yamakawa,[98] so we omit detailed discussion, referring only to some more recent work on the approximation of the distribution of end-to-end distance based on moments. This problem is of interest for short chains, where the central limit theorem is not relevant. The second general approach refers to the elucidation of properties of the so-called wormlike chains first studied in the context of light scattering from polymer solutions by Kratky and Porod,[103] and shortly thereafter more extensively, from the mathematical side, by Daniels.[104]

A simple representation of a polymethylene chain is shown in Fig. 10, which serves to define the bond angles θ_i and the bond rotation angles φ_i. The bond angles are generally fixed while the bond rotation angles are random because of thermal motion, but are restricted to some degree by the interaction potential between atoms. The mean squared end-to-end distance can be written

$$\langle R_n^2 \rangle = \sum_{i=1}^{n} \langle r_i^2 \rangle + 2 \sum_{1 \leqslant i < j \leqslant n} \sum \langle \mathbf{r}_i \cdot \mathbf{r}_j \rangle \tag{3.39}$$

If the bond length is equal to l and the angles θ_i are random and identically distributed so that $\langle \cos \theta \rangle = 0$, then $\langle \mathbf{r}_i \cdot \mathbf{r}_j \rangle = 0$ and $\langle R_n^2 \rangle = nl^2$. If $\theta_i = \theta$ for all i and the φ_i are uniformly distributed, then

$$\langle \mathbf{r}_i \cdot \mathbf{r}_{i+j} \rangle = l^2 \cos^{|j|} \theta \tag{3.40}$$

and the sums in Eq. (3.39) can be evaluated, leading to

$$\langle R_n^2 \rangle = nl^2 \left(\frac{1 + \cos \theta}{1 - \cos \theta} \right) - 2l^2 \cos \theta \frac{1 - \cos^n \theta}{(1 - \cos \theta)^2}$$

$$\sim nl^2 \left(\frac{1 + \cos \theta}{1 - \cos \theta} \right) \tag{3.41}$$

a result given by several authors.[95] When φ is not uniformly distributed the asymptotic form for $\langle R_n^2 \rangle$ is to be replaced by[15]

$$\langle R_n^2 \rangle \sim nl^2 \left(\frac{1 + \cos \theta}{1 - \cos \theta} \right) \left(\frac{1 + \langle \cos \varphi \rangle}{1 - \langle \cos \varphi \rangle} \right) \tag{3.41a}$$

More elaborate calculations allow one to take the bond rotation angles into account, and a parallel analysis can be given for the radius of gyration. Jernigan and Flory[102] have discussed, at some length, the mathematics involved in calculating moments of the end-to-end distance of polymer chains. Since moments can, in theory, be calculated fairly easily, the question that suggests itself is how well one can approximate the actual probability density in terms of these moments. Nagai[128] and Jernigan and Flory[101] were the first to approximate the probability density of realistic chains in terms of an expansion in a series of Hermite polynomials. Letting $p_n(\mathbf{r})$ be the required probability density of end-to-end distance, Jernigan and Flory transform to a dimensionless variable $\zeta = 3r^2/(nl^2)$ and set

$$w_n(\zeta) = \left(\frac{nl^2}{3} \right)^{3/2} p_n(\mathbf{r}) \tag{3.42}$$

for chains that are spherically symmetric by virtue of averaging over all possible conformations of the first bond. They then expand

$$w_n(\zeta) = \frac{1}{(2\pi)^{3/2}} \exp \left(- \frac{\zeta^2}{2} \right) \sum_{j=0}^{\infty} \frac{h_j H_j(\zeta)}{j! \zeta} \tag{3.43}$$

where the $H_j(\zeta)$ are Hermite polynomials orthogonal to the weight function multiplying the infinite series. The central-limit result corresponds to $h_1 = 1$, $h_2 = h_3 = \cdots = 0$. By symmetry, only odd terms in the series survive. The expansion in orthogonal polynomials is useful because the coefficients h_j are expressible as a linear combination of moments. Results of this kind of approximation were generated and compared to the exactly known distribution of the freely jointed chain, with good agreement obtained using terms up to $\langle R^6 \rangle$ for a 10-bond chain. It was also found that for more realistic polymer chains the convergence of the series in Eq. (3.43) was extremely slow. This difficulty also appeared in the work of Yoon and Flory[129] on polymethylene chains. These authors found that the expansion in Hermite functions did not give good results for short chains, and that longer chains ($n \geqslant$ 50) were well represented by the freely jointed chain density function with an appropriately chosen bond length. Jernigan and Weiss[130] suggested an expansion in terms of polynomials orthogonal with respect to the density function for the freely jointed chain. They did not present results for any specific chain, so that the utility of this method is not presently known.

Another rather interesting approach to this problem is that taken by Fixman and Alben,[131] who derive an approximation to the $p_n(r)$ where $r = (\mathbf{r} \cdot \mathbf{r})^{1/2}$. The authors choose $p_n(r)$ to maximize the entropy

$$S = -4\pi \int_0^L r^2 p_n(r) \ln p_n(r)\, dr \qquad (3.44)$$

where L is the maximum allowed length. This function is to be maximized subject to constraints that may be imposed somewhat arbitrarily. Fixman and Alben in particular chose to impose equality between the first two nonvanishing moments ($\langle r^2 \rangle$ and $\langle r^4 \rangle$) from an exact calculation and those from the approximate $p_n(r)$. This leads to the form

$$p_n(r) = C \exp(-ar^2 - br^4) \qquad (3.45)$$

where a and b are determined from the moments and C is a normalizing constant. The results were tested for a rotational-isomer‡ model on a tetrahederal lattice. Good agreement is obtained with more exact calculations for

---

‡The rotational-isomer model, frequently used in polymer physics, and exhaustively described by Volkenshtein[15] and Flory,[97] approximates the probability density function of the bond rotation angle,

$$p(\varphi) = \frac{\exp[-U(\varphi)/(kT)]}{\displaystyle\int_{-\pi}^{\pi} \exp[-U(\beta)/(kT)]\, d\beta}$$

by a function defined for a small number, generally three, of discrete values of φ. It can also allow for dependence on more than one value of φ, that is, one can also have a joint densities like $p(\varphi_i, \varphi_{i+1})$.

values of n down to 12. A later paper by Fixman and Skolnick[132] starts from an approximation analogous to that in Eq. (3.45):

$$p_n^{(0)}(r) = C \exp\left[-ar^2 - (br^2)^s\right] \qquad (3.46)$$

where s is a disposable parameter. Correction terms to this formula are obtained by expanding the true distribution function as

$$p_n(r) = p_n^{(0)}(r) \sum_{j=0}^{M} a_j r^{2j} \qquad (3.47)$$

where the sum can be calculated in terms of the polynomials orthogonal to $p_n^{(0)}(r)$. The improved approximation performs well except at the origin and at $r \simeq r_{max}$, where successive approximations tend to oscillate. This oscillation can be reduced to some extent by introducing smoothing, but it is not clear that more accurate values are thereby obtained. Freire and Fixman[133] and Freire and Rodrigo[134] have taken a different approach to the general problem that takes angular dependence into account. The function $p_n(\mathbf{r})$ for a realistic model of a polymer chain depends only on r because the configuration of the first two bonds includes an average over all orientations. If this average is not taken, $p_n(\mathbf{r})$ will have some angular dependence which decreases as the number of bonds increases. The angular dependence can be incorporated by expanding $p_n(\mathbf{r})$ as

$$p_n(\mathbf{r}) = \sum_{l=0}^{\infty} \sum_{m=-l}^{l} F_{lm}(r) Y_{lm}(\theta, \varphi) \qquad (3.48)$$

where the $Y_{lm}(\theta, \varphi)$ are spherical harmonics that depend on the spherical coordinates of r. The coefficients appearing in the expansion of $F_{lm}(r)$ are to be expressed in terms of generalized moments

$$M_{lm}^{(k)} = 4\pi \int_0^{\infty} F_{lm}(r) r^{k+2} \, dr \qquad (3.49)$$

whose calculation has been formalized by Fixman and Skolnick.[132] Since these moments can be calculated exactly, Freire and Rodrigo suggest an expansion of the type shown in Eq. (3.47). The approximation technique just described has been applied to short polymethylene chains so as to compare results with those of Yoon and Flory[129] for $n = 10$ and 20 bonds. The series generated by Freire and Rodrigo led to curves that appear to be more realistic than those of Yoon and Flory by virtue of remaining nonnegative on a wider set of intervals. The authors take this to be a sign of better convergence of their series. A detailed comparison of the estimated radial density function and Monte Carlo results showed good agreement for $n = 10$ except in the neighborhood of the peak. The results for $n = 20$ are in good agreement over the entire curve.

Clearly, much remains to be done in assessing the best way of using moment data to generate approximations to the underlying densities. Other ideas that might be pursued in this regard include different choices of the weighting function (the gaussian for the expansion in Hermite polynomials, the freely jointed chain density, or the density generated by the maximum-entropy analysis). Another possibility is that of trying to improve the convergence of the series. For example, one might try replacing the series by equivalent Padé approximations.

4. The Wormlike Chain

One of the most widely used configurational models in polymer physics is that of the wormlike chain, originally introduced by Kratky and Porod[103] in an analysis of light scattering from polymer molecules. The idea behind this model can be understood by referring to Eq. (3.41). One wants to model a chain of fixed length and gradual curvature. Let the total length be $L = nl$, and let $l \to 0$ and $n \to \infty$, keeping L fixed. To keep the curve smooth we must also let $\theta \to 0$, and we shall do this in such a way that

$$\frac{2l}{\theta^2} \to a \tag{3.50}$$

where a is a constant known as the persistence length of the chain. In these limits Eq. (3.41) becomes

$$\langle r^2 \rangle = 2a \left[L - a + a \exp\left(-\frac{L}{a} \right) \right] \to \begin{cases} 2aL, & L/a \to \infty \\ L^2, & L/a \to 0 \end{cases} \tag{3.51}$$

as given in the original formulation by Kratky and Porod. Daniels[104] derived a partial differential equation for the probability density function for the end-to-end distance, and this was followed by a slightly simpler analysis by Hermans and Ullman.[105] More recent analyses of configurational properties of wormlike chains have tended to use path-integral methods originally introduced by Saito, Takahashi, and Yunoki.[135] A good introduction to the use of path integrals for the study of configurational problems in polymers is contained in the review by Freed.[136]

To look at the ideas involved in formulating the equation for the wormlike chain, we follow Hermans and Ullman in deriving an equation for the two-dimensional case (Daniel's derivation is a closely related alternative). Consider a chain as in Fig. 11, let Q be an arbitrary point on the chain with coordinates (x, y), and let the tangent to the curve at Q make an angle φ with respect to the x axis. Further, let s be the length of chain from the origin to Q, and let $\psi_{\Delta s}(\varphi)$ be the probability density for the change in φ when the length of curve changes from s to $s + \Delta s$. This function will be assumed to be symmetric in φ and to have the property

$$\int_{-\pi}^{\pi} \varphi^2 \psi_{\Delta s}(\varphi) \, d\varphi = 2\lambda \Delta s + o(\Delta s) \qquad (3.52)$$

for $\Delta s \to 0$. If $f(x, y, \varphi; s)$ is the probability density for the chain coordinates to be (x, y) and the tangent angle to be φ when the total length from the origin to Q is s, then f satisfies

$$f(x, y, \varphi; s + \Delta s) = \int_{-\pi}^{\pi} f(x - \Delta s \cos \varphi, y - \Delta s \sin \varphi, \varphi - \Delta \varphi; s)$$
$$\times \psi_{\Delta s}(\Delta \varphi) \, d\Delta \varphi \qquad (3.53)$$

Passing to the limit $\Delta s = 0$ and using Eq. (3.52), we find that f satisfies the equation

$$\frac{\partial f}{\partial s} = \lambda \frac{\partial^2 f}{\partial \varphi^2} - \cos \varphi \frac{\partial f}{\partial x} - \sin \varphi \frac{\partial f}{\partial y} \qquad (3.54)$$

This is the equation that would describe the brownian motion of rods whose centers have a translational velocity $\dot{x} = \cos \varphi$, $\dot{y} = \sin \varphi$. In three dimensions

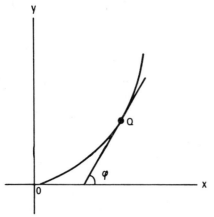

Fig. 11.   Diagram that defines the angle φ used in the text. The length of the curve from 0 to Q is s.

one can follow similar steps to find an equation for the probability density $f(r, \theta; s)$ for the simultaneous occurrence of an end-to-end distance r and angle θ between the direction of the chain at Q and the end-to-end radius vector $\mathbf{r}$. The equation is

$$\frac{\partial f}{\partial s} = \frac{\lambda}{\sin\theta} \frac{\partial}{\partial\theta}\left(\sin\theta \frac{\partial f}{\partial\theta}\right) + \frac{\sin\theta}{r} \frac{\partial f}{\partial\theta} - \cos\theta \frac{\partial f}{\partial r} \qquad (3.55)$$

A convenient solution in closed form has not been found for this equation [nor for Eq. (3.54) for that matter], but Gobush et al.[137] and Yamakawa and Stockmayer[138] have calculated asymptotic corrections to the gaussian limiting behavior starting from Eq. (3.55).

A second approach to the derivation of results for the wormlike chain is the path-integral approach of Saito, Takahashi, and Yunoki.[135] We briefly summarize this technique, following Yamakawa's exposition.[98] Let $\mathbf{r}(s)$ be the vector from the origin to an arbitrary point on the random walk, where s is the arc length along the curve. We note that $\mathbf{u}(s) = \partial\mathbf{r}/\partial s$ is a unit vector, since

$$ds = ds\left[\left(\frac{\partial x}{\partial s}\right)^2 + \left(\frac{\partial y}{\partial s}\right)^2 + \left(\frac{\partial z}{\partial s}\right)^2\right]^{1/2} \qquad (3.56)$$

and it is tangent to the curve. The potential energy per unit length in a bent rod is inversely proportional to the square of the radius of curvature R, which leads to the relation

$$U = \frac{\varepsilon}{2}\int_0^s \left(\frac{\partial\mathbf{u}}{\partial s'}\right)^2 ds' = \frac{\varepsilon}{2}\int_0^s \left(\frac{\partial^2\mathbf{r}}{\partial s'^2}\right)^2 ds' \qquad (3.57)$$

where ε is the bending force constant. The configurational partition function for the random walk is therefore

$$Z(\mathbf{u}, s) = \int \exp\left\{-\frac{\varepsilon}{2kT}\int_0^s \left(\frac{\partial\mathbf{u}}{\partial s'}\right)^2 ds'\right\} d\{\mathbf{r}\} \qquad (3.58)$$

where the integration over $\mathbf{r}$ is over all configurations with $\mathbf{u}$ fixed. One can infer from this representation using the technique introduced by Feynman[139] and Kac[140] that Z satisfies the equation

$$\frac{\partial Z}{\partial s} = \lambda \nabla_{\mathbf{u}}^2 Z \qquad (3.59)$$

where $\lambda = kT/(2\varepsilon)$. This equation is to be solved subject to the constraint that $u^2 = 1$. A solution can be written as an infinite series of spherical harmonics.[98] The probability density for finding the random walk with polar angles (θ, φ) at arc length s given that $(\theta, \varphi) = (0,0)$ at s' is found to be

$$p(\theta, \varphi; s \mid 0,0; s') = \frac{1}{4\pi} \sum_{n=0}^{\infty} (2n+1) P_n(1) P_n(\cos\theta) e^{-\lambda n(n+1)|s-s'|}$$

(3.60)

where the $P_n(x)$ are Legendre polynomials. For the continuous chain the mean squared end-to-end length when the total arc length is equal to L is

$$\langle r^2 \rangle = \int_0^L \int_0^L \langle \mathbf{u}(s_1) \cdot \mathbf{u}(s_2) \rangle \, ds_1 ds_2$$

(3.61)

but the average $\langle \mathbf{u}(s_1) \cdot \mathbf{u}(s_2) \rangle$ can be found to be

$$\langle \mathbf{u}(s_1) \cdot \mathbf{u}(s_2) \rangle = \int_0^\pi \int_0^{2\pi} \cos\theta \, p(\theta, \varphi; s_2 \mid 0,0, s_1) \sin\theta \, d\theta \, d\varphi$$

(3.62)

Substituting this result into Eq. (3.61), we find that $\langle r^2 \rangle$ takes the form of Eq. (3.51) provided that we make the identification $\lambda = (2a)^{-1}$. A more detailed calculation suffices to confirm that $\langle r^4 \rangle$ also agrees with the value calculated by more elementary means. Some implications of using more complicated forms of the potential energy than (3.57) have been discussed by Freed[136] and Yamakawa and Fujii.[141] The analysis becomes considerably more intricate, but the model of Yamakawa and Fujii leads to good agreement with parameters calculated for polymethylmethacrylate chains by Yoon and Flory.[129] The theory has been further elaborated in great detail by Yamakawa and his collaborators,[142] including a study of many of the physical properties of interest in polymer physics. There are many studies of the brownian motion of polymer chains, as well as of hydrodynamic properties such as viscosity, diffusion, and sedimentation. Reasons of space and personal interests preclude our discussing these, but a good collection of reprints of classic papers on these topics can be found in References 143 and 144. A further topic of some current interest is that of reptation. This concept was introduced by de Gennes[145] and refers to random walks in a space with fixed obstacles. This snakelike motion is the origin of the word. In effect the polymer moves in a tube defined by network entanglements. Edwards and his collaborators have also developed a theory of random walks with entanglements,[146] but the theories all contain approximations of one sort or another and the correlation with experimental data is not always good.[147]

5. Polymer-Chain Adsorption at a Surface

The problems described to this point have mainly required dealing with nonlattice random walks. When a physical boundary is present in a solution of polymers, an important class of problems arises that have been dealt with by many authors using both lattice and nonlattice random walks. We describe several of these in the remainder of this section. The general problem area is that of the characterization of polymer adsorption on surfaces. For chains in the vicinity of a surface there is competition between the energy gained by establishing additional contacts with the surface and the entropy lost because chain configurations are eliminated by the presence of the surface. The evaluation of this tradeoff has been the subject of several investigations.

Silberberg[148] gave the first correct formulation of this problem in the case of a discrete lattice model, and Pouchlý[149] did the same for the continuous random-flight limit. These authors recognized that the presence of an impenetrable surface could be represented by an absorbing boundary condition at that surface. The central problem in a model of polymer-chain adsorption at a solution surface is to enumerate or classify n-step configurations according to the number of visits to a surface layer adjacent to the absorbing surface and to assign a Boltzmann statistical weight $e^{\theta n}$ to the class of configurations which visit that surface layer n times. Energy of interaction effects of polymer chain segments with a surface can be treated in a more straightforward manner in a lattice model, and an extensive variety of formulations and solutions of this problem rapidly appeared.[67, 150-153] Correspondingly complete solutions of the adsorption problem in the random-flight limit have appeared more recently in two limiting cases. Lépine and Caillé[154] treated the case of a "weakly adsorbed" chain following a formulation of de Gennes[155]; and Wiegel[156] treated the case of a weak but long-range electrostatic interaction between monomer units of a chain and a charged surface. In the solution of the lattice model of adsorption, to be described more fully below, it was found that as the Boltzmann weighting factor e^{θ} increases, the polymer chain collapses discontinuously (at a critical value e^{θ_c}) into the surface layer. The description of such discontinuous behavior on the basis of a diffusion equation clearly presents technical difficulties in the case $e^{\theta} \gg e^{\theta_c}$, where the entire chain is practically confined to a few layers near the solution surface. This difficulty motivated de Gennes's treatment in the limiting case of a "weakly adsorbed" chain where e^{θ} is close to e^{θ_c}.

To give some idea of the method used to treat a lattice model of adsorption, we summarize the approach used by Rubin.[67] His method can be adapted to treat adsorption on a long rigid rod represented by the z axis in a cubic lattice,[157] and adsorption of a chain polymer constrained between two parallel planes[158] as well as several other applications.[72] Consider a random walk with steps to nearest neighbors only, on a cubic lattice. The starting point is assumed to be in the surface layer labeled $k = 0$. Successive layers in the solution are labelled $k = 1, 2, \ldots$. In order to eliminate from consideration walks which pass out of the solution, layer $k = -1$ is assumed to be an absorbing or trapping layer. This artifice has been used many times.[149,159] The description of the model in terms of visits to layers concentrates attention on the one-dimensional aspect of the model. Thus, for layers $k > 0$, the probability of stepping to an adjacent layer is $a/2$, and the probability of stepping laterally in the same layer (i.e., remaining in layer k) is $1 - a$. The probability, $P_{n+1}(k|0)$, of being in layer k at the $n + 1$th step, having started in layer 0, satisfies the nearest-neighbor recurrence relations

$$P_{n+1}(0|0) = \frac{a}{2} P_n(1|0) + (1 - a) P_n(0|0)$$

$$P_{n+1}(k|0) = \frac{a}{2} P_n(k+1|0) + (1 - a) P_n(k|0) + \frac{a}{2} P_n(k-1|0) \quad (3.63)$$

with the additional relation for the trapping layer

$$P_{n+1}(-1|0) = \frac{a}{2} P_n(0|0) + P_n(-1|0) \quad (3.64)$$

and

$$P_0(k|0) = \delta_{k,0} \quad (3.65)$$

The foregoing recurrence equations can be written in matrix form as

$$\mathbf{P}_{n+1}^{(0)} = \mathbf{A} \mathbf{P}_n^{(0)} \quad (3.66)$$

where the consecutive components of $\mathbf{P}_n^{(0)}$ are $P_n(0|0), P_n(1|0), P_n(2|0), \ldots$. The initial state and the transition matrix are, respectively,

$$\mathbf{P}_0^{(0)} = \begin{pmatrix} 1 \\ 0 \\ 0 \\ \vdots \\ 0 \end{pmatrix}$$

$$\mathbf{A} = \begin{pmatrix} 1-a & a/2 & 0 & 0 & \cdots \\ a/2 & 1-a & a/2 & 0 & \cdots \\ 0 & a/2 & 1-a & a/2 & \cdots \\ 0 & 0 & a/2 & 1-a & \cdots \\ \vdots & \vdots & \vdots & \vdots & \end{pmatrix} \quad (3.67)$$

Rubin[67] devised a formal procedure to introduce a weighting factor for each visit to layer 0: multiply the right-hand side of Eq. (3.66) on the left by the diagonal matrix

$$\mathbf{W} = \begin{pmatrix} e^{\theta} & & & & \\ & 1 & & 0 & \\ & & 1 & & \\ & 0 & & 1 & \\ & & & & \ddots \end{pmatrix} \tag{3.68}$$

The recurrence equations for the weighted probabilities are then given by the equation

$$\mathscr{P}_{n+1}^{(0)} = \mathbf{WA}\mathscr{P}_n^{(0)} \tag{3.69}$$

with

$$\mathscr{P}_0^{(0)} = \mathbf{WP}_0^{(0)} = \begin{pmatrix} e^{\theta} \\ 0 \\ 0 \\ \vdots \end{pmatrix} \tag{3.70}$$

providing a weight e^{θ} for starting in layer 0. The weighted probability of being in the kth layer at the nth step is given by the kth component of $\mathscr{P}_n^{(0)}$, which is

$$\begin{aligned} \mathscr{P}_n(k|0) &= \left[(\mathbf{WA})^n \right]_{k,0} e^{\theta} \\ &= e^{\theta} \sum_{k_1} \cdots \sum_{k_n} (\mathbf{WA})_{kk_1} (\mathbf{WA})_{k_1 k_2} \ldots (\mathbf{WA})_{k_n 0} \end{aligned} \tag{3.71}$$

Equation (3.71) is a compact representation for the weighted sum of probabilities of all n-step random-walk paths which start in layer 0 and terminate in layer k, in the presence of an absorbing layer at -1. Each random-walk path is weighted by a power of e^{θ} given by the number of steps of that path in layer 0. The above procedure is easily generalized to weight or count visits to each one of a set of layers or of a set of lattice points in D dimensions.

The formal expression in Eq. (3.71) is not particularly useful. However, it is possible to obtain generating functions for these weighted probabilities using the technique referred to earlier for treating the effect of traps on lattice random walks. Asymptotic properties of the weighted probabilities for large n can be obtained from the generating functions. Essentially the same technique was used by Rubin[160] in studying the effect of isotope defects on the transmission of disturbances in harmonic crystals. It can be verified from the definition of $P_n(k|0)$, Eq. (3.71), that the weighted average number of steps in an n-step walk which lie in the surface layer is

$$\nu(\theta, n) = \frac{d}{d\theta} \sum_{k=0}^{\infty} P_n(k|0) \qquad (3.72)$$

The generating function obtained for the sum of the weighted probabilities in (3.72), which plays the role of a partition function, is[67]

$$\Gamma(0, z) = \sum_{n=0}^{\infty} z^n \sum_{k=0}^{\infty} P_n(k|0)$$

$$= \frac{1}{1-z} \frac{1 + \frac{1}{2}az(I_1 - I_0)}{1 + \frac{1}{2}azI_1 - (1 - e^{-\theta})I_0} \qquad (3.73)$$

where

$$I_k = \frac{1}{2\pi} \int_{-\pi}^{\pi} \frac{\cos k\varphi \, d\varphi}{1 - z(1 - a + a\cos\varphi)}$$

$$= \{(1-z)[1 - (1-2a)z]\}^{-1/2}$$

$$\times \left[\frac{1 - (1-a)z - \{(1-z)[1 - (1-2a)z]\}^{1/2}}{az} \right]^{|k|} \qquad (3.74)$$

is the perfect lattice propagator. In the limit $n \gg 1$, the average number of steps in the surface layer is a discontinuous function of θ:

$$\nu(\theta, n) = \begin{cases} [1 - \exp(\theta - \theta_c)]^{-1}, & \theta < \theta_c \\[2mm] 2\exp(-\theta_c)\left(\dfrac{2n}{a\pi}\right)^{1/2}, & \theta = \theta_c \\[2mm] n\left\{1 - \dfrac{1}{2(1 - e^{-\theta})}\left[1 - \left\{1 + \dfrac{a}{1-a}\dfrac{1}{e^\theta - 1}\right\}^{-1/2}\right]\right\}, & \theta > \theta_c \end{cases}$$

$$(3.75)$$

where $\theta_c = \ln[2/(2-a)]$. According to this result, for sufficiently large energies of attraction to the surface layer ($\theta > \theta_c$) the average number of adsorbed monomer units is proportional to the number of monomer units in the chain, while for $\theta = 0$, the value of $\nu(0, \infty)$ is $2/a$. Moreover, when θ is very large

$$\nu(\theta, n) \sim \begin{cases} n, & a \neq 0 \\ n/2, & a = 0 \end{cases} \qquad (3.76)$$

Thus, in this limit ($\theta \gg 1$), the random walk lies flat in the surface layer.

Other weighted averages for an adsorbed polymer chain have been evaluated in the limit $n \gg 1$, such as the mean squared displacement normal and parallel to the surface layer,[166] and the average number of steps in layer $k > 0$. We content ourselves here with indicating a connection between the weighted probability, $\mathcal{P}_n(0|0)$, and the probability of first return to the starting point, discussed in Section II.B. The generating function for the weighted probability of starting in layer 0 and arriving in layer 0 at the nth step is

$$h_0(z) = \sum_{n=0}^{\infty} \mathcal{P}_n(0|0) z^n$$

$$= e^{\theta} \left\{ 1 - e^{\theta} \left[1 - I_0^{-1} - \tfrac{1}{2} az \frac{I_1}{I_0} \right] \right\}^{-1}. \qquad (3.77)$$

It is clear from the form of Eq. (3.77) for $h_0(z)$ that when the right-hand side is expanded as a power series in e^{θ}, the coefficient of $e^{2\theta}$ is the generating function of the probability of first return to surface layer $k = 0$ (which is located next to the absorbing layer $k = -1$),

$$F(0, z) = 1 - I_0^{-1} - 0.5 az I_1 I_0^{-1} \qquad (3.78)$$

The probability of eventual return to layer 0 is [using Eq. (3.74)]

$$F(0, 1) = 1 - \frac{a}{2} \qquad (3.79)$$

an obvious result for a one-dimensional random walk between layers once it is recognized that, starting in layer 0, there is a probability $a/2$ of stepping to the left (and into the absorbing layer at -1) and a probability $1 - a/2$ of certain return to layer 0.

We have already mentioned that in the continuous random-flight limit de Gennes[155] has proposed a method for treating the case of a weakly "adsorbed chain." He proposed that the appropriate boundary condition at the adsorbing surface is the mixed boundary condition

$$\left(\frac{\partial G}{\partial x} + hG \right)_{x=0} = 0 \qquad (3.80)$$

where $G(x, n)$ is the solution of the random-flight equation

$$\frac{\partial G}{\partial n} = \frac{l^2}{6} \frac{\partial^2 G}{\partial x^2} \qquad (3.81)$$

de Gennes drew the analogy between this treatment of the adsorbed state of the random-flight chain and an analogous treatment of the bound state of the deuteron. We have addressed the question of the appropriate boundary condition for the random-flight equation in a different way.[162] Starting with the random-walk equations (3.69), we have assumed that not only is the surface layer modified by a factor e^{θ}, but each of the first k layers are modified by this factor. We then go to the random-flight limit, in which the lattice equations become

$$\frac{\partial G}{\partial n} = D \frac{\partial^2 G}{\partial x^2} + \lambda(x) G \qquad (3.82)$$

where

$$D = \lim_{\Delta n, \Delta x \to 0} \frac{1}{2} \frac{(\Delta x)^2}{\Delta n}$$

$$\lambda(x) = \begin{cases} \lambda, & 0 \leqslant x \leqslant k \\ 0, & x > k \end{cases} \qquad (3.83)$$

and

$$\lambda' = \lim_{\Delta n, \theta \to 0} \frac{\theta}{2n} \qquad (3.84)$$

and the absorbing boundary condition $G(0, n) = 0$ applies at the solution surface, $x = 0$. In addition, $G(x, n)$ and $\partial G / \partial x$ are assumed to be continuous at $x = k$. When $\lambda > 0$ in the region near the solution surface, this region acts as an effective source of probability. When $\lambda < 0$, the region acts as a sink. The condition $\lambda > 0$ (< 0) corresponds to $\theta > 0$ (< 0), that is, it is energetically advantageous (disadvantageous) for the polymer segments to lie in the surface region. A similar form for the random-flight equation (3.82) has been suggested by Pouchlý.[149] We have solved Eq. (3.82) for the starting condition $G(x, 0) = \delta(x - x_0)$ where $x_0 > k$, using the method of Laplace transforms. We have then compared this solution with the one obtained for the analogous problem proposed by de Gennes. That is, we solve Eq. (3.81) with the mixed boundary condition (3.80) and the initial condition $G(x, 0) = \delta(x - x_0)$. Lépine and Caillé[154] have obtained a solution to these equations using the Laplace transform. In our notation their solution for the

Laplace transform of $G(x, n)$ is

$$\Gamma(x, s) = \frac{1}{2(sD)^{1/2}} \left\{ \exp\left[-\left(\frac{s}{D}\right)^{1/2} |x - x_0| \right] \right.$$

$$\left. - F_0(s) \exp\left[-\left(\frac{s}{D}\right)^{1/2} (x + x_0 - 2k) \right] \right\} \qquad (3.85)$$

where

$$F_0(s) = \frac{1 + h(s/D)^{1/2}}{1 - h(s/D)^{1/2}} \qquad (3.86)$$

[Equation (3.85) can be inverted, and the expression for $G(x, n)$ is qualitatively similar to the solution of the lattice equations (3.69).] The solution of Eq. (3.82) corresponding to (3.85) is identical in form with (3.85) except that $F_0(s)$ is replaced by

$$F_1(s) = \frac{\left(1 - \frac{\lambda}{s}\right)^{1/2} - \tanh\left[k\left(\frac{s - \lambda}{D^2}\right)^{1/2} \right]}{\left(1 - \frac{\lambda}{s}\right)^{1/2} + \tanh\left[k\left(\frac{s - \lambda}{D^2}\right)^{1/2} \right]} \qquad (3.87)$$

If $\pi/2 < (\lambda/D^2)^{1/2} < 3\pi/2$, the asymptotic $(n \gg 1)$ form of $G(x, n)$ obtained with Eq. (3.87) in (3.85) is indistinguishable from the asymptotic form of $G(x, n)$ obtained with (3.86) in (3.85). Thus, for this special model, we see how to establish a correspondence between the solution of Eq. (3.82) with a source term and an absorbing boundary condition at $x = 0$ and the solution of Eq. (3.81) with the mixed boundary condition proposed by de Gennes.

The study of the influence of boundaries on chain configurations continues to be an active area of research. In the case where boundaries are energetically neutral ($\lambda = 0$), the random-flight diffusion equation with an absorbing boundary has been used to treat polymer-chain configurations in a number of different bounded regions. Casassa and Tagami[124] treated the sphere and the cylinder, and Lauritzen and DiMarzio[163] the wedge. Discrete random-walk models of adsorption have been analyzed in which immediate reversals are not allowed. Rubin[164] treated the 4-choice simple cubic lattice, and Hoeve[165] has recently treated the 5-choice simple cubic lattice where the relative probability of the fifth choice to the other four is an adjustable parameter associated with chain "stiffness." In other generalizations, Rubin[166] has treated three-dimensional aspects of chain configurations and calculated the mean squared end-to-end distance of a lattice random walk which starts on an adsorbing surface. DiMarzio and Rubin[158] also considered a polymer chain between two parallel solution surfaces and further analyzed configurations in terms of runs of similar steps. That is, a configuration is characterized by alternating sequences of steps which lie on or off a surface. This type of decomposition of a random walk had been introduced earlier by Motomura and Matuura[152] and Roe[151] in the single-surface adsorption problem. The mean squared end-to-end distance of a lattice random walk in which return to the starting point is excluded has also been analyzed by the foregoing method.[160] In addition to these investigations, some references to related work may be found in a recent review by Dickenson and Lal.[167]

B. Multistate Random Walks

One major idea that has appeared in one guise or another over a number of years is that of the multistate random walk, that is, one in which the random walker can be in one of a number of states and the properties of the random walk depend on the state.

The earliest reference to such random walks appears to have been in a study of diffusion in a chromatographic column by Giddings and Eyring.[168] The Giddings–Eyring model makes the assumption that a molecule in a chromatographic column can be in either a mobile or a trapped phase. In the original formulation the velocity is assumed to be constant in the mobile phase and zero in the trapped phase. In this model the statistical properties of the time for a molecule to traverse the column are of interest. Extensions of this theory were developed by Weiss,[169] and an application of the Giddings–Eyring model to affinity chromatography was suggested recently by Denizot and Delaage.[170, 171] The two-state diffusion process was first discussed by Cann, Kirkwood, and Brown[172] and by Bak[173] as a model for certain types of electrophoresis, by Meiboom[174] in the analysis of NMR experiments of isomerizing molecules, by Friedman and Ben-Naim[175] in the calculation of transport coefficients in electrolyte solutions, by Weiss[176] in the context of brownian motion, and by Singwi and Sjölander[177] in the analysis of the self-diffusion of water. More recently, Lindenberg and Cukier[178] used the theory of two-state random walks to study molecular rotational dynamics. Landman, Montroll, and Shlesinger[179] developed the theory of n-state random walks to treat a problem related to the surface diffusion of atom clusters, and several applications of these ideas have been made by Landman and Shlesinger[180] to diffusion in crystals, following a suggestive analysis by Reed and Ehrlich.[181]

The theory will be developed following the analysis by Weiss.[182] Let us consider a random walker capable of being in any one of k distinct states, and let us assume that the duration of stay in each state is a random variable. Suppose that the probability density of displacement of the random walker in state i is $p_i(\mathbf{r}, t)$, where t is the sojourn time of a single stay in that state. If $\psi_i(t)$ is the probability density for the time of a single sojourn in state i, then the probability density for displacement during a single complete sojourn in state i for a time t is $f_i(\mathbf{r}, t) = p_i(\mathbf{r}, t)\psi_i(t)$. The factorized CTRW with no correlation between displacement and time can be described in terms of a two-state random walk if we make the particular choices $f_1(\mathbf{r}, t) = \delta(\mathbf{r})\psi(t)$, $f_2(\mathbf{r}, t) = p(\mathbf{r})\delta(t)$. One can also develop a theory in which a factorization assumption is unnecessary, that is, in which $p_i(\mathbf{r}, t)$ is a probability density of displacement and duration of stay in state i. The theory is quite similar to that described below and will not be given here. In the case specified by the factorization assumption we shall calculate an expression for the Fourier–Laplace transform of $P_i(\mathbf{r}, t)$, the probability density for being at r in state i at time t for the general two-state random walk. In order to do this we must specify how transitions between states take place. We assume that such transitions occur by means of a time-independent Markov chain with probabilities a_{ij} for making a transition from state i to state j. The transition matrix will be denoted by $\mathbf{a}$. Let $f_i^0(\mathbf{r}, t)$ be the joint probability density for the displacement $\mathbf{r}$ and sojourn time during the first occurrence when it is in state i. Let $\eta_i(\mathbf{r}, t)d^D\mathbf{r}\,dt$ be the probability for a sojourn in state i to end during the time interval $(t, t + dt)$, the random walker at that time being in $(\mathbf{r}, \mathbf{r} + d^D\mathbf{r})$, and let $\eta_i^0(\mathbf{r}, t)d^D\mathbf{r}\,dt$ be the analogous quantity for the first sojourn. The probability that the first state to occur at $t = 0$ is state i will be denoted by λ_i. We let $F_i(\mathbf{r}, t)d^D\mathbf{r}$ be the probability that a displacement of a random walker in state i is in $(\mathbf{r}, \mathbf{r} + d^D\mathbf{r})$ in time t, conditional on the sojourn in i being longer than t, and let $F_i^0(\mathbf{r}, t)d^D\mathbf{r}$ the same quantity for the initial sojourn.

With these definitions we can write an integral equation for $P_i(\mathbf{r}, t)$ in terms of the $F_i(\mathbf{r}, t)$:

$$P_i(\mathbf{r}, t) = \lambda_i F_i^0(\mathbf{r}, t) + \sum_{l=1}^{k} a_{li} \int_0^t d\tau \int_{-\infty}^{\infty} d^D\rho\,\eta_l(\boldsymbol{\rho}, \tau) F_i(\mathbf{r} - \boldsymbol{\rho}, t - \tau)$$

$$(3.88)$$

To complete the calculation we must give the recipe for finding $F_i(\mathbf{r}, t)$ in terms of the given $f_i(\mathbf{r}, t)$, and secondly, show how the $\eta_i(\mathbf{r}, t)$ are to be calculated. The first relation is easily seen to be

$$F_i(\mathbf{r}, t) = p_i(\mathbf{r}, t)\Psi_i(t) \qquad (3.89)$$

where $\Psi_i(t) = \int_t^\infty \psi_i(\tau) \, d\tau$. Finally, by enumeration one can write the following equation for the $\eta_i(r, t)$:

$$\eta_i(\mathbf{r}, t) = \lambda_i f_i^0(\mathbf{r}, t) + \sum_{l=1}^{k} a_{li} \int_0^t d\tau \int_{-\infty}^{\infty} d^D \rho \eta_l(\boldsymbol{\rho}, \tau) f_i(\mathbf{r} - \boldsymbol{\rho}, t - \tau)$$

(3.90)

Since both Eqs. (3.88) and (3.90) are in convolution form, they can be reduced to algebraic relations by introducing the Fourier–Laplace transform, which for an arbitrary function $g(\mathbf{r}, t)$ is defined to be

$$g^*(\boldsymbol{\omega}, s) = \int_0^\infty dt \int_{-\infty}^{\infty} \cdots \int d^D \mathbf{r} \, g(\mathbf{r}, t) \exp(i\boldsymbol{\omega} \cdot \mathbf{r} - st)$$

(3.91)

Then it follows from Eq. (3.90) that the $\eta_i^*(\boldsymbol{\omega}, s)$ satisfy

$$\eta_i^*(\boldsymbol{\omega}, s) - \sum_{l=1}^{k} a_{li} \eta_l^*(\boldsymbol{\omega}, s) f_i^*(\boldsymbol{\omega}, s) = \lambda_i f_i^{0*}(\boldsymbol{\omega}, s)$$

(3.92)

and the $P_i^*(\boldsymbol{\omega}, s)$ can be expressed in terms of the $\eta_i^*(\boldsymbol{\omega}, s)$ as

$$P_i^*(\boldsymbol{\omega}, s) = \lambda_i F_i^{0*}(\boldsymbol{\omega}, s) + \sum_{l=1}^{k} \eta_l^*(\boldsymbol{\omega}, s) a_{li} F_i^*(\boldsymbol{\omega}, s)$$

(3.93)

These relations can also be rewritten in matrix form, as was done by Landman, Montroll, and Shlesinger.[179] The theoretical development in Eqs. (3.88)–(3.93), which generalizes that of Reference 182, is also more general than that of Landman, Montroll, and Shlesinger in that it allows for the possibility that the first sojourn can have different statistical properties from those of subsequent sojourns. This would not generally have any effect on asymptotic results, but might be important in a calculation of transient properties.

Some interesting and simple asymptotic properties can be derived from the preceding formalism by specializing to the case of $k = 2$ states. Without loss of generality one can always choose the transition matrix to have 0's on the diagonal. Therefore for the two-state case we have

$$\mathbf{a} = \begin{pmatrix} 0 & 1 \\ 1 & 0 \end{pmatrix}$$

(3.94)

as the only possible transition matrix. If we let $P(\mathbf{r}, t) = P_1(\mathbf{r}, t) + P_2(\mathbf{r}, t)$ [i.e., $P(\mathbf{r}, t) d^D \mathbf{r}$ is the probability that starting from the origin at $t = 0$ the random walker is in $(\mathbf{r}, \mathbf{r} + d^D \mathbf{r})$ at time t], then

$$P^*(\omega, s) = \lambda_1 F_1^{0*} + \lambda_2 F_2^{0*}$$

$$+ \frac{1}{1 - f_1^* f_2^*} \{ \lambda_1 \eta_1^{0*} (F_1^* f_2^* + F_2^*) + \lambda_2 \eta_2^{0*} (F_2^* f_1^* + F_1^*) \}$$

$$(3.95)$$

One can show that at sufficiently large values of t/μ, where $\mu = \mu_1 + \mu_2$ is the average time spent in a single cycle comprising a sojourn in states 1 and 2, that $P(\mathbf{r}, t)$ approaches a gaussian form when the average displacements, the associated variances, and average sojourn times are finite. This can be done either by expanding Eq. (3.95) for small $\omega [= (\boldsymbol{\omega} \cdot \boldsymbol{\omega})^{1/2}]$ and $|s|$, or by appeal to the central limit theorem. The gaussian distribution is completely specified by the first two moments of displacement. Let $\langle r_i \rangle$ and $\langle r_i r_j \rangle$ be the indicated space averages taken over a cycle consisting of complete sojourns in states 1 and 2, let $\sigma^2(T)$ be the variance of a complete cycle time $[\sigma^2(T_1) + \sigma^2(T_2)$, where T_i is the time spent in a sojourn in state i], and let $\rho_{i,T}$ be the mixed space–time covariance defined by

$$\rho_{i,T} = \sum_{j-1}^{2} \{ \langle r_i t_j \rangle - \langle r_i \rangle \mu_j \} \qquad (3.96)$$

in which $\langle r_i t_j \rangle$ is the mixed moment

$$\langle r_i t_j \rangle = \int_0^\infty t \, dt \int_{-\infty}^{\infty} \cdots \int r_i f_j(\mathbf{r}, t) \, d^D \mathbf{r} \qquad (3.97)$$

Then, for large t/μ, the asymptotic expressions for average displacements and the associated variances and covariances are

$$\langle r_i(t) \rangle \sim \langle r_i \rangle \frac{t}{\mu}$$

$$\langle r_i(t) r_j(t) \rangle - \langle r_i(t) \rangle \langle r_j(t) \rangle \sim \left[\langle r_i r_j \rangle - \langle r_i \rangle \langle r_j \rangle + \sigma^2(T) \frac{\langle r_i \rangle \langle r_j \rangle}{\mu^2} \right.$$

$$\left. - \frac{1}{\mu} (\langle r_i \rangle \rho_{j,T} + \langle r_j \rangle \rho_{i,T}) \right] \frac{t}{\mu} \qquad (3.98)$$

where $\langle r_i(t) \rangle$ is the average displacement in direction i at time t. It is interesting to note that mixed space–time correlations appear in this last equation in addition to the pure spatial and temporal moments. Although the results in Eq. (3.98) have been derived here in the context of random walks, similar results for sojourn times were first given by Smith.[183] A development of the theory of multistate random walks and its relation to master equations has recently been given by van Kampen.[184] In particular, van Kampen discusses the conditions under which it is legitimate to approximate the nonmarkovian theory by a markovian description.

Several known results can be derived directly from Eq. (3.98). We have already mentioned that the CTRW can be subsumed under the present formalism. In the simplest version of the CTRW, since both functions $f_i(\mathbf{r}, t)$ can be factored into a function of $\mathbf{r}$ multiplied by a function of t, the $\rho_{i, T}$ are zero. If we specialize Eq. (3.98) to 1-D, we find that

$$\langle r(t) \rangle \sim \langle r \rangle \frac{t}{\mu}$$

$$\sigma^2(r(t)) \sim \left[\sigma^2 + \sigma^2(T) \left(\frac{\langle r \rangle}{\mu} \right)^2 \right] \frac{t}{\mu} \tag{3.99}$$

as first stated by Shlesinger.[50] A second application is to the calculation of the asymptotic distribution of moments and residence time in a given state for a two-state system. If state 1 is the one of interest, then we can set $f_1(r, t) = \delta(r - t)\psi_1(t)$ and $f_2(r, t) = \delta(r)\psi_2(t)$. Since $f_1(r, t)$ is not separable, the space–time correlations need to be taken into account and we find that

$$\langle r_1(t) \rangle \sim \mu_1 \frac{t}{\mu}$$

$$\sigma_1^2(t) \sim \left(\sigma_1^2 \mu_2^2 + \sigma_2^2 \mu_1^2 \right) \frac{t}{\mu^3} \tag{3.100}$$

This result was first given by Takacs[185] using a different method. Finally, consider the Giddings–Eyring[168] model for transport in a chromatographic column with two phases. If we suppose that the molecules move according to a diffusion process in each case, that is,

$$f_i(r, t) = \frac{1}{(4\pi D_i t)^{1/2}} \psi_i(t) \exp\left[-\frac{(r - v_i t)^2}{4 D_i t} \right] \tag{3.101}$$

and define average speeds and average diffusion constants by

$$\bar{v} = \frac{v_1 \langle t \rangle_1 + v_2 \langle t \rangle_2}{\langle t \rangle_1 + \langle t \rangle_2}$$

$$\bar{D} = \frac{D_1 \langle t \rangle_1 + D_2 \langle t \rangle_2}{\langle t \rangle_1 + \langle t \rangle_2} \tag{3.102}$$

then Eq. (3.98) implies that the moments of displacement are

$$\langle r(t) \rangle \sim \bar{v}t$$

$$\sigma^2(t) \sim 2\bar{D}t + \left[\sigma_1^2 (v_1 - \bar{v})^2 + \sigma_2^2 (v_2 - \bar{v})^2 \right] \frac{t}{\langle t \rangle_1 + \langle t \rangle_2} \tag{3.103}$$

where σ_i^2 is the variance of a single sojourn time in state i. Thus one can define an effective diffusion constant

$$D_{\text{eff}} = \bar{D} + \frac{1}{2} \frac{\sigma_1^2 (v_1 - \bar{v})^2 + \sigma_2^2 (v_2 - \bar{v})^2}{\langle t \rangle_1 + \langle t \rangle_2} \tag{3.104}$$

which implies that even if the diffusion in each state is zero (as is the case in the original Giddings–Eyring formulation) there will still appear to be diffusion because of the random time spent in each of the states. Similar results can be derived by starting from a more detailed enumeration of the dynamics using a Langevin equation to describe the motion in each state. In a recent paper on applications of the theory to affinity chromatography, Weiss[169] developed asymptotic results for the multistate analogue of the Giddings–Eyring model.

Landman and Shlesinger[180] have considered more complicated multistate models for random walks on lattices with periodic defects. A simpler and more accurate treatment of this problem has been given by Scher and Wu.[186] Such models have been used to study diffusion of clusters on crystalline surfaces. Persistent random walks, as discussed earlier in Section II.B, can also be regarded as multistate random walks, and the theory developed along similar lines to that discussed earlier. Haus and Kehr and Shlesinger[187] have recently treated some effects in the theory of ac conductivity in solids using the theory of multistate random walks. In particular, these authors calculated the diffusion constant for the motion of some simple multistate hopping models.

Another type of multistate walk has been studied by Silver, Shuler, and Lindenberg,[188] Shuler,[189] and Seshadri, Lindenberg, and Shuler.[190] To define the nature of this walk we assume that it takes place on a 2-D square lattice. In the usual definition of a random walk with transitions only to nearest neighbors on a square lattice, the random walker can make a transition in the x direction followed by one in the y direction, at any point of the lattice. In the simplest version of this sparse random walk, motion in the x direction can be followed by motion in the y direction at a subset of points, and similarly for $y-x$ transitions. References 188 and 189 contain detailed calculations of properties of random walks on periodically and randomly connected lattices, derived by rather involved manipulations of generating functions. Reference 189 is more heuristic appealing to asymptotic properties of Markov chains. We follow the reasoning in the paper by Shuler[189] and for simplicity consider a 2-D random walk on a lattice which can be described by a unit cell in which there are B_y points on any x axis which allow a transition from the x to the y direction, and B_x points on any y axis that allow the transition from y motion to x motion. On a simple square (nonsparse) lattice, $B_x = B_y = 1$. We assume that at a lattice point at which a transition can be made only in the x (or y) direction, the probability of moving to either nearest neighbor is $\frac{1}{2}$. At a point at which transitions can be made in either the x or the y direction, the probability of moving to either nearest point in the x direction is $\frac{1}{2}p_x$ and in the y direction $\frac{1}{2}p_y$. Let $\bar{S}_x(n)$ and $\bar{S}_y(n)$ be the expected number of steps made in the x and y directions respectively in an n-step walk. If we let $\lambda = p_x/p_y$ the Shuler assumption is

$$\lim_{n \to \infty} \frac{\bar{S}_x(n)}{\bar{S}_y(n)} = \lambda \frac{B_x}{B_y} \qquad (3.105)$$

One can show that for symmetric n-step walks with steps to nearest neighbors, $\langle x^2(n) \rangle = \bar{S}_x(n)$ and $\langle y^2(n) \rangle = \bar{S}_y(n)$, so that from Eq. (3.105)

$$\lim_{n \to \infty} \frac{\langle x^2(n) \rangle}{\langle y^2(n) \rangle} = \lambda \frac{B_x}{B_y} \qquad (3.106)$$

For the symmetric and isotropic random walks with one lattice point per unit cell the above ratio is equal to 1. To derive the expected number of points visited in an n-step walk, Shuler writes $s_n(k) = f(k)s_n(1)$, where $k = B_x/B_y$, $f(k)$ is a function to be determined, and $s_n(1) = \pi n / \ln n$ as in Eq. (2.126). He further assumes that $s_n(k) = a[\langle x^2(n) \rangle \langle y^2(n) \rangle]^{1/2}$, where a is a constant and the factor in brackets is the analogue of area. We can combine these last two relations, finding thereby

$$f(k) = \left[\frac{\langle x^2(n) \rangle \langle y^2(n) \rangle}{\langle x^2(1) \rangle \langle y^2(1) \rangle} \right]^{1/2}$$

$$\sim \frac{2(\lambda B_x / B_y)^{1/2}}{1 + \lambda B_x / B_y} \qquad (3.107)$$

Although the derivation of this result is heuristic, it has been confirmed by much more complicated techniques.[190] Other asymptotic properties can be found in a similar spirit, and the results extended to nearest-neighbor random walks in 3-D. It would be of considerable interest to find the limits of validity of the *Ansatz* of Eq. (3.105). In particular, it appears difficult to extend the analysis to random walks with jumps to distant (i.e., not necessarily nearest) neighbors using the analysis in Reference 190, yet we feel that the assumption of nearest-neighbor walks is unnecessarily restrictive.

Recently Shuler and Mohanty[191] have made an interesting contribution in which they showed that the model just discussed can be used to derive the Maxwell–Rayleigh[192] result for the effective conductivity of a set of spheres of conductivity σ_1 embedded either randomly or periodically in a medium of conductivity σ_m. So far it is not known whether one can recover higher-order approximations from the same formalism.

Some generalizations of multistate models that involve random walks on disordered lattices will be mentioned in the next section. At this point it should be mentioned that two kinds of multistate random walks have been discussed in this section. In the first of these the position of the random walker and the state were independent, and in the second the internal state is defined by the location of the random walker. No application has arisen in which it is natural for these two classes to be combined, and no analysis of this case has been undertaken.

C. Solid-State Physics

It is natural to study diffusion phenomena in terms of continuum models, and the analysis of polymer configurations requires a combination of lattice and continuum models. By virtue of their structure, problems that arise in solid state physics are most naturally phrased in terms of lattice random walks, and a battery of the techniques discussed in Sections II.B to II.D have been used in their solution. In this section we shall describe several problems that have been of interest in recent years. As in other sections on applications, we fall far short of completeness in our discussion of these models. We begin with a short description of analyses of correlations in the random-walk description of the diffusion of atoms in crystals. This will be followed by a description of several analyses of trapping in random walks, that is, random walks that occur in the presence of absorbing sites on a lattice. This model arises in a number of applications and has been the subject of numerous recent investigations. Finally, we outline a number of studies on energy conduction in solids.

1. Correlated Diffusion Models

One of the first applications of the techniques associated with random walks to problems in solid-state physics is that required for taking account of correlations in calculating diffusion constants for the migration of atoms through crystals. When, for example, a radioactive tracer makes a jump in a lattice, it creates a vacancy at the site just left. Consequently, the tracer atom tends to make its next jump back to the vacancy, rather than making a jump according to an isotropic set of probabilities. This asymmetry introduces correlations into the random-walk picture. A similar effect occurs in interstitial diffusion. These correlations were first described by Bardeen and Herring,[193] with related analysis by Compaan and Haven[194] appearing shortly thereafter. Good reviews of the subject are to be found in the book by Manning[195] and a review article by Le Claire.[196] To see what is involved, let us assume that a random walker, or tracer atom, makes jumps at a rate $1/\Delta t$ on a lattice, the average absolute displacement in a single step being ΔL. We shall write $\mathbf{r}_i = \Delta L \boldsymbol{\rho}_i$ where $\boldsymbol{\rho}_i$ is a unit vector in the direction of step i. The diffusion constant for such a random walk is then defined to be

$$D = \lim_{n \to \infty} \frac{1}{6n\,\Delta t} \left\langle \left(\sum_{i=1}^{n} \mathbf{r}_i \right)^2 \right\rangle = \lim_{n \to \infty} \frac{(\Delta L)^2}{6\Delta t} \left(1 + \frac{2}{n} \sum_{i<j=1}^{n} \langle \boldsymbol{\rho}_i \cdot \boldsymbol{\rho}_i \rangle \right)$$

$$= D_0 f \tag{3.108}$$

where $D_0 = (\Delta L)^2/(6\Delta t)$ and f is the correlation factor defined by the terms in parentheses. If, for example, the motion of the tracer atom constitutes a translationally invariant process, we have $\langle \boldsymbol{\rho}_i \cdot \boldsymbol{\rho}_j \rangle = t_1^{|i-j|}$ and the resulting sum in Eq. (3.108) can be evaluated exactly, leading to the expression

$$f = \frac{1 + t_1}{1 - t_1} \tag{3.109}$$

Most of the papers on the subject of correlations in crystalline diffusion deal with the calculation of f for different crystal structures.[195,196] The derivations of correlations factors found in most of the literature make the assumption that only a single vacancy, that created by the jump of the tracer atom, is important. Somewhat more difficult questions are raised when there is a finite concentration of vacancies. The most comprehensive analysis of such problems has been given by Ishioka and Koiwa,[197] who were able to check their results against simulations by Benoist, Bocquet, and Lafore.[198] The analysis by Ishioka and Koiwa is complicated, and the results involve approximations, so that the results are valid for low concentrations only ($\leqslant 3\%$). The questions that are raised in this area of research are of considerable theoretical interest, but most of the practically interesting questions may be in the low-concentration region where Ishioka's and Koiwa's analysis appears to be satisfactory.

2. Trapping Models

A second area of solid-state physics in which random-walk models play a central role is that of trapping. We describe two types of trapping models: those in which traps occupy random positions, and those in which the traps occur on sublattices. The first analysis of the lattice trapping problem is that due to Rosenstock,[199] who developed a theory of luminescent emission for an organic solid with traps. The model that he describes represents the solid as a lattice composed of two types of sites, hosts and traps. An incident photon is absorbed at a host site and excites it; after a time t the excitation energy is transferred to one of its nearest neighbors unless, with probability α per step, it is emitted as luminescence. When the new site is a trap, the walk ends and luminescence occurs at some later time. Rosenstock first calculates the probability of host luminescence at the nth step of the random walk in terms of S_n, the number of distinct sites visited. If c is the probability that a given site is a trap, the probability of luminescence at the nth step is

$$Q_n = \alpha(1-\alpha)^{n-1}(1-c)^{S_n} \qquad (3.110)$$

If we suppose that the random walk is a separable CTRW with $\psi(t) = (1/T)\exp(-t/T)$, then the probability density of the time to the nth step is

$$\psi_n(t) = \frac{1}{n!} \frac{t^n}{T^{n+1}} \exp\left(-\frac{t}{T}\right) \qquad (3.111)$$

If luminescence occurs immediately when an exciton appears at a host site, the mean intensity of illumination at time t is

$$\langle Q(t) \rangle = \alpha \sum_{n=1}^{\infty} \frac{(1-\alpha)^{n-1}}{n!} \frac{t^n}{T^{n+1}} \langle (1-c)^{S_n} \rangle e^{-t/T} \qquad (3.112)$$

Although this sum is difficult to evaluate in general, one can get an approximation to it by first observing that c, the concentration of traps, is generally quite small. For example, Powell and Kepler[200] have measured this concentration in anthracene and tetracene, finding $c = 10^{-6}$ and 8.3×10^{-5}, respectively. These small numbers make plausible the approximation

$$\langle (1-c)^{S_n} \rangle \sim (1-c)^{\langle S_n \rangle} \qquad (3.113)$$

Since, in 3-D, $S_n \sim n/P(0;1)$, the series in Eq. (3.112) can be summed exactly to give a result in closed form for $\langle Q(t) \rangle$.

Rosenstock[201] has studied further ramifications of the assumption embodied in Eq. (3.113), but it is clear that it can, at best, be valid for small c, since the asymptotic form for $\langle S_n \rangle$ is used. One quantity that is of interest is the expected trapping time. This is given by

$$\langle n_T \rangle = \sum_{j=1}^{\infty} \langle (1-c)^{S_j} \rangle \qquad (3.114)$$

and the probability of surviving till step n is

$$P_n = \langle (1-c)^{S_n} \rangle \qquad (3.115)$$

The series in Eq. (3.114) can be summed in closed form in 3-D with the Rosenstock approximation in Eq. (3.113), leading to the result

$$\langle n_T \rangle = \frac{(1-c)^{1-F}}{1-(1-c)^{1-F}} \qquad (3.116)$$

where F is the probability of return to the origin for the particular lattice. In 1-D and 2-D the equivalent results cannot be found in closed form. Recently Weiss[202] has investigated the validity of the Rosenstock approximation in 3-D by using the results of Jain and his collaborators,[60] which show that S_n has an asymptotic gaussian distribution with mean equal to $(1-F)n$ and variance given in Eq. (2.127). The result of the investigation was that the Rosenstock approximation is useful for $c \leq 0.05$ but that it leads to errors of the order of 10% in the expected time till trapping for larger concentrations.

A second approach to the study of trapping has been suggested by Montroll[203] in a study in which traps are embedded in a periodic sublattice, rather than randomly as in Rosenstock's work. This periodic sublattice defines a unit cell and allows the possibility of reducing the random walk on the entire lattice to one performed on the unit cell, provided that only the time till trapping is of interest (rather than a quantity like displacement from a starting point). The degree to which the periodic trap model approximates to the random trap model has not been investigated. It is obvious, however, that the approximation can only be a good one at low concentrations of trapping sites.

If trapping sites are sufficiently rare, we can assume that there is exactly one such site in every unit cell. If we further assume that the incident photons are absorbed at any site with a uniform probability, then we are left with a first-passage-time problem for an exciton that moves as a random walk. As a further approximation it will be assumed that the random walker steps to nearest neighbors only, with uniform jump probabilities. The average number of steps before trapping in different numbers of dimensions with N sites per unit cell is found, for square or simple cubic lattices and steps to nearest neighbors only, to be[203]

1-D:
$$\langle n_T \rangle = \frac{N(N+1)}{6}$$

2-D:
$$\langle n_T \rangle \sim \frac{1}{\pi} N \ln N + 0.195 N + O(1) \tag{3.117}$$

3-D:
$$\langle n_T \rangle \sim 1.52 N + O(N^{1/2})$$

for large N. Montroll gives the estimate of N as being in the range 250–500, corresponding to experimental results of Emerson and Arnold.[204] Pearlstein[205] found the closest agreement with experiment to be that given by the 2-D result, suggesting that the excitons travel on the surface. Montroll[206] has generalized the preceding theory to allow the possibility of an arbitrary number of trapping sites per unit cell, but the calculations become increasingly complicated with the number of sites assumed. Sanders, Ruijgrok, and Ten Bosch[207] have derived results for finite lattices in 2-D without making the assumption of periodic boundary conditions. They solved the resulting equations numerically. These authors distinguish between two possibilities: (a) the time spent on any site is inversely proportional to the number of neighboring sites (so that an edge site differs from an interior site), and (b) the time spent on any site is a constant. If the time spent on an interior site is τ, the origin $(0,0)$ is an interior point of the lattice, and the number of neighbors of the point (r, s) is $n(r, s)$, then the equation for the expected time to trapping starting from (r, s) is

$$T(r, s) = \frac{4\tau + \sum_{r', s'} T(r', s')}{n(r, s)} \tag{3.118}$$

where the sum is taken over sites (r', s') that are nearest neighbors to (r, s). The mean time to trapping is then given by an average of the $T(r, s)$. Case (b) leads to an equation similar to Eq. (3.118) except that $\tau/n(r, s)$ is replaced by τ. Similar equations are obtained for the case in which one assumes that the exciton can decay spontaneously before reaching a trapping site. Knox[208] also analyzed the same equations numerically, but not as completely as in the paper just cited.

Kopelman and his collaborators (see the many citations in Reference 209) have investigated by simulation the efficiency of energy transfer in heterogeneous lattices. For example, Argyrakis and Kopelman[209] studied random walks on lattices which have two types of sites in addition to trapping sites: A sites which allow the random walker to occupy them, and B sites which do not. Therefore the random walk takes place on random islands of A sites embedded in a sea of forbidden B sites. It would appear to be almost impossible to derive any analytical results for these very complicated models. Recent studies by Hatlee and Kozak[210] investigate, by simulation, the effect of changing the number of dimensions on the statistics of trapping times. These investigations were motivated by a suggestion of Adam and Delbrück[211] that organisms might optimize efficiency in reaching a target by moving in a 2-D rather than in a 3-D space. Hatlee and Kozak showed that for a fixed concentration of trapping sites the mean trapping time decreases as the number of dimensions increases, while for a fixed number of sites the mean trapping time increases with the number of dimensions. They also investigated the effects of different boundary conditions on the first two moments of trapping time. Hatlee et al.[212] discuss some applications of these ideas to chemical reactions.

Two subjects related to energy transport that have been studied intensively in recent years have been trapping and the kinetics of motion in disordered structures. Many of the techniques for studying these problems are related, and we therefore discuss them together. One of the early phenomenological approaches to these problems starts from a master-equation description. Hemenger, Lindenberg, and Pearlstein[213] have treated several cases of exciton quenching on lattices by using a random-walk formalism in the approximation in which a master equation describes the kinetics of the process. The theory makes the assumption that quencher–host interactions are weak. In 1-D, in the case of fully incoherent excitons (the physics of which was first discussed by Förster[214] and Dexter[215]), one finds that the occupation probability $\rho_n(t)$ for site n satisfies a master equation that typically can be written

$$\dot{\rho}_n = F(\rho_{n+1} - 2\rho_n + 2\rho_{n-1}) - \left(S_{jn} + \tau^{-1}\right)\rho_n \qquad (3.119)$$

where F is the rate constant for energy transfer between neighboring sites, S_{jn} is the rate of quenching by a single quencher at site j, and τ is the average exciton lifetime in the absence of quenching. A slightly different master equation describes the kinetics of fully coherent Frenckel excitons, but the mathematical development of the theory is quite similar. If one takes the Laplace transform of Eq. (3.119), rewriting it as

$$(s - F\Delta^2)\rho_n^*(s) = \rho_n(0) - (S_{jn} + \tau^{-1})\rho_n^*(s) \qquad (3.120)$$

where $\Delta^2 a_n = a_{n+1} - 2a_n + a_{n-1}$, then one can find a formal solution for the $\rho_n(t)$ in terms of the Green's functions for a variety of boundary conditions and use it to discuss different models of exciton propagation in quenched linear chains.

The models just described have been called phenomenological, since they have not been derived systematically from a physical analysis. Grover and Silbey[216] derived a more complicated master equation based on a density-matrix analysis of an exciton–phonon system described by a model hamiltonian. The form of this equation used by Hemenger, Lindenberg, and Pearlstein[213] is in terms of a function‡ $\rho_{nm}(t)$ [where $\rho_{nn}(t)$ is the probability that an exciton is at site n at time t] is

$$\begin{aligned}
\dot{\rho}_{nm} = & \ iJ(\rho_{n+1,m} + \rho_{n-1,m} - \rho_{n,m+1} - \rho_{n,m-1}) - F(\mu+1)\rho_{nm} \\
& + F\delta_{nm}[(\mu-1)\rho_{nm} + \rho_{n+1,m+1} + \rho_{n-1,m-1}] \\
& + F(\delta_{n,m+1} + \delta_{n,m-1})\rho_{nm} \qquad (3.121)
\end{aligned}$$

In this equation J is the renormalized pairwise interaction energy in units of $\hbar$, F is the nonlocal exciton scattering rate constant, and $F(\mu-1)$ is a local exciton-scattering rate constant. Equation (3.121) describes both coherent and incoherent excitons, but the former case does not include exciton–phonon interactions. One can show that $\langle n^2 \rangle$, the mean squared displacement of an exciton initially at $n=0$, is

$$\langle n^2 \rangle = \sum_{n=-\infty}^{\infty} n^2 \rho_{nn} = \left(2F + \frac{4J^2}{F_0}\right)t - \frac{4J^2}{F_0^2}(1 - e^{-F_0 t}) \qquad (3.122)$$

‡This function is defined by $\rho_{nm}(t) = \langle 0 | A_0 \langle A_n^\dagger(t) A_m(t) \rangle A_i^\dagger | 0 \rangle$, where $A_n^\dagger$ and A_m are exciton creation and annihilation operators, the inner brackets denote an average over the free phonon ensemble, and the outer brackets denote an average over the electronic and vibrational ground state of the system.

where $F_0 = F(\mu + 2)$. This result is valid for an infinite chain. Grover and Silbey[216] and Munn[217] have derived a somewhat more general expression for $\langle n^2 \rangle$, but the important feature of Eq. (3.122) is that in the limit $J/F_0 \to 0$ we have $\langle n^2 \rangle = 2Dt$, where $D = F + 2J^2/F_0$. This would be the result for a pure diffusion process, and describes so-called Förster–Dexter excitons. In the opposite limit, $J/F_0 \to \infty$, $\langle n^2 \rangle$ is proportional to t for $F_0 t \gg 1$, but for shorter times the behavior of $\langle n^2 \rangle$ is more complicated. Hemenger, Lindenberg, and Pearlstein[213] have described the kinetics of exciton quenching in great detail, starting from Eq. (3.121) together with an approximation that preserves the physical properties of interest but leads to a tractable mathematical development.

Greer[218] has considered the related problem of exciton transport in linear crystals with substitutional impurities occurring in a finite segment of the crystal. More recently Dlott, Fayer, and Wieting[219] have analyzed exciton transport in linear crystals with randomly distributed impurities, by assuming that the exciton can become "caged" between two impurity scatterers. Thus, exciton transport is modeled as a random walk between linear cages that are embedded in a lattice of cages. The random-walk formalism is used because the probability that an exciton has not been trapped in time t is proportional to $(1 - c)^{S(t)}$, where c is the probability that a given site is a trap, and $S(t)$ is the number of distinct sites visited in time t. The analogous problem in higher dimensions is considerably more difficult because of the complicated trapping structures that can occur. All of the authors cited find that impurities play a significant role in the dynamics of exciton transport.

Helman and Funabashi,[220] Shlesinger,[221] and Klafter and Silbey[222] have developed very similar theories for trapping, using a CTRW approach. All of these studies use the assumption that migration of impurities takes place on a lattice, and that the concentration of traps is much smaller than that of the impurities. This enables them to use the single-trapping-site approximation. When the trap is located at $\mathbf{r} = \mathbf{0}$, the rate at which particles are trapped is

$$k(t) = \sum_{\mathbf{r} \neq 0} \bar{F}(\mathbf{r}, t) \qquad (3.123)$$

where $\bar{F}(\mathbf{r}, t)$ is the probability density for the time to trapping of an impurity initially located at $\mathbf{r}$. For a separable CTRW we can express $\bar{F}(\mathbf{r}, t)$ as

$$\bar{F}(\mathbf{r}, t) = \sum_{n=0}^{\infty} F_n(\mathbf{r})\psi_n(t) \qquad (3.124)$$

where $\psi_n(t)$ is the probability density for the sum of n hopping times. Using the fact that $\mathcal{L}\{\psi_n(t)\} = [\psi^*(s)]^n$ together with Eq. (2.100), one easily shows that the Laplace transform of $k(t)$ is given by

$$k^*(s) = \frac{1}{[1 - \psi^*(s)]P(\mathbf{0}, \psi^*(s))} - 1 \tag{3.125}$$

The extension of this result to the situation where trapping occurs with probability $f < 1$ is immediate. Funabashi[223] has shown that Eq. (3.125) is to be replaced by

$$k^*(s, f) = k^*(s)\frac{fP(\mathbf{0}, \psi^*(s))}{1 - f + fP(\mathbf{0}, \psi^*(s))} \tag{3.126}$$

where $k^*(s)$ is given by Eq. (3.125). It is not feasible to use these expressions in the development of a transient theory, but one can certainly find the asymptotic time dependence of $k(t)$ by using Tauberian theorems as in Section II.B. One finds that in 1-D, when the steps have finite variance, $k(t) \sim t^{-1/2}$, while in 3-D, $k(t) \sim$ constant. Furthermore, Funabashi[223] has shown that in 1-D the coefficient of $t^{-1/2}$ in the asymptotic expansion is independent of the trapping probability f, at least partially justifying the assumption by Dlott, Fayer, and Wieting[219] that $f = 1$.

3. Nonlattice Models

The models described so far make the assumption that both the trapping sites and the random walk take place on a lattice. In many contexts this assumption is inappropriate, or at least its utility as an approximation needs closer examination. Haan and Zwanzig[224] have considered a continuum model for the migration of electronic excitation between molecules randomly distributed in space. The study was motivated by fluorescence quenching in a dilute solution of chlorophyll, as was that of Montroll.[203] When a molecule is excited by the absorption of a photon, the molecule can reemit a photon, or the excitation can jump to another molecule by inductive resonance, or it can jump to a trap from which reemission is impossible. The question raised by this phenomenon is whether a diffusion equation can be used to describe it, or whether the random distribution of trapping sites plays a significant qualitative role. To analyze this random-trapping system, Haan and Zwanzig locate N transfer sites at $\mathbf{r}_1, \mathbf{r}_2, \ldots, \mathbf{r}_N$ and let $p_j(t)$ be the probability that the excitation is at $\mathbf{r}_j$ at time t. They then assume that the $p_j(t)$ satisfy a master equation

$$\dot{p}_j(t) = -\frac{p_j(t)}{\tau} + \sum_k w_{jk} p_k - \sum_k w_{kj} p_j \tag{3.127}$$

where τ is a lifetime for the decay of the excitation by trapping (fluorescence) and the rate constants for transitions from $\mathbf{r}_j$ to $\mathbf{r}_k$ are assumed to be those proposed by Förster[214]:

$$w_{jk} = \frac{1}{\tau}\left(\frac{R_0}{r_{jk}}\right)^6 \tag{3.128}$$

The decay term in Eq. (3.127) can be eliminated by transforming to a new set of dependent random variables $p_j' = p_j \exp(t/\tau)$, which leads to a more usual form of the master equation. Since the resulting equation is linear, it can be solved in terms of the associated Green's function, which we denote by $G(\mathbf{r}, t)$. Let the Fourier–Laplace transform of G be denoted by $G^*(\omega, s)$:

$$G^*(\omega, s) = \int_0^\infty e^{-st}\,dt \int_{-\infty}^\infty G(\mathbf{r}, t)\exp(i\omega \cdot \mathbf{r})\,d^3\mathbf{r} \tag{3.129}$$

Since isotropy is assumed, $G^*(\omega, s) = G^*(\omega, s)$, where ω is the magnitude of ω. A generalized diffusion coefficient $\hat{D}(\omega, s)$ can be defined by the relation

$$G^*(\omega, s) = \left[s + \omega^2 \hat{D}(\omega, s)\right]^{-1} \tag{3.130}$$

For ordinary diffusion the inverse Fourier–Laplace transform of $\hat{D}(\omega, s)$ is a constant, D. Otherwise, $P(\mathbf{r}, t)$, the density function for the excitation satisfies a generalized master equation:

$$\frac{\partial P(\mathbf{r}, t)}{\partial t} = \nabla_r^2 \int_{-\infty}^\infty d^3\mathbf{r}' \int_0^t D(\mathbf{r}-\mathbf{r}', t-t')P(\mathbf{r}', t')\,dt' \tag{3.131}$$

and, for example, the Laplace transform of the mean squared displacement is

$$\int_0^\infty e^{-st}\langle r^2(t)\rangle\,dt = \frac{6\hat{D}(0, s)}{s^2} \tag{3.132}$$

The transformed Green's function can be written in terms of the matrix defined by Eq. (3.127).

If the density is denoted by $\rho = N/V$, where V is the volume containing the N randomly distributed sites, then Haan and Zwanzig introduce a novel scaling argument which allows them to conclude that if D is considered as a function of ρ, ω, and s, then

$$\hat{D}(\rho, 0, s) = \lambda^4 \hat{D}(\lambda^{-3}\rho, \lambda^{-1}\omega, \lambda^{-6}s) \tag{3.133}$$

for λ a scale factor. In particular, in the limit $\omega = 0$ we can conclude that

$$\hat{D}(\rho,0,s) = s^{2/3}\hat{D}(s^{-1/2}\rho,0,1) \qquad (3.134)$$

A subsequent argument shows that at short times

$$\langle r^2(t) \rangle \sim R_0^2 \left(\frac{t}{\tau}\right)^{1/3}\left[2.98C\left(\frac{t}{\tau}\right)^{1/2} + 0.33C^2\frac{t}{\tau} + \cdots\right] \qquad (3.135)$$

where $C = \frac{4}{3}\pi\rho R_0^3$. Clearly this is not the same behavior as would be predicted by diffusion theory. The authors do conjecture that diffusion theory is valid at high concentrations or at long times, but the transition to diffusive behavior does not follow from their approximations. Gochanour, Andersen, and Fayer[225] continued this approach to transport in disordered media by developing a diagrammatic method for approximating the Green's function needed to calculate $\langle r^2(t) \rangle$. They found a difference in the second term on the right-hand side of Eq. (3.135) and derive a self-consistent approximation, indicating an approach to diffusion behavior at sufficiently large times. Godzik and Jortner[226] advanced this approach to energy transport by introducing an averaged Dyson-equation approximation. More interestingly, they show[226,227] that a CTRW with a coupled displacement–pausing-time density reproduces exactly the results of the Dyson-equation approximation. The joint density for displacement and pausing time, $p(\mathbf{r}, t)$, is that suggested by Scher and Lax[22]:

$$p(\mathbf{r}, t) = \theta w(r)\exp[-tw(r)]\Phi(t) \qquad (3.136)$$

where θ is the probability that r is an accessible impurity site, $w(r)$ is the continuum form of w_{ij} shown in Eq. (3.128), and

$$\Phi(t) = \left\langle \exp\left(-t\sum_i w_{1i}\right)\right\rangle \qquad (3.137)$$

is the probability that the initial excitation is still localized at the donor site (here denoted by $i = 1$) at time t. The short-time results of this theory agree in form with Eq. (3.135) (although the coefficients differ from those given by Haan and Zwanzig[224] and Gochanour, Andersen, and Fayer[225]), and the long-term results are consistent with diffusion. Many authors[228] have considered the problem of finding approximations to the function $\Phi(t)$ in Eq. (3.137) in different cases, deriving the effects of these choices on diffusive properties characterized by the diffusion coefficient and $\langle r^2(t) \rangle$. At this time we cannot say for sure which theory for these phenomena will be most worthwhile to pursue. Only a combination of simulation and experiment will be able to resolve conflicting results, as all of the theories contain some type of approximation. A different approach to some of these problems will be described later, in Section III.C.5.

Another variation of some of these techniques is taken by Klafter and Silbey[222] in a study of energy transport in 1-D molecular crystals. They use a separable CTRW model on a lattice in which $\psi(t)$ is assumed to be $\lambda e^{-\lambda t}$. The disorder is incorporated into their model by assuming that λ is itself random. The averaged value of $\psi(t)$ is then

$$\bar{\psi}(t) = \int_0^\infty \lambda e^{-\lambda t} \rho(\lambda)\, d\lambda \qquad (3.138)$$

with Laplace transform

$$\bar{\psi}^*(s) = \int_0^\infty \frac{\lambda \rho(\lambda)}{\lambda + s}\, d\lambda \qquad (3.139)$$

in which $\rho(\lambda)$ is the probability density of λ. For triplet excitation of the molecular crystals in which intermolecular interactions are dominated by exchange or superexchange, it has been shown[229] that when the mean separation between impurities and the range of interaction are large with respect to the lattice spacing, $\rho(\lambda)$ goes like $\lambda^{-\alpha}$ for small λ. For this dependence the asymptotic analysis of $\bar{\psi}^*(s)$ in Eq. (3.139) leads to the result that $k(t) \sim t^{-(1+\alpha)/2}$ for $t \to \infty$. Klafter and Silbey[230] further related the asymptotic behavior of $\langle r^2(t) \rangle$ and the localization function $\langle p(\mathbf{0}, t) \rangle$ to the behavior of $\rho(\lambda)$ near $\lambda = 0$. These following formulas are also implied by results of Shlesinger.[50] They find that in any number of dimensions, when:

$$\langle \lambda^{-1} \rangle = \int_0^\infty \frac{\rho(\lambda)}{\lambda}\, d\lambda < \infty, \qquad \langle r^2(t) \rangle \sim \frac{t}{\langle \lambda^{-1} \rangle} \qquad (3.140a)$$

$$\rho(\lambda) \sim \lambda^{-\alpha}, \quad 0 \leqslant \alpha < 1, \qquad \langle r^2(t) \rangle \sim t^{1-\alpha} \qquad (3.140b)$$

$$\rho(\lambda) \sim \text{constant}, \quad \lambda \to 0, \qquad \langle r^2(t) \rangle \sim \frac{t}{\ln t} \qquad (3.140c)$$

In contrast, the behavior of $\langle p(\mathbf{0}, t) \rangle$ depends on dimension. When $\langle \lambda^{-1} \rangle$ exists, $\langle p(\mathbf{0}, t) \rangle \sim \langle \lambda^{-1} \rangle^{-1} t^{-D/2}$, where D is the number of dimensions. When $\rho(\lambda) \sim \lambda^{-\alpha}$, for λ near 0 one finds that in various dimensions $\langle p(\mathbf{0}, t) \rangle$ takes various forms:

1-D: $\langle p(0, t) \rangle \sim t^{-(1-\alpha)/2}$

2-D: $\langle p(\mathbf{0}, t) \rangle \sim t^{\alpha-1} \ln t$ $\qquad (3.141)$

3-D: $\langle p(\mathbf{0}, t) \rangle \sim t^{\alpha-1}.$

When $\rho(\lambda) \sim$ constant for small λ, they find that in 1-D $\langle p(0, t) \rangle \sim$ $(\ln t / t)^{1/2}$. Klafter and Silbey also derive results for $\langle r^2(t) \rangle$ from a scaling argument. More will be said about Eq. (3.140) in Section III.C.5.

Helman and Funabashi[220] and Shlesinger[221] have developed theories of electron scavenging in glasses, based on the CTRW formalism. The experiments that motivate development of the theory measure the optical absorption spectra of electrons injected by pulse radiolysis into aqueous or organic glasses.[231] Specifically, one is interested in calculating the rate of disappearance of electrons that interact weakly either with radiation products or with scavenger molecules that can be added to the system. Shlesinger's model considers a dilute-limit approximation in which a single electron and M scavengers are randomly distributed in a unit cell. He treats the randomness of the electron localization sites by assuming that they occur on a regular lattice but that there is a distribution of hopping times—that is, he deals with a CTRW. He essentially calculates $k(t)$ as given in Eq. (3.123), and shows that by assuming that the probability density of hopping times is

$$\psi(t) = a\alpha t^{\alpha - 1} \exp(-a\alpha t^\alpha) \tag{3.142}$$

He is able to reproduce data[231] in the literature that indicate that $\log_{10} \log_{10}$ $[N(0)/\langle N(t) \rangle]$ plots as a straight line as a function of $\log_{10} t$. In this last relation $N(t)$ is the average number of undecayed reactants left after time t.

4. Transport in Disordered Structures

One of the active lines of research in solid-state physics has been that of elucidating features of energy transport in disordered structures. To study this phenomena one must drop the lattice picture, since the essence of the problem is the lack of long-range structural order. The random potential that arises because of the disorder creates localized states. Carriers then move between localized centers by hopping. Trap-controlled hopping occurs when a carrier temporarily resides in a trap between hops. A considerable amount of experimental information is available on different modes of energy transport in amorphous semiconductors, as discussed recently by Mort and Knights.[232] Scher and Lax[22] have made important contributions to the problem of hopping transport in amorphous solids, using random-walk methodology. Starting from the Nyquist relation they show that if $\mathbf{r}(t)$ is the position of a carrier at time t, the frequency-dependent conductivity can be expressed as $\sigma(\omega) = [ne^2/(kT)]D(\omega)$, where n is the density of effective carriers, and $D(\omega)$ is the function

$$D(\omega) = -\frac{\omega^2}{6} \int_0^\infty e^{-i\omega t} \langle [\mathbf{r}(t) - \mathbf{r}(0)]^2 \rangle \, dt$$

$$= -\frac{\omega^2}{6} \sum_{\mathbf{r}} \sum_{\mathbf{r}_0} (\mathbf{r} - \mathbf{r}_0)^2 P^*(\mathbf{r}, \omega | \mathbf{r}_0, 0) f(\mathbf{r}_0) \tag{3.143}$$

where $f(r_0)$ is the distribution of initial positions and

$$P^*(\mathbf{r}, \omega \,|\, \mathbf{r}_0, 0) = \int_0^\infty e^{-i\omega t} P(\mathbf{r}, t \,|\, \mathbf{r}_0, 0)\, dt \qquad (3.144)$$

Thus, once a form for the transition probabilities $P(\mathbf{r}, t \,|\, \mathbf{r}_0, 0)$ is specified, one can derive an expression for the experimentally accessible conductivity from Eq. (3.143). For $P^*(r, \omega \,|\, r_0, 0)$ Scher and Lax chose the form that results from a CTRW model with $t = 0$ coinciding with the beginning of a sojourn time. With the assumption that $r_0 = 0$, and that the probability density for the jump time is $\psi(t)$, Scher and Lax find

$$D(\omega) = \frac{\sigma^2}{6} \frac{i\omega \psi^*(i\omega)}{1 - \psi^*(i\omega)} \qquad (3.145)$$

where σ^2 is the variance (assumed finite) associated with a single jump, and $\psi^*(s)$ is the Laplace transform of $\psi(t)$. The calculation is then completed by following an analysis by Thomas, Hopfield, and Augustyniak[233] allowing the authors to find an expression for $\psi(t)$.

Tunaley[25] pointed out that the assumption that the first transition has the same statistical characteristics as later ones is crucial to the derivation of $D(\omega)$. If one makes the assumption that $t = 0$ occurs at a random time between jumps, then Tunaley finds $D(\omega)$ to be independent of ω. A simple way of seeing this is to consider the case where $\psi(t) = \lambda \exp(-\lambda t)$, for which $\psi_0(t) = \psi(t)$. In that case Eq. (3.145) does not depend on which assumption is made, and $D(\omega)$ is found to be equal to $\sigma^2/6$ independent of ω. Tunaley's objection, while mathematically correct, does not reflect the physics correctly. The crux of Scher and Lax's argument is the replacement of the effect of randomly located impurities by a regular lattice of impurities with a distribution of hopping times, that is, a CTRW. Lax and Scher[234] and Kumar and Heinrichs[235] have examined the details of the replacement carefully, concluding that Tunaley's objection is invalid and that Scher and Lax's development is the correct one. An interesting point made by Kumar and Heinrichs is that the configuration averaging implies that the CTRW describing hopping conduction must be nonseparable. This was taken as an assumption by Scher and Lax in their original treatment.[22] Further light was shed on random-walk modeling of conductivity in disordered structures in a later paper by Kumar and Heinrichs,[236] in which they relax the assumption that transition rates between different sites are independent of time as explicitly stated in Eq. (3.137). Instead, they assume that the time between successive jumps is a renewal process in which the probability density for the first jump is not necessarily equal to that for subsequent jumps (the assumption of constant rates implies equality of these two densities). Specifically, they replace Eq. (3.137) by a more general $\Phi(\mathbf{r}, t)$. The theory developed by Kumar and Heinrichs leads to a generalization of the formula for ac conductivity and somewhat better agreement with data[237] on dc conductivity than does the original theory of Scher and Lax. The residual discrepancy is due, at least in part, to approximations made in the configuration averaging. This question was also examined in a diagrammatic analysis by Moore,[238] who showed that the Scher–Lax averaging amounts to summing over selected classes of diagrams. McInnes, Butcher, and Clark[239] have recently calculated statistical properties of ac conductivity for a hopping system in 3-D, starting from a transport equation, and comparing results with those derived from the CTRW-based theory. They found that results of the random-walk model agreed well with their simulation results except at very high or very low frequencies. There is an extensive literature of conductivity calculations for related problems that does not use the random-walk formalism,[240] but so far the approach pioneered by Scher and Lax appears to be a fruitful approach to this class of problems. Additional support is lent to this conclusion in a paper by Klafter and Silbey,[27] in which they show by using Zwanzig's projection-operator formalism for configuration averaging[241] that $P(\mathbf{r}, t)$ satisfies a generalized master equation and therefore also a CTRW, as assumed by Scher and Lax. However, it must be emphasized that transport modeling by CTRWs represents an approximation whose limits of validity are not fully known at the present time.

One of the most successful of the applications of random-walk techniques is the Scher–Montroll[242] model of anomalous charge transport in amorphous solids. In the experiments described by this model, one excites the surface of a film of specially prepared amorphous material with a light source, creating pairs of charge carriers. The electric field due to the charge deposited during the material preparation causes one of the carrier components to move

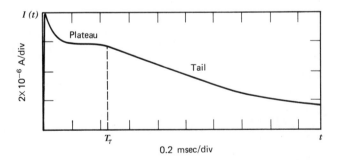

Fig. 12. Dispersive transient photocurrent trace, $I(t)$, measured on As_2Se_3 by Scharfe.[243] The features of this trace are incompatible with the prediction of a diffusion model.

through the film to the other face, where it can be detected. Surprisingly, the experimentally measured current deviates considerably from that expected from a diffusion or random-walk theory. This finding has been made on several materials.[243] A typical transient photocurrent trace is shown in Fig. 12. It is characterized by a current spike, a plateaulike region, and then a long tail. On the basis of diffusion one would expect the ratio of average displacement to standard deviation of displacement to satisfy $\langle r(t) \rangle / \sigma(t) \sim t^{1/2}$. What is found experimentally is that $\langle r(t) \rangle / \sigma(t) \sim$ constant. Figure 13 illustrates the behavior expected from a theory with propagation at uniform speed. Diffusion would simply round the shoulder and base of the curve. A typical set of experimental data is shown in Fig. 14 on a logarithmic scale. The measurements are plotted on a log–log scale. They were made by Dr. G. Pfister, and kindly lent us by Dr. H. Scher, both of the Xerox Corporation. The remarkable feature of these data is that they fall on a universal curve. Further, the curves appear to consist of two straight segments joined by a curvilinear part. Scher and Montroll, wishing to preserve the random-walk model because it has a qualitative correspondence to the physics of the process, assume that a CTRW describes the motion of the holes, but $\psi(t)$

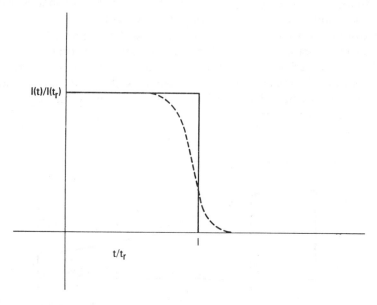

Fig. 13. Idealized transient current trace for charge injected at one face and measured at the opposite face, when the velocity is constant (solid line). The dashed line indicates the influence of diffusion.

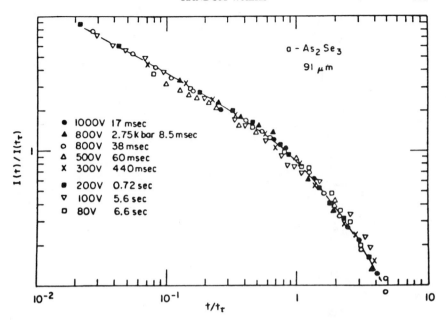

Fig. 14. A log–log plot of $I(t)/I(t_r)$ against t/t_r for a-As$_2$Se$_3$ for the range of transit times indicated. The measurements were made by Pfister, and the solid curve represents the prediction of the theory of Scher and Montroll.

has the large-t form

$$\psi(t) \sim Ct^{-(1+\alpha)}, \qquad 0 < \alpha < 1 \qquad (3.146)$$

The transport process can be modeled as one in which there are a number of trapping sites in the amorphous material and the charge carrier jumps from site to site. It is implicitly assumed that the time spent in any site is greater than the time spent in making intersite jumps. When $\psi(t)$ has the form given in Eq. (3.146), the graph of $P(r, t)$ as a function of r shows a peak at small r that decreases slowly as a function of t, and a greatly skewed portion due to the random walkers that are initially trapped for a long period of time. This behavior is exemplified by the curves in Fig. 15, also furnished by Dr. Scher. Shlesinger[50] has shown [Eq. (2.89) of this chapter] that when the average jump size is finite and unequal to zero, both $\langle r(t) \rangle$ and $\sigma(t)$ are proportional to t; in the absence of boundaries, this leads to the conclusion that $\langle r(t) \rangle / \sigma(t)$ is constant, in agreement with the experimental results.

The results just described are valid in the absence of boundaries. The experiments themselves include boundaries, since the holes injected at one face

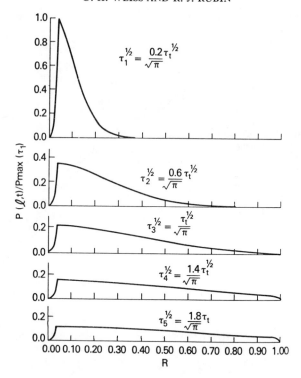

Fig. 15. A typical propagator for a carrier packet with an absorbing barrier at $R = 1$, and $\psi(t) \sim t^{-3/2}$ for large t. If propagation were by means of biased diffusion one would expect the propagator, at early times, to look gaussian.

of the film move, under the influence of an electric field, to the opposite face, where the current is measured. The influence of the face at which holes are injected is negligible because of the field, but the collecting face must be modeled as an absorbing barrier. Scher and Montroll[242] derived an equation to calculate the state probability in the presence of an absorbing barrier. The equation is strictly correct only for (lattice) random walks restricted to nearest-neighbor transitions, but is probably asymptotically correct for any random walk with a finite expected step length that is small relative to the distance to be traversed. Let us assume that the random walker starts from $r = 0$ in 1-D, and that there is an absorbing boundary at $r = R$. Let $P_0(r, t)$ be the state probability in the absence of boundaries, let $P(r, t)$ be the state probability in the presence of boundaries, and let $F(r, t)$ be the first-passage-time probability for arrival at r. Then $P(r, t)$ satisfies

$$P(r, t) = P_0(r, t) - \int_0^t F(R, \tau) P(r - R, t - \tau) \, d\tau \qquad (3.147)$$

That is, to reach $r < R$ in the presence of an absorbing boundary at R, one must subtract all those random walks that reach the boundary point (in a boundary-free walk) and return to r at time t. Thus, $F(R, t)$ can also be calculated only for the free-space case. Equation (3.147) is strictly valid only for nearest-neighbor walks, because more general random walks need not stop at R but may make a transition to some point $r > R$ at time t and then return to $r < R$ at a later time without ever landing at $r = R$. Assuming the validity of that equation, we can take the Laplace transform of Eq. (3.147) with respect to time, finding that

$$P^*(r, s) = P_0^*(r, s) - \frac{P_0^*(R, s)}{P_0^*(0, s)} P^*(r - R, s) \qquad (3.148)$$

where we have used the continuous-time analogue of Eq. (2.100) to replace the quantity $F^*(R, s)$. In the transport experiments one measures an induced current. This current, $I(t)$, is proportional to the instantaneous velocity, which may be defined as

$$\langle v(t) \rangle = \frac{d\langle r(t) \rangle}{dt} \qquad (3.149)$$

thereby establishing the relation between the random walk and experimental parameters. When $\psi(t)$ has the asymptotic form of Eq. (3.146), Shlesinger's results show that the current $I(t)$ falls off as $t^{-(1-\alpha)}$ as long as there is negligible interaction between the random walkers and the absorbing boundary, so that his result for $\langle v(t) \rangle$ can be expected to hold only at sufficiently short times. The asymptotic form for $\langle v(t) \rangle$ must be found from Eq. (3.148). Multiplying both sides of that equation by $r - R$ and summing over all r, one finds

$$\langle r^*(s) \rangle = \langle r^*(s) \rangle_0 [1 - F^*(R, s)] \qquad (3.150)$$

When $\psi^*(s) \sim 1 - As^\alpha$ for small $|s|$, and both the expected step length μ and σ^2 are finite, one can show that

$$F^*(R, s) \sim \exp\left\{ \frac{\mu R}{\sigma^2} \left(1 - \sqrt{1 + \frac{2A\sigma^2 s^\alpha}{\mu^2}} \right) \right\}$$

$$\sim 1 - as^\alpha + bs^{2\alpha} + o(s^{2\alpha}) \qquad (3.151)$$

where a and b are constants easily calculated from the small-$|s|$ expansion of the exponential. Since $\langle r^*(s)\rangle_0 \sim s^{-(1+\alpha)}$, it follows that

$$\langle v^*(s)\rangle \sim \frac{a}{s} - \frac{b}{s^{1-\alpha}} + \cdots \tag{3.152}$$

from which one can infer that at very long times $\langle v(t)\rangle$ goes like $t^{-(1+\alpha)}$. Therefore, the theory predicts that the current at early times falls off like $t^{-(1-\alpha)}$, and at late times like $t^{-(1+\alpha)}$. If one plots $\log I(t)$ against $\log t$, then there should be two straight line segments and the sum of their slopes should be -2. This behavior is in substantial agreement with experiment on several materials,[242] as exemplifed in Fig. 14. McLean and Ausman[244] suggested some easily calculated approximations for the current $I(t)$.

Workers in the general area of transport in the presence of traps tend to draw a distinction between hopping transport and multiple-trapping models. Multiple trapping has the effect of randomly stopping the motion of the moving particles, followed by a later release similar to the motion of molecules through a chromatographic column.[168] Schmidlin[245] has developed the theory of multiple trapping using a continuum approach. The following exposition is based on that of Pfister and Scher.[246] We point out, however that Schmidlin[245] was the first to establish the correspondence between the multiple-trapping formalism and that for CTRWs.

The multiple-trapping formalism is expressed in terms of $c(\mathbf{r}, t)$, the concentration of mobile carriers at $\mathbf{r}$; $c_i(\mathbf{r}, t)$, the concentration of carriers in traps of type i at $\mathbf{r}$; μ, the (constant) mobility of the carriers; g, the local photogeneration rate; and E, the applied electric field. Since the transport equations are linear, the theory can be developed in terms of either probabilities or concentrations, the same formalism applying in both cases. If the total concentration of carriers, both mobile and trapped, is denoted by $\rho(\mathbf{r}, t)$, so that $\rho = c + \Sigma_i c_i$, the kinetics of the process are described by

$$\frac{\partial \rho}{\partial t} = g - \mu \mathbf{E} \cdot \nabla c$$
$$\frac{\partial c_i}{\partial t} = \omega_i c - W_i c_i \tag{3.153}$$

where g and E are taken to be constant, and ω_i and W_i are rate constants for trapping and release. One can solve Eq. (3.153) by introducing Laplace transforms of all of the variables. We find from the second line

$$c_i^*(\mathbf{r}, s) = \frac{\omega_i}{s + W_i} c^*(\mathbf{r}, s) \tag{3.154}$$

or

$$c^*(\mathbf{r}, s) = Q^*(s)\rho^*(\mathbf{r}, s) \tag{3.155}$$

where

$$Q^*(s) = \left(1 + \sum_i \frac{\omega_i}{s + W_i}\right)^{-1} \qquad (3.156)$$

But the relation (3.155) allows us to write

$$c(\mathbf{r}, t) = \int_0^t Q(t - \tau)\rho(\mathbf{r}, \tau)\, d\tau \qquad (3.157)$$

so that

$$\frac{\partial \rho}{\partial t} = g - \mu \mathbf{E} \cdot \nabla \int_0^t Q(t - \tau)\rho(\mathbf{r}, t)\, d\tau \qquad (3.158)$$

Leal Ferreira[247] has shown that the continuum limit of the generalized master equation, neglecting diffusion as we have done in writing Eq. (3.158), takes the form of this last equation with a kernel, $\varphi(t)$, in place of $Q(t)$. Since a generalized master equation can be written for every CTRW, it follows that the two formulations are equivalent.

In a recent paper Scher, Alexander, and Montroll[248] have used a combination of some of the ideas so far described to discuss field-induced trapping as a probe of anisotropy in certain molecular crystals (e.g., TTF-TCNQ). In nearly 1-D crystals free charge tends to move along a preferred or stack axis under the influence of an applied field **E**. There may, however, be jumps perpendicular to the axis as well. One can measure a trapping time, $\tau(E)$, for this system. When **E** is applied parallel to the stack axis, $\tau(E)$ is proportional to $1/E$, while if **E** is applied perpendicular to it, $\tau(E)$ is independent of E. Thus, measurements of trapping time can be used to estimate anisotropy. The authors consider this phenomenon as an asymmetric random walk in the presence of traps, the asymmetry being due to the electric field. When **E** is applied along the stack axis the two hopping rates in that direction are chosen to be $w_1 \exp(\pm \beta)$, where β is proportional to $E/(kT)$. Let the traps be randomly distributed, the probability that any lattice site is a trap being c. The probability that a charge is trapped by time t is

$$1 - (1 - c)^{S(t)} \sim cS(t) \qquad (3.159)$$

where $S(t)$ is the distinct number of sites visited in time t, and c is assumed small. To calculate $\tau(E)$ the authors set $c\langle S(\tau)\rangle = 1$, corresponding to making the expected trapping probability close to certainty, and calculate τ from this relation. To do so they calculate asymptotic properties of the Green's functions for anisotropic random walks and compare the results to experimental data of Haarer and Möhwald.[249] The fit to the data is shown in Fig.

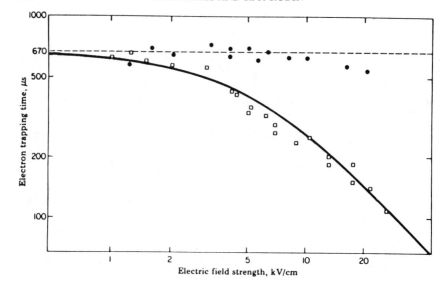

Fig. 16.   A double-logarithmic plot of $\tau(E)$ as a function of E, for **E** parallel to the stack axis ($\square$) and perpendicular to it ($\bullet$). The solid lines are theoretical predictions of the theory by Scher, Alexander, and Montroll. The data are those of Haarer and Möhwald.

16, and is evidently a very good one, suggesting that the random-walk model does embody significant features of the physics.

5. Random Walk in 1-D with Disordered Rate Constants

The study of random walks in 1-D with random rate constants, as a model for percolation, is a recent development, the earliest papers being those of Bernasconi, Alexander, and Orbach.[250] The random walk is described by a master equation

$$\dot{P}_r = W_{r,r-1}(P_{r-1} - P_r) + W_{r,r+1}(P_{r+1} - P_r) \qquad (3.160)$$

where the W's are rate constants that are identically distributed random variables. A first problem that is reasonably simple to solve and has been discussed by several authors[250,251] is that in which

$$W_{r,r+1} = \begin{cases} W & \text{with probability } c \\ 0 & \text{with probability } 1-c \end{cases}$$

This formulation is solvable because the lattice, in 1-D, separates into isolated islands of excitable points. As long as the random walk is to nearest neighbors only, the solution need only be found for a finite lattice, and at the end an average can be taken over all sizes of lattice. Bernasconi et al.[250] and others[252] show that the probability that an initial excitation will survive till at least time t can be written

$$\bar{P}(t) = A + BI(t) \qquad (3.161)$$

where A and B are constants and $I(t)$ is

$$I(t) = c \sum_{N=2}^{\infty} (1-c)^{N-1} \sum_{l=1}^{N-1} \exp\left[2\tau\left(\cos\frac{\pi l}{N} - 1\right)\right] \qquad (3.162)$$

where $\tau = Wt$ is a dimensionless time. The function $I(t)$ has a rather complicated behavior as a function of time which can be summarized as going like $\exp(-\tau)$ at earliest times, like $\tau^{-1/2}$ at intermediate times, and like $\exp(-b\tau^{1/3})$ at long times, where b is constant.

The generalization of these results to the case where the W's are all non-zero presents considerable difficulties even in 1-D. The resemblance of the problem to that of disorder in harmonic lattices has prompted several analyses using some of the same tools developed for disordered lattices. Some of this work has been reviewed recently by Alexander et al.,[253] so that we need only give a brief discussion of the major results. One starts by assuming that the W's have a (possibly) continuous probability density $\rho(w)$. If $P_r^*(s)$ denotes the Laplace transform of $P_r(t)$, Eq. (3.160) is equivalent to

$$W_{r-1,r}(P_r^* - P_{r-1}^*) + W_{r+1,r}(P_r^* - P_{r+1}^*) + sP_r^* = \delta_{r,0} \qquad (3.163)$$

The quantity of greatest interest is $\langle P_0^*(s) \rangle$, that is, the configuration averaged value of P_0^*. A formal solution to Eq. (3.163) can be written

$$P_0^*(s) = (s + G_+ + G_-)^{-1} \qquad (3.164)$$

where G_+ is the continued fraction

$$G_+ = \cfrac{1}{\cfrac{1}{W_{0,1}} + \cfrac{1}{s + \cfrac{1}{\cfrac{1}{W_{1,2}} + \cdots}}} \qquad (3.165)$$

and G_- is defined similarly in terms of the $W_{r+1,r}$. The probability density for the G's can be shown, following Dyson,[254] to satisfy the integral equation

$$f_s(g) = \int_0^\infty f_s(\gamma)\, d\gamma \int_0^\infty \rho(w)\delta\left[\gamma - \left(\frac{1}{w} + \frac{1}{s+\gamma}\right)^{-1}\right] dw \quad (3.166)$$

If the function $f_s(g)$ is known, $\langle P_0^*(s)\rangle$ can be expressed as

$$\langle P_0^*(s)\rangle = \int_0^\infty f_s(\gamma)\, d\gamma \int_0^\infty f_s(\gamma')\frac{d\gamma'}{s+\gamma+\gamma'} \quad (3.167)$$

All of the results in Eqs. (3.164)–(3.167) are exact. As in most other investigations described by us, the asymptotic results corresponding to $t \to \infty$ or $s \to 0$ are the only ones that have been calculated. For $s \to 0$ Bernasconi, Schneider, and Wyss[255] show that asymptotically $f_s(g)$ can be written as

$$f_s(g) = \frac{1}{\varepsilon(s)} h\left(\frac{g}{\varepsilon(s)}\right) \quad (3.168)$$

where both the scale factor $\varepsilon(s)$ and the scaling function $h(x)$ depend on $\rho(w)$. Results have been obtained for the three classes characterized in Eq. (3.140). [Notice that $\rho(\lambda)$ in that equation is $\rho(w)$ in the present case.] One finds that

$$\langle P_0(t)\rangle \sim \begin{cases} t^{-1/2}, & \text{case a} \\ t^{-(1-\alpha)/(2-\alpha)}, & \text{case b} \\ \left(\dfrac{\ln t}{t}\right)^{1/2}, & \text{case c} \end{cases} \quad (3.169)$$

The calculation of $\langle r^2(t)\rangle$ is more difficult in that it requires a knowledge of all of the $\langle P_r(t)\rangle$. To find these, Bernasconi, Schneider, and Wyss[255] propose a scaling hypothesis of the form $\langle P_r^*(s)\rangle \sim \langle P_0^*(s)\rangle F(r/\mathcal{S}(s))$ for $s \to 0$. It follows from this representation that

$$\langle D^*(s)\rangle = \frac{s^2}{2}\sum_{r=-\infty}^{\infty} r^2\langle P_r^*(s)\rangle$$

$$\sim D_0\left(s\langle P_0^*(s)\rangle^2\right)^{-1} \quad (3.170)$$

where D_0 is a calculable constant. From these results it has been shown that

$$\langle r^2(t)\rangle \sim \begin{cases} t, & \text{case a} \\ t^{2(1-\alpha)/(2-\alpha)}, & \text{case b} \\ \dfrac{t}{\ln t}, & \text{case c} \end{cases} \quad (3.171)$$

The validity of the scaling hypothesis on which these results are based is confirmed by simulations.[255,256] It should be noted that the result in case b is not in agreement with those obtained from theories based on the CTRW,[230] the dependence in those theories being $t^{1-\alpha}$ rather than as in Eqs. (3.169) and (3.171). Present evidence appears to favor the correctness of the equations derived by Bernasconi et al., since they are based on the analysis of an exact rather than a phenomenological model. However, the analysis does contain a scaling approximation which requires further scrutiny.

A slightly different approach to these problems, the so-called effective-medium approximation has been used by Bernasconi and his collaborators.[257] In this approach the solution to Eq. (3.167) for $f_s(g)$ is approximated as

$$f_s(g) = \delta(g - g_{\text{eff}}(s)) \qquad (3.172)$$

The function $g_{\text{eff}}(s)$ is calculated from a self-consistency equation which takes the form

$$g_{\text{eff}}(s) = \int_0^\infty \rho(w)\left(\frac{1}{w} + \frac{1}{s + g_{\text{eff}}(s)}\right)^{-1} dw \qquad (3.173)$$

In this approximation one can find $\langle P_r^*(s)\rangle$ rather simply as

$$\langle P_r^*(s)\rangle = \langle P_0^*(s)\rangle\left(\frac{g_{\text{eff}}}{g_{\text{eff}} + s}\right)^{|r|}$$

$$\langle P_0^*(s)\rangle = (s + 2g_{\text{eff}})^{-1} \qquad (3.174)$$

The small-$|s|$ behavior of $\langle P_0^*(s)\rangle$ can be calculated from Eq. (3.173) and is in agreement with those found in cases a and b in Eq. (3.169). Alexander et al.[253] discuss the limitations of the effective-medium approximation, pointing out that it neglects fluctuations and correlations and therefore cannot be pushed much further than the calculation of the asymptotic quantities just described. They also point out that the CTRW methodology that subsumes disorder into a single pausing-time density $\psi(t)$ is also an effective-medium approximation. However, the $\psi(t)$ derived by Scher and Lax[22] differs from that obtained by Bernasconi et al., whence the slight discrepancy between asymptotic forms for $\langle P_0(t)\rangle$ and $\langle r^2(t)\rangle$ obtained from the two theories.

There is clearly much more work to be done on this class of problems. In 1-D one is fortunate in having the explicit representation in Eq. (3.164) as a starting point for any investigation. Even so, the resulting theory is quite difficult. The 3-D theory cannot start from an explicit solution, and its analysis may have to proceed in terms of more *ad hoc* approximations.

We mention, in passing, a final class of random-walk problems that is beginning to receive some attention, namely random walks on lattices with inequivalent sites.[258] A particularly difficult model is one in which multiple occupancy of a given site is prohibited. This problem in 1-D was discussed by Huber[259] for an alternating sublattice $ABABA\ldots$. If $P_i^A(t)$ denotes the probability that site i on the A sublattice is occupied (similarly for the B sublattice), then Huber analyzes the random walk described by the equations

$$\dot{P}_i^A = -W_{AB}P_i^A\left(2 - P_i^B - P_{i-1}^B\right) + W_{BA}\left(1 - P_i^A\right)\left(P_i^B + P_{i-1}^B\right)$$
$$\dot{P}_i^B = -W_{BA}P_i^B\left(2 - P_i^A - P_{i+1}^A\right) + W_{AB}\left(1 - P_i^B\right)\left(P_i^A + P_{i-1}^A\right)$$
$$(3.175)$$

The nonlinear terms are due to the restriction on multiple occupancy, and the W_{AB} and W_{BA} are rate constants. Huber does not solve this last equation exactly; however, he shows that the solution to the linearized equation leads to good accuracy at sufficiently large times but is in error at short times. Many ramifications of this model remain open for investigation and may be of some importance in the elucidation of properties of superionic conductors.

D. Models of the Motion of Microorganisms

One of the more recent applications of random-walk theory has been to the interpretation of data on the motion of microorganisms on surfaces. Early experiments by Gail and Boone,[260] Peterson and Noble,[261] Berg and Brown,[262] and Macnab and Koshland[263] indicated that the motion of such microorganisms could be adequately modeled in terms of two- and three-dimensional random walks. The underlying biological problem that motivates development of a quantitative analysis is that of elucidating the mechanism of chemotaxis, that is, motion that is biased by the presence of a chemical attractant.

The random walk of bacteria can roughly be described as nearly linear trajectories interrupted by periods in which the bacteria appear to "tumble," remaining at the same spot, followed by the choice of a new direction for the next linear segment. A schematic diagram of such a random walk is shown in Fig. 17. At least two phenomenological mechanisms have been identified to characterize microorganism motion in the presence of a chemoattractant. The first of these was suggested by data of Berg and Brown[262] on the motion of flagellated $E.\ coli\ K_{12}$ bacteria. They found that the time spent in the linear trajectories is random and has a negative-exponential distribution. They and Mcnab and Koshland[264] suggested that the time spent in a linear segment moving towards the chemoattractant source would be

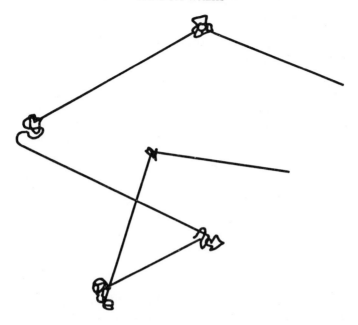

Fig. 17.   Typical bacterial path shown as a succession of straight line segments and "tumbles."

longer, on average, than the time spent moving away from that source. In this picture the turn angles separating successive linear segments are uniformly distributed in $(0, 2\pi)$. A second, and possibly more obvious, mechanism is one in which the turn-angle distribution is influenced by the presence of a chemoattractant. This mechanism has been found to operate in polymorphonuclear leukocytes.[265]

Patlak[266] appears to have been the first to develop a theory of random walks with persistence and external bias in the context of problems of biological motion. His theory allows only for weak biasing mechanisms, and his results are expressed in terms of a Fokker–Planck equation. Nossal and Weiss[267] developed a more elaborate theory as a model for the time-sensing chemotactic mechanism mentioned earlier, and Nossal[268] developed a theory along the same lines for the angle-bias mechanism. In both cases the authors present asymptotic results for the first two moments, relying on the central limit theorem to suggest a gaussian distribution of the distance wandered from the starting point. The first model is characterized by the following assumptions. At $t = 0$ the bacterium is assumed to start from the origin, and the first linear segment has an angle with respect to a fixed axis that is uniformly distributed in $(0, 2\pi)$. The motion is characterized as follows:

1. The motion along any linear segment has a constant random speed independent of direction. The first and second moments of the speed are $\bar{v}$ and $\overline{v^2}$ respectively.
2. Two successive line segments form an angle θ, which is characterized by a probability density $p(\theta)$, independent of orientation with respect to the chemoattractant source.
3. If φ is the angle between a space-fixed axis and a straight line segment, the time spent traversing that segment has a negative-exponential density, that is,

$$p_\varphi(t) = \frac{1}{T(\varphi)} \exp\left(-\frac{t}{T(\varphi)}\right) \tag{3.176}$$

Thus, the mean time, $T(\varphi)$, expresses the preference for moving towards the chemoattractant.

The history of the random walks can be described in terms of the path traversal times $\tau_1, \tau_2, \ldots, \tau_{n(t)}$, where $n(t)$ is the number of segments generated in time t; the initial turn angle θ_0 formed by the first linear segment with respect to a space-fixed axis; and the subsequent turn angles $\theta_1, \theta_2, \ldots, \theta_{n(t)-1}$ with respect to the successive straight line segments. The nth straight line segment is inclined at an angle $\zeta_n = \theta_0 + \theta_1 + \cdots + \theta_{n-1}$ with respect to the fixed axis. Hence at time t the displacement in the x direction is

$$x_n(t) = \sum_{i=1}^{n} v_i \tau_i \cos \zeta_i \tag{3.177}$$

where

$$\sum_{i=1}^{n} \tau_i = t \tag{3.178}$$

Notice that no time has been assigned to the tumbling period in this calculation. This additional factor could easily be accommodated. If we denote by $E_{\tau,v}\{x_n(t)\}$ the average of $x_n(t)$ with respect to the v's and the travel times, it is not too difficult to show that

$$\int_0^\infty e^{-st} E_{\tau,v}\{x_n(t)\} \, dt = \frac{\bar{v}}{s} F_1 F_2 \cdots F_{n-1}(1 - F_n)(G_1 + G_2 + \cdots + G_n) \tag{3.179}$$

where

$$F_j = \left[1 + sT(\zeta_j)\right]^{-1}, \qquad G_j = T(\zeta_j) F_j \cos \zeta_j \qquad (3.180)$$

When the sum over n is performed on Eq. (3.179), one finds that

$$\sum_{n=1}^{\infty} \int_0^{\infty} e^{-st} E_{\tau,v}\{x_n(t)\}\, dt = \frac{\bar{v}}{s}(G_1 + F_1 G_2 + F_1 F_2 G_3 + \cdots)$$

$$(3.181)$$

Under the assumption that θ_0 is uniformly distributed in $(0, 2\pi)$ and that the θ_i are identically distributed random variables, the angular averages can be performed, leading to

$$\int_0^{\infty} e^{-st}\langle x(t)\rangle\, dt = \frac{\bar{v}}{2\pi s} \int_{-\pi}^{\pi} U(\varphi, s)\, d\varphi \qquad (3.182)$$

where $U(\varphi, s)$ is the solution to the integral equation

$$U(\varphi, s) = G(\varphi, s) + F(\varphi, s) \int_{-\pi}^{\pi} U(\varphi + \theta, s) p(\theta)\, d\theta \qquad (3.183)$$

in which $F(\varphi, s)$ and $G(\varphi, s)$ are defined as in Eq. (3.180). Although this last equation cannot be solved in closed form, it is possible to perform an analysis for $|s| \to 0$ that enables us to determine the behavior of $\langle x(t)\rangle$ for $t \to \infty$. We find that

$$\langle x(t)\rangle \sim \bar{v}\frac{\alpha_1}{\alpha_0}t, \qquad \langle y(t)\rangle \sim \bar{v}\frac{\beta_1}{\beta_0}t \qquad (3.184)$$

where

$$\alpha_j = \frac{1}{2\pi} \int_{-\pi}^{\pi} T(\varphi) \cos j\varphi\, d\varphi, \qquad \beta_j = \frac{1}{2\pi} \int_{-\pi}^{\pi} T(\varphi) \sin j\varphi\, d\varphi$$

$$(3.185)$$

A slightly more elaborate analysis leads to an expression for the variances.[267] Although the turn-angle distribution doesn't appear in the lowest-order term for the displacement, it does appear in the more complicated expression for the variances as well as in the correction terms for the displacement. In this model the bias depends primarily on $T(\varphi)$. As mentioned earlier, Nossal[268] developed a parallel theory applicable when the bias is due mainly to the turn-angle distribution $p(\theta)$. The theory has been applied to data obtained by Zigmond[265,269] on polymorphonuclear leukocytes. Another experimental study attempting to correlate data on human granulocytes with random-walk parameters is that of Hall and Peterson,[270] using statistical tests developed by Hall.[271] Lovely and Dahlquist[272] have given a systematic exposition of techniques for the reduction of data to useful statistical parameters. Their work can be regarded as the three-dimensional generalization of the theory of Nossal and Weiss.[267,268] Although we have mentioned two mechanisms producing bias, other possibilities are by no means excluded. It would seem desirable to develop more formal statistical tests to distinguish between different possibilities.

There has been considerable work on the mathematics of random walks by microorganisms. The most recent comprehensive work along this line is a paper by Alt[273] that attempts to relate the kinetic equations to theoretical models for the underlying mechanisms. Alt shows that under suitable restrictions he can obtain the diffusion equations first discussed by Patlak[266] and Keller and Segel.[274] Alt's paper has an excellent bibliography on the general subject. Another good source of material is a recent volume[275] and in particular the review by Nossal.[276] At this time a considerable amount of experimental effort is being expended in estimating random-walk parameters for different living systems. It is clear that further development of experimental techniques in the biological area will lead to the consideration of new and different questions in the mathematics of random walks.

ADDENDUM

Since we completed the body of this chapter, a number of papers have appeared as preprints or in the chemical-physics literature that extend many of the results discussed by us. While we cannot be exhaustive in mentioning this literature, several of the papers may be of considerable interest and will therefore be described briefly. Most relate to the general area of energy transfer modeled as a random walk on either regular or disordered lattices.

Blumen and Zumofen[277] have discussed several aspects of random walks on homogeneous lattices in the context of the energy transfer. They present useful numerical results for the expected number of distinct sites visited and the expected number of returns to the origin by an n-step random walk. The results are developed for the three cubic lattices with nearest-neighbor random walks, as well as for the diamond and edge lattices. These all lead to expansions of $\langle S_n \rangle$, the expected number of distinct sites, and $\langle M_n \rangle$, the expected number of returns to the origin, of the form

$$\langle S_n \rangle \sim an + bn^{1/2} + c$$
$$\langle M_n \rangle \sim a' - b'n^{-1/2} + O(n^{-3/2}) \tag{A.1}$$

where the a's, b's, and c's are positive constants. Similar results to these are also obtained by simulation for random walks with steps to further neighbors, having the form $p(r) = Ar^{-s}$, where A and s are constants and $r^2 = \mathbf{r} \cdot \mathbf{r}$. The authors then proceed to consider diffusion models (in continuous space) for the continuous analogue of $\langle M_n \rangle$. They show that if $\langle M(t) \rangle$ is the expected time spent in a sphere of radius R, then

$$\langle M(t) \rangle \sim \frac{R^2}{2D} - \frac{R^3}{3\sqrt{\pi} \, D^{3/2}} t^{-1/2} + \frac{R^5}{60\sqrt{\pi D^5}} t^{-3/2} + \cdots \qquad (A.2)$$

(Parenthetically it should be noticed that the problem of the occupation time of a sphere in any number of dimensions has been analyzed in the mathematics literature by Ciesielski and Taylor.[278]) Blumen and Zumofen then show that the results from the continuum approach can be brought into coincidence with Eq. (A.1) for discrete lattices, to the order shown, by choosing $\frac{4}{3}\pi R^3$ to be equal to the reciprocal density of lattice points.

Machta[279] has recently introduced the notion of the renormalization group into calculations of long-time and large-distance properties of 1-D random walks with static disorder. Machta analyzes a random walk on a line in which distances between successive lattice points are random variables. The results of his calculation indicate that the properties of the random walk can be described by a generalized diffusion equation with a $t^{-3/2}$ decay in the memory kernel. One can also derive this result by means of perturbation theory, as both Machta and Zwanzig[280] have shown. Whether the renormalization-group technique leads to sensible results in higher dimensions is not presently known but is being investigated. Another approach to problems involving disorder is that of the coherent-medium approximation developed and applied by Odagaki and Lax[252,281] in a series of papers. This approximation starts by writing as an expression for the Laplace transform of the transition probabilities on a disordered lattice

$$P^*(\mathbf{r}, s | \mathbf{r}_0, 0) = \mathcal{L}\{P(\mathbf{r}, t | \mathbf{r}_0, 0)\} = \langle \mathbf{r} | (s\mathbf{I} - \mathbf{H})^{-1} | \mathbf{r}_0 \rangle \qquad (A.3)$$

where $\mathbf{H}$ is the matrix containing random elements. To evaluate the conductivity, according to Eq. (3.143) one needs to configuration-average. This can be done by writing

$$\langle (s\mathbf{I} - \mathbf{H})^{-1} \rangle = (s\mathbf{I} - \mathbf{\Sigma})^{-1} \qquad (A.4)$$

where $\mathbf{\Sigma}$ is a deterministic matrix whose properties are calculated by either perturbation theory or some other type of approximation. Several applications have been made of these ideas: to hopping conduction in 1-D chains, to the bond percolation problem in 1-D and 3-D systems, and to a lattice model for impurity conduction in doped semiconductors.[281] The results obtained by this approximation appear to be in good agreement with experimental data for the latter problem.

The CTRW has been used extensively to describe the kinetics of motion in disordered materials, as discussed in Section 8. Halpern[282] has recently presented a series of approximations to the description of disordered-medium random walks that can be described as a hierarchy of CTRWs. One starts from the master equation

$$\frac{dP(\mathbf{r}, t | \mathbf{r}_0)}{dt} = \sum_{\mathbf{r}'} W(\mathbf{r}, \mathbf{r}') P(\mathbf{r}', t | \mathbf{r}_0) - P(\mathbf{r}, t | \mathbf{r}_0) \sum_{\mathbf{r}'} W(\mathbf{r}', \mathbf{r}) \quad (A.5)$$

and takes the Laplace transform with respect to time to find

$$\left(s + \sum_{\mathbf{r}'} W(\mathbf{r}', \mathbf{r}) \right) P^*(\mathbf{r}, s | \mathbf{r}_0) - \sum_{\mathbf{r}'} W(\mathbf{r}, \mathbf{r}') P^*(\mathbf{r}', s | \mathbf{r}_0) = \delta_{\mathbf{r}, \mathbf{r}_0} \quad (A.6)$$

This may be rewritten in matrix form as

$$\mathbf{P} = \mathbf{G}_0 + \mathbf{G}_0 \mathbf{V} \mathbf{P} = \mathbf{G}_0 + \mathbf{P} \mathbf{V} \mathbf{G}_0 \quad (A.7)$$

where

$$\mathbf{G}_0 = (s\mathbf{I} + \mathbf{\Lambda})^{-1}, \qquad (\mathbf{\Lambda})_{\mathbf{r}, \mathbf{r}'} = \delta_{\mathbf{r}, \mathbf{r}'} \sum_{\mathbf{r}''} W(\mathbf{r}, \mathbf{r}'')$$

$$\mathbf{P} = \left(P^*(\mathbf{r}, s | \mathbf{r}_0) \right) \quad (A.8)$$

$$\mathbf{V} = \left(W(\mathbf{r}, \mathbf{r}') \right)$$

Equation (A.7) is an exact transcription of Eq. (A.5), and forms a starting point for the introduction of various approximations based on decoupling the possible averages in different ways. As an example, Halpern writes

$$\mathbf{P} = \mathbf{G}_0 \sum_{j=0}^{n-1} (\mathbf{V}\mathbf{G}_0)^j + \mathbf{P}(\mathbf{V}\mathbf{G}_0)^n \quad (A.9)$$

which is exact. The CTRW(n) approximation is obtained by taking configurational averages as follows:

$$\langle \mathbf{P}^{(n)} \rangle = \sum_{j=0}^{n-1} \langle \mathbf{G}_0(\mathbf{V}\mathbf{G}_0)^j \rangle + \langle \mathbf{P}^{(n)} \rangle \langle (\mathbf{V}\mathbf{G}_0)^n \rangle \quad (A.10)$$

so that the last term on the right is not averaged exactly. The usual CTRW approximation corresponds to CTRW(1) in this hierarchy. The decoupling procedure always satisfies the requirement of conservation of probabilities, and the CTRW(n) approximation allows one to study systems with correlations between successive steps. Halpern used this approximation technique to analyze two model systems, for which he calculated the ac conductivity. He concludes on the basis of these examples that the CTRW (1) approximation generally leads to qualitatively, but not necessarily quantitatively, correct results. At this time we can only conclude that the idea is an interesting one which may lead to valuable results, but much more testing on specific systems is needed for a more definitive assessment.

A scattering formalism for studying diffusion in the presence of traps has been used by Huber and his collaborators[283] in the context of fluorescence in the presence of traps, and more recently by Bixon and Zwanzig.[284] In both cases one expects approximations derived by this formalism to be valid at low concentrations. Bixon and Zwanzig model the traps as N randomly located spheres, while the diffusing particles are described by a diffusion equation subject to boundary conditions $c = 0$ on the surface and in the interior of the trapping spheres. If the diffusion equation is subjected to a Laplace transform, the solution can be written in terms of the Green's function $G_0 = (s - D\nabla^2)^{-1}$, a set of functions $C_j^*(\mathbf{r}, s)$, $j = 1, 2, \ldots, N$, and a set of operators $\{\mathbf{t}_j\}$, as

$$C^*(\mathbf{r}, s) = G_0 C(\mathbf{r}, 0) + \sum_{j=1}^{N} G_0 \mathbf{t}_j C_j^*(\mathbf{r}, s) \qquad (A.11)$$

where

$$C_j^*(\mathbf{r}, s) = G_0 C(\mathbf{r}, 0) + \sum_{k(\neq j)=1}^{N} G_0 \mathbf{t}_k C_k^*(\mathbf{r}, s) \qquad (A.12)$$

The $\mathbf{t}$ operators can be calculated through the analysis of diffusion in the presence of a single trap. The configuration-averaged concentrations $\langle C^*(\mathbf{r}, s) \rangle$ are the quantities of interest. In the low-concentration regime Bixon and Zwanzig use the approximation $\langle \mathbf{t}_k C_k^* \rangle \simeq \langle \mathbf{t}_k \rangle \langle C_k^* \rangle$. In this approximation one finds that at very long times the concentration decays as $t^{-3/2}$. Thus, no simple effective transport equation can be used to derive results for this system. Ghosh and Huber use a variant of this approach to discuss fluorescence in the presence of a high concentration of traps, but the approximation would seem to be questionable except at low concentrations.[283]

496   G. H. WEISS AND R. J. RUBIN

In conclusion, much more detailed simulation studies will be needed to sort out the areas of validity of different approximation techniques in the study of energy transport in disordered media. We expect to see much work along these lines in the very near future.

Acknowledgments

We are exceedingly grateful to several workers who read, commented, and corrected all or parts of this work while it was in preparation. These include Drs. Frank den Hollander, Robert Jernigan, Katja Lindenberg, Ralph Nossal, Michael Shlesinger, and Kurt Shuler. We are also indebted to Dr. Harvey Scher for several useful discussions as well as the use of figures from several of his publications.

References

1. I. Todhunter, *History of the Mathematical Theory of Probability*, Chelsea, New York, reprint, 1949.
2. F. Spitzer, *Principles of Random Walk*, Springer-Verlag, New York, 2nd ed., 1976.
3. K. Pearson, *Nature* **72**, 294 (1905).
4. K. Pearson, *A Mathematical Theory of Random Migration*, Draper's Company Research Memoirs, Biometrics Series, #3, 1906.
5. L. Bachelier, *Ann. Sci. Ecole Norm. Sup.* **17**, 21 (1900).
6. C. Domb and M. S. Green, *Phase Transitions and Critical Phenomena*, Vols. 1–6, Academic, New York, 1973–77.
7. J. C. Kluyver, *Kon. Akad. Wet. Amst.* **8**, 325 (1906): G. N. Watson, *A Treatise on the Theory of Bessel Functions*, Cambridge University Press, Cambridge, 1966; J. A. Greenwood and D. Durand, *Ann. Math. Stat.* **26**, 233 (1955).
8. Lord Rayleigh, *Philos. Mag.* **37**, 321 (1919).
9. S. Chandrasekhar, *Rev. Mod. Phys.* **15**, 1 (1943).
10. G. Polya, *Math. Ann.* **84**, 149 (1921).
11. G. I. Taylor, *Proc. London Math. Soc.* **20**, 196 (1921).
12. W. Kuhn, *Kolloid Z.* **68**, 2 (1934).
13. W. Kuhn and F. Grün, *Kolloid Z.* **101**, 248 (1942).
14. L. R. G. Treloar, *Trans. Faraday Soc.* **42**, 77 (1946).
15. M. V. Volkenshtein, *Configurational Statistics of Polymeric Chains*, Interscience, New York, 1959.
16. R. Kubo, *J. Phys. Soc. Jpn.* **2**, 47 (1947); **4**, 319 (1949).
17. M. Kac, *Probability and Related Topics in Physical Sciences*, Interscience, New York, 1959.
18. H. E. Daniels, *Proc. R. Soc. Edinburgh Sect. A* **LXIII**, 290 (1952).
19. B. V. Gnedenko and A. N. Kolmogoroff, *Limit Distributions for Sums of Independent Random Variables*, Addison-Wesley, Reading, Mass., rev. ed., 1968.
20. E. W. Montroll, *J. Soc. Ind. Appl. Math.* **4**, 241 (1956); in *Proceedings of the Sixteenth Symposium on Applied Mathematics*, 1964, p. 193.
21. E. W. Montroll and G. H. Weiss, *J. Math. Phys.* **6**, 167 (1965).
22. H. Scher and M. Lax, *Phys. Rev.* **137**, 4491, 4502 (1973).
23. H. Scher and C. H. Wu, *Proc. Natl. Acad. Sci.* **78**, 22 (1981).

24. D. R. Cox, *Renewal Theory*, Wiley, New York, 1962.

25. J. K. E. Tunaley, *Phys. Rev. Lett.* **33**, 1037 (1974); *J. Stat. Phys.* **11**, 397 (1974); **14**, 461 (1976).

26. V. M. Kenkre, E. W. Montroll, and M. F. Shlesinger, *J. Stat. Phys.* **9**, 45 (1973).

27. J. Klafter and R. Silbey, *Phys. Rev. Lett.* **44**, 55 (1980).

28. D. Bedeaux, K. Lindenberg, and K. E. Shuler, *J. Math. Phys.* **12**, 2116 (1971).

29. F. W. Schmidlin, *Phys. Rev. B* **16**, 2362 (1977).

30. J. Noolandi, *Phys. Rev. B* **16**, 4474 (1977).

31. K. W. Kehr and J. W. Haus, *Physica* **93A**, 412 (1978).

32. W. J. Shugard and H. Reiss, *J. Chem. Phys.* **65**, 2827 (1976).

33. B. J. Alder, in L. Garrido, P. Seglar, P. J. Shepherd, Ed., *Stochastic Processes in Nonequilibrium Systems*, Springer-Verlag, Berlin, 1978, p. 168.

34. W. E. Alley and B. J. Alder, *Phys. Rev. Lett.* **43**, 653 (1979).

35. S. Goldstein, *Quart. J. Mech. Appl. Math.* **IV**, 129 (1950).

36. J. Wishart and H. O. Hirschfeld, *J. London Math. Soc.* **11**, 227 (1936); E. W. Montroll, *J. Chem. Phys.* **41**, 2256 (1950); C. M. Tchen, *J. Chem. Phys.* **20**, 214 (1952); J. Gillis, *Proc. Cambridge Philos. Soc.* **51**, 639 (1955); C. Domb and M. E. Fisher, *Proc. Cambridge Philos. Soc.* **54**, 48 (1958); S. Seth, *J. Roy. Stat. Soc. B* **25**, 394 (1963); A. J. Allnutt and R. Fürth, *Proc. Roy. Soc. Edinb.* **69**, 1 (1970); J. U. Keller, *Z. Naturf.* **26**, 1939 (1971); G. C. Jain, *Can. Math. Bull.* **14**, 341 (1971); *ibid.* **16**, 389 (1973); A. D. Proudfoot and D. G. Lampard, *J. Appl. Prob.* **9**, 436 (1972); M. Kac, *Rocky Mountain J. Math.* **4**, 497 (1974); S. Corrsin, *Adv. Geophys.* **18A**, 25 (1974); M. Gordon, J. A. Torkington, and S. B. Ross-Murphy, *Macromolecules* **10**, 1090 (1977); D. J. Daley, *Proc. Cambridge Philos. Soc.* **86**, 115 (1979); R. B. Nain and K. Sen, *J. Appl. Prob.* **17**, 253 (1980); S. Fujita, Y. Okamura, and J. T. Chen, *J. Chem. Phys.* **72**, 3993 (1981).

37. G. H. Weiss, *J. Stat. Phys.* **15**, 157 (1976).

38. U. Landman, M. F. Shlesinger, and E. W. Montroll, *Proc. Natl. Acad. Sci.* **74**, 430 (1977).

39. K. Lindenberg and K. E. Shuler, *J. Math. Phys.* **12**, 633 (1971).

40. K. Lindenberg, *J. Stat. Phys.* **10**, 485 (1974).

41. M. F. Shlesinger, J. Klafter, and Y. M. Wong, *J. Stat. Phys.* **27**, 499 (1982).

42. G. Doetsch, *Theorie und Anwendungen der Laplace Transformation*, Dover reprint, New York, 1945.

43. C. Giddings, *Dynamics of Chromatography*, M. Dekker, New York, 1965; E. Glueckauf, K. H. Barker, and G. P. Kitt, *Discuss. Faraday Soc.* **7**, 199 (1949); E. Glueckauf, *Trans. Faraday Soc.* **51**, 34, 1540 (1955); H. C. Thomas, *J. Am. Chem. Soc.* **60**, 1664 (1944); *Ann. N.Y. Acad. Sci.* **49**, 161 (1948); L. Lapidus and N. R. Amundsen, *J. Phys. Chem.* **56**, 984 (1952); G. K. Ackers, *J. Biol. Chem.* **242**, 3026 (1967).

44. P. M. Morse and H. Feshbach, *Methods of Theoretical Physics*, McGraw-Hill, New York, 1953.

45. P. Levy, *Calcul des Probabilités*, Gauthier-Villars, Paris, 1925.

46. W. Feller, *An Introduction to Probability Theory and its Applications*, Vol. 2, Wiley, New York, 2nd ed., 1971.

47. E. W. Montroll and B. J. West, in *Fluctuation Phenomena*, North Holland, Amsterdam, 1979, p. 134.

48. J. Gillis and G. H. Weiss, *J. Math. Phys.* **11**, 1308 (1970).

49. D. R. Holt and E. L. Crow, *J. Res. Natl. Bur. Stand.* **77B**, 3 (1973); J. K. E. Tunaley, *J. Stat. Phys.* **12**, 1 (1975).

50. M. F. Shlesinger, *J. Stat. Phys.* **10**, 421 (1974).

51. P. J. Brockwell and B. M. Brown, *Z. Wahrsch. Verw. Geb.* **45**, 213 (1978).

52. E. W. Montroll and H. Scher, *J. Stat. Phys.* **9**, 101 (1973).

53. S. Foldes and G. Gabor, *Discrete Math.* **24**, 103 (1978).

54. G. N. Watson, *Quart. J. Math. Oxf.* **10**, 266 (1939); M. L. Glasser, *J. Res. Natl. Bur. Stand.* **B80**, 313 (1976); M. L. Glasser and I. J. Zucker, *Proc. Natl. Acad. Sci.* **74**, 1800 (1977).

55. K. Lindenberg, V. Seshadri, K. E. Shuler, and G. H. Weiss, *J. Stat. Phys.* **23**, 11 (1980).

56. A. Dvoretzky and P. Erdös, in *Proceedings of the Second Berkeley Symposium*, Vol. 33, Univ. of California Press, Berkeley, 1951.

57. G. H. Vineyard, *J. Math. Phys.* **4**, 1191 (1963).

58. G. H. Hardy, *Divergent Series*, Oxford University Press, Oxford, 1949.

59. G. H. Weiss and M. F. Shlesinger, *J. Stat. Phys.* **27**, 355 (1982).

60. N. C. Jain and S. Orey, *Isr. J. Math.* **6**, 373 (1968); N. C. Jain, *Z. Wahrsch. Verw. Geb.* **16**, 270 (1970); N. C. Jain and W. E. Pruitt, *J. Analyse Math.* **24**, 369 (1969); in *Proceedings of the Sixth Berkeley Symposium*, Vol. III, University of California Press, Berkeley, (1971), p. 31.

61. R. J. Beeler and J. A. Delaney, *Phys. Rev.* **130**, 926 (1963).

62. R. J. Beeler, *Phys. Rev. A* **134**, 1396 (1964).

63. P. Erdös and S. J. Taylor, *Acta Math. Acad. Sci.* **11**, 137 (1960).

64. A. A. Maradudin, E. W. Montroll, G. H. Weiss, R. Herman, and H. W. Milnes, *Mem. Acad. Roy. Belgique* **14**: 7 (1960); J. Mahanty, *Proc. Phys. Soc. (London)* **88**, 1011 (1966); J. S. Byrnes, S. Podgor, and W. W. Zachary, *Proc. Camb. Phil. Soc.* **66**, 377 (1969); G. Iwata, *Nat. Sci. Rep. Ochanomizu Univ.* **20**, 13 (1969); S. Katsura, T. Morita, S. Inawashiro, and Y. Abe, *J. Math. Phys.* **12**, 895 (1971); F. T. Hioe, *J. Math Phys.* **19**, 1064 (1978); D. J. Daley, *J. Appl. Prob.* **16**, 45 (1979).

65. G. S. Joyce, *J. Phys.* **C4**, L53 (1971); *J. Phys.* **A5**, L65 (1972); *J. Math. Phys.* **12**, 1390 (1971); *Phil. Trans. Roy. Soc.* **A273**, 585 (1973).

66. R. J. Rubin and G. H. Weiss, *J. Math. Phys.* **23**, 250 (1982).

67. R. J. Rubin, *J. Chem. Phys.* **43**, 2392 (1965).

68. B. W. Conolly, *SIAM Rev.* **13**, 81 (1971).

69. D. A. Darling and M. Kac, *Trans. Am. Math. Soc.* **84**, 444 (1957).

70. M. Kac, *Trans. Am. Math. Soc.* **84**, 459 (1957).

71. E. W. Montroll, *J. Phys. Soc. Jpn. Suppl.* **26**, 6 (1969); *J. Math. Phys.* **10**, 753 (1969).

72. R. J. Rubin, *J. Math. Phys.* **8**, 576 (1967).

73. G. H. Weiss, *J. Math. Phys.* **22**, 562 (1981).

74. K. E. Shuler, H. Silver, and K. Lindenberg, *J. Stat. Phys.* **15**, 393 (1976).

75. E. Schrödinger, *Phys. Z.* **16**, 289 (1915).

76. N. G. van Kampen and I. Oppenheim, *J. Math. Phys.* **13**, 842 (1972).

77. D. R. Cox and H. D. Miller, *The Theory of Stochastic Processes*, Wiley, New York, 1965.

78. M. J. Lighthill, *Fourier Analysis and Generalized Functions*, Cambridge U.P., Cambridge, 1964.

79. U. Grenander and G. Szegö, *Toeplitz Forms and Their Applications*, Univ. of Calif. Press, Berkeley, 1958; M. E. Fisher and R. E. Hartwig, *Adv. Chem. Phys.* **XV**, 333 (1969).

80. F. L. Spitzer and C. J. Stone, *Ill. J. Math.* **4**, 253 (1960).

81. H. Kesten, *Ill. J. Math.* **5**, 246, 267 (1961).

82. D. Gutkowitz-Krusin, I. Procaccia, and J. Ross, *J. Stat. Phys.* **19**, 525 (1978). G. H. Weiss, *J. Stat. Phys.* **21**, 369 (1980).

83. A. Wald, *Sequential Analysis*, Wiley, New York, 1947.

84. H. Daniels, *Proc. Camb. Phil. Soc.* **XXXVII**, 244 (1941).

85. W. Feller, *Ann. Math. Stat.* **22**, 427 (1951).

86. D. A. Darling and A. J. F. Siegert, *Ann. Math. Stat.* **24**, 624 (1953).

87. G. H. Weiss and R. J. Rubin, *J. Stat. Phys.* **14**, 333 (1976).

88. R. J. Rubin, J. Mazur, and G. H. Weiss, *Pure Appl. Chem.* **46**, 143 (1976).

89. P. Erdös and M. Kac, *Bull. Am. Math. Soc.* **52**, 292 (1946).

90. M. Donsker, *Mem. Am. Math. Soc.* **6**, 1951.

91. R. J. Rubin and J. Mazur, *Macromolecules* **10**, 139 (1977).

92. G. H. Weiss and R. J. Rubin, *J. Stat. Phys.* **22**, 97 (1980).

93. S. H. Northrup and J. T. Hynes, *J. Stat. Phys.* **18**, 91 (1978).

94. Y. Zvirin and R. Shinnar, *Water Res.* **10**, 765 (1976).

95. H. Eyring, *Phys. Rev.* **39**, 746 (1932); E. Guth and H. Mark, *Monatsh. Chem.* **65**, 93 (1935); W. Kuhn, *Kolloid Z.* **76**, 258 (1936); **87**, 3 (1939).

96. T. M. Birshtein and O. B. Ptitsyn, *Conformations of Macromolecules*, Wiley, New York, 1966.

97. P. J. Flory, *Statistical Mechanics of Chain Molecules*, Wiley, New York, 1969.

98. H. Yamakawa, *Modern Theory of Polymer Solutions*, Harper & Row, San Francisco, 1971.

99. N. L. Johnson, *Technometrics* **8**, 303 (1966); R. Barakat, *J. Phys.* **A6**, 796 (1973).

100. S. Dvorak, *J. Phys.* **A5**, 78, 85 (1972).

101. R. L. Jernigan and P. J. Flory, *J. Chem. Phys.* **50**, 4185 (1969).

102. R. L. Jernigan and P. J. Flory, *J. Chem. Phys.* **50**, 4178 (1969).

103. A. Kratky and G. Porod, *Rec. Trav. Chim.* **68**, 1106 (1949).

104. H. E. Daniels, *Proc. R. Soc. Edinburgh* **A63**, 290 (1952).

105. J. J. Hermans and R. Ullman, *Physica* **18**, 951 (1952).

106. C. A. Hollingsworth, *J. Chem. Phys.* **16**, 544 (1948); **17**, 97 (1949); **20**, 1580 (1952).

107. R. J. Rubin, J. Mazur, and G. H. Weiss, *Pure Appl. Chem.* **46**, 143 (1976).

108. R. J. Rubin and G. H. Weiss, *Macromolecules* **11**, 1046 (1978).

109. A. Katchalsky, O. Kuenzle, and W. Kuhn, *J. Polym. Sci.* **5**, 283 (1950).

110. C. Domb, *J. Phys.* **A14**, 219 (1981).

111. B. H. Zimm, R. S. Stein, and P. Doty, *Polym. Bull.* **1**, 90 (1945).

112. M. Fixman, *J. Chem. Phys.* **36**, 306 (1962); W. C. Forsman and R. E. Hughes, *J. Chem. Phys.* **38**, 2118 (1963); **42**, 2829 (1965); **44**, 1716 (1966); R. F. Hoffman and W. C. Forsman, *J. Chem. Phys.* **50**, 2316 (1969); **52**, 2222 (1970); S. G. Gupta, and W. C. Forsman, *Macromolecules* **5**, 779 (1972); **6**, 285 (1973); **7**, 853 (1974); S. G. Gupta and W. C. Forsman, *J. Chem. Phys.* **65**, 201 (1976); B. E. Eichinger, *Macromolecules* **13**, 1 (1980).

113. H. Fujita and T. Norisuye, *J. Chem. Phys.* **52**, 1115 (1970).

114. S. R. Coriell and J. L. Jackson, *J. Math. Phys.* **8**, 1276 (1965).

115. P. J. Flory and S. Fisk, *J. Chem. Phys.* **44**, 2243 (1966).

116. K. Solč, *J. Chem. Phys.* **55**, 335 (1971); *Macromolecules* **5**, 705 (1972); **6**, 378 (1973); K. Solč and W. Gobush, *Macromolecules* **7**, 814 (1974).

117. K. Solč and W. H. Stockmayer, *J. Chem. Phys.* **54**, 2726 (1971).

118. D. E. Kranbuehl, P. H. Verdier, and J. M. Spencer, *J. Chem. Phys.* **59**, 3861 (1973).

119. P. H. Verdier, *J. Comp. Phys.* **4**, 204 (1969).

120. G. H. Weiss and R. J. Rubin, *J. Stat. Phys.* **14**, 333 (1976).

121. J. Mazur and R. J. Rubin, *J. Chem. Phys.* **60**, 341 (1974); R. J. Rubin and J. Mazur, *J. Chem. Phys.* **63**, 5362 (1975).

122. R. J. Rubin and G. H. Weiss, *Macromolecules* **10**, 332 (1977).

123. H. E. Daniels, *Proc. Camb. Phil. Soc.* **37**, 244 (1941).

124. E. F. Casassa and Y. Tagami, *Macromolecules* **2**, 14 (1969).

125. C. M. Guttman and E. A. DiMarzio, *Macromolecules* **3**, 681 (1970).

126. E. F. Casassa, *Macromolecules* **9**, 182 (1976).

127. P. Debye, *Ann. Phys.* **46**, 809 (1915).

128. K. Nagai, *J. Chem. Phys.* **38**, 924 (1963).

129. D. Y. Yoon and P. J. Flory, *J. Chem. Phys.* **61**, 5358, 5366 (1974); *Polym.* **16**, 645 (1975).

130. R. L. Jernigan and G. H. Weiss, *Polym. Prepr. Am. Chem. Soc. Div. Polym. Chem.* **14**, 214 (1973).

131. M. Fixman and R. Alben, *J. Chem. Phys.* **58**, 1553 (1973).

132. M. Fixman and J. Skolnick, *J. Chem. Phys.* **65**, 1700 (1976).

133. J. Freire and M. Fixman, *J. Chem. Phys.* **69**, 634 (1978).

134. J. J. Freire and M. M. Rodrigo, *J. Chem. Phys.* **72**, 6376 (1980).

135. N. Saito, K. Takahashi, and Y. Yunoki, *J. Phys. Soc. Jpn* **22**, 219 (1967).

136. K. F. Freed, *Adv. Chem. Phys.* **XXII**, 1 (1972).

137. W. Gobush, H. Yamakawa, W. H. Stockmayer, and W. S. Magee, *J. Chem. Phys.* **57**, 2839 (1972).

138. H. Yamakawa and W. H. Stockmayer, *J. Chem. Phys.* **57**, 2843 (1972).

139. R. P. Feynman, *Phys. Rev.* **80**, 440 (1950).

140. M. Kac, in *Proceedings of the Second Berkeley Symposium*, 1951, p. 189.

141. H. Yamakawa and M. Fujii, *J. Chem. Phys.* **64**, 5222 (1976).

142. H. Yamakawa, M. Fujii, and J. Shimada, *J. Chem. Phys.* **65**, 2371 (1976); M. Fujii and H. Yamakawa, *J. Chem. Phys.* **66**, 2578 (1977); H. Yamakawa and M. Fujii, *J. Chem. Phys.* **66**, 2584 (1977); J. Shimada and H. Yamakawa, *J. Chem. Phys.* **67**, 344 (1977); H. Yamakawa, J. Shimada, and M. Fujii, *J. Chem. Phys.* **68**, 4722 (1978); H. Yamakawa and J. Shimada, *J. Chem. Phys.* **70**, 609 (1979); H. Yamakawa, M. Fujii, and J. Shimada, *J. Chem. Phys.* **71**, 1611 (1978); H. Yamakawa, J. Shimada, and K. Nagasaka, *J. Chem. Phys.* **71**, 3573 (1979); M. Fujii and H. Yamakawa, *J. Chem. Phys.* **72**, 6005 (1980); J. Shimada and H. Yamakawa, *J. Chem. Phys.* **73**, 4037 (1980).

143. D. McIntyre and F. Gornick, Eds., *Light Scattering from Dilute Polymer Solutions*, Gordon & Breach, New York, 1964.

144. J. J. Hermans, Ed., *Polymer Solution Properties: Vol. 1. Statistics and Thermodynamics*; *Vol. 2. Hydrodynamics and Light Scattering*, Dowden, Hutchinson, and Ross, Stroudsburg, Pa., 1978.

145. P. de Gennes, *J. Chem. Phys.* **55**, 572 (1971); *Macromolecules* **9**, 587 (1976); *Macromolecules* **11**, 852 (1978); *Scaling Concepts in Polymer Physics*, Cornell U.P. Ithaca, 1979.

146. S. F. Edwards and J. W. Grant, *J. Phys.* **A6**, 1169, 1186, 1670 (1973); M. Doi and S. F. Edwards, *J. Chem. Soc. Faraday Trans. II* **74**, 1789, 1802, 1818 (1978).

147. J. Klein, *Contemp. Phys.* **20**, 611 (1979).

148. A. Silberberg, *J. Phys. Chem.* **66**, 1872 (1962).

149. J. Pouchlý, *Coll. Czech. Chem. Comm.* **28**, 1804 (1963).

150. C. A. J. Hoeve, E. A. DiMarzio, and P. Peyser, *J. Chem. Phys.* **42**, 2558 (1965); E. A. DiMarzio and F. L. McCrackin, *J. Chem. Phys.* **43**, 539 (1965); C. A. J. Hoeve, *J. Chem. Phys.* **43**, 3007 (1965).

151. R. J. Roe, *Proc. Natl. Acad. Sci.* **53**, 50 (1965); *J. Chem. Phys.* **43**, 1591; **44**, 4264 (1965).

152. K. Motomura and R. Matuura, *J. Chem. Phys.* **50**, 1281 (1969); K. Motomura, Y. Moroi, and R. Matuura, *Bull. Chem. Soc. Jpn.* **44**, 1243, 1248 (1971); K. Motomura and R. Matuura, *Mem. Fac. Sci. Kyushu Univ.* **6**, 97 (1968); K. Motomura, *J. Chem. Phys.* **51**, 4681 (1969).

153. D. Chan, D. J. Mitchell, B. W. Ninham, and L. R. White, *J. Faraday Soc. II* **71**, 235 (1975).

154. H. Lépine and A. Caillé, *Can. J. Phys.* **56**, 403 (1978).

155. P. G. de Gennes, *Rep. Prog. Phys.* **32**, 187 (1969).

156. F. W. Wiegel, *J. Phys.* **A10**, 299 (1977).

157. R. J. Rubin, *J. Chem. Phys.* **44**, 2130 (1966).

158. E. A. DiMarzio and R. J. Rubin, *J. Chem. Phys.* **55**, 4318 (1971); D. Chan, B. Davies, and P. Richmond, *J. Faraday Soc. II* **72**, 1584 (1976); A. M. Skvortsov, A. A. Gorbunov, Y. B. Zhuling, and T. M. Birshtein, *Polym. Sci. USSR* **20**, 919 (1979).

159. E. A. DiMarzio, *J. Chem. Phys.* **42**, 2101 (1965); E. A. DiMarzio and F. L. McCrackin, *J. Chem. Phys.* **43**, 539 (1965); A. Silberberg, *J. Chem. Phys.* **46**, 1105 (1967); E. F. Casassa, *Polym. Lett.* **5**, 773 (1967); P. Richmond and M. Lal, *Chem. Phys. Lett.* **24**, 594 (1974); A. K. Dolan and S. F. Edwards, *Proc. R. Soc. Edinburgh Sect. A* **337**, 509 (1974); D. J. Meier, *J. Phys. Chem.* **71**, 1861 (1967); F. Th. Hesselink, *J. Phys. Chem.* **73**, 3488 (1969); **75**, 65 (1971); F. Th. Hesselink, A. Vrij, and J. Th. G. Overbeek, *J. Phys. Chem.* **75**, 2094 (1971); M. E. van Kreveld, *Polym. Phys.* **13**, 2253 (1975).

160. R. J. Rubin, *J. Math. Phys.* **9**, 2252 (1968).

161. R. J. Rubin, *J. Chem. Phys.* **44**, 2130 (1966).

162. R. J. Rubin and G. H. Weiss, in preparation.

163. J. I. Lauritzen and E. A. DiMarzio, *J. Res. Natl. Bur. Stand.* **83B**, 381 (1981).

164. R. J. Rubin, *J. Res. Natl. Bur. Stand.* **69B**, 301 (1965).

165. C. A. J. Hoeve, *J. Polym. Sci. Polym. Symp.* **61**, 389 (1977).

166. R. J. Rubin, *J. Res. Natl. Bur. Stand.* **70B**, 237 (1966).

167. E. Dickenson and M. Lal, *Adv. Mol. Relaxation Interactions* **17**, 1 (1980).

168. J. C. Giddings and H. Eyring, *J. Phys. Chem.* **59**, 416 (1955); J. C. Giddings, *J. Chem. Phys.* **26**, 169 (1957); **31**, 1462 (1959); *J. Chromatogr.* **3**, 443 (1960).

169. G. H. Weiss, *Sep. Sci.* **5**, 51 (1970).

170. F. C. Denizot and M. A. Delaage, *Proc. Natl. Acad. Sci.* **12**, 4840 (1975).

171. G. H. Weiss, *Sep. Sci. Technol.* **16**, 75 (1981).

172. J. R. Cann, J. G. Kirkwood, and R. A. Brown, *Arch. Biochem. Biophys.* **72**, 37 (1957).

173. T. A. Bak, *Contributions to the Theory of Chemical Kinetics*, Munksgaard, Copenhagen, 1959.

174. S. Meiboom, *J. Chem. Phys.* **34**, 1 (1961).

175. H. L. Friedman and A. Ben-Naim, *J. Chem. Phys.* **34**, 1 (1961).

176. G. H. Weiss, *J. Stat. Phys.* **8**, 221 (1973).

177. K. S. Singwi and A. Sjölander, *Phys. Rev.* **119**, 863 (1960).

178. K. Lindenberg and R. I. Cukier, *J. Chem. Phys.* **62**, 3271 (1975).

179. U. Landman, E. W. Montroll, and M. F. Shlesinger, *Proc. Natl. Acad. Sci.* **74**, 430 (1977).

180. U. Landman and M. F. Shlesinger, *Phys. Rev. Lett.* **41**, 1174 (1978); *Solid State Comm.* **27**, 939 (1978); *Phys. Rev. B* **19**, 6207, 6220 (1979).

181. D. E. Reed and G. Ehrlich, *J. Chem. Phys.* **64**, 4616 (1976).

182. G. H. Weiss, *J. Stat. Phys.* **15**, 157 (1976).

183. W. L. Smith, *Proc. R. Soc. Edinburgh Sect. A* **232**, 6 (1955); *J. R. Stat. Soc.* **20**, 243 (1958).

184. N. G. van Kampen, *Physica* **96A**, 435 (1979).

185. L. Takacs, *Acta Math. Hung.* **8**, 169 (1957).

186. H. Scher and C. H. Wu, *Proc. Natl. Acad. Sci.* **78**, 22 (1981).

187. J. W. Haus, K. W. Kehr, *Solid St. Comm.* **26**, 753 (1978); *J. Phys. Chem. Sol.* **40**, 1019 (1979); M. F. Shlesinger, *Solid St. Comm.* **32**, 1207 (1979).

188. H. Silver, K. E. Shuler, and K. Lindenberg, in U. Landman, Ed., *Statistical Mechanics and Statistical Methods in Theory and Application*, Plenum, New York, 1977, p. 463.

189. K. E. Shuler, *Physica* **95A**, 12 (1979).

190. V. Seshadri, K. Lindenberg, and K. E. Shuler, *J. Stat. Phys.* **21**, 517 (1979).

191. K. E. Shuler and U. Mohanty, *Proc. Natl. Acad. Sci.* **78**, 6576 (1981).

192. J. C. Maxwell, *A Treatise on Electricity & Magnetism*, Vol. 1, Clarendon Press, Oxford, 1892; Lord Rayleigh, *Philos. Mag.* **34**, 481 (1892).

193. J. Bardeen and C. Herring, in W. Shockley, Ed., *Imperfections in Nearly Perfect Crystals*, Wiley, New York, 1952, p. 261.

194. K. Compaan and Y. Haven, *Trans. Faraday Soc.* **52**, 786 (1956); **54**, 1498 (1958).

195. J. R. Manning, *Diffusion Kinetics for Atoms in Crystal*, Van Nostrand, Princeton, 1968.

196. A. D. Le Claire, in W. Jost, Ed., *Physical Chemistry: An Advanced Treatise*, Vol. 10, Academic, New York, 1970, p. 261.

197. S. Ishioka and M. Koiwa, *Philos. Mag.* **A41**, 385 (1980).

198. P. Benoist, J. L. Bocquet, and P. Lafore, *Acta Metall.* **25**, 265 (1977).

199. H. B. Rosenstock, *Phys. Rev.* **187**, 1166 (1969).

200. R. C. Powell and R. G. Kepler, *Phys. Rev. Lett.* **22**, 636 (1969).

201. H. B. Rosenstock, *J. Math. Phys.* **11**, 487 (1970); **21**, 1643 (1980); H. B. Rosenstock and C. L. Marquardt, *Phys. Rev. B* **22**, 5797 (1980).

202. G. H. Weiss, *Proc. Natl. Acad. Sci.* **77**, 4391 (1980).

203. E. W. Montroll, *J. Math. Phys.* **10**, 753 (1969).

204. R. Emerson and W. Arnold, *J. Gen. Physiol.* **15**, 391 (1932).

205. R. M. Pearlstein, *J. Chem. Phys.* **56**, 2431 (1972).

206. E. W. Montroll, *J. Phys. Soc. Jpn. Suppl.* **26**, 6 (1969).

207. J. W. Sanders, T. W. Ruijgrok, and J. J. Ten Bosch, *J. Math. Phys.* **12**, 534 (1971).

208. R. S. Knox, *J. Theor. Biol.* **21**, 244 (1968).

209. J. Hoshen and R. Kopelman, *J. Chem. Phys.* **65**, 2817 (2976); *Phys. Rev. B* **14**, 3438 (1976); P. Argyrakis and R. Kopelman, *J. Theor. Biol.* **73**, 205 (1978).

210. M. D. Hatlee and J. J. Kozak, *Phys. Rev. B* **21**, 1400 (1980); *Proc. Natl. Acad. Sci.* **78**, 972 (1981).

211. G. Adam and M. Delbrück, in A. Rich, and N. Davidson, Eds., *Structural Chemistry and Molecular Biology*, Freeman, San Francisco, 1968, p. 198.

212. M. D. Hatlee, J. J. Kozak, G. Rothenberg, P. P. Infelta, and M. Gratzel, *J. Phys. Chem.* **84**, 1508 (1980).

213. K. Lindenberg, R. P. Hemenger, and R. M. Pearlstein, *J. Chem. Phys.* **56**, 4852 (1972); R. P. Hemenger, R. M. Pearlstein, and K. Lindenberg, *J. Math. Phys.* **13**, 1056 (1972); R. P. Hemenger, K. Lindenberg, and R. M. Pearlstein, *J. Chem. Phys.* **60**, 3271 (1974).

214. T. Förster, *Ann. Phys.* **2**, 55 (1948); *Z. Naturforsch.* **A4**, 321 (1949).

215. D. L. Dexter, *J. Chem. Phys.* **21**, 836 (1953).

216. M. K. Grover and R. Silbey, *J. Chem. Phys.* **52**, 2099 (1970); **54**, 4843 (1971).

217. R. W. Munn, *J. Chem. Phys.* **58**, 3230 (1973).

218. W. L. Greer, *J. Chem. Phys.* **60**, 744 (1974).

219. D. D. Dlott, M. D. Fayer, and R. D. Wieting, *J. Chem. Phys.* **67**, 3808 (1977); R. D. Wieting, M. D. Fayer, and D. D. Dlott, *J. Chem. Phys.* **69**, 1996 (1978).

220. W. P. Helman and K. Funabashi, *J. Chem. Phys.* **66**, 5790 (1977); **70**, 4813 (1979).

221. M. F. Shlesinger, *J. Chem. Phys.* **70**, 4813 (1979).

222. J. Klafter and R. Silbey, *J. Chem. Phys.* **72**, 843 (1980).

223. K. Funabashi, *J. Chem. Phys.* **72**, 3123 (1980).

224. S. W. Haan and R. Zwanzig, *J. Chem. Phys.* **68**, 1879 (1978).

225. C. R. Gochanour, H. C. Andersen, and M. D. Fayer, *J. Chem. Phys.* **70**, 4254 (1979).

226. K. Godzik and J. Jortner, *J. Chem. Phys.* **72**, 4471 (1980).

227. K. Godzik and J. Jortner, *Chem. Phys. Lett.* **63**, 428 (1979).

228. A. I. Burshtein, *Sov. Phys. JETP* **35**, 882 (1972); L. D. Zusman, *Sov. Phys. JETP* **46**, 347 (1977); A. Blumen and G. Zumofen, *Chem. Phys. Lett.* **70**, 387 (1980); K. Allinger and A. Blumen, *J. Chem. Phys.* **72**, 4608 (1980); A. Blumen, J. Klafter, and R. Silbey, *J. Chem. Phys.* **72**, 5320 (1980).

229. G. Theodorou and M. H. Cohen, *Phys. Rev. B* **19**, 1561 (1978).

230. J. Klafter and R. Silbey, *J. Chem. Phys.* **72**, 849 (1980).

231. M. S. Matheson and L. M. Dorfman, *Pulse Radiolysis*, MIT Press, Cambridge, Mass., 1969; J. H. Baxendale and P. H. G. Sharpe, *Chem. Phys. Lett.* **39**, 401 (1976); J. R. Miller, *J. Phys. Chem.* **82**, 2143 (1978).

232. J. Mort and J. Knights, *Nature* **290**, 659 (1981).

233. D. G. Thomas, J. J. Hopfield, and W. M. Augustyniak, *Phys. Rev.* **140**, 202 (1965).

234. M. Lax and H. Scher, *Phys. Rev. Lett.* **39**, 781 (1977).

235. A. A. Kumar and J. Heinrichs, *J. Phys. C* **13**, 2131 (1980).

236. A. A. Kumar and J. Heinrichs, *J. Phys. C* **13**, 5971 (1980).

237. H. Kahlert, *J. Phys. C* **9**, 491 (1976).

238. E. J. Moore, *J. Phys. C* **7**, 339 (1974).

239. J. A. McInnes, P. N. Butcher, and D. L. Clark, *Philos. Mag.* **B41**, 1 (1980).

240. M. Pollak and T. H. Geballe, *Phys. Rev.* **122**, 1742 (1961); A. Miller and E. Abrahams, *Phys. Rev.* **120**, 745 (1960); P. N. Butcher, *J. Phys. C* **5**, 1817 (1972), C**7**, 879, 2645 (1974); P. N. Butcher and K. J. Hayden, *Philos. Mag.* **36**, 657 (1977); P. N. Butcher and J. A. McInnes, *Philos. Mag.* **B37**, 249 (1978); P. N. Butcher and P. L. Morys, *J. Phys. C* **6**, 2147 (1973); J. A. McInnes and P. N. Butcher, *Philos. Mag.* **B39**, 1 (1979); S. Kirkpatrick, *Rev. Mod. Phys.* **45**, 574 (1973); B. Movaghar, D. Miller, and K. H. Benneman, *J. Phys. F* **4**, 687 (1974); B. Movaghar and D. Miller, *J. Phys. F* **5**, 261 (1975); B. Movaghar, B. Pohlmann, and G. W. Sauer, *Phys. Stat. Sol. (b)* **97**, 533 (1980); B. Movaghar, B. Pohlmann, and W. Schirmacher, *Philos. Mag.* **B41**, 49 (1980); *Solid State Commun.* **34**, 451 (1980);

B. Movaghar and G. W. Sauer, *Solid State Commun.* **35**, 841 (1980); B. Movaghar, *J. Phys. C* **13**, 4915 (1980); S. Kivelson, *Phys. Rev. B* **21**, 5755 (1980).

241. R. Zwanzig, in W. E. Townsend, and J. Down, Eds., *Lectures in Theoretical Physics III*, Interscience, New York, 1961, p. 106.

242. H. Scher and E. W. Montroll, *Phys. Rev. B* **12**, 2455 (1975).

243. M. E. Scharfe, *Phys. Rev. B* **2**, 5025 (1970); D. M. Pai and M. E. Scharfe, *J. Non-Cryst. Sol.* **8–10**, 752 (1972); J. Mort and A. I. Lakatos, *J. Non-Cryst. Sol.* **4**, 117 (1970); H. Seki, in J. Stuke and W. Brenig, Eds., *Proceedings of the Fifth International Conference on Amorphous and Liquid Semiconductors*, Taylor and Francis, London, 1974, p. 105; G. Pfister and H. Scher, *Phys. Rev. B* **15**, 2062 (1977); G. Pfister, *Phys. Rev. Lett.* **36**, 271 (1976).

244. F. B. McLean and G. A. Ausman, *Phys. Rev. B* **15**, 1052 (1977).

245. F. W. Schmidlin, *Phys. Rev. B* **16**, 2362 (1977); *Philos. Mag.* **B41**, 535 (1980).

246. G. Pfister and H. Scher, *Adv. Phys.* **27**, 747 (1978).

247. G. F. Leal Ferreira, *Phys. Rev. B* **16**, 4719 (1977).

248. H. Scher, S. Alexander, and E. W. Montroll, *Proc. Natl. Acad. Sci.* **77**, 3758 (1980).

249. D. Haarer and H. Möhwald, *Phys. Rev. Lett.* **34**, 1447 (1975).

250. J. Bernasconi, S. Alexander, and R. Orbach, *Phys. Rev. Lett.* **41**, 185 (1978); S. Alexander, J. Bernasconi, and R. Orbach, *Phys. Rev. B* **17**, 4311 (1978); *J. Phys. (Paris) Suppl.* **39**, C6-706.

251. C. Domb, A. A. Maradudin, E. W. Montroll, and G. H. Weiss, *Phys. Rev.* **115**, 18, 24 (1959).

252. J. Heinrichs, *Phys. Rev. B* **22**, 3093 (1980); T. Odagaki and M. Lax, *Phys. Rev. Lett.* **45**, 847 (1980).

253. S. Alexander, J. Bernasconi, W. R. Schneider, and R. Orbach, *Rev. Mod. Phys.* **53**, 175 (1981).

254. F. J. Dyson, *Phys. Rev.* **92**, 1331 (1953).

255. J. Bernasconi, W. R. Schneider, and W. Wyss, *Z. Phys.* **B37**, 175 (1980).

256. P. M. Richards and R. L. Renken, *Phys. Rev. B* **21**, 3740 (1980).

257. J. Bernasconi, S. Alexander, and R. Orbach, *Phys. Rev. Lett.* **41**, 185 (1978); J. Bernasconi, H. U. Beyeler, S. Strassler, and S. Alexander, *Phys. Rev. Lett.* **42**, 819 (1979); J. Bernasconi and H. U. Beyeler, *Phys. Rev. B* **21**, 3745 (1980).

258. P. M. Richards, *Phys. Rev. B* **16**, 1393 (1977); P. A. Fedders, *Phys. Rev. B.* **17**, 40 (1978); P. A. Fedders and P. M. Richards, *Phys. Rev. B* **21**, 377 (1980).

259. D. L. Huber, *Phys. Rev. B* **15**, 533 (1977).

260. M. H. Gail and C. W. Boone, *Biophys. J.* **10**, 980 (1970).

261. S. C. Peterson and P. B. Noble, *Biophys. J.* **12**, 1048 (1972).

262. H. C. Berg and D. A. Brown, *Nature* **239**, 500 (1972); *Antibiot. Chemother.* **19**, 55 (1974).

263. R. M. Nacnab and D. E. Koshland, Jr., *Proc. Natl. Acad. Sci.* **69**, 2509 (1972).

264. R. M. Macnab and D. E. Koshland, Jr., *J. Mechanochem. Cell Motil.* **2**, 141 (1973).

265. S. H. Zigmond, *Nature* **249**, 450 (1974).

266. C. S. Patlak, *Bull. Math. Biophys.* **15**, 311, 431 (1953).

267. R. Nossal and G. H. Weiss, *J. Stat. Phys.* **10**, 245 (1974); *J. Theor. Biol.* **47**, 103 (1974).

268. R. Nossal, *Math. Biosci.* **31**, 121 (1976).

269. R. Nossal and S. H. Zigmond, *Biophys. J.* **16**, 1171 (1976).

270. R. L. Hall and S. C. Peterson, *Biophys. J.* **25**, 1979 (1979).

271. R. L. Hall, *J. Math. Biol.* **4**, 327 (1977).

272. P. S. Lovely and F. W. Dahlquist, *J. Theor. Biol.* **50**, 477 (1975).

273. W. Alt, *J. Math. Biol.* **9**, 147 (1980).

274. E. F. Keller and L. A. Segel, *J. Theor. Biol.* **30**, 225 (1971).

275. W. Jäger, H. Röst, and P. Tautu, Eds., *Biological Growth and Spread. Mathematical Theories and Applications*, Springer, Heidelberg, 1980.

276. R. Nossal, in Ref. 275, p. 410.

277. A. Blumen and G. Zumofen, *J. Chem. Phys.* **75**, 892 (1981).

278. Z. Ciesielski and S. J. Taylor, *Trans. Am. Math. Soc.* **103**, 434 (1962).

279. J. Machta, *Phys. Rev. B* **24**, 5260 (1981).

280. R. Zwanzig, *J. Stat. Phys.* **28**, 127 (1982).

281. T. Odagaki and M. Lax, *Phys. Rev. B* **25**, 2301, 2307 (1982); M. Lax and T. Odagaki, in R. Burridge, S. Childress, and G. Papanicolau, Eds., *Proceedings of the Conference on Macroscopic Properties of Disordered Media*, Springer-Verlag, New York 1982 p. 148

282. V. Halpern, *J. Phys. C* **14**, 3195, 3208 (1981).

283. D. L. Huber, *Phys. Rev. B* **20**, 2307 5333 (1979); K. K. Ghosh, J. Hegarty, and D. L. Huber, *Phys. Rev. B* **22**, 2837 (1980); K. K. Ghosh and D. L. Huber, *Phys. Rev. B* **23**, 4441 (1981).

284. M. Bixon and R. Zwanzig, *J. Chem. Phys.* **75**, 2354 (1981).

AUTHOR INDEX

Numbers in parentheses are reference numbers and indicate that the author's work is referred to although his name is not mentioned in the text. Numbers in italic show the pages on which the complete references are listed.

Lasaga, A. C., 70(130), 72(130), 105(130), *155*

Laurinc, V., 190(34), 191(34, 36), 193(34, 36), 197(42), 199(34), 200(45), 203(53), 205(36, 42), 208(34), 225(42), 227(60), 237(34), 238(34), *260*

Lauritzen, J. I., 455(163), *501*

Lawson, R. D., 3(14), *111*

Lax, M., 369(22), 473(22), 475(22), 476(22, 234), 485(252), 487(22), 493(252, 281), 494(281), *496, 503–505*

LeBlanc, F. J., 307(138), 309(138), 322(138), 327–330(138), 333(138), *359*

LeBreton, P. R., 350(242), *362*

Leckenby, R. E., 273(71), 284(71), *357*

Le Claire, A. D., 464(196), *502*

Lee, A., 350(243), *362*

Lee, J. K., 280(75), 281(75), 284(75), *357*

Lee, S. T., 324(162), 336(198), *360, 361*

Lee, Y. T., 266(11–16, 19–21, 26), 273(15), 274(13, 15), 277(11), 280(12, 75, 77), 281(12, 13, 15, 26, 75, 77, 93), 282(12, 13), 283(12, 13, 98), 284(15, 75, 77, 98), 285(15, 98), 286(15), 287(99), 288(11), 290(14), 293(14, 26), 299(16, 19, 21), 300(21), 302(14, 26), 304(26), 305(26), 310(26), 322(12), 334(26), 344(16), 345(16), 347(19), 348(19), 350(20), 351(20), *355–358*

Lepine, H., 449(154), 454(154), *501*

Leroi, G. E., 288(100), *358*

Levy, D. H., 265(5–8), *355*

Levy, P., 384(45), *497*

Liao, C.-L., 70(123), 82(123), 89(123), 95–97(123), 103(123), 104(123), 107(123), *114*

Lieb, E. H., 90(142), 98(142), *115*

Liepmann, H. W., 271(67), *357*

Lifshitz, E. M., 9(57), 24(81), 38(81), *113*

Lighthill, M. J., 411(78), *498*

Lindenberg, K., 373(28), 377(39, 40), 395(55), 409(74), 456(178), 462(188, 190), 463(190), 468–470(213), *496, 497, 498, 502, 503*

Lindgren, I., 195(39), 234(39, 72), 235(39, 72), 236–239(72), 240(39), *260, 261*

Lindholm, E., 327(164), 354(249), *360, 362*

Linn, S. H., 266(23–25, 27, 29, 30), 276(24), 278(24), 280(24), 288(25), 289(25), 290(25, 30), 291(25, 30), 292(30), 293(24, 29, 30), 294(29), 295(29), 297–299(29), 302(24, 29,

30), 303(30, 127), 304(30, 129), 306(129), 307–309(24), 311(24, 30), 313(24), 314(24, 27), 315(24), 316(23), 317(23, 27), 318–320(27), 321(23, 24, 27), 323(24), 325(27), 326–329(24), 330(24, 129), 331(24, 129), 333(129), 334(24, 29), 335(129), 344(129), *356, 359*

Lippert, E., Jr., 291(114), *358*

Lipscomb, W. N., 291(114), *358*

Lischka, H., 255(93), *261*

Liston, S. K., 333(180), 334(185), 335(185), 339(185), *360*

Litlewood, D. E., 85(140), *115*

Liverman, M. G., 265(4), *355*

Lloyd, D. R., 334(192), 339–342(192), *361*

Long, J., 345(206), *361*

Longuet-Higgins, H. C., 119(4, 5), 122(5), 131(4, 5), 173(4), *179, 180*

Lorents, D. C., 281(83), 283(83), *357*

Lossing, F. P., 350(244, 245), *362*

Lovely, P. S., 492(272), *505*

Lucas, N. J. D., 119(6), 157(6), 162(6), *180*

Lührmann, K. H., 182(8), 190(8), *259*

McCloskey, K. E., 334(181), *360*

McClure, D. J., 305(134), *359*

McCrackin, F. L., 449(150), 450(159), *501*

McCulloh, K. E., 348(223), 349(223), *361*

McDowell, C. A., 324(162), *360*

Macek, J., 15(74), *113*

MacFarlane, M. H., 3(14), *111*

Machta, J., 493(279), *505*

McInnes, J. A., 477(239, 240), *503*

McIntyre, D., 448(143), *500*

McIver, R. T., Jr., 347(215), *361*

McLachlan, A. D., 74(135), 80(135), *115*

McLean, F. B., 482(244), *504*

Macnab, R. M., 488(263, 264), *504*

Madden, R. P., 27(86), 30(86), *113*

Maeda, K., 350(235), *362*

Magee, W. S., 447(137), *500*

Mahan, B. H., 266(11–16, 19–21, 26), 273(15), 274(13, 15), 277(11), 280(12), 281(12, 13, 15, 26), 282(12, 13), 283(12, 13), 284–286(15), 288(11), 290(14), 293(14, 26), 299(16, 19, 21), 300(21), 302(14, 26), 304(26), 305(26), 310(26), 322(12), 334(26), 344(16), 345(16), 347(19), 348(19), 350(20), 351(20), 355, *356*

Mahanty, J., 402(64), *498*

SUBJECT INDEX

521